50 1800 1850 1900 1950

Rumford Kelvin

Watt Rankine

Joule

Carnot Planck

Clausius

Clapeyron Carathéodory

Nernst

Helmholtz

Gibbs

Einstein

Maxwell Bose

Lagrange Boltzmann

Laplace Fermi

Gay-Lussac Dirac

Young Van der Waals

Dalton Sackur

Avogadro Tetrode

Dulong Debye

Petit Born

Coulomb Wien

Faraday Jeans

Ampere Rayleigh

Volta Stefan

Ohm Hertz Pauli

Gauss Curie

Oersted Bohr

Henry Schrodinger

Washington Lincoln F.D. Roosevelt

Beethoven Brahms Shostakovich

Keats Dostoevsky Faulkner

Goya Manet Picasso

750 1800 1850 1900 1950

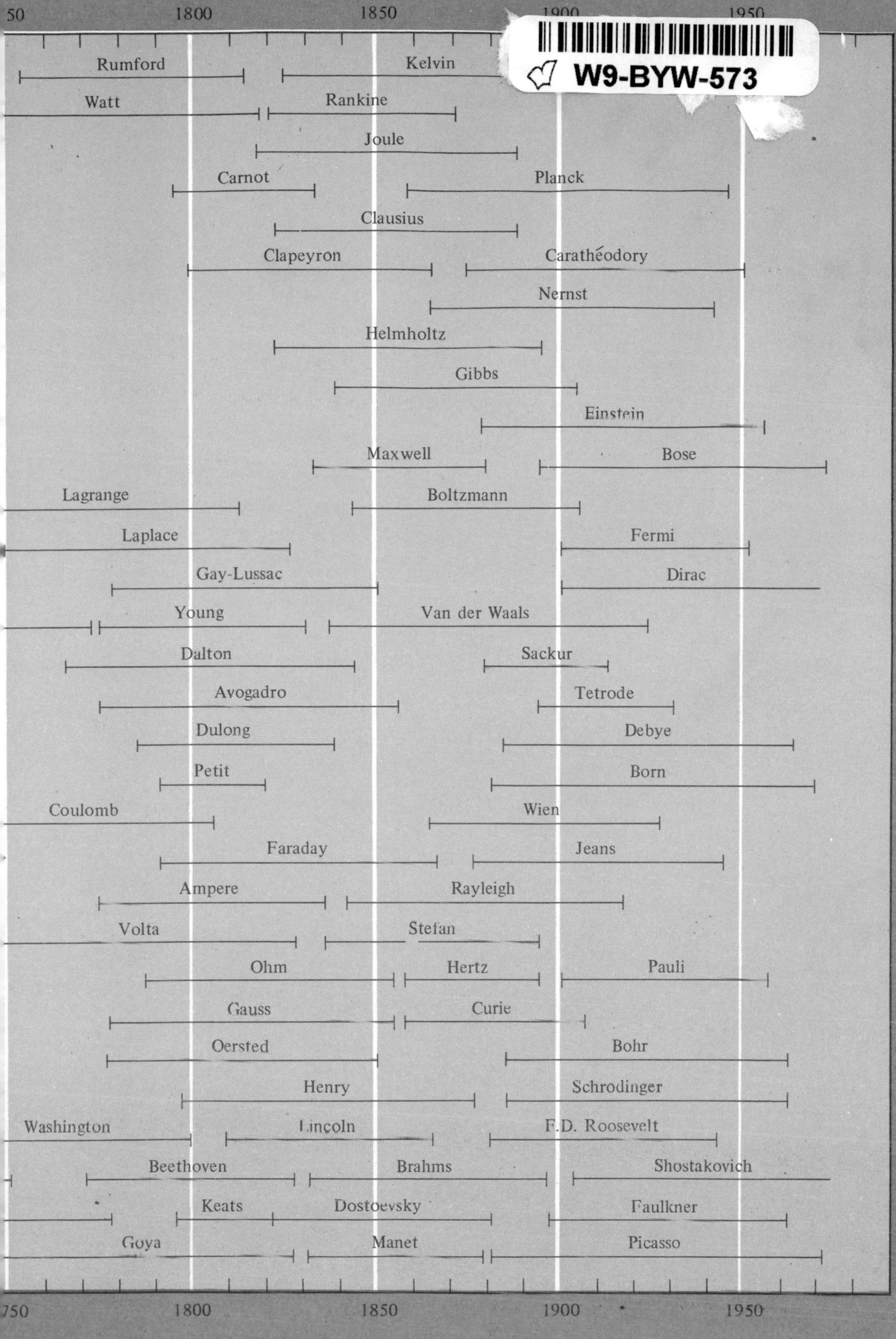

Thermodynamics, Kinetic Theory, and Statistical Thermodynamics

Francis W. Sears
Professor Emeritus, Dartmouth College

Gerhard L. Salinger
Associate Professor of Physics, Rensselaer Polytechnic Institute

Thermodynamics, Kinetic Theory, and Statistical Thermodynamics

THIRD EDITION

Addison-Wesley Publishing Company
Reading, Massachusetts, Menlo Park, California • London • Amsterdam
Don Mills, Ontario • Sydney

This book is in the
ADDISON-WESLEY PRINCIPLES OF PHYSICS SERIES.

 Printed in the United States of America. Published simultaneously in Canada. Library of Congress Catalog Card No. 74-2851.

ISBN 0-201-06894-X
FGHIJKLMNO-HA-89876543210

Preface

This text is a major revision of *An Introduction to Thermodynamics, Kinetic Theory, and Statistical Mechanics* by Francis W. Sears. The general approach has been unaltered and the level remains much the same, perhaps being increased somewhat by greater coverage. The text is still considered useful for advanced undergraduates in physics and engineering who have some familiarity with calculus.

The first eight chapters are devoted to a presentation of classical thermodynamics without recourse to either kinetic theory or statistical mechanics. We feel it is important for the student to understand that if certain macroscopic properties of a system are determined experimentally, all the properties of the system can be specified without knowing anything about the microscopic properties of the system. In the later chapters we show how the microscopic properties of the system can be determined by using the methods of kinetic theory and statistical mechanics to calculate the dependence of the macroscopic properties of a system on thermodynamic variables.

The presentation of many topics differs from the earlier text. Non-PVT systems are introduced in the second chapter and are discussed throughout the text. The first law is developed as a definition of the difference in the internal energy of a system between two equilibrium states as the work in an adiabatic process between the states and in which the kinetic and potential energy of the system do not change. The heat flow is then the difference between the work in any process between two equilibrium states and the work in an adiabatic process between the same states. Care is taken to explain the effects of changes in kinetic and potential energy as well. After the discussion of the first law, various examples are presented to show which properties of the system can be determined on the basis of this law alone.

The statement that "in every process taking place in an isolated system the entropy of the system either increases or remains constant" is used as the second law. It is made plausible by a series of examples and shown to be equivalent to the "engine" statements and the Carathéodory treatment. Thermodynamic potentials are presented in greater detail than in the earlier text. A new potential F^* is introduced to make consistent the thermodynamic and statistical treatments of processes in which the potential energy of a system changes. The discussion of open systems, added in Chapter 8, is necessary for the new derivation of statistics.

Kinetic theory of gases is treated in Chapters 9 and 10. Although the coverage appears to be reduced from the previous edition, the remaining material is discussed from the point of view of statistics in Chapter 12.

The derivation of the distribution functions for the various types of statistics is completely different from previous editions. Discrete energy levels are assumed from the outset. The number of microstates belonging to each macrostate is calculated in the conventional manner for Bose-Einstein, Fermi-Dirac and Maxwell-Boltzmann statistics. The entropy is shown to be proportional to the natural logarithm of the total number of microstates available to the system and not to the number of microstates in the most probable macrostate. The distribution of particles among energy levels is determined without the use of Lagrange multipliers and Stirling's approximation, by calculating the change in the total number of microstates when a particle in a particular energy level is removed from the system. The logarithm of this change is proportional to the change of entropy of the system.

Only the single-particle partition function is introduced and it is used to derive the thermodynamic properties of systems. The coverage is much the same as the earlier text except that it is based entirely on discrete levels. The chapter on fluctuations has been omitted.

The number of problems at the end of each chapter has been expanded. Some of the problems would become tedious if one did not have access to a small calculator. The International System (SI) has been adopted throughout. Thus the units are those of the MKS system and are written, for example, as J kilomole^{-1} K^{-1} for specific heat capacity.

The section on classical thermodynamics can be used for a course lasting one quarter. For a one-semester course it can be used with either the chapters on kinetic theory or statistical thermodynamics, but probably not both, unless only classical statistics are discussed, which can be done by using the development given in the sections on Bose-Einstein statistics and taking the limit that $g_j \gg N_j$.

We appreciate the helpful comments of the reviewers of the manuscript, especially L. S. Lerner and C. F. Hooper, who also gave part of the manuscript a field test. One of us (GLS) wishes to thank his colleagues at Rensselaer for many helpful discussions. J. Aitken worked all the problems and checked the answers. Phyllis Kallenburg patiently retyped many parts of the manuscript with great accuracy and good humor. The encouragement of our wives and tolerance of our children helped considerably in this undertaking. Criticisms from teachers and students will be welcomed.

Norwich, Vermont
Troy, New York
October 1974

F.W.S.
G.L.S.

Contents

1 Fundamental concepts

2 Equations of state

3 The first law of thermodynamics

4 Some consequences of the first law

5 Entropy and the second law of thermodynamics

6 Combined first and second laws

7 Thermodynamic potentials

8 Applications of thermodynamics to simple systems

1

Fundamental concepts

1–1 SCOPE OF THERMODYNAMICS

Thermodynamics is an experimental science based on a small number of principles that are generalizations made from experience. It is concerned only with *macroscopic* or large-scale properties of matter and it makes no hypotheses about the small-scale or *microscopic* structure of matter. From the principles of thermodynamics one can derive general relations between such quantities as coefficients of expansion, compressibilities, specific heat capacities, heats of transformation, and magnetic and dielectric coefficients, especially as these are affected by temperature. The principles of thermodynamics also tell us which few of these relations must be determined experimentally in order to completely specify *all* the properties of the system.

The actual magnitudes of quantities like those above can be calculated only on the basis of a molecular model. The *kinetic theory* of matter applies the laws of mechanics to the individual molecules of a system and enables one to calculate, for example, the numerical value of the specific heat capacity of a gas and to understand the properties of gases in terms of the law of force between individual molecules.

The approach of *statistical thermodynamics* ignores the detailed consideration of molecules as individuals and applies statistical considerations to find the distribution of the very large number of molecules that make up a macroscopic piece of matter over the energy states of the system. For those systems whose energy states can be calculated by the methods of either quantum or classical physics, both the magnitudes of the quantities mentioned above and the relations between them can be determined by quite general means. The methods of statistics also give further insight into the concepts of entropy and the principle of the increase of entropy.

Thermodynamics is complementary to kinetic theory and statistical thermodynamics. Thermodynamics provides *relationships* between physical properties of any system once certain measurements are made. Kinetic theory and statistical thermodynamics enable one to calculate the magnitudes of these properties for those systems whose energy states can be determined.

The science of thermodynamics had its start in the early part of the nineteenth century, primarily as a result of attempts to improve the efficiencies of steam engines, devices into which there is an input in the form of heat, and whose output is mechanical work. Thus as the name implies, thermodynamics was concerned with both *thermal* and mechanical, or *dynamical*, concepts. As the subject developed and its basic laws were more fully understood, its scope became broader. The principles of thermodynamics are now used by engineers in the design of internal combustion engines, conventional and nuclear power stations, refrigeration and air-conditioning systems, and propulsion systems for rockets, missiles, aircraft, ships, and land vehicles. The sciences of physical chemistry and chemical physics consist in large part of the applications of thermodynamics to chemistry and chemical equilibria. The production of extremely low temperatures, in the neighborhood of absolute zero, involves the application of thermodynamic principles to systems of molecular and nuclear magnets. Communications, information theory, and even certain biological processes are examples of the broad areas in which the thermodynamic mode of reasoning is applicable.

In this book we shall first develop the principles of thermodynamics and show how they apply to a system of any nature. The methods of kinetic theory and statistics are then discussed and correlated with those of thermodynamics.

1–2 THERMODYNAMIC SYSTEMS

The term *system*, as used in thermodynamics, refers to a certain portion of the Universe within some closed surface called the *boundary* of the system. The boundary may enclose a solid, liquid, or gas, or a collection of magnetic dipoles, or even a batch of radiant energy or photons in a vacuum. The boundary may be a real one, like the inner surface of a tank containing a compressed gas, or it may be imaginary, like the surface bounding a certain mass of fluid flowing along a pipe line and followed in imagination as it progresses. The boundary is not necessarily fixed in either shape or volume. Thus when a fluid expands against a piston, the volume enclosed by the boundary increases.

Many problems in thermodynamics involve interchanges of energy between a given system and other systems. Any systems which can interchange energy with a given system are called the *surroundings* of that system. A system and its surroundings together are said to constitute a *universe*.

If conditions are such that no energy interchange with the surroundings can take place, the system is said to be *isolated*. If no matter crosses the boundary, the system is said to be *closed*. If there is an interchange of matter between system and surroundings, the system is *open*.

1–3 STATE OF A SYSTEM. PROPERTIES

The *state* of a thermodynamic system is specified by the values of certain experimentally measurable quantities called *state variables* or *properties*. Examples of properties are the temperature of a system, the pressure exerted by it, and the volume it occupies. Other properties of interest are the magnetization of a magnetized body, the polarization of a dielectric, and the surface area of a liquid.

Thermodynamics deals also with quantities that are not properties of any system. Thus when there is an interchange of energy between a system and its surroundings, the energy transferred is not a property of either the system or its surroundings.

Those properties of a system in a given state that are proportional to the mass of a system are called *extensive*. Examples are the total volume and the total energy of a system. Properties that are independent of the mass are called *intensive*. Temperature, pressure, and density are examples of intensive properties.

The *specific* value of an extensive property is defined as the ratio of the value of the property to the mass of the system, or as its value *per unit mass*. We shall use capital letters to designate an extensive property and lower case letters for the corresponding specific value of the property. Thus the total volume of a system is represented by V and the specific volume by v, and

$$v = \frac{V}{m}.$$

The specific volume is evidently the reciprocal of the density ρ, defined as the mass per unit volume:

$$\rho = \frac{m}{V} = \frac{1}{v}.$$

Since any extensive property is proportional to the mass, the corresponding specific value is independent of the mass and is an *intensive* property.

The ratio of the value of an extensive property to the number of moles of a system is called the *molal specific value* of that property. We shall use lower case letters also to represent molal specific values. Thus if n represents the number of moles of a system, the molal specific volume is

$$v = \frac{V}{n}.$$

Note that in the MKS system, the term "mole" implies kilogram-mole or kilomole, that is, a mass in kilograms numerically equal to the molecular weight. Thus one kilomole of O_2 means 32 kilograms of O_2.

No confusion arises from the use of the same letter to represent both the volume per unit mass, say, and the volume per mole. In nearly every equation in which such a quantity occurs there will be some other quantity which indicates which specific volume is meant, or, if there is no such quantity, the equation will hold equally well for either.

In many instances, it is more convenient to write thermodynamic equations in terms of specific values of extensive properties, since the equations are then independent of the mass of any particular system.

1–4 PRESSURE

The stress in a continuous medium is said to be a *hydrostatic pressure* if the force per unit area exerted on an element of area, either within the medium or at its surface, is (a) normal to the element and (b) independent of the orientation of the element. The stress in a fluid (liquid or gas) at rest in a closed container is a hydrostatic pressure. A solid can be subjected to a hydrostatic pressure by immersing it in a liquid in which it is insoluble and exerting a pressure on the liquid. The pressure P is defined as the magnitude of the force per unit area and the unit of pressure in the MKS system is 1 *newton* per square meter* (1 N m^{-2}). A pressure of

* Sir Isaac Newton, English mathematician (1642–1727).

exactly 10^5 N m^{-2} (= 10^6 dyne cm^{-2}) is called 1 *bar*, and a pressure of 10^{-1} N m^{-2} (= 1 dyne cm^{-2}) is 1 *microbar* (1 μ bar).

A pressure of 1 *standard atmosphere* (atm) is defined as the pressure produced by a vertical column of mercury exactly 76 cm in height, of density $\rho = 13.5951$ g cm^{-3}, at a point where g has its standard value of 980.665 cm s^{-2}. From the equation $P = \rho g h$, we find

$$1 \text{ standard atmosphere} = 1.01325 \times 10^6 \text{ dyne cm}^{-2} = 1.01325 \times 10^5 \text{ N m}^{-2}.$$

Hence 1 standard atmosphere is very nearly equal to 1 bar, and 1 μ bar is very nearly 10^{-6} atm.

A unit of pressure commonly used in experimental work at low pressures is 1 *Torr* (named after Torricelli*) and defined as the pressure produced by a mercury column exactly 1 millimeter in height, under the conditions above; therefore 1 Torr = 133.3 N m^{-2}.

1–5 THERMAL EQUILIBRIUM AND TEMPERATURE. THE ZEROTH LAW

The concept of temperature, like that of force, originated in man's sense perceptions. Just as a force is something we can correlate with muscular effort and describe as a push or a pull, so temperature can be correlated with the sensations of relative hotness or coldness. But man's temperature sense, like his force sense, is unreliable and restricted in range. Out of the primitive concepts of relative hotness and coldness there has developed an objective science of thermometry, just as an objective method of defining and measuring forces has grown out of the naive concept of a force as a push or a pull.

The first step toward attaining an objective measure of the temperature sense is to set up a criterion of *equality* of temperature. Consider two metal blocks A and B, of the same material, and suppose that our temperature sense tells us that A is warmer than B. If we bring A and B into contact and surround them by a thick layer of felt or glass wool, we find that after a sufficiently long time has elapsed the two feel equally warm. Measurements of various properties of the bodies, such as their volumes, electrical resistivities, or elastic moduli, would show that these properties changed when the two bodies were first brought into contact but that eventually they became constant also.

Now suppose that two bodies of *different* materials, such as a block of metal and a block of wood, are brought into contact. We again observe that after a sufficiently long time the measurable properties of these bodies, such as their volumes, cease to change. However, the bodies will not feel equally warm to the touch, as evidenced by the familiar fact that a block of metal and a block of wood,

* Evangelista Torricelli, Italian physicist (1608–1647).

both of which have been in the same room for a long time, do not feel equally warm. This effect results from a difference in thermal conductivities and is an example of the unreliability of our temperature sense.

The feature that is common in both instances, whether the bodies are of the same material or not, is that an end state is eventually reached in which there are no further observable changes in the measurable properties of the bodies. This state is then defined as one of *thermal equilibrium*.

Observations such as those described above lead us to infer that all ordinary objects have a physical property that determines whether or not they will be in thermal equilibrium when placed in contact with other objects. This property is called *temperature*. If two bodies *are* in thermal equilibrium when placed in contact, then by definition their temperatures are equal. Conversely, if the temperatures of two bodies are equal, they will be in thermal equilibrium when placed in contact. A state of thermal equilibrium can be described as one in which the temperature of the system is the same at all points.

Suppose that body A, say a metal block, is in thermal equilibrium with body B, also a metal block. The temperature of B is then equal to the temperature of A. Suppose, furthermore, that body A is also separately in thermal equilibrium with body C, a wooden block, so that the temperatures of C and A are equal. It follows that the temperatures of B and C are equal; but the question arises, and it can only be answered by experiment, what will actually happen when B and C are brought in contact? Will they be in thermal equilibrium? We find by experiment that they *are*, so that the definition of equality of temperature in terms of thermal equilibrium is self-consistent.

It is not immediately obvious that because B and C are both in thermal equilibrium with A, that they are necessarily in thermal equilibrium with each other. When a zinc rod and a copper rod are dipped in a solution of zinc sulfate, both rods come to *electrical* equilibrium with the solution. If they are connected by a wire, however, it is found that they are *not* in electrical equilibrium with each other, as evidenced by an electric current in the wire.

The experimental results above can be stated as follows:

When any two bodies are each separately in thermal equilibrium with a third, they are also in thermal equilibrium with each other.

This statement is known as the *zeroth law of thermodynamics*, and its correctness is tacitly assumed in every measurement of temperature. Thus if we want to know when two beakers of water are at the same temperature, it is unnecessary to bring them into contact and see whether their properties change with time. We insert a thermometer (body A) in one beaker of water (body B) and wait until some property of the thermometer, such as the length of the mercury column in a glass capillary, becomes constant. Then by definition the thermometer has the same temperature as the water in this beaker. We next repeat the procedure with the other beaker of water (body C). If the lengths of the mercury columns are the same, the temperatures of B and C are equal, and experiment shows that if the two beakers are brought into contact, no changes in their properties take place.

Note that the thermometer used in this test requires no calibration—it is only necessary that the mercury column stand at the same point in the capillary. Such an instrument can be described as a *thermoscope.* It will indicate equality of temperature without determining a numerical value of temperature.

Although a system will eventually come to thermal equilibrium with its surroundings if these are kept at constant temperature, the *rate* at which equilibrium is approached depends on the nature of the boundary of the system. If the boundary consists of a thick layer of a thermal insulator such as glass wool, the temperature of the system will change very slowly, and it is useful to imagine an ideal boundary for which the temperature would not change at all. A boundary that has this property is called *adiabatic*, and a system enclosed in an adiabatic boundary can remain permanently at a temperature different from that of its surroundings without ever coming to thermal equilibrium with them. The ideal adiabatic boundary plays somewhat the same role in thermodynamics as the ideal frictionless surface does in mechanics. Although neither actually exists, both are helpful in simplifying physical arguments and both are justified by the correctness of conclusions drawn from arguments making use of them.

Although we have not as yet defined the concept of *heat*, it may be said at this point that an ideal adiabatic boundary is one across which the flow of heat is zero, even when there is a difference in temperature between opposite surfaces of the boundary.

At the opposite extreme from an adiabatic boundary is a *diathermal* boundary, composed of a material which is a good thermal conductor such as a thin sheet of copper. The temperature of a system enclosed in a diathermal boundary very quickly approaches that of its surroundings.

1–6 EMPIRICAL AND THERMODYNAMIC TEMPERATURE

To assign a numerical value to the temperature of a system, we first select some one system, called a *thermometer*, that has a *thermometric property* which changes with temperature and is readily measured. An example is the volume V of a liquid, as in the familiar liquid-in-glass thermometer. The thermometers used most widely in precise experimental work, however, are the *resistance thermometer* and the *thermocouple*.

The thermometric property of the resistance thermometer is its resistance R. For good sensitivity, the *change* in the thermometric property of a thermometer, for a given change in temperature, should be as large as possible. At temperatures that are not too low, a resistance thermometer consisting of a fine platinum wire wound on an insulating frame is suitable. At extremely low temperatures, the resistivity of platinum changes only slightly with changes in temperature, but it has been found that arsenic-doped germanium makes a satisfactory resistance thermometer down to very low temperatures.

The thermocouple consists of an electrical circuit shown in its simplest form in Fig. 1–1(a). When wires of any two unlike metals or alloys are joined so as to form a complete circuit, it is found that an emf $\mathscr{E}$ exists in the circuit whenever the junctions A and B are at different temperatures, and this emf is the thermometric property of the couple. To measure the emf, a galvanometer or potentiometer must be inserted in the circuit, and this introduces a pair of junctions at the points where the instrument leads are connected. If these leads are of the same material, usually copper, and if both of these junctions are at the same temperature, called the *reference temperature*, the emf is the same as in a simple circuit, one of whose junctions is at the reference temperature. Figure 1–1(b) shows a typical thermocouple circuit. Junctions B and C are kept at some known reference temperature, for example by inserting them in a Dewar flask* containing ice and water. Junction A, the *test junction*, is placed in contact with the body whose temperature is to be determined.

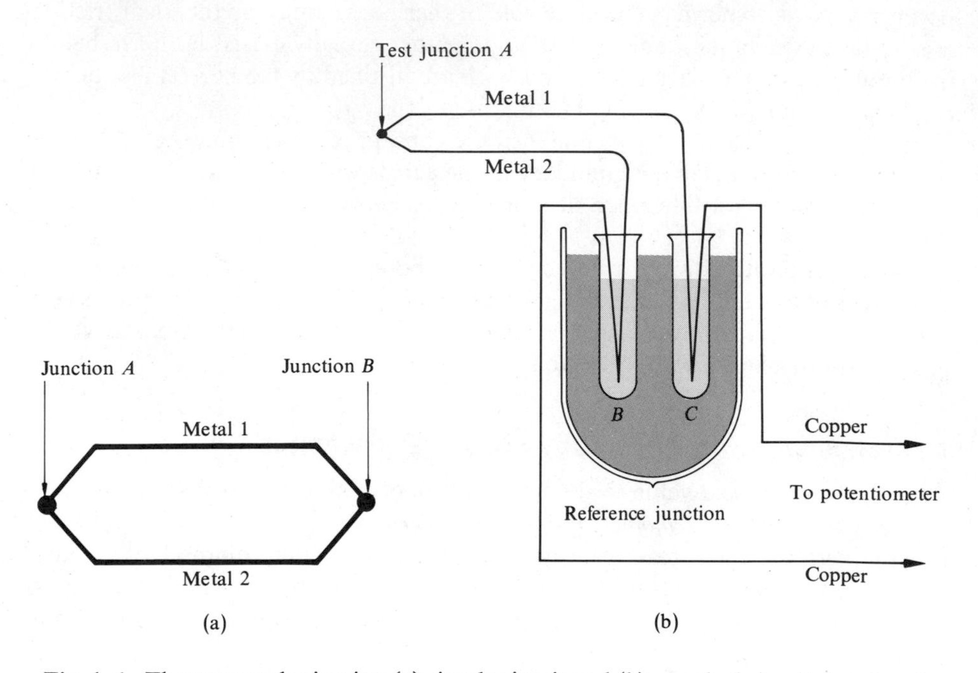

Fig. 1–1 Thermocouple circuits: (a) simple circuit and (b) practical circuit showing the test junction and the reference junction.

* A Dewar flask is a double-walled container. The space between the walls is evacuated to keep heat from entering or leaving the contents of the container. It was invented by Sir James Dewar, British chemist (1848–1923).

Another important type of thermometer, although it is not suitable for routine laboratory measurements, is the *constant volume gas thermometer*, illustrated schematically in Fig. 1-2. The gas is contained in bulb C and the pressure exerted by it can be measured with the open tube mercury manometer. As the temperature of the gas increases, the gas expands, forcing the mercury down in tube B and up in tube A. Tubes A and B communicate through a rubber tube D with a mercury reservoir R. By raising R, the mercury level in B may be brought back to a reference mark E. The gas is thus kept at constant volume. Gas thermometers are used mainly in bureaus of standards and in some university research laboratories. The materials, construction, and dimensions differ in various laboratories and depend on the nature of the gas and the temperature range to be covered.

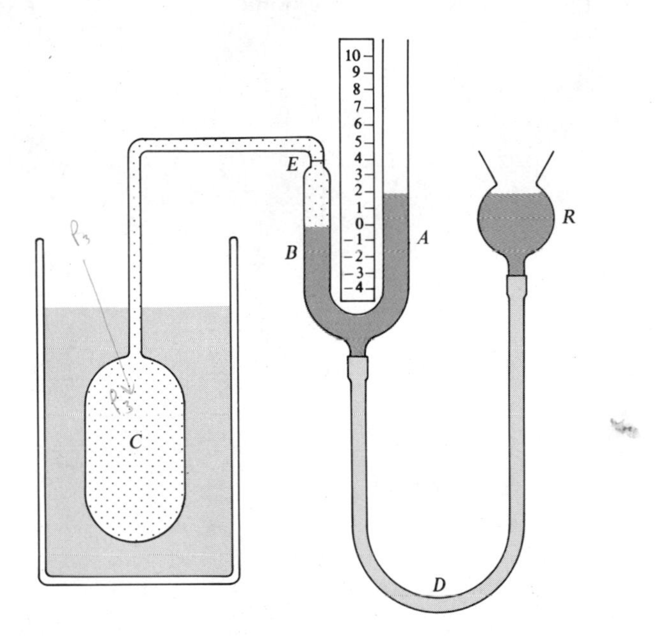

Fig. 1-2 The constant-volume gas thermometer.

Let X represent the value of any thermometric property such as the emf $\mathscr{E}$ of a thermocouple, the resistance R of a resistance thermometer, or the pressure P of a fixed mass of gas kept at constant volume, and θ the *empirical temperature* of the thermometer or of any system with which it is in thermal equilibrium. The ratio of two empirical temperatures θ_2 and θ_1, as determined by a particular thermometer, is defined as equal to the corresponding ratio of the values of X:

$$\frac{\theta_2}{\theta_1} = \frac{X_2}{X_1}.$$

The next step is to arbitrarily assign a numerical value to some one temperature called the *standard fixed point*. By international agreement, this is chosen to be the *triple point* of water, the temperature at which ice, liquid water, and water vapor coexist in equilibrium. We shall see in Section 8–2 that the three states of any substance can coexist at only one temperature.

To achieve the triple point, water of the highest purity which has substantially the isotopic composition of ocean water is distilled into a vessel like that shown schematically in Fig. 1–3. When all air has been removed, the vessel is sealed off. With the aid of a freezing mixture in the inner well, a layer of ice is formed around the well. When the freezing mixture is removed and replaced with a thermometer, a thin layer of ice is melted nearby. So long as the solid, liquid, and vapor coexist in equilibrium, the system is at the triple point.

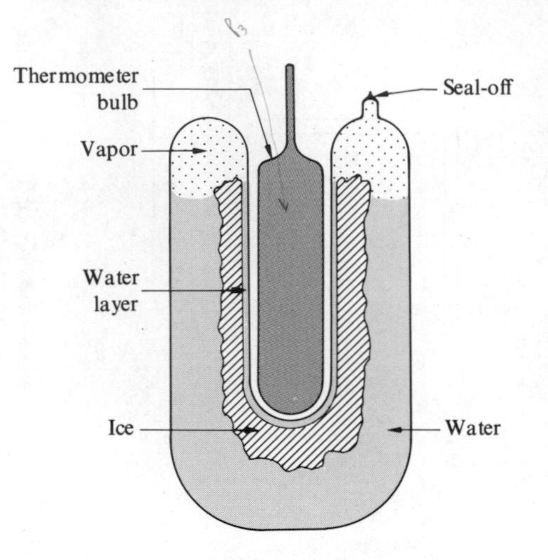

Fig. 1–3 Triple-point cell with a thermometer in the well, which melts a thin layer of ice nearby.

If we now assign some arbitrary value θ_3 to the triple point temperature, and let X_3 represent the corresponding value of the thermometric property of a thermometer, the empirical temperature θ when the value of the thermometric property is X, is given by

$$\frac{\theta}{\theta_3} = \frac{X}{X_3},$$

or

$$\theta = \theta_3 \frac{X}{X_3}. \tag{1–1}$$

Table 1–1 lists the values of the thermometric properties of each of four different thermometers at a number of temperatures, and the ratio of the property at each temperature to its value at the triple point. The first thermometer is a copper-constantan thermocouple, the second is a platinum resistance thermometer, the third is a constant volume hydrogen thermometer filled to a pressure of 6.80 atm at the triple point, and the fourth is also a constant volume hydrogen thermometer but filled to a lower pressure of 1.00 atm at the triple point. Values of the thermometric properties are given at the normal boiling point (NBP) of nitrogen, the normal boiling point of oxygen, the normal sublimation point (NSP) of carbon dioxide, the triple point of water, the normal boiling point of water, and the normal boiling point of tin.

Table 1–1 Comparison of thermometers

System	(Cu–Constantan) $\mathscr{E}$, mV	$\dfrac{\mathscr{E}}{\mathscr{E}_3}$	(Pt) R, ohms*	$\dfrac{R}{R_3}$	(H_2, V const) P, atm	$\dfrac{P}{P_3}$	(H_2, V const) P, atm	$\dfrac{P}{P_3}$
N_2 (NBP)	0.73	0.12	1.96	0.20	1.82	0.27	0.29	0.29
O_2 (NBP)	0.95	0.15	2.50	0.25	2.13	0.31	0.33	0.33
CO_2 (NSP)	3.52	0.56	6.65	0.68	4.80	0.71	0.72	0.72
H_2O (TP)	$\mathscr{E}_3 = 6.26$	1.00	$R_3 = 9.83$	1.00	$P_3 = 6.80$	1.00	$P_3 = 1.00$	1.00
H_2O (NBP)	10.05	1.51	13.65	1.39	9.30	1.37	1.37	1.37
Sn (NMP)	17.50	2.79	18.56	1.89	12.70	1.87	1.85	1.85

We see that a complication arises. The ratio of the thermometric properties, at each temperature, is different for all four thermometers, so that for a given value of θ_3, the empirical temperature θ is different for all four. The agreement is closest, however, for the two hydrogen thermometers and it is found experimentally that constant volume gas thermometers using different gases agree more and more closely with each other, the lower the pressure P_3 at the triple point. This is illustrated in Fig. 1–4, which shows graphs of the ratio P_s/P_3 for four different constant volume gas thermometers plotted as function of the pressure P_3. The pressure P_s is that at the normal boiling point of water (the steam point). Experimental measurements cannot, of course, be made all the way down to zero pressure, P_3, but the extrapolated curves all intersect the vertical axis at a common point at which $P_s/P_3 = 1.3660$. At any other temperature, the extrapolated graphs also intersect at a (different) common point, so that *all constant volume gas thermometers agree* when their readings are extrapolated to zero pressure P_3. We therefore define the *empirical gas temperature* θ_g as

$$\theta_g = \theta_3 \times \lim_{P_3 \to 0} \left(\frac{P}{P_3}\right)_V, \tag{1–2}$$

* Georg S. Ohm, German physicist (1787–1854).

the subscript V indicating that the pressures are measured at constant volume. Temperatures defined in this way are therefore independent of the properties of any particular gas, although they do depend on the characteristic behavior of gases as a whole and are thus not entirely independent of the properties of a particular material.

There remains the question of assigning a numerical value to the triple-point temperature θ_3. Before 1954, gas temperatures were defined in terms of *two* fixed points: the normal boiling point of pure water (the *steam point*) and the equilibrium temperature of pure ice and air-saturated water at a pressure of 1 atmosphere (the *ice point*). (The triple-point and ice-point temperatures are not exactly the same because the pressure at the triple point is not 1 atm, but is the vapor pressure of water, 4.58 Torr, and the ice is in equilibrium with pure water, not air-saturated water. This is discussed further in Section 7-6.)

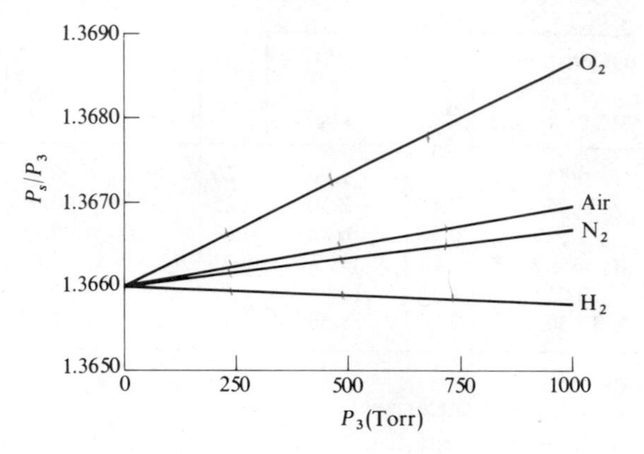

Fig. 1-4 Readings of a constant-volume gas thermometer for the temperature of condensing steam, when different gases are used at various values of P_3.

If the subscripts s and i designate values at the steam and ice points, the gas temperatures θ_s and θ_i were defined by the equations

$$\frac{\theta_s}{\theta_i} = \left(\frac{P_s}{P_i}\right)_V, \qquad \theta_s - \theta_i = 100 \text{ degrees.}$$

(The pressure ratio is understood to be the limiting value extrapolated to zero pressure.) When these equations are solved for θ_i, we have

$$\theta_i = \frac{100P_i}{P_s - P_i} = \frac{100}{(P_s/P_i) - 1}. \tag{1-3}$$

The best experimental value of the ratio P_s/P_i was found to be 1.3661. (This differs slightly from the limiting value of the ratio P_s/P_3 of 1.3660 in Fig. 1-4 because the temperature of the triple point is slightly larger than that of the ice point.) Hence from Eq. (1–3),

$$\theta_i = \frac{100}{1.3661 - 1} = 273.15 \text{ degrees},$$

and from the defining equations for θ_s and θ_i

$$\theta_s = 373.15 \text{ degrees}.$$

The triple point temperature θ_3 is found by experiment to be 0.01 degree above the ice point, so the best experimental value of θ_3 is

$$\theta_3 = 273.16 \text{ degrees}.$$

In order that temperatures based on a single fixed point, the triple point of water, shall agree with those based on two fixed points, the ice and steam points, the triple point temperature is assigned the value

$$\theta_3 \equiv 273.16 \text{ degrees } (\textit{exactly}).$$

Hence

$$\theta_g = 273.16 \times \lim_{P_3 \to 0} \left(\frac{P}{P_3}\right)_V. \tag{1–4}$$

It will be shown in Section 5–2 that, following a suggestion made by Lord Kelvin*, one can define the ratio of two temperatures on the basis of the second law of thermodynamics in a way that is completely independent of the properties of any particular material. Temperatures defined in this way are called *absolute* or *thermodynamic* temperatures and are represented by the letter T. We shall show later that thermodynamic temperatures are equal to gas temperatures as defined above. Since all thermodynamic equations are best expressed in terms of thermodynamic temperature, we shall use, from now on, the symbol T for temperature, understanding that it can be measured experimentally with a gas thermometer.

It has been customary for many years to speak of a thermodynamic temperature as so many "degrees kelvin," abbreviated deg K or °K. The word "degree" and the degree symbol have now been dropped. The unit of temperature is called 1 kelvin (1 K), just as the unit of energy is called 1 joule (1 J)†, and we say, for example, that the triple point temperature is 273.16 kelvins (273.16 K). The unit of temperature is thus treated in the same way as the unit of any other physical quantity. Thus we can write finally, accepting for the present that $T = \theta_g$,

$$T = 273.16 \text{ K} \times \lim_{P_3 \to 0} \left(\frac{P}{P_3}\right)_V. \tag{1–5}$$

Celsius‡ temperature t (formerly known as centigrade temperature) is defined by the equation

$$t = T - T_i, \tag{1–6}$$

* William Thomson, Lord Kelvin, Scottish physicist (1824–1907).

† James P. Joule, British physicist (1818–1889).

‡ Anders Celsius, Swedish astronomer (1701–1744).

where T_i is the thermodynamic temperature of the ice point, equal to 273.15 K. The *unit* employed to express Celsius temperature is the degree Celsius (°C), which is equal to the kelvin. Thus at the ice point, where $T = T_i$, $t = 0°C$; at the triple point of water, where $T = 273.16$ K, $t = 0.01°C$; and at the steam point, $t = 100°C$. A *difference* in temperature is expressed in kelvins; it may also be expressed in degrees Celsius (deg C).

The Rankine* and Fahrenheit† scales, commonly used in engineering measurements in the United States, are related in the same way as the Kelvin and Celsius scales. Originally these scales were defined in terms of two *fixed* points, with the difference between the steam point and ice point temperatures taken as 180 degrees instead of 100 degrees. Now they are defined in terms of the Kelvin scale through the relation

$$1\ \mathrm{R} = \frac{5}{9}\ \mathrm{K}\ \text{(exactly)}. \tag{1–7}$$

Thus the thermodynamic temperature of the ice point is

$$T_i = \frac{9}{5}\frac{\mathrm{R}}{\mathrm{K}} \times 273.15\ \mathrm{K} = 491.67\ \mathrm{R}.$$

Fahrenheit temperature t is defined by the equation

$$t = T - 459.67\ \mathrm{R}, \tag{1–8}$$

where T is the thermodynamic temperature expressed in rankines. The unit of Fahrenheit temperature is the degree Fahrenheit (°F), which is equal to the rankine. Thus at the ice point, where $T = T_i = 491.67$ R, $t = 32.00°F$ and at the steam

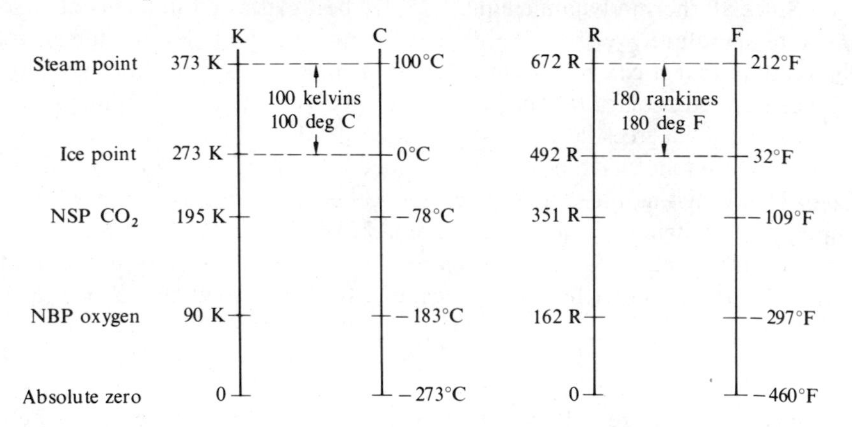

Fig. 1–5 Comparison of Kelvin, Celsius, Rankine, and Fahrenheit temperatures. Temperatures have been rounded off to the nearest degree.

* William J. M. Rankine, Scottish engineer (1820–1872).

† Gabriel D. Fahrenheit, German physicist (1686–1736).

point $t = 212.00°F$. A temperature *difference* is expressed in rankines; it may also be expressed in degrees Fahrenheit (deg F). These scales are no longer used in scientific measurements. Some Kelvin, Celsius, Rankine, and Fahrenheit temperatures are compared in Fig. 1–5.

1–7 THE INTERNATIONAL PRACTICAL TEMPERATURE SCALE

To overcome the practical difficulties of direct determination of thermodynamic temperature by gas thermometry and to unify existing national temperature scales, an International Temperature Scale was adopted in 1927 by the Seventh General Conference on Weights and Measures. Its purpose was to provide a practical scale of temperature which was easily and accurately reproducible and which gave as nearly as possible thermodynamic temperatures. The International Temperature Scale was revised in 1948, in 1960, and most recently in 1968. It is now known as the International Practical Temperature Scale of 1968 (IPTS–68).

International Practical Kelvin Temperature is represented by the symbol T_{68}, and International Practical Celsius Temperature by the symbol t_{68}. The relation between T_{68} and t_{68} is

$$t_{68} = T_{68} - 273.15 \text{ K}.$$

The units of T_{68} and t_{68} are the kelvin (K) and the degree Celsius (°C), as in the case of the thermodynamic temperature T and the Celsius temperature t.

The IPTS–68 is based on assigned values to the temperatures of a number of reproducible equilibrium states (fixed points) and on standard instruments calibrated at those temperatures. Within the limits of experimental accuracy, the temperatures assigned to the fixed points are equal to the best experimental values in 1968 of the *thermodynamic* temperatures of the fixed points. Interpolation between the fixed-point temperatures is provided by formulas used to establish the relation between indications of the standard instruments and values of International Practical Temperature. Some of these equilibrium states, and values of the International Practical Temperatures assigned to them, are given in Table 1–2.

Table 1–2 Assigned temperatures of some of the fixed points used in defining the International Practical Temperature Scale of 1968 (IPTS–68)

Fixed point	T_{68} (K)	t_{68} (°C)
Triple point of hydrogen	13.81	−259.34
Boiling point of neon	27.102	−246.048
Triple point of oxygen	54.361	−218.789
Triple point of water	273.16	0.01
Boiling point of water	373.15	100
Freezing point of zinc	692.73	419.58
Freezing point of silver	1235.08	961.93
Freezing point of gold	1337.58	1064.43

The standard instrument used from 13.81 K to 630.74°C is a platinum resistance thermometer. Specified formulas are used for calculating International Practical Temperature from measured values of the thermometer resistance over temperature ranges in this interval, the constants in these formulas being determined by measuring the resistance at specified fixed points between the triple point of hydrogen and the freezing point of zinc.

In the range from 630.74°C to 1064.43°C, the standard instrument is a thermocouple of platinum and an alloy of platinum and 10% rhodium. The thermocouple is calibrated by measuring its emf at a temperature of 630.74°C as determined by a platinum resistance thermometer, and at the normal freezing points of silver and of gold.

At temperatures above the freezing point of gold, (1337.58 K or 1064.43°C) International Practical Temperature is determined by measuring the spectral concentration of the radiance of a black body and calculating temperature from the Planck* law of radiation (see Section 13–2). The freezing point of gold, 1337.58 K is used as a reference temperature, together with the best experimental value of the constant c_2 in the Planck law of radiation given by

$$c_2 = 0.014388 \text{ m K}.$$

For a complete description of the procedures to be followed in determining IPTS-68 temperatures, see the article in *Metrologia*, Vol. 5, No. 2 (April 1969). The IPTS-68 is not defined below a temperature of 13.8 K. A description of experimental procedures in this range can be found in "Heat and Thermodynamics," 5th ed., by Mark W. Zemansky (McGraw-Hill).

1–8 THERMODYNAMIC EQUILIBRIUM

When an arbitrary system is isolated and left to itself, its properties will in general change with time. If initially there are temperature differences between parts of the system, after a sufficiently long time the temperature will become the same at all points and then the system is in *thermal equilibrium.*

If there are variations in pressure or elastic stress within the system, parts of the system may move, or expand or contract. Eventually these motions, expansions, or contractions will cease, and when this has happened we say that the system is in *mechanical* equilibrium. This does not necessarily mean that the pressure is the same at all points. Consider a vertical column of fluid in the earth's gravitational field. The pressure in the fluid decreases with increasing elevation, but each element of the fluid is in mechanical equilibrium under the influence of its own weight and an equal upward force arising from the pressure difference between its upper and lower surfaces.

Finally, suppose that a system contains substances that can react chemically. After a sufficiently long time has elapsed, all possible chemical reactions will have taken place, and the system is then said to be in *chemical equilibrium.*

* Max K. E. L. Planck, German physicist (1858–1947).

A system which is in thermal, mechanical, and chemical equilibrium is said to be in *thermodynamic equilibrium.* For the most part, we shall consider systems that are in thermodynamic equilibrium, or those in which the departure from thermodynamic equilibrium is negligibly small. Unless otherwise specified, the "state" of a system implies an equilibrium state. In this discussion it is assumed that the system is not divided into portions such that the pressure, for example, might be different in different portions, even though the pressure in each portion would approach a constant value.

1–9 PROCESSES

When any of the properties of a system change, the state of the system changes and the system is said to undergo a *process.* If a process is carried out in such a way that at every instant the system departs only infinitesimally from an equilibrium state, the process is called *quasistatic* (i.e., almost static). Thus a quasistatic process closely approximates *a succession of equilibrium states.* If there are finite departures from equilibrium, the process is *nonquasistatic.*

Consider a gas in a cylinder provided with a movable piston. Let the cylinder walls and the piston be adiabatic boundaries and neglect any effect of the earth's gravitational field. With the piston at rest, the gas eventually comes to an equilibrium state in which its temperature, pressure, and density are the same at all points. If the piston is then suddenly pushed down, the pressure, temperature, and density immediately below the piston will be increased by a finite amount above their equilibrium values, and the process is not quasistatic. To compress the gas quasistatically, the piston must be pushed down very slowly in order that the processes of wave propagation, viscous damping, and thermal conduction may bring about at all instants a state which is essentially one of both mechanical and thermal equilibrium.

Suppose we wish to increase the temperature of a system from an initial value T_1 to a final value T_2. The temperature *could* be increased by enclosing the system in a diathermal boundary and maintaining the surroundings of the system at the temperature T_2. The process would not be quasistatic, however, because the temperature of the system near its boundary would increase more rapidly than that at internal points, and the system would not pass through a succession of states of thermal equilibrium. To increase the temperature quasistatically, we must start with the surroundings at the initial temperature T_1 and then increase this temperature sufficiently slowly so that at all times it is only infinitesimally greater than that of the system.

All actual processes are nonquasistatic because they take place with finite differences of pressure, temperature, etc., between parts of a system. Nevertheless, the concept of a quasistatic process is a useful and important one in thermodynamics.

Many processes are characterized by the fact that some property of a system remains constant during the process. A process in which the volume of a system is constant is called *isovolumic* or *isochoric*. If the pressure is constant, the process is called *isobaric* or *isopiestic*. A process at constant temperature is called *isothermal*.

A process carried out by a system enclosed by an adiabatic boundary is an *adiabatic* process. As stated earlier, such a process can also be described as one in which there is no flow of heat across the boundary. Many actual processes, such as a single stroke of the piston of an internal combustion engine, are very nearly adiabatic simply because the process takes place in such a short time that the flow of heat into or out of the system is extremely small. A process can also be made adiabatic by adjusting the temperature of the surroundings as the process proceeds so that this temperature is always equal to that of the system.

A *reversible* process can be defined as one whose "direction" can be reversed by an infinitesimal change in some property of the system. Thus if the temperature of a system within a diathermal boundary is always slightly lower than that of its surroundings, there will be a flow of heat from the surroundings into the system; whereas if the temperature of the system is slightly greater than that of the surroundings, there will be a flow of heat in the opposite direction. Such a process is therefore *reversible* as well as *quasistatic*.

If there is a finite temperature difference between system and surroundings, the direction of the heat flow cannot be reversed by an *infinitesimal* change in temperature of the system, and the process is *irreversible* as well as nonquasistatic. Suppose, however, that the boundary of the system is nearly, but not completely adiabatic, so that the heat flow is very small even with a finite difference in temperature. The system is then very nearly in thermal equilibrium at all times and the process is quasistatic although it is not reversible.

The slow compression or expansion of a gas in a cylinder provided with a piston is quasistatic, but if there is a force of sliding friction, f, between piston and cylinder when the piston is in motion, the process is not reversible. The force exerted on the piston by the gas when the gas is expanding differs by $2f$ from its value when the gas is being compressed. Therefore the direction of motion of the piston can be reversed only by a finite change in gas pressure. All reversible processes are necessarily quasistatic, but a quasistatic process is not necessarily reversible. The terms *reversible* and *irreversible* have a deeper significance also, which can only be brought out after a discussion of the second law of thermodynamics.

PROBLEMS

1–1 State whether or not classical thermodynamic reasoning alone can be used to determine (a) the average velocity of the molecules of a gas; (b) the relation between the pressure dependence of the specific heat capacity of a solid and the temperature dependence of its volume; (c) the magnitude of the magnetic moment of a gas; (d) the relation between the pressure and temperature of electromagnetic radiation in a cavity; (e) the magnitude of the specific heat capacity of a solid. Briefly justify your answers.

1–2 Which of the following quantities are extensive and which are intensive? (a) The magnetic moment of a gas. (b) The electric field E in a solid. (c) The length of a wire. (d) The surface tension of an oil film.

1–3 The density of water in cgs units is 1 g cm^{-3}. Compute (a) the density in MKS units; (b) the specific volume in m^3 kg^{-1}; (c) the MKS molal specific volume. (d) Make the same computations for air whose density is 0.00129 g cm^{-3}. The mean molecular weight of air is 29; that is, the mass of 1 kilomole of air is 29 kg.

1–4 Estimate the pressure you exert on the floor when standing. Express the answer in atmospheres and in Torr.

1–5 One standard atmosphere is defined as the pressure produced by a column of mercury exactly 76 cm high, at a temperature of 0°C, and at a point where $g = 980.665$ cm s^{-2}. (a) Why do the temperature and the acceleration of gravity have to be specified in this definition? (b) Compute the pressure in N m^{-2} produced by a column of mercury of density 13.6 g cm^{-3}, 76 cm in height at a point where $g = 980$ cm s^{-2}.

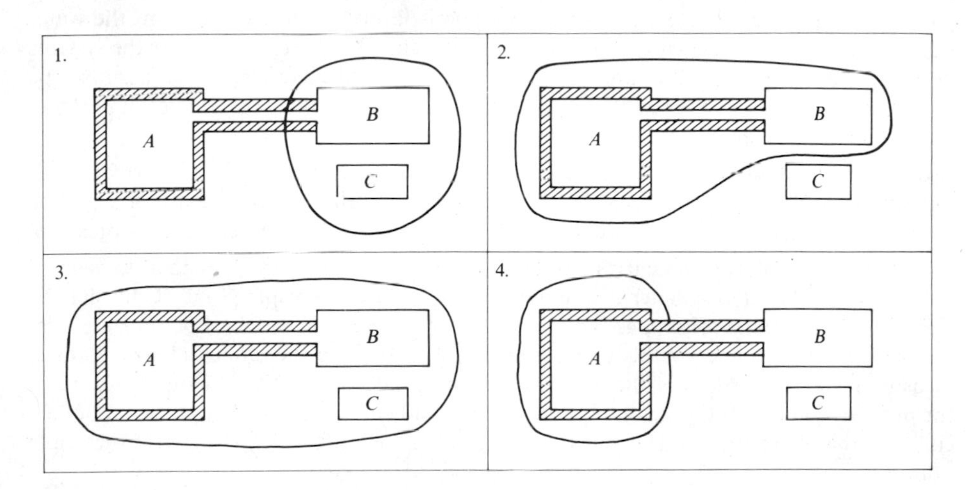

Figure 1–6

1–6 Two containers of gas are connected by a long, thin, thermally insulated tube. Container A is surrounded by an adiabatic boundary, but the temperature of container B can be varied by bringing it into contact with a body C at a different temperature. In Fig. 1–6, these systems are shown with a variety of boundaries. Which figure represents (a) an open system enclosed by an adiabatic boundary; (b) an open system enclosed by a diathermal boundary; (c) a closed system enclosed by a diathermal boundary; (d) a closed system enclosed by an adiabatic boundary.

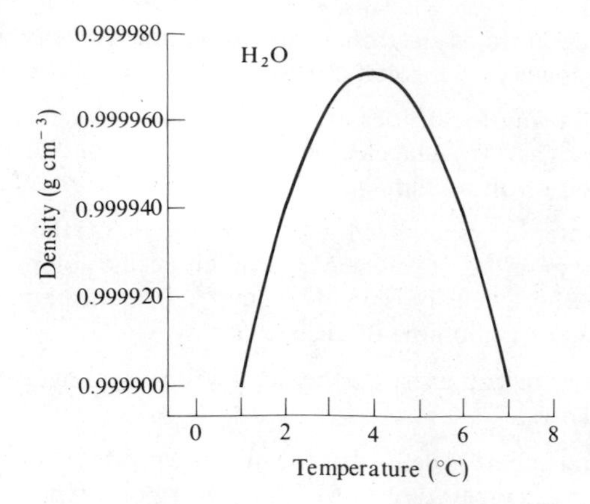

Figure 1–7

1–7 A water-in-glass thermoscope is to be used to determine if two separated systems are in thermal equilibrium. The density of water, shown in Fig. 1–7, is the thermometric parameter. Suppose that when the thermoscope is inserted into each system, the water rises to the same height, corresponding to a density of 0.999945 g cm^{-3}. (a) Are the systems necessarily in thermal equilibrium? (b) Could the height of the water in the thermoscope change if the systems are brought into thermal contact? (c) If there is a change in part (b), would the height increase or decrease?

1–8 Using the data of Table 1–1, find the empirical temperature of the normal sublimation point of CO_2 as measured by the thermocouple, the platinum thermometer, the hydrogen thermometer at high pressure, and the hydrogen thermometer at low pressure.

1–9 The length of the mercury column in a certain mercury-in-glass thermometer is 5.00 cm when the thermometer is in contact with water at its triple point. Consider the length of the mercury column as the thermometric property X and let θ be the empirical temperature determined by this thermometer. (a) Calculate the empirical temperature, measured when the length of the mercury column is 6.00 cm. (b) Calculate the length of the mercury column at the steam point. (c) If X can be measured with a precision of 0.01 cm, can this thermometer be used to distinguish between the ice point and the triple point?

1–10 A temperature t^* is defined by the equation

$$t^* = a\theta^2 + b,$$

where a and b are constants, and θ is the empirical temperature determined by the mercury-in-glass thermometer of the previous problem. (a) Find the numerical values of a and b, if $t^* = 0$ at the ice point and $t^* = 100$ at the steam point. (b) Find the value of t^* when the length of the mercury column $X = 7.00$ cm. (c) Find the length of the mercury column when $t^* = 50$. (d) Sketch t^* versus X.

1–11 Suppose a numerical value of 100 is assigned to the steam point temperature, and that the ratio of two temperatures is defined as the limiting ratio, as $P_3 \to 0$, of the corresponding pressures of a gas kept at constant volume. Find (a) the best experimental value of the ice point temperature on this scale, and (b) the temperature interval between the ice and steam points.

1–12 Suppose that a numerical value of exactly 492 is assigned to the ice point temperature, and that the ratio of two temperatures is defined as the limiting ratio, as $P_i \to 0$, of the corresponding pressures of a gas kept at constant volume. Find (a) the best experimental value of the steam point temperature on this scale, and (b) the temperature interval between the ice and steam points.

1–13 The pressure of an ideal gas kept at constant volume is given by the equation

$$P = AT$$

where T is the thermodynamic temperature and A is a constant. Let a temperature T^* be defined by

$$T^* = B \ln CT$$

where B and C are constants. The pressure P is 0.1 atm at the triple point of water. The temperature T^* is 0 at the triple point and T^* is 100 at the steam point. (a) Find the values of A, B, and C. (b) Find the value of T^* when $P = 0.15$ atm. (c) Find the value of P when T^* is 50. (d) What is the value of T^* at absolute zero? (e) Sketch a graph of T^* versus the Celsius temperature t for $-200°\text{C} < t < 200°\text{C}$.

1–14 When one junction of a thermocouple is kept at the ice point, and the other junction is at a Celsius temperature t, the emf $\mathscr{E}$ of the thermocouple is given by a quadratic function of t:

$$\mathscr{E} = \alpha t + \beta t^2.$$

If $\mathscr{E}$ is in millivolts, the numerical values of α and β for a certain thermocouple are found to be

$$\alpha = .50, \qquad \beta = -1 \times 10^{-3}.$$

(a) Compute the emf when $t = -100°\text{C}$, $200°\text{C}$, $400°\text{C}$, and $500°\text{C}$, and sketch a graph of $\mathscr{E}$ versus t. (b) Suppose the emf is taken as a thermometric property and that a temperature scale t^* is defined by the linear equation

$$t^* = a\mathscr{E} + b.$$

Let $t^* = 0$ at the ice point, and $t^* = 100$ at the steam point. Find the numerical values of a and b and sketch a graph of $\mathscr{E}$ versus t^*. (c) Find the values of t^* when $t = -100°\text{C}$, $200°\text{C}$, $400°\text{C}$, and $500°\text{C}$, and sketch a graph of t^* versus t over this range. (d) Is the t^* scale a Celsius scale? Does it have any advantage or disadvantages compared with the IPTS scale?

1–15 The thermodynamic temperature of the normal boiling point of nitrogen is 77.35 K. Calculate the corresponding value of (a) the Celsius, (b) the Rankine, and (c) the Fahrenheit temperature.

1–16 The thermodynamic temperature of the triple point of nitrogen is 63.15 K. Using the data of the preceding problem, what is the temperature difference between the boiling point and the triple point of nitrogen on (a) the Kelvin, (b) the Celsius, (c) the Rankine, and (d) the Fahrenheit scales? Include the proper unit in each answer.

1–17 A mixture of hydrogen and oxygen is isolated and allowed to reach a state of constant temperature and pressure. The mixture is exploded with a spark of negligible energy and again allowed to come to a state of constant temperature and pressure. (a) Is the initial state an equilibrium state? Explain. (b) Is the final state an equilibrium state? Explain.

1–18 (a) Describe how a system containing two gases can be in mechanical but not in thermal or chemical equilibrium. (b) Describe how a system containing two gases can be in thermal but not in mechanical or chemical equilibrium. (c) Describe how a system containing two gases can be in thermal and mechanical equilibrium but not in chemical equilibrium.

1–19 On a graph of volume versus temperature draw and label lines indicating the following processes, each proceeding from the same initial state T_0, V_0: (a) an isothermal expansion; (b) an isothermal compression; (c) an isochoric increase in temperature.

1–20 Give an example of (a) a reversible isochoric process; (b) a quasistatic, adiabatic, isobaric process; (c) an irreversible isothermal process. Be careful to specify the system in each case.

1–21 Using the nomenclature similar to that in the previous problem, characterize the following processes. (a) The temperature of a gas, enclosed in a cylinder provided with a frictionless piston, is slowly increased. The pressure remains constant. (b) A gas, enclosed in a cylinder provided with a piston, is slowly expanded. The temperature remains constant. There is a force of friction between the cylinder wall and the piston. (c) A gas enclosed in a cylinder provided with a frictionless piston is quickly compressed. (d) A piece of hot metal is thrown into cold water. (Assume that the system is the metal which neither contracts nor expands.) (e) A pendulum with a frictionless support swings back and forth. (f) A bullet is stopped in a target.

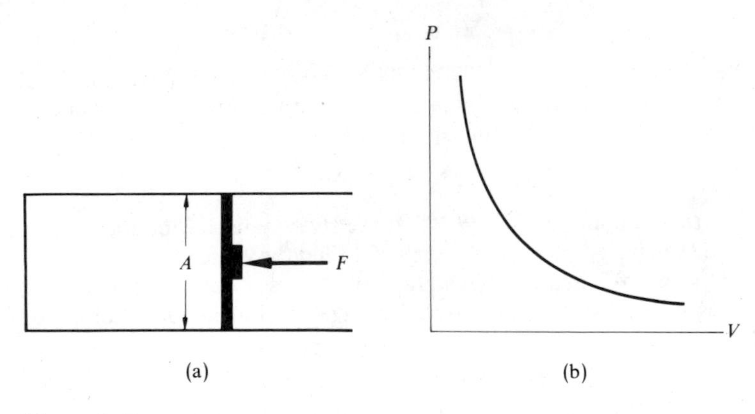

Figure 1–8

1–22 A gas is enclosed in a cylinder provided with a piston of area A, as in Fig. 1–8(a). The relation between the pressure and volume of the gas, at a constant temperature T, is shown in Fig. 1–8(b). On a similar figure sketch graphs of the ratio of the external force F to the area A, F/A, as a function of V, as the gas is (a) slowly compressed, and (b) slowly expanded at the temperature T. There is a force of sliding friction f between the piston and the cylinder.

2

Equations of state

2–1 EQUATIONS OF STATE

It is found by experiment that only a certain minimum number of the properties of a pure substance can be given arbitrary values. The values of the remaining properties are then determined by the nature of the substance. For example, suppose that oxygen gas is allowed to flow into an evacuated tank, the tank and its contents being kept at a thermodynamic temperature T. The volume V of the gas admitted is then fixed by the volume of the tank and the mass m of gas is fixed by the amount which we allow to enter. Once T, V, and m have been fixed, the pressure P is determined by the nature of oxygen and cannot be given any arbitrary value. It follows that there exists a certain relation between the properties P, V, T, and m which can be expressed in general as

$$f(P, V, T, m) = 0. \tag{2–1}$$

This relation is known as the *equation of state* of the substance. If *any* three of the properties are fixed, the fourth is determined.

In some instances, properties in addition to those listed above are necessary to completely describe the state of a system and these properties must be included in the equation of state. Examples are the area and surface tension of a liquid-vapor surface, the magnetization and flux density in a magnetic material, and the state of charge of an electrolytic cell. For the present, however, we shall consider only systems whose state can be completely described by the properties P, V, T, and m.

The equation of state can be written in a form which depends only on the *nature* of a substance, and not on *how much* of the substance is present, if all extensive properties are replaced by their corresponding specific values, per unit mass or per mole. Thus if the properties V and m are combined in the single intensive property $v = V/m$, the equation of state becomes

$$f(P, v, T) = 0. \tag{2–2}$$

The equation of state varies from one substance to another. In general, it is an extremely complicated relation and is often expressed as a converging power series. A general idea of the nature of the function is often better conveyed by presenting the data in graphical form.

2–2 EQUATION OF STATE OF AN IDEAL GAS

Suppose one has measured the pressure, volume, temperature, and mass of a certain gas, over wide ranges of these variables. Instead of the actual volume V, we shall use the molal specific volume, $v = V/n$. Let us take all the data collected at a given temperature T, calculate for each individual measurement the ratio Pv/T, and plot these ratios as ordinates against the pressure P as abscissa. It is found experimentally that these ratios all lie on a smooth curve, whatever the temperature, but that the ratios at different temperatures lie on different curves.

The data for carbon dioxide are plotted in Fig. 2–1, for three different temperatures. The remarkable feature of these curves is (a) that they all converge to exactly the same point on the vertical axis, whatever the temperature, and (b) that the curves for all other gases converge to exactly the same point. This common limit of the ratio Pv/T, as P approaches zero, is called the *universal gas constant* and is denoted by R. The unit of Pv/T is

$$1(\text{N m}^{-2})(\text{m}^3\text{ kilomole}^{-1})(\text{K}^{-1}) = 1(\text{N m})(\text{kilomole}^{-1}\text{ K}^{-1}) = 1\text{ J kilomole}^{-1}\text{ K}^{-1},$$

and the value of R in this system is

$$R = 8.3143 \times 10^3\text{ J kilomole}^{-1}\text{ K}^{-1}.$$

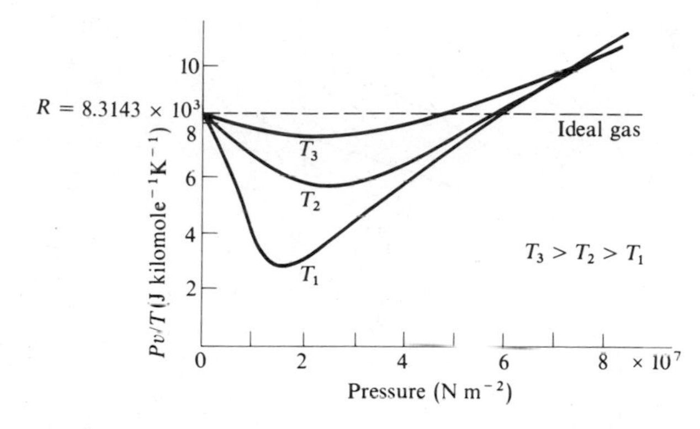

Fig. 2–1 The limiting value of Pv/T is independent of T for all gases. For an ideal gas, Pv/T is constant.

It follows that at sufficiently low pressures we can write, for all gases,

$$Pv/T = R, \qquad \text{or} \quad Pv = RT.$$

It is convenient to postulate an *ideal gas* for which, by definition, the ratio Pv/T is exactly equal to R at all pressures and temperatures. The equation of state of an ideal gas is therefore

$$Pv = RT, \tag{2-3}$$

or, since $v = V/n$,

$$PV = nRT. \tag{2-4}$$

For an ideal gas, the curves in Fig. 2–1 coalesce into a single horizontal straight line at a height R above the pressure axis.

2–3 *P-v-T* SURFACE FOR AN IDEAL GAS

The equation of state of a PvT system defines a *surface* in a rectangular coordinate system in which P, v, and T are plotted along the three axes. A portion of this surface for an ideal gas is shown in Fig. 2–2. Every possible equilibrium state of an ideal gas is represented by a point on its P-v-T surface, and every point on the surface represents a possible equilibrium state. A quasistatic process, i.e., a succession of equilibrium states, is represented by a line on the surface. The full lines in Fig. 2–2 represent processes at constant temperature, or *isothermal* processes. The dotted lines represent *isochoric* processes, and the dashed lines, *isobaric* processes.

Figures 2–3(a) and 2–3(b) are projections of the lines in Fig. 2–2 onto the P-v and P-T planes.

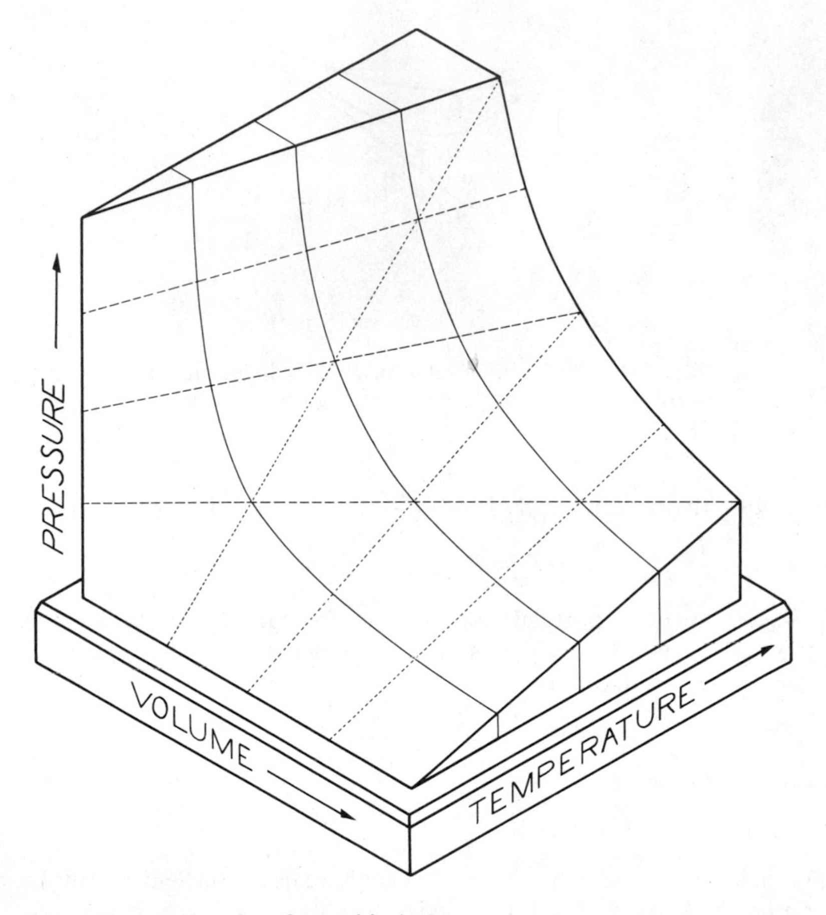

Fig. 2–2 P-v-T surface for an ideal gas.

In an isothermal process, for a fixed mass of an ideal gas,

$$Pv = RT = \text{constant}. \tag{2–5}$$

Robert Boyle*, in 1660, discovered experimentally that the product of the pressure and volume is very nearly constant for a fixed mass of a real gas at constant temperature. This fact is known as *Boyle's law*. It is, of course, exactly true for an ideal gas, by definition. The curves in Fig. 2–3(a) are graphs of Eq. (2–5) for different temperatures and hence for different values of the constant. They are equilateral hyperbolas.

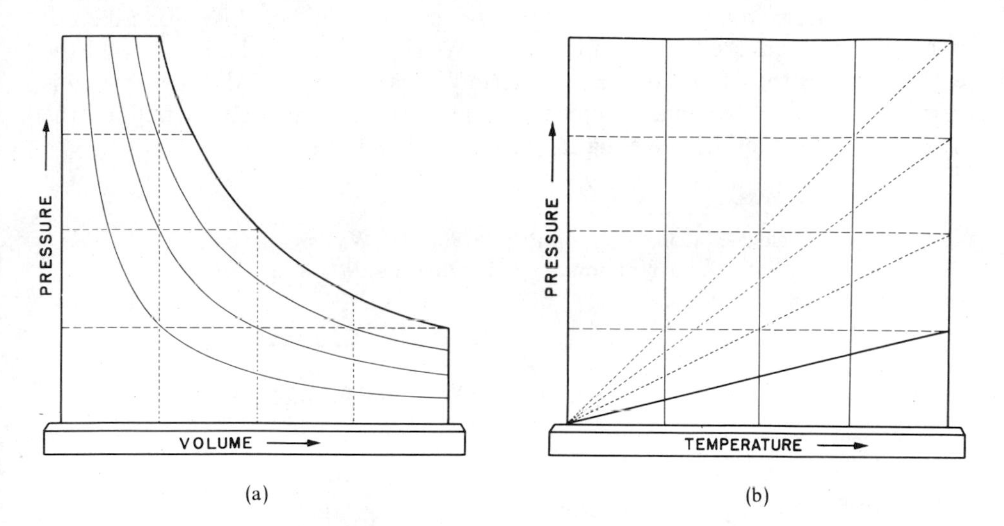

Fig. 2–3 Projections of the ideal gas *P-v-T* surface onto (a) the *P-v* plane, and (b) the *P-T* plane.

In a process at constant volume, for a fixed mass of an ideal gas,

$$P = \left(\frac{nR}{V}\right)T = \text{constant} \times T. \tag{2–6}$$

That is, the pressure is a linear function of the temperature T. The dotted lines in Fig. 2–3(b) are graphs of Eq. (2–6) for different volumes and hence different values of the constant.

If the pressure of a fixed mass of an ideal gas is constant,

$$V = \left(\frac{nR}{P}\right)T = \text{constant} \times T, \tag{2–7}$$

and the volume is a linear function of the temperature at constant pressure.

* Robert Boyle, British chemist (1627–1691).

2–4 EQUATIONS OF STATE OF REAL GASES

Many equations have been proposed which describe the P-v-T relations of real gases more accurately than does the equation of state of an ideal gas. Some of these are purely empirical, while others are derived from assumptions regarding molecular properties. Van der Waals*, in 1873, derived the following equation:

$$\left(P + \frac{a}{v^2}\right)(v - b) = RT. \tag{2–8}$$

The quantities a and b are constants for any one gas but differ for different gases. Some values are listed in Table 2–1. We shall show in Chapter 10 that the term a/v^2 arises from the existence of intermolecular forces and that the term b is proportional to the volume occupied by the molecules themselves, but for the present we shall consider the equation as an empirical one.

Table 2–1 Constants a and b in van der Waals equation. P in N m^{-2}, v in m^3 kilomole^{-1}, T in kelvins, $R = 8.31 \times 10^3$ J kilomole^{-1} K^{-1}.

Substance	a (J m^3 kilomole^{-2})	b (m^3 kilomole^{-1})
He	3.44×10^3	0.0234
H_2	24.8	.0266
O_2	138	.0318
CO_2	366	.0429
H_2O	580	.0319
Hg	292	.0055

At sufficiently large specific volumes, the term a/v^2 becomes negligible in comparison with P, and b becomes negligible in comparison with v. The van der Waals equation then reduces to the equation of state of an ideal gas, which any equation of state must do at large specific volumes.

Figure 2–4 is a diagram of a portion of the P-v-T surface of a van der Waals gas, and Fig. 2–5 is a projection of a number of isotherms onto the P-v plane.

---

* Johannes D. van der Waals, Dutch physicist (1837–1923).

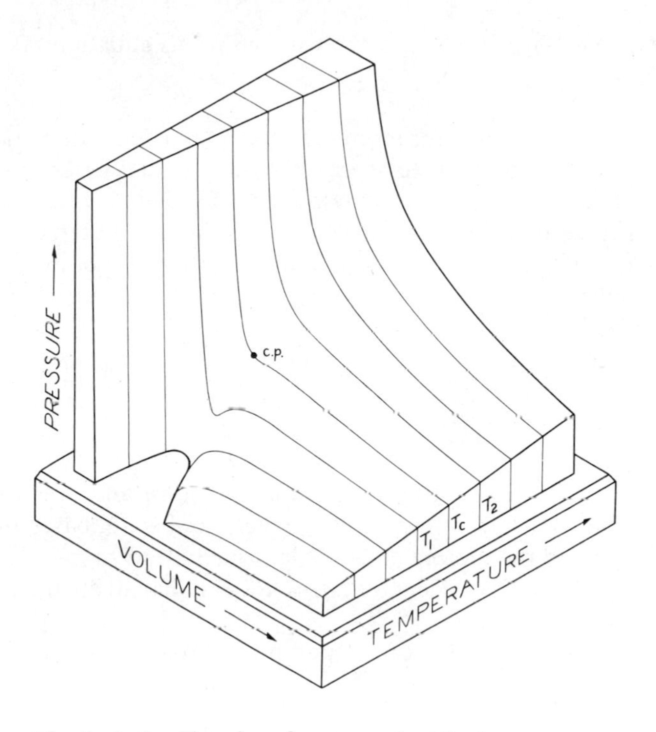

Fig. 2–4 *P-v-T* surface for a van der Waals gas.

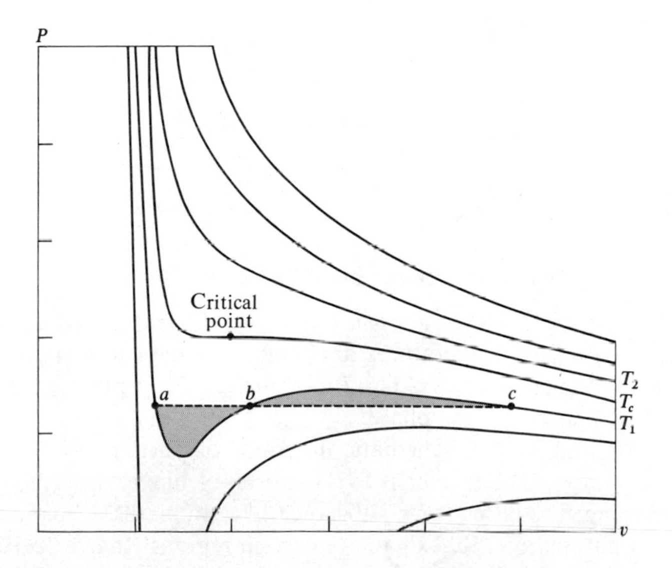

Fig. 2–5 Isotherms of a van der Waals gas.

When expanded in powers of v, the van der Waals equation takes the form

$$Pv^3 - (Pb + RT)v^2 + av - ab = 0. \tag{2-9}$$

It is therefore a cubic in v and for given values of P and T has three roots, of which only one need be real. For low temperatures, such as that lettered T_1 in Fig. 2–5, three positive real roots exist over a certain range of values of P. As the temperature increases, the three real roots approach one another, and at the temperature T_c they become equal. Above this temperature only one real root exists for all values of P. The significance of the point lettered c.p. and of the dotted line abc, will be explained in Section 2–5.

Another useful form of the equation of state of a real gas is

$$Pv = A + \frac{B}{v} + \frac{C}{v^2} + \cdots, \tag{2-10}$$

where A, B, C, etc., are functions of the temperature and are called the *virial coefficients*. Theoretical derivations of the equation of state, based on an assumed law of force between the molecules of a gas, usually lead to an equation in virial form. For an ideal gas, it is evident that $A = RT$ and that all other virial coefficients are zero.

The van der Waals equation can be put in virial form as follows. We first write it as

$$Pv = RT\left(1 - \frac{b}{v}\right)^{-1} - \frac{a}{v}.$$

By the binomial theorem,

$$\left(1 - \frac{b}{v}\right)^{-1} = 1 + \frac{b}{v} + \frac{b^2}{v^2} + \cdots.$$

Hence

$$Pv = RT + \frac{RTb - a}{v} + \frac{RTb^2}{v^2} + \cdots, \tag{2-11}$$

and for a van der Waals gas, $A = RT$, $B = RTb - a$, $C = RTb^2$, etc.

2–5 *P-v-T* SURFACES FOR REAL SUBSTANCES

Real substances can exist in the *gas phase* only at sufficiently high temperatures and low pressures. At low temperatures and high pressures transitions occur to the *liquid phase* and the *solid phase*. The P-v-T surface for a pure substance includes these phases as well as the gas phase.

Figures 2–6 and 2–7 are schematic diagrams of portions of the P-v-T surface for a real substance. The former is for a substance like carbon dioxide that contracts on freezing, the latter for a substance like water that expands on freezing. Study of the figures shows that there are certain regions (that is, certain ranges of the variables) in which the substance can exist in a single phase only. These are the regions lettered solid, liquid, and gas or vapor. (The distinction between a gas and a vapor will be discussed shortly.) In other regions, labeled solid-liquid, solid-vapor, and liquid-vapor, two phases can exist simultaneously in equilibrium, and along a line called the *triple line*, all three phases can coexist. As with the *P-v-T* surface for an ideal gas, any line on the surface represents a possible quasistatic process, or a succession of equilibrium states. The lines in Figs. 2–6 and 2–7 represent isothermal processes.

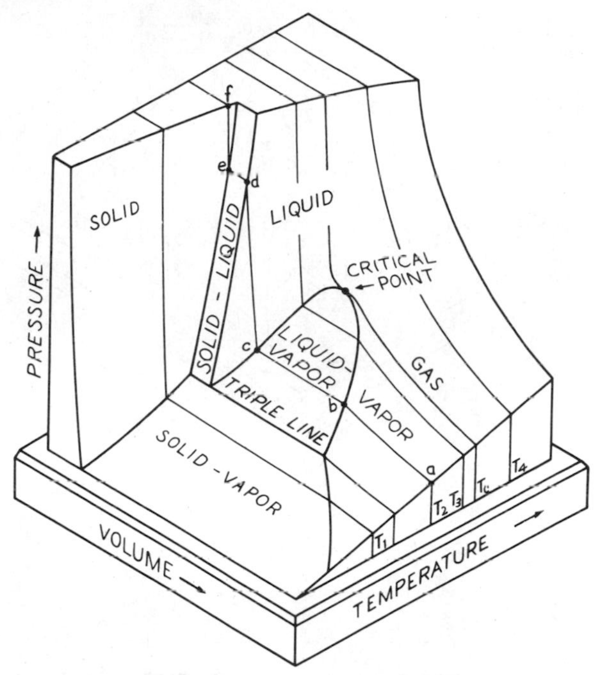

Fig. 2–6 *P-v-T* surface for a substance that contracts on freezing.

Those portions of a surface at which two phases can coexist are *ruled surfaces*. That is, a straight edge parallel to the v-axis makes contact with the surface at all points. Hence when the surfaces in Figs. 2–6 and 2–7 are projected onto the *P-T* plane, these surfaces project as lines. The projection of the surface in Fig. 2–6 onto the *P-T* plane is shown in Fig. 2–8(a), and that of the surface in Fig. 2–7 is shown in Fig. 2–9(a). The lines corresponding to values of pressure and temperature at which the solid and vapor phases, and the liquid and vapor phases, can coexist, always slope upward to the right. The line representing the equilibrium between solid and liquid slopes upward to the right in Fig. 2–8, but upward to the left in Fig. 2–9. We shall show in Section 7 6 that the former is characteristic of all substances that contract on freezing, the latter of substances (like water) that expand on freezing.

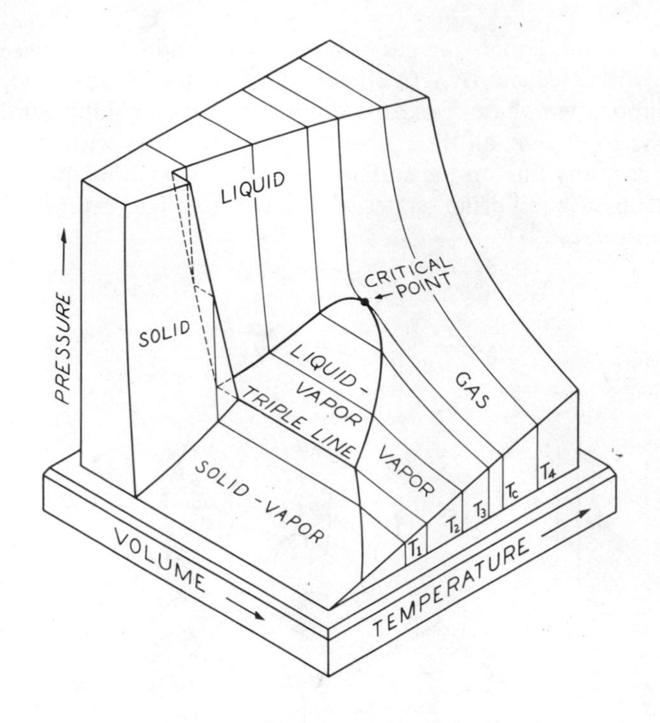

Fig. 2–7 *P-v-T* surface for a substance that expands on freezing.

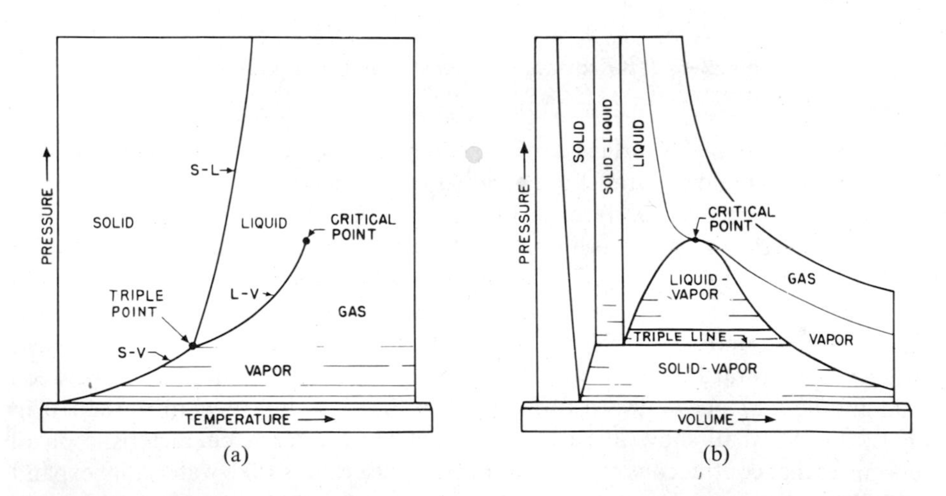

(a) (b)

Fig. 2–8 Projections of the surface in Fig. 2–6 onto (a) the *P-T* plane and (b) the *P-v* plane.

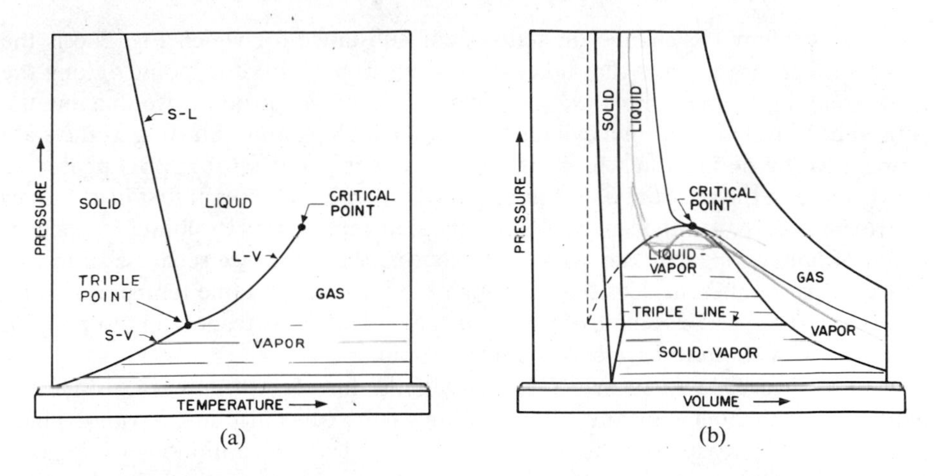

Fig. 2–9 Projections of the surface in Fig. 2–7 onto (a) the P-T plane and (b) the P-v plane.

The triple lines in Figs. 2–6 and 2–7 project as a point, called the *triple point*, in the P-T diagram. Triple-point data for a few common substances are given in Table 2–2. The triple-point temperature of water is the standard fixed point to which is assigned the arbitrary temperature of 273.16 K.

The projections of the surfaces in Figs. 2–6 and 2–7 onto the P-v plane are shown in Figs. 2–8(b) and 2–9(b). The surfaces can also be projected onto the v-T plane, but this projection is rarely used since all essential features of the surface can be shown in the first two projections.

Table 2–2 Triple-point data

Substance	Temperature, (K)	Pressure, (Torr)
Helium (4) (λ point)	2.186	38.3
Hydrogen (normal)	13.84	52.8
Deuterium (normal)	18.63	128
Neon	24.57	324
Nitrogen	63.18	94
Oxygen	54.36	1.14
Ammonia	195.40	45.57
Carbon dioxide	216.55	3880
Sulfur dioxide	197.68	1.256
Water	**273.16**	4.58

Let us follow the changes in state of the substance for which Fig. 2–6 is the *P-v-T* surface in a process that takes the system from point *a* to point *f* along the isothermal line at the temperature T_2. To carry out this process, we imagine the substance to be enclosed in a cylinder with a movable piston. Starting at the state represented by point *a*, at which the substance is in the gas (or vapor) phase, we slowly increase the pressure on the piston. The volume decreases at first in a manner approximating that of an ideal gas. When the state represented by point *b* is reached, drops of liquid appear in the cylinder.* That is, the substance separates into two phases of very different densities, although both are at the same temperature and pressure. The specific volume of the vapor phase is that corresponding to point *b*, and that of the liquid phase corresponds to point *c*.

With further decrease in volume, along the line *bc*, the pressure does not increase but remains constant. The fraction of the substance in the vapor phase continuously decreases and the fraction in the liquid phase continuously increases. In this part of the process, where liquid and vapor can exist in equilibrium, the vapor is called a *saturated vapor* and the liquid a *saturated liquid.* (The adjective "saturated" is an unfortunate one, for it brings to mind the concept of a "saturated solution," that is, one in which the concentration of a dissolved substance is a maximum. There is nothing dissolved in a saturated vapor; the substance that "precipitates" out with decreasing volume is not a solute but the same substance as that of which the vapor is composed.)

The pressure exerted by a *saturated* vapor or liquid is called the *vapor pressure.* The vapor pressure is evidently a function of temperature, increasing as the temperature increases. The curve lettered *L-V* in Fig. 2–8(a), the projection of the liquid-vapor surface onto the *P-T* plane, is the *vapor pressure curve.* The general shape of this curve is the same for all substances, but the vapor pressure at a given temperature varies widely from one substance to another. Thus at a temperature of 20°C, the vapor pressure of mercury is 0.0012 Torr, that of water is 17.5 Torr, and that of CO_2 is 42,960 Torr.

Let us now return to the isothermal compression process. At point *c* in Fig. 2–6 the substance is entirely in the liquid phase. To decrease the volume from that at point *c* to that at point *d*, a very large increase in pressure is required, since liquids are not very compressible. At point *d*, the substance again separates into two phases. Crystals of the solid begin to develop, with a specific volume corresponding to point *e*, and the pressure remains constant while both liquid and solid phases are present. The substance is entirely in the solid phase at point *e* and the volume decreases only slightly with further increase in pressure unless other forms of the solid can exist. Ice is an example of the latter case, where at least seven different forms have been observed at extremely high pressures, as illustrated in Fig. 2–10.

If the volume of the system is now slowly increased, all of the changes described above proceed in the opposite direction.

---

* However, see Section 7–5 for a further discussion of this phenomenon.

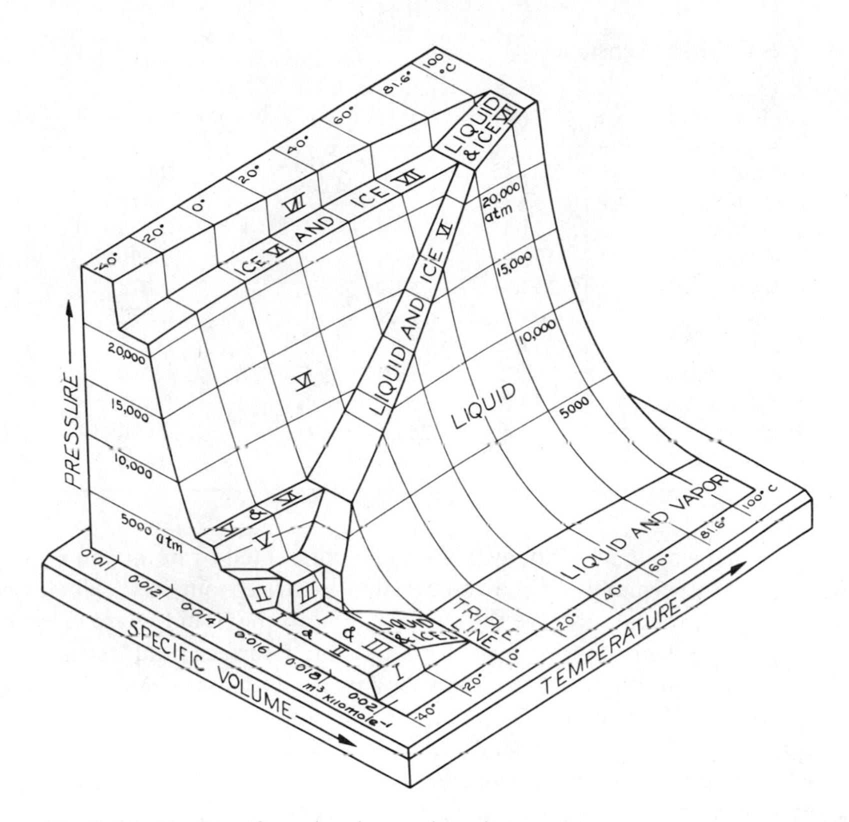

Fig. 2–10 *P-v-T* surface showing various forms of ice.

It will be seen from a study of Fig. 2–6 that if a compression process like that just described were carried out at a higher temperature, such as T_3, a higher pressure and a smaller specific volume would be required before a phase change from vapor to liquid commenced, and that when the substance was completely liquefied, its specific volume would be somewhat larger than that at the lower temperature. At the particular temperature lettered T_c, called the *critical temperature*, the specific volumes of saturated liquid and vapor become equal. Above this temperature, no separation into two phases of different densities occurs in an isothermal compression from a large volume. (That is, the liquid phase does not separate out. Separation into a gas and solid phase may occur at sufficiently high pressures.) The common value of the specific volumes of saturated liquid and vapor at the critical temperature is called the *critical specific volume*, v_c, and the corresponding pressure the *critical pressure*, P_c. The point on the *P-v-T* surface whose coordinates at P_c, v_c, and T_c is the *critical point*. The critical constants for a number of substances are given in Table 2–3.

Table 2–3 Critical constants

Substance	T_c(K)	P_c(N m^{-2})	v_c(m^3 kilomole^{-1})
Helium 4	5.25	1.16×10^5	0.0578
Helium 3	3.34	1.15	0.0726
Hydrogen	33.3	12.8	0.0650
Nitrogen	126.2	33.6	0.0901
Oxygen	154.8	50.2	0.078
Ammonia	405.5	111.0	0.0725
Freon 12	384.7	39.7	0.218
Carbon dioxide	304.2	73.0	0.094
Sulfur dioxide	430.7	77.8	0.122
Water	647.4	209.0	0.056
Carbon disulfide	552	78	0.170

Suppose that a system originally in the state represented by point *a* in Fig. 2–11 is compressed isothermally. If the compression is carried out in a cylinder with transparent walls, we can observe the condensation to the liquid phase commence at the point where the isotherm meets the liquid-vapor surface, and we can see the liquid phase grow in amount while the vapor phase decreases. At the state represented by point *b* we would be sure that the substance in the cylinder was wholly in the liquid phase. On the other hand, we could start with the substance in the same state (point *a*) and carry out the process represented by the line from *a* to *b* that curves around the critical point. (This process, of course, is not isothermal.) The end state of the system is the same in both processes but at no time in the second process did the substance separate into two phases. Nevertheless, it would certainly be described as a liquid at the end of the second process as well as at the end of the first. It has all the properties of a liquid; i.e., it is a fluid of high density (small specific volume) and small compressibility (the pressure increases rapidly for small decreases in volume), but its properties change *continuously* from those associated with a vapor, at point *a*, to those associated with a liquid, at point *b*. It is therefore possible to convert a vapor to a liquid without going through the process of "condensation," but no sharp dividing line can be drawn separating the portion of the *P-v-T* surface labeled "liquid" from that labeled "gas" or "vapor."

So far we have used the terms "gas" and "vapor" without distinguishing between them; and the distinction is, in fact, an artificial and unnecessary one. The term "vapor" is usually applied to a gas in equilibrium with its liquid (i.e., a *saturated* vapor) or to a gas at a temperature below its critical temperature, but the properties of a "vapor" differ in no essential respect from those of a "gas."

When the temperature of a gas at a given pressure is greater than the saturation temperature at this pressure, the gas is said to be "superheated" and is called a "superheated vapor." Thus "superheated" is synonymous with "nonsaturated."

Note that the term does not necessarily imply a high temperature. The saturation temperature of nitrogen at a pressure of 0.8 bar (its partial pressure in the earth's atmosphere) is -197.9°C, so that the nitrogen in the earth's atmosphere is always superheated.

One may wonder whether or not the edges of the solid-liquid surface approach one another as do those of the liquid-vapor surface, and if there is another critical point for the solid-liquid transition. No such point has ever been observed; i.e., there is always a finite difference in specific volume or density between the liquid and solid phases of a substance at the same temperature and pressure. This does not exclude the possibility of such critical points existing at extremely high pressures.

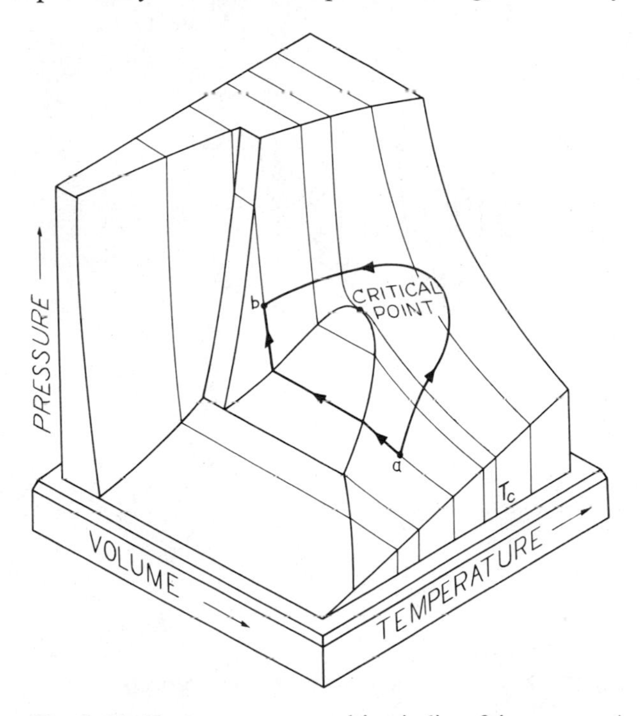

Fig. 2–11 Two processes resulting in liquefying a gas. A phase separation is observed in the isothermal process, but not in the other process.

Now consider the phase changes in an isobaric process. Suppose we have a vessel of liquid open to the atmosphere at a pressure P_1, in the state represented by point a in Fig. 2–12. If the temperature is increased at constant pressure, the representative point moves along an isobaric line to point b. When point b is reached, the system separates into two phases, one represented by point b and the other by point c. The specific volume of the vapor phase is much greater than that of the liquid, and the volume of the system increases greatly. This is the familiar phenomenon of boiling. If the vessel is open, the vapor diffuses into the atmosphere. Thus the temperature T_b at which a liquid boils is merely that temperature at which its vapor pressure is equal to the external pressure, and the vapor pressure curve in Fig. 2–8(a) can also be considered as the *boiling point curve*. If the substance diagramed in Fig. 2–12 is water (actually the solid-liquid line for water slopes in the opposite direction) and the pressure P_1 is 1 atmosphere, the corresponding temperature T_b is 373 K. The vapor pressure curve always slopes upward to the right, so that an increase in external pressure always results in an elevation of the boiling temperature, and vice versa.

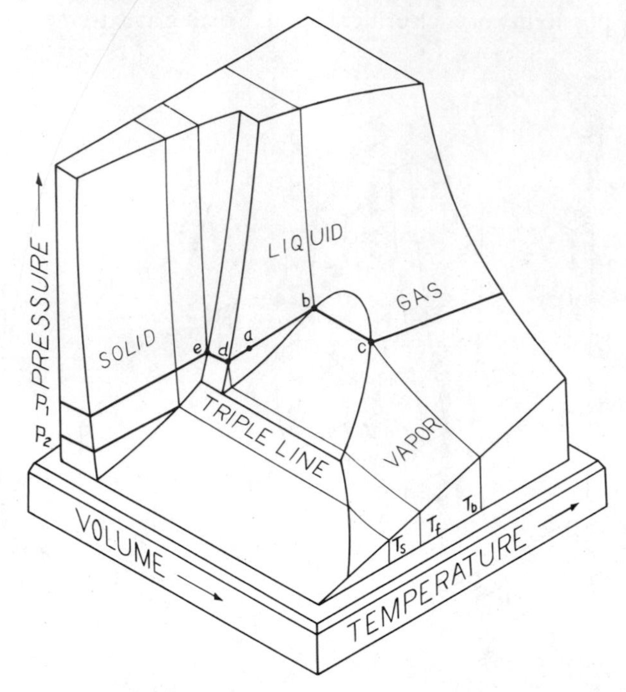

Fig. 2–12 Phase changes in an isobaric process.

If, starting with the liquid at point *a* in Fig. 2–12, the temperature is lowered while the pressure is kept constant, the representative point moves along the isobaric line to point *d*. At this point, the system again separates into two phases, one represented by point *d* and the other by point *e*. For a substance like that represented in Fig. 2–12, the specific volume of the solid is less that that of the liquid, and the volume decreases. The process is that of *freezing*, and evidently the solid-liquid equilibrium line in a *P-T* diagram like Fig. 2–8 is the *freezing point curve*, and at the pressure P_1 the freezing temperature is T_f. If the solid-liquid equilibrium line slopes upward to the right as in Fig. 2–12, an increase in pressure raises the freezing point, and vice versa.

It is evident from a study of Fig. 2–12 that the liquid phase cannot exist at a temperature lower than that of the triple point, or at a pressure less than that at the triple point. If the pressure is less than that at the triple point, say the value P_2, the substance can exist in the solid and vapor phases only, or both can exist in equilibrium. The transition from one to the other takes place at the *temperature of sublimation* T_s. Thus the solid-vapor equilibrium curve is also the *sublimation point curve*.

For example, the triple-point temperature of CO_2 is $-56.6°C$ and the corresponding pressure is 5.2 bar. Liquid CO_2 therefore cannot exist at atmospheric pressure. When heat is supplied to solid CO_2 (dry ice) at atmospheric pressure, it

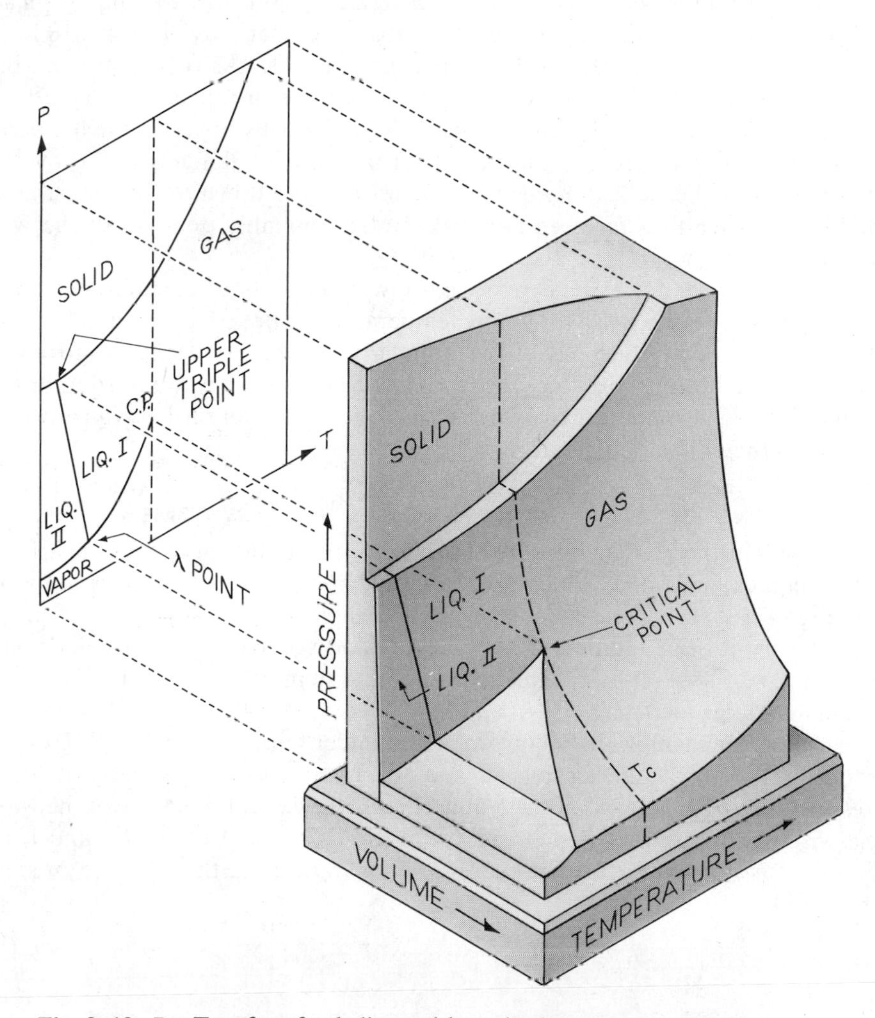

Fig. 2–13 *P-v-T* surface for helium with projection onto the *P-T* plane.

sublimes and changes directly to the vapor phase. Liquid CO_2 can, of course, exist at room temperature, provided the pressure is sufficiently high. This material is commonly stored in steel tanks which when "full" contain mostly liquid and a small amount of vapor (both, of course, saturated). The temperature is room temperature if the tank has been standing in the room, and the pressure is that of the ordinate of the vapor pressure curve at room temperature.

Figure 2–13 is a schematic diagram of the P-v-T surface of ordinary helium (of mass number 4). This substance exhibits a unique behavior at low temperatures in the neighborhood of 2 K. The critical temperature and pressure are 5.25 K and 2.29 bar respectively. When helium in the vapor phase is compressed isothermally at temperatures between 5.25 K and 2.18 K, it condenses to a liquid phase called helium I. When the vapor is compressed at temperatures below 2.18 K, a liquid phase called helium II, which is superfluid, results. As is evident from the diagram, He I and He II can coexist in equilibrium over a range of temperatures and pressures, and He I can be converted to He II either by lowering the temperature, provided the pressure is not too great, or by reducing the pressure, provided the temperature is below 2.18 K. He II remains a liquid down to the lowest temperatures that have thus far been attained, and presumably does so all the way down to absolute zero.

Solid helium cannot exist at pressures lower than about 25 bar, nor can it exist in equilibrium with its vapor at any temperature or pressure. Helium has two triple points, at one of which (called the lambda-point or λ-point) the two forms of liquid are in equilibrium with the vapor; while at the other they are in equilibrium with the solid. It is interesting to note that the solid phase can exist at temperatures greater than the critical temperature.

2–6 EQUATIONS OF STATE OF OTHER THAN *P-v-T* SYSTEMS

The principles of thermodynamics are of general applicability and are not restricted to gases, liquids, and solids under a uniform hydrostatic pressure. Depending on the nature of a system, we may be interested in intensive-extensive pairs of properties other than, or in addition to, the pressure and volume of the system. Whatever its nature, however, the *temperature* of a system is always a fundamental thermodynamic property.

Consider, for example, a metal wire or rod under tension. The length L of the wire depends both on the tension $\mathscr{F}$ and the temperature T, and the relation expressing the length in terms of these quantities is the equation of state of the wire. If the wire is not stretched beyond its proportional limit of elasticity, and if its temperature is not too far from a reference temperature T_0, the equation of state of the wire is

$$L = L_0\left[1 + \frac{\mathscr{F}}{YA} + \alpha(T - T_0)\right], \tag{2–12}$$

where L_0 is length under zero tension at the temperature T_0, Y is the isothermal stretch modulus (Young's* modulus), A is the cross-sectional area, and α is the coefficient of linear expansion, or the *linear expansivity.* In this example, the intensive variable is the tension $\mathscr{F}$ and the extensive variable is the length L.

The magnetic moment M of a paramagnetic material, within which there is a uniform magnetic field of intensity $\mathscr{H}$, depends both on $\mathscr{H}$ and the temperature T. Except at extremely low temperatures and in large fields, the magnetic moment can be represented with sufficient accuracy by the equation

$$M = C_C \frac{\mathscr{H}}{T}, \qquad (2\text{–}13)$$

where C_C, a constant characteristic of a given material, is called the *Curie*† *constant.* This relation is known as *Curie's law.* The magnetic moment M is an extensive variable and the field intensity $\mathscr{H}$ is an intensive variable.

The total dipole moment P of a dielectric in an external electric field E is given by a similar equation:

$$P = \left(a + \frac{b}{T}\right)E. \qquad (2\text{–}14)$$

The surface film of a liquid can be considered a thermodynamic system, although it is not a closed system because as the surface area of a given mass of liquid is changed, molecules move from the liquid into the film, or vice versa. The intensive property of interest is the surface tension σ, which may be defined as the force per unit length exerted by the film on its boundary. The corresponding extensive property is the area of the film, but unlike the systems considered thus far (and unlike a stretched rubber membrane) the surface tension is independent of the area of the film and depends only on its temperature. The surface tension of all liquids decreases with increasing temperature and becomes zero at the critical temperature T_c (see Section 8–4). To a first approximation, the surface tension can be represented by the equation

$$\sigma = \sigma_0\left(\frac{T_c - T}{T_c - T_0}\right), \qquad (2\text{–}15)$$

where σ_0 is the surface tension at a reference temperature T_0.

Another thermodynamic system, and one that is of great importance in physical chemistry, is the electrolytic cell. The electromotive force $\mathscr{E}$ of the cell is the intensive property of interest, and the corresponding extensive property is the charge Z, whose absolute value is of no importance but whose *change* in any process equals the quantity of charge flowing past a point in a circuit to which the cell is connected, and which is proportional to the number of moles reacting in the cell in the process. An electrolytic cell resembles a surface film in that the emf of a given

* Thomas Young, British physicist (1773–1829).

† Pierre Curie, French physicist (1859–1906).

cell depends only on the temperature and not on the charge Z. The emf can be represented by a power series in the temperature and is usually written as

$$\mathscr{E} = \mathscr{E}_{20} + \alpha(t - 20°) + \beta(t - 20°)^2 + \gamma(t - 20°)^3, \tag{2–16}$$

where t is the Celsius temperature, $\mathscr{E}_{20}$ is the emf at 20°C, and α, β, and γ are constants depending on the materials composing the cell.

2–7 PARTIAL DERIVATIVES. EXPANSIVITY AND COMPRESSIBILITY

The equation of state of a *PVT* system is a relation between the values of pressure, volume, and temperature for any equilibrium state of the system. The equation defines a surface in a rectangular coordinate system, and Fig. 2–14 represents schematically the *P-V-T* surface of a solid or liquid. (The vertical scale is greatly exaggerated.) The volume increases with increasing temperature if the pressure is constant, and decreases with increasing pressure if the temperature is constant. The surface in Fig. 2–14 corresponds to the surfaces lettered "solid" or "liquid" in Figs. 2–6 and 2–7, except that in Fig. 2–14 the volume axis is vertical and the pressure axis is horizontal.

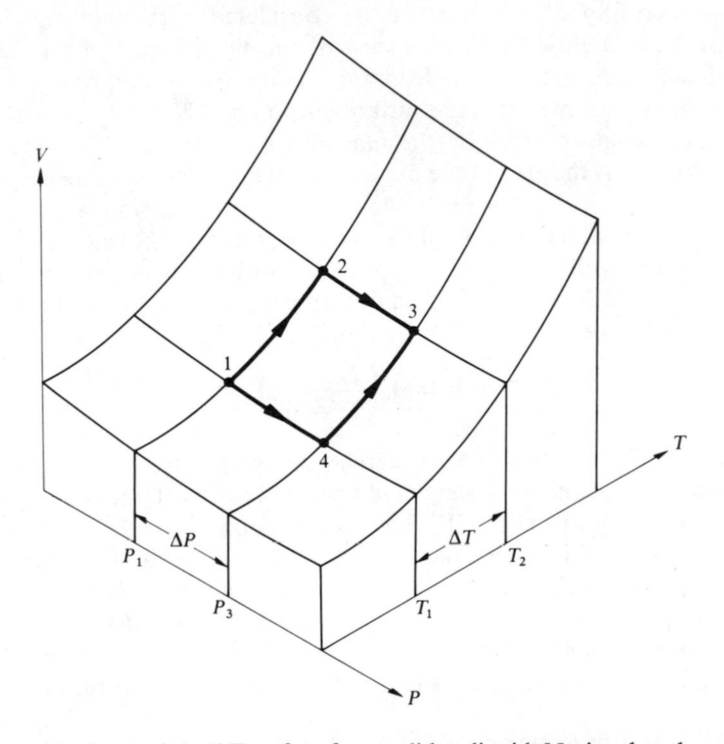

Fig. 2–14 A *P-V-T* surface for a solid or liquid. Notice that the *V* axis is now vertical and has been greatly exaggerated.

If the equation of state is solved for V, thus expressing V as a function of the *two* independent variables P and T, the value of V corresponds to the vertical height of the surface above the P-T plane, at any given pair of values of P and T.

Instead of specifying the height of the surface above the P-T plane, at any point, the surface can be described by giving its *slope* at any point. More specifically, we can specify the slope, at any point, of the lines of intersection of the surface with planes of constant pressure and of constant temperature.

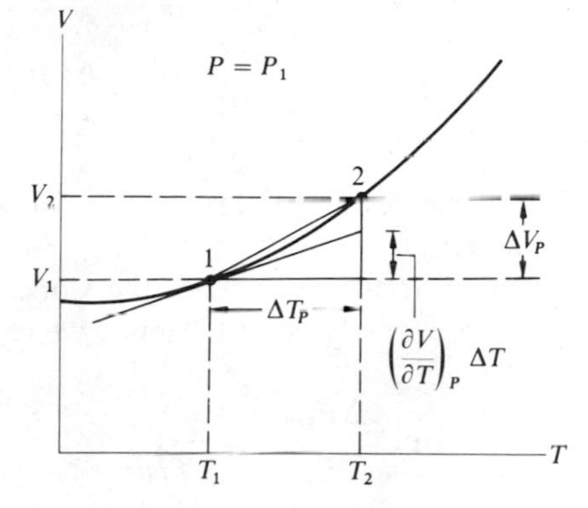

Fig. 2–15 The intersection of the surface of Fig. 2–14 with the v-T plane at pressure P_1.

The curve in Fig. 2–15 is a graph of the intersection of the surface in Fig. 2–14 with the plane at which the pressure has the constant value P_1. That is, it is a graph of the volume V as a function of the temperature T, for the isobaric curve along which the pressure equals P_1. The *slope* of this curve at any point means the slope of the *tangent* to the curve at that point, and this is given by the derivative of V with respect to T at the point. In Fig. 2–15, the tangent has been constructed at point 1, at which the temperature is T_1 and the pressure is P_1. However, the volume V is a function of P as well as of T, and since P is constant along the curve, the derivative is called the *partial derivative of V with respect to T at constant pressure* and is written:

$$\text{Slope of tangent} = \left(\frac{\partial V}{\partial T}\right)_P.$$

If the equation of state is known, expressing V as a function of T and P, the partial derivative is calculated in the same way as an ordinary derivative of a function of a *single* variable, except that P is considered constant. Thus if the system is an ideal gas, for which $V = nRT/P$, the quantity nR/P is considered constant and

$$\left(\frac{\partial V}{\partial T}\right)_P = \frac{nR}{P}.$$

In mathematics, the partial derivative would be written simply as $(\partial V/\partial T)$. In thermodynamics, the subscript P is included because, as we shall see later, a PVT system has many other properties in addition to pressure, volume, and temperature, and the volume can be expressed in terms of any two of these. The subscript not only indicates that P is held constant, but that V is to be expressed in terms of P and T.

Point 2, in Figs. 2–14 and 2–15, is a second point on the isobaric curve at which the volume is V_2 and the temperature is T_2. The slope of the *chord* from point 1 to point 2 is

$$\text{Slope of chord} = \frac{V_2 - V_1}{T_2 - T_1} = \frac{\Delta V_P}{\Delta T_P},$$

where again the subscript P indicates that the pressure is constant. The slope of the chord is not equal to the slope of the tangent, but if point 2 is taken closer and closer to point 1, so that ΔT_P approaches zero, the slopes of the chord and the tangent become more and more nearly equal. Hence we can say

$$\lim_{\Delta T_P \to 0} \frac{\Delta V_P}{\Delta T_P} = \left(\frac{\partial V}{\partial T}\right)_P. \tag{2–17}$$

Another point of view is the following. Suppose the volume of the system were to increase with temperature, not along the actual curve but along the *tangent* at point 1. The increase in volume when the temperature was increased by ΔT_P would then be represented by the length of the intercept of the tangent on the vertical line through point 2, or it would be given by

$$\left(\frac{\partial V}{\partial T}\right)_P \Delta T_P,$$

the product of the slope of the tangent line, $(\partial V/\partial T)_P$, and the base ΔT_P.

As can be seen from Fig. 2–15, the intercept is not equal to ΔV_P, but the two approach equality as ΔT_P approaches zero. Then

$$\lim_{\Delta T_P \to 0} \left(\frac{\partial V}{\partial T}\right)_P \Delta T_P = \Delta V_P, \tag{2–18}$$

which is the same as Eq. (2–17). Hence if we let dV_P and dT_P represent the *limiting* values of ΔV_P and ΔT_P as $\Delta T_P \to 0$, we can write

$$dV_P = \left(\frac{\partial V}{\partial T}\right)_P dT_P. \tag{2–19}$$

Instead of giving the value of the slope itself at any point, it is convenient to give the value of the slope, $(\partial V/\partial T)_P$, divided by the volume V at the point. The quotient is called the *coefficient of volume expansion* of the material, or its *expansivity* β, defined as

$$\beta \equiv \frac{1}{V}\left(\frac{\partial V}{\partial T}\right)_P. \tag{2–20}$$

Thus for an ideal gas,

$$\beta = \frac{1}{V}\frac{nR}{P} = \frac{1}{T}, \tag{2–21}$$

and the expansivity depends only on the temperature and is equal to the reciprocal of the temperature. The unit of expansivity is evidently 1 *reciprocal kelvin* (1 K^{-1}).

Equation (2–20) can also be written in terms of specific volumes:

$$\beta = \frac{1}{v}\left(\frac{\partial v}{\partial T}\right)_P. \tag{2–22}$$

It follows from Eq. (2–20) that for two closely adjacent states of a system at the same pressure,

$$\beta = \frac{1}{V}\frac{dV_P}{dT_P} = \frac{dV_P/V}{dT_P}. \tag{2–23}$$

The expansivity can therefore be described as the limiting value of the *fractional* increase in volume, dV_P/V, per unit change in temperature at constant pressure.

The *mean* expansivity $\bar{\beta}$ over a finite temperature interval between T_1 and T_2 is defined as

$$\bar{\beta} = \frac{(V_2 - V_1)/V_1}{T_2 - T_1} = \frac{1}{V_1}\frac{\Delta V_P}{\Delta T_P}. \tag{2–24}$$

That is, the mean expansivity equals the slope of the *chord* shown in Fig. 2–15, $\Delta V_P/\Delta T_P$, divided by the volume V_1.

Since both the slope of an isobar and the volume V will in general vary from point to point, the expansivity is a function of both temperature and pressure. Figure 2–16 shows how the expansivity β of copper varies with temperature at a constant pressure of 1 atm, from absolute zero up to a temperature of 1200 K. The *ordinate* of this graph, at any temperature, is equal to the *slope* of a graph of V versus T, as in Fig. 2–15, divided by the volume. A particularly interesting feature of the graph in Fig. 2–16 is that the expansivity approaches zero as the temperature approaches zero. Other metals show a similar variation.

Figure 2–17 shows how the expansivity of mercury varies with pressure at a constant temperature of 0°C. Note that the origin of the scale of β, in Fig. 2–17, does not appear in the diagram; the expansivity changes only slightly with changes in pressure, even for pressures as great as 7000 atm.

Liquid water has a maximum density and a minimum specific volume at a temperature of 4°C. In the temperature interval between 0°C and 4°C its specific volume *decreases* with increasing temperature and its expansivity is *negative*, while at 4°C it is zero.

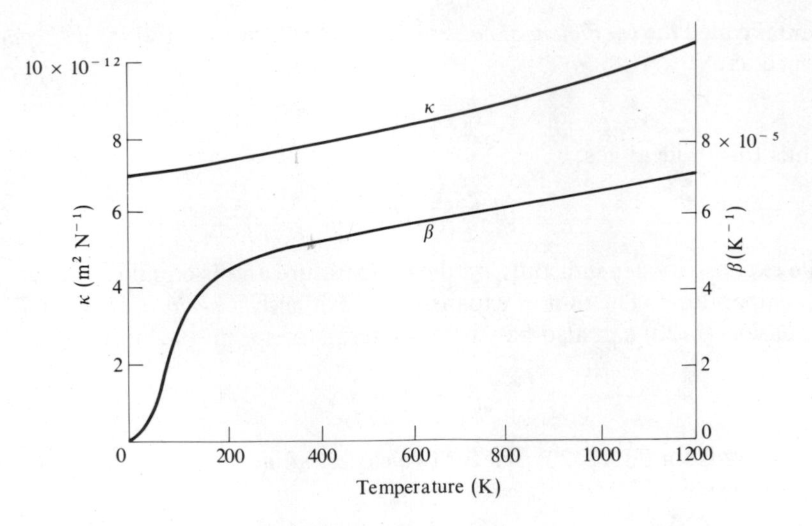

Fig. 2–16 Compressibility κ and expansivity β of copper as functions of temperature at a constant pressure of 1 atm.

Tables of properties of materials usually list values of the *linear* expansivities α of solids, related to β by the equation

$$\beta = 3\alpha. \tag{2–25}$$

Tabulated values are ordinarily *mean* values, over a temperature interval near room temperature and at atmospheric pressure, and provide only a very incomplete description of the complicated dependence of volume on temperature and pressure.

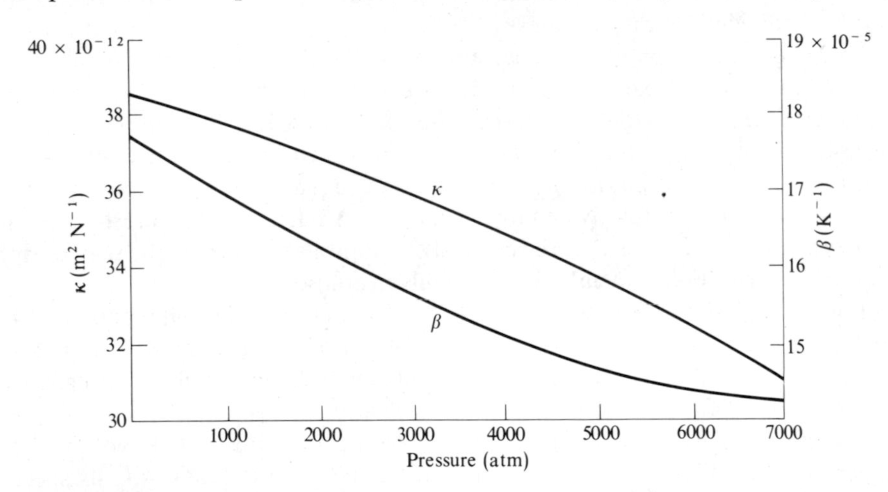

Fig. 2–17 Compressibility κ of and expansivity β of mercury as functions of pressure at a constant temperature of 0°C.

Consider next the change in volume of a material when the pressure is changed at constant temperature, for example, when the state of the system in Fig. 2-14 is changed from point 2 to point 3, along the isothermal curve at temperature T_2. It should be evident without a detailed discussion that the slope of the tangent line to an isothermal curve at any point is given by

$$\text{Slope of tangent} = \left(\frac{\partial V}{\partial P}\right)_T.$$

Hence if dV_T and dP_T represent the limiting values of the volume and pressure differences between two neighboring states at the same temperature,

$$dV_T = \left(\frac{\partial V}{\partial P}\right)_T dP_T. \tag{2-26}$$

For an ideal gas, considering T constant, we have

$$\left(\frac{\partial V}{\partial P}\right)_T = -\frac{nRT}{P^2}.$$

The *isothermal compressibility* κ of a material is defined in the same way as its expansivity, namely, as the slope of an isothermal curve at any point, divided by the volume

$$\kappa \equiv -\frac{1}{V}\left(\frac{\partial V}{\partial P}\right)_T. \tag{2-27}$$

The negative sign is included because the volume always *decreases* with increasing pressure at constant temperature so that $(\partial V/\partial P)_T$ is inherently negative. The compressibility itself is therefore a positive quantity. The unit of compressibility is the reciprocal of the unit of pressure, and in the MKS system it is 1 square meter per newton (1 m^2 N^{-1}).

For an ideal gas,

$$\kappa = -\frac{1}{V}\left(-\frac{nRT}{P^2}\right) = \frac{1}{P}. \tag{2-28}$$

The *mean* compressibility $\bar{\kappa}$ is defined as

$$\bar{\kappa} = -\frac{1}{V_1}\frac{\Delta V_T}{\Delta P_T}.$$

The compressibility of a material, like its expansivity, is in general a function of both temperature and pressure. A graph of κ versus T for copper is given in Fig. 2-16, and a graph of κ versus P for mercury in Fig. 2-17.

In the preceding discussion, we have considered two states at the same pressure, such as states 1 and 2 in Fig. 2-14, or two states at the same temperature, such as states 2 and 3. Suppose, however, that two states of a system are neither at the same pressure nor at the same temperature, such as states 1 and 3 in Fig. 2-14. The volume difference between the states depends only on the states and is independent of any particular process by which the system is taken from one state to the other. Let us therefore take the system from state 1 to state 3, first along path 1-2, at constant pressure P_1, followed by path 2-3 at constant temperature T_2. The volume difference ΔV between the states then equals the sum of the volume change ΔV_P in process 1-2 and the change ΔV_T in process 2-3. In the limit as ΔP_T and ΔT_P approach zero, by Eqs. (2–19) and (2–26) the volume difference dV is

$$dV = \left(\frac{\partial V}{\partial T}\right)_P dT + \left(\frac{\partial V}{\partial P}\right)_T dP. \tag{2–29}$$

In terms of β and κ,

$$dV = \beta V\, dT - \kappa V\, dP, \tag{2–30}$$

or

$$\frac{dV}{V} = \beta\, dT - \kappa\, dP. \tag{2–31}$$

Now instead of considering that the partial derivatives of V (or the quantities β and κ) can be calculated if the equation of state is known, we reverse this point of view. That is, if β and κ have been measured experimentally and are known as functions of temperature and pressure, we can find the equation of state by *integrating* Eq. (2–30) or (2–31). Thus suppose we had found experimentally, for a gas at low pressure, that $\beta = 1/T$ and $\kappa = 1/P$. Then from Eq. (2–31),

$$\frac{dV}{V} - \frac{dT}{T} + \frac{dP}{P} = 0,$$

$$\ln V - \ln T + \ln P = \ln\,(\text{constant}),$$

and

$$\frac{PV}{T} = \text{constant},$$

which is the equation of state of an ideal gas if we identify the constant as nR.

If Eq. (2–30) is integrated from some reference state V_0, P_0, T_0, to some arbitrary state V, P, T, we obtain

$$\int_{V_0}^{V} dV = V - V_0 = \int_{T_0}^{T} \beta V\, dT - \int_{P_0}^{P} \kappa V\, dP.$$

The volume change of a solid or liquid is relatively small when the pressure and temperature are changed and to a first approximation we can consider V to be constant and equal to V_0 in the integrals on the right. If β and κ can also be considered constant, then

$$V = V_0[1 + \beta(T - T_0) - \kappa(P - P_0)]. \tag{2–32}$$

Therefore measurements of the expansivity and compressibility, plus a knowledge of the values of V_0, P_0, and T_0 in the reference state, are sufficient to determine the equation of state of a solid or liquid, subject to the approximations above.

2–8 CRITICAL CONSTANTS OF A VAN DER WAALS GAS

As another example of the use of partial derivatives in thermodynamics, we show how they are used to determine the critical constants of a van der Waals gas. In spite of the relative simplicity of the van der Waals equation, a van der Waals gas exhibits a critical point and its P-v-T surface has features that correspond to the liquid–vapor region of a real gas. The point of coincidence of the three real values of v for a van der Waals gas is its critical point (see Figs. 2–4 and 2–5). At temperatures below the critical temperature, the van der Waals isotherms do not exhibit the horizontal portion along which the liquid and vapor phases of a real gas can coexist. One can, however, justify the construction of the horizontal line *abc* in Fig. 2–5 by drawing it as a pressure such that the shaded areas are equal. Points *a* and *c* then correspond respectively to the specific volumes of saturated liquid and vapor.

Since an isotherm represents those equilibrium states at which the temperature is constant, the slope of an isothermal curve projected on the P-v plane is given by $(\partial P/\partial v)_T$. An examination of Fig. 2–5 will show that at the critical point not only is the slope zero, but since the isotherm is concave upward at the left of this point and concave downward at the right, the critical point is also a point of inflection. Hence at this point,

$$\left(\frac{\partial P}{\partial v}\right)_T = 0, \qquad \text{and} \qquad \left(\frac{\partial^2 P}{\partial v^2}\right)_T = 0. \tag{2–33}$$

One of the useful properties of the van der Waals equation is that it may be solved for P, and hence partial derivatives of P are easily calculated. We find

$$P = \frac{RT}{v - b} - \frac{a}{v^2}.$$

Hence

$$\left(\frac{\partial P}{\partial v}\right)_T = -\frac{RT}{(v-b)^2} + \frac{2a}{v^3},$$

$$\left(\frac{\partial^2 P}{\partial v^2}\right)_T = \frac{2RT}{(v-b)^3} - \frac{6a}{v^4}.$$

When $T = T_c$, the critical temperature, and $v = v_c$, the critical volume, each of the expressions above is zero. Solving the two equations simultaneously for v_c and T_c, and inserting these values in the original equation, we get

$$P_c = \frac{a}{27b^2}, \qquad v_c = 3b, \qquad T_c = \frac{8a}{27Rb}. \tag{2–34}$$

These equations are commonly used to determine the values of a and b for a particular gas, in terms of measured values of the critical constants. However, there are three equations for the two unknowns a and b, hence these are over-determined. That is, we find from the second of the equations above that

$$b = \frac{v_c}{3},$$

while from simultaneous solution of the first and third equations,

$$b = \frac{RT_c}{8P_c}. \tag{2-35}$$

When experimental values of P_c, v_c, and T_c are inserted in the two preceding equations, we do not obtain the same value for b. In other words, it is not possible to fit a van der Waals P-v-T surface to that of a real substance at the critical point. Any two of the variables may be made to coincide, but not all three. Since the critical volume is more difficult to measure accurately than the critical pressure and temperature, the latter two are used to determine the values of a and b in Table 2-1.

Another way of comparing the van der Waals equation with the equation of state of a real substance is to compare the values of the quantity Pv/RT at the critical point. For a van der Waals gas,

$$\frac{P_c v_c}{RT_c} = \frac{3}{8} = 0.375, \tag{2-36}$$

and according to the van der Waals equation this ratio should have the value 3/8 for *all* substances at the critical point. (For an ideal gas, of course, the ratio equals unity.) Experimental values are given in Table 2-4. The two are not equal, although the discrepancies are not large.

Table 2-4 Experimental values of $P_c v_c/RT_c$

Substance	$P_c v_c/RT_c$
He	0.327
H_2	0.306
O_2	0.292
CO_2	0.277
H_2O	0.233
Hg	0.909

The van der Waals equation can be put in a form that is applicable to any substance by introducing the *reduced* pressure, volume, and temperature, that is, the ratios of the pressure, volume, and temperature to the critical pressure, volume, and temperature:

$$P_r = \frac{P}{P_c}, \qquad v_r = \frac{v}{v_c}, \qquad T_r = \frac{T}{T_c}. \tag{2-37}$$

Combining these equations with Eqs. (2–34) and (2–8), the van der Waals equation, we get

$$\left(P_r + \frac{3}{v_r^2}\right)(3v_r - 1) = 8T_r. \tag{2–38}$$

The quantities a and b have disappeared and the same equation applies to any van der Waals gas. The critical point has the coordinates 1, 1, 1, in a P_r-v_r-T_r diagram. Equation (2–38) is called the *law of corresponding states*. It is a "law," of course, only to the extent that real gases obey the van der Waals equation. Two different substances are said to be in "corresponding states" if their pressures, volumes, and temperatures are the same fraction (or multiple) of the critical pressure, volume, and temperature of the two substances.

2–9 RELATIONS BETWEEN PARTIAL DERIVATIVES

We have shown in Section 2–7 that the volume difference dV between two neighboring equilibrium states of a system can be written

$$dV = \left(\frac{\partial V}{\partial T}\right)_P dT + \left(\frac{\partial V}{\partial P}\right)_T dP.$$

It is assumed in this equation that the volume V is expressed as a function of T and P. But we can also consider that the pressure P is expressed as a function of V and T, and by the same reasoning as above we can write

$$dP = \left(\frac{\partial P}{\partial T}\right)_V dT + \left(\frac{\partial P}{\partial V}\right)_T dV.$$

Let us now eliminate dP between the preceding equations and collect coefficients of dV and dT. The result is

$$\left[1 - \left(\frac{\partial V}{\partial P}\right)_T \left(\frac{\partial P}{\partial V}\right)_T\right] dV = \left[\left(\frac{\partial V}{\partial P}\right)_T \left(\frac{\partial P}{\partial T}\right)_V + \left(\frac{\partial V}{\partial T}\right)_P\right] dT.$$

This equation must hold for *any* two neighboring equilibrium states. In particular, for two states at the same temperature but having different volumes, $dT = 0$, $dV \neq 0$, and to satisfy the equation above we must have

$$1 - \left(\frac{\partial V}{\partial P}\right)_T \left(\frac{\partial P}{\partial V}\right)_T = 0,$$

or

$$\left(\frac{\partial V}{\partial P}\right)_T = \frac{1}{(\partial P/\partial V)_T}. \tag{2–39}$$

Similarly, since we can have $dV = 0$, $dT \neq 0$, it must be true that

$$\left(\frac{\partial V}{\partial P}\right)_T \left(\frac{\partial P}{\partial T}\right)_V + \left(\frac{\partial V}{\partial T}\right)_P = 0. \tag{2–40}$$

By combining Eqs. (2–39) and (2–40) the preceding equation can be put in the more symmetrical form,

$$\left(\frac{\partial V}{\partial P}\right)_T\left(\frac{\partial P}{\partial T}\right)_V\left(\frac{\partial T}{\partial V}\right)_P = -1. \tag{2–41}$$

Note that in this equation the denominator in any partial derivative becomes the numerator of the next, and that the symbols V, P, T occur cyclically in each of the partial derivatives.

To illustrate the use of the preceding equations, suppose we wish to calculate the increase in pressure when the temperature of a system is increased but the system is not allowed to expand. That is, we wish to have the value of the partial derivative $(\partial P/\partial T)_V$. Having measured the expansivity and compressibility of a material, it is not necessary to perform a third series of experiments to find the dependence of pressure on temperature at constant volume. It follows from Eq. (2–41) that

$$\left(\frac{\partial P}{\partial T}\right)_V = -\frac{(\partial V/\partial T)_P}{(\partial V/\partial P)_T} = -\frac{\beta V}{-\kappa V} = \frac{\beta}{\kappa}, \tag{2–42}$$

and the desired partial derivative is the ratio of the expansivity to the compressibility. The larger the expansivity and the smaller the compressibility, the greater the increase in pressure for a given increase in temperature.

The pressure change in a finite change in temperature at constant volume is

$$\int_{P_1}^{P_2} dP = P_2 - P_1 = \int_{T_1}^{T_2} \frac{\beta}{\kappa}\, dT,$$

and if β and κ can be considered constant,

$$P_2 - P_1 = \frac{\beta}{\kappa}(T_2 - T_1),$$

a relation that can also be obtained from Eq. (2–32) by setting $V = V_0$.

Throughout the foregoing, we have considered only a PVT system so as to give the analysis a physical rather than simply a mathematical basis. Let us now rewrite the important equations in a more general form. Suppose we have any three variables satisfying the equation

$$f(x, y, z) = 0.$$

Then Eqs. (2–39) and (2–41) become

$$\left(\frac{\partial x}{\partial y}\right)_z = \frac{1}{(\partial y/\partial x)_z}, \tag{2–43}$$

$$\left(\frac{\partial x}{\partial y}\right)_z\left(\frac{\partial y}{\partial z}\right)_x\left(\frac{\partial z}{\partial x}\right)_y = -1. \tag{2–44}$$

The letters x, y, and z can be associated with any of the three variables whose values specify the state of any system.

2–10 EXACT DIFFERENTIALS

Since the volume difference between two equilibrium states of a system is independent of the nature of any process between the states, we can also evaluate the volume difference between states 1 and 3 in Fig. 2–14 along the path 1-4-3. In our earlier derivation, in which we used path 1-2-3, the pressure along portion 1-2 had the constant value P_1 and the temperature along portion 2-3 had the constant value T_2. We therefore write Eq. (2–29) explicitly as

$$dV_{1\text{-}2\text{-}3} = \left(\frac{\partial V}{\partial T}\right)_{P_1} dT + \left(\frac{\partial V}{\partial P}\right)_{T_2} dP.$$

Along path 1-4-3,

$$dV_{1\text{-}4\text{-}3} = \left(\frac{\partial V}{\partial P}\right)_{T_1} dP + \left(\frac{\partial V}{\partial T}\right)_{P_3} dT.$$

Since these volume changes are the same, it follows that

$$\frac{\left[\left(\frac{\partial V}{\partial T}\right)_{P_3} - \left(\frac{\partial V}{\partial T}\right)_{P_1}\right]}{dP} = \frac{\left[\left(\frac{\partial V}{\partial P}\right)_{T_2} - \left(\frac{\partial V}{\partial P}\right)_{T_1}\right]}{dT}. \tag{2–45}$$

In the limit, as dP and dT approach zero, we can consider that the partial derivative $(\partial V/\partial T)_{P_3}$ is evaluated at point *4*, and the partial derivative $(\partial V/\partial T)_{P_1}$ is evaluated at point *1*, which is at the same temperature as point *4*. The numerator on the left side of Eq. (2–45) is therefore the *change* in the value of this partial derivative when the pressure is changed by dP, from P_1 to P_3, at constant temperature. When divided by dP, the quotient is the *rate* of change with pressure, at constant temperature, of the partial derivative $(\partial V/\partial T)_P$, or, it is the so-called *mixed second partial derivative* of V with respect to P and T and is written

$$\left[\frac{\partial}{\partial P}\left(\frac{\partial V}{\partial T}\right)_P\right]_T \quad \text{or} \quad \frac{\partial^2 V}{\partial P\, \partial T}.$$

In the same way, the right side of Eq. (2–45) is

$$\left[\frac{\partial}{\partial T}\left(\frac{\partial V}{\partial P}\right)_T\right]_P \quad \text{or} \quad \frac{\partial^2 V}{\partial T\, \partial P}.$$

We therefore have the important result that

$$\frac{\partial^2 V}{\partial P\, \partial T} = \frac{\partial^2 V}{\partial T\, \partial P}. \tag{2–46}$$

That is, the value of the mixed second partial derivative is *independent of the order of differentiation.*

Note that the preceding result is true only if the volume difference dV between states 1 and 3 is the same for *all* processes between the states. A differential for which this is true is called an *exact* differential. The differentials of all *properties* of a system—such as volume, pressure, temperature, magnetization, etc.—are exact. In fact, this criterion can be considered the definition of a thermodynamic property. A quantity whose differential is *not* exact is not a thermodynamic property. Later on, when we consider energy interchanges between a system and its surroundings, we shall encounter quantities whose differentials are not exact and which are therefore *not* properties of a system.

Still another point of view is the following. The volume difference between any two arbitrary states of a system can be found by summing or integrating the infinitesimal volume changes dV along any arbitrary path between the states. Thus if V_1 and V_2 are the volumes in the two states,

$$\int_{V_1}^{V_2} dV = V_2 - V_1, \tag{2–47}$$

and the value of the integral is *independent of the path.*

It follows that if the path is *cyclic*, so that points 1 and 2 coincide, $V_2 = V_1$, $V_2 - V_1 = 0$, and

$$\oint dV = 0, \tag{2–48}$$

where the symbol $\oint$ means that the integral is evaluated around a closed path.

Conversely, if the integral of a differential between two arbitrary states is independent of the path, the integral around any closed path is zero and the differential is *exact*.

A test as to whether or not a differential is exact can be determined as follows. The exact differential dV can be written

$$dV = \left(\frac{\partial V}{\partial T}\right)_P dT + \left(\frac{\partial V}{\partial P}\right)_T dP.$$

The partial derivatives are the *coefficients* of the differentials dT and dP; and as we have shown, the partial derivative with respect to P of the coefficient of dT is equal to the partial derivative with respect to T of the coefficient of dP. In general, if for any three variables x, y, z, we have a relation of the form

$$dz = M(x, y)\,dx + N(x, y)\,dy,$$

the differential dz is exact if

$$\frac{\partial M}{\partial y} = \frac{\partial N}{\partial x}. \tag{2–49}$$

PROBLEMS

2–1 The table below lists corresponding values of the pressure and specific volume of steam at the three temperatures of 700°F, 1150°F and 1600°F. Without converting to MKS units, compute the ratio Pv/T at each temperature and pressure; and for each temperature plot these ratios as a function of pressure. Estimate the extrapolated value of Pv/T as P approaches zero, and find the value of R in J kilomole^{-1} K^{-1}.

P	t = 700°F	t = 1150°F	t = 1600°F
lb in^{-2}	v ft^3 lb^{-1}	v ft^3 lb^{-1}	v ft^3 lb^{-1}
500	1.304	1.888	2.442
1000	0.608	0.918	1.215
2000	.249	.449	0.601
3000	.0984	.289	.397
4000	.0287	.209	.294
5000	.0268	.161	.233

2–2 (a) Estimate as accurately as you can from Fig. 2–1 the molal specific volume of CO_2 at a pressure of 3×10^7 N m^{-2} and a temperature T_1. Assume $T_1 = 340$ K. (b) At this pressure and temperature, how many kilomoles of CO_2 are contained in a tank of volume 0.5 m^3? (c) How many kilomoles would the tank contain if CO_2 were an ideal gas?

2–3 A cylinder provided with a movable piston contains an ideal gas at a pressure P_1, specific volume v_1, and temperature T_1. The pressure and volume are simultaneously increased so that at every instant P and v are related by the equation

$$P = Av,$$

where A is a constant. (a) Express the constant A in terms of the pressure P_1, the temperature T_1, and the gas constant R. (b) Construct the graph representing the process above in the P-v plane. (c) Find the temperature when the specific volume has doubled, if $T_1 =$ 200 K.

2–4 The U-tube in Fig. 2–18, of uniform cross section 1 cm^2, contains mercury to the depth shown. The barometric pressure is 750 Torr. The left side of the tube is now closed at the top, and the right side is connected to a good vacuum pump. (a) How far does the mercury level fall in the left side and (b) what is the final pressure of the trapped air? The temperature remains constant.

2–5 The left side of the U-tube in Fig. 2–18 is closed at the top. (a) If the initial temperature is 300 K, find the temperature T at which the air column at the left is 60 cm long. The barometric pressure remains constant at 750 Torr. (b) Sketch the isotherms at 300 K and at the temperature T, in the P-v plane, and show the curve representing the process through which the gas in the left side of the U-tube is carried as its temperature increases.

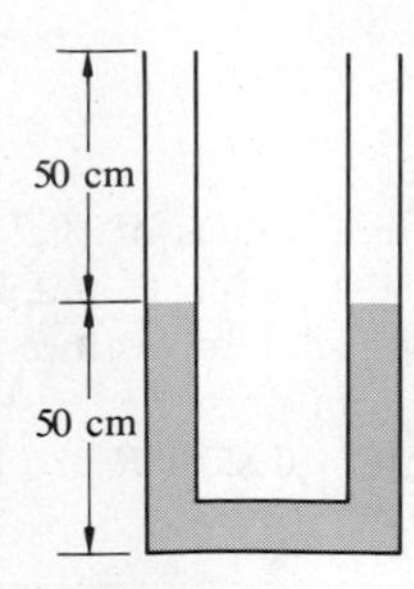

Figure 2–18

2–6 The J-shaped tube, of uniform cross section, in Fig. 2–19 contains air at atmospheric pressure. The barometric height is h_0. Mercury is poured into the open end, trapping the air in the closed end. What is the height h of the mercury column in the closed end when the open end is filled with mercury? Assume that the temperature is constant and that air is an ideal gas. Neglect any effect of the curvature at the bottom. As a numerical example, let $h_0 = 0.75$ m, $h_1 = 0.25$ m, $h_2 = 2.25$ m.

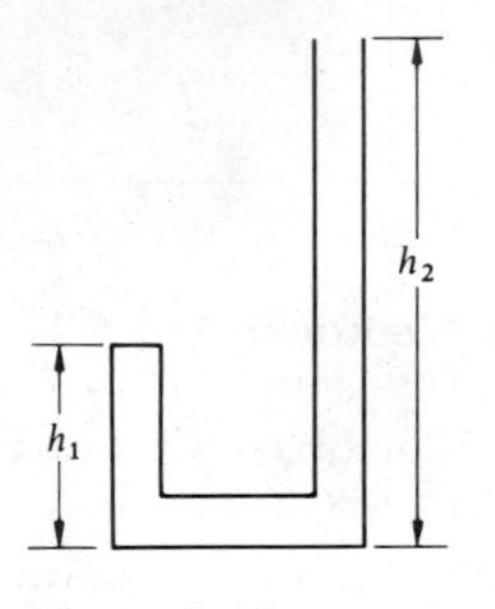

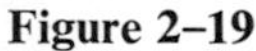

Figure 2–19

2–7 If n moles of an ideal gas can be pumped through a tube of diameter d at 4 K, what must be the diameter of the tube to pump the same number of moles at 300 K?

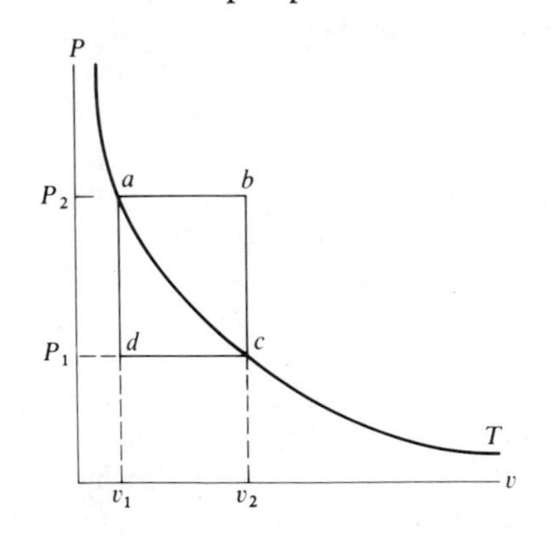

Figure 2–20

2–8 Figure 2–20 shows five processes, $a - b, b - c, c - d, d - a$ and $a - c$, plotted in the P-v plane for an ideal gas in a closed system. Show the same processes (a) in the P-T plane, (b) in the T-v plane. (c) Locate the four points of intersection of the lines on the P-v-T surface in Fig. 2–2 that correspond to a, b, c, and d in Fig. 2–20.

2–9 In Fig. 2–20, let $P_2 = 10 \times 10^5$ N m^{-2}, $P_1 = 4 \times 10^5$ N m^{-2}, $v_1 = 2.5$ m^3 kilomole^{-1}. Find (a) the temperature T, (b) the specific volume v_2, (c) the temperatures at points b and d, (d) the actual volume V at point a if the system consists of 4 kilomoles of hydrogen, (e) the mass of the hydrogen.

2–10 A tank of volume 0.5 m^3 contains oxygen at an absolute pressure of 1.5×10^6 N m^{-2} and a temperature of 20°C. Assume that oxygen behaves like an ideal gas. (a) How many kilomoles of oxygen are there in the tank? (b) How many kilograms? (c) Find the pressure if the temperature is increased to 500°C. (d) At a temperature of 20°C, how many kilomoles can be withdrawn from the tank before the pressure falls to 10 percent of the original pressure?

2–11 A quantity of air is contained in a cylinder provided with a movable piston. Initially the pressure of the air is 2×10^7 N m^{-2}, the volume is 0.5 m^3 and the temperature is 300 K. Assume air is an ideal gas. (a) What is the final volume of the air if it is allowed to expand isothermally until the pressure is 1×10^7 N m^{-2}, the piston moving outward to provide for the increased volume of the air? (b) What is the final temperature of the air if the piston is held fixed at its initial position and the system is cooled until the pressure is 1×10^7 N m^{-2}? (c) What are the final temperature and volume of the air if it is allowed to expand isothermally from the initial conditions until the pressure is 1.5×10^7 N m^{-2} and then it is cooled at constant volume until the pressure is 1×10^7 N m^{-2}? (d) What are the final temperature and volume of the air if an isochoric cooling to 1.5×10^7 N m^{-2} is followed by an isothermal expansion to 1×10^7 N m^{-2}? (e) Plot each of these processes on a T-V diagram.

2–12 A volume V at temperature T contains n_A moles of ideal gas A and n_B moles of ideal gas B. The gases do not react chemically. (a) Show that the total pressure P of the system is given by

$$P = p_A + p_B \tag{2–50}$$

where p_A and p_B are the pressures that each gas would exert if it were in the volume alone. The quantity p_A is called the partial pressure of gas A and Eq. (2–50) is known as Dalton's* law of partial pressures. (b) Show that $p_A = x_A P$ where x_A is the fraction of moles of A in the system.

2–13 In all so-called diatomic gases, some of the molecules are dissociated into separated atoms, the fraction dissociated increasing with increasing temperature. The gas as a whole thus consists of a diatomic and a monatomic portion. Even though each component may act as an ideal gas, the mixture does not, because the number of moles varies with the temperature. The degree of dissociation δ of a diatomic gas is defined as the ratio of the mass m_1 of the monatomic portion to the total mass m of the system

$$\delta = m_1/m.$$

* John Dalton, British chemist (1766–1844).

(a) Show that the equation of state of the gas is

$$PV = (\delta + 1)(m/M_2)RT,$$

where M_2 is the molecular "weight" of the diatomic component. Assume that the gas obeys Dalton's law (see Problem 2–12). (b) The table below lists measured values of the ratio PV/m, for iodine vapor, at three different temperatures. Compute and show in a graph the degree of dissociation as a function of temperature.

t(°C)	800	1000	1200
$\dfrac{PV}{m}$, J kg^{-1}	3.72×10^4	5.08×10^4	7.30×10^4

2–14 A vessel contains CO_2 at a temperature of 137°C. The specific volume is 0.0700 m^3 kilomole^{-1}. Compute the pressure in N m^{-2} (a) from the ideal gas equation, (b) from the van der Waals equation. (c) Calculate the ratio Pv/T, in J kilomole^{-1} K^{-1}, for the two pressures found above, and compare with the experimental value as read from Fig. 2–1 assuming T_2 = 137°C.

2–15 A cylinder provided with a piston contains water vapor at a temperature of −10°C. From a study of Fig. 2–10, describe the changes that take place as the volume of the system is decreased isothermally. Make a graph of the process in the P-v plane, approximately to scale.

2–16 The critical constants of CO_2 are given in Table 2–3. At 299 K the vapor pressure is 66×10^5 N m^{-2} and the specific volumes of the liquid and the vapor are, respectively, 0.063 and 0.2 m^3 kilomole^{-1}. At the triple point, T = 216 K, $P = 5.1 \times 10^5$ N m^{-2}, and the specific volumes of the solid and liquid are respectively 0.029 and 0.037 m^3 kilomole^{-1}. (a) Construct as much as you can of the P-v diagram for CO_2 corresponding to Fig. 2–5. (b) One mole of solid CO_2 is introduced into a vessel whose volume varies with pressure according to the relation $P = 7 \times 10^7\ V$, where V is in m^3 and P in N m^{-2}. Describe the change in the contents of the vessel as the temperature is slowly increased to 310 K.

2–17 Show that $\beta = 3\alpha$ for an isotropic solid.

2–18 (a) Show that the coefficient of volume expansion can be expressed as

$$\beta = -\frac{1}{\rho}\left(\frac{\partial \rho}{\partial T}\right)_P,$$

where ρ is the density. (b) Show that the isothermal compressibility can be expressed as

$$\kappa = \frac{1}{\rho}\left(\frac{\partial \rho}{\partial P}\right)_T.$$

2–19 The temperature of a block of copper is increased from 400 K to 410 K. What change in pressure is necessary to keep the volume constant? Obtain the necessary numerical data from Fig. 2–16.

2–20 Design a mercury-in-glass thermometer for use near room temperature. The length of the mercury column should change one centimeter per deg C. Assume that the volume expansivity of mercury is equal to $2 \times 10^{-4}\ \text{K}^{-1}$ and is independent of temperature near room temperature and that the expansivity of glass is essentially zero.

2–21 (a) Show that the coefficient of volume expansion of a van der Waals gas is

$$\beta = \frac{Rv^2(v - b)}{RTv^3 - 2a(v - b)^2}.$$

(b) What is the expression for β if $a = b = 0$ (ideal gas)?

2–22 (a) Show that the compressibility of a van der Waals gas is

$$\kappa = \frac{v^2(v - b)^2}{RTv^3 - 2a(v - b)^2}.$$

(b) What is the expression for κ if $a = b = 0$?

2–23 An approximate equation of state is $P(v - b) = RT$. (a) Compute the expansivity and the compressibility for a substance obeying this equation of state. (b) Show that the corresponding equations for a van der Waals gas (see Problems 2–21 and 2–22) reduce to the expressions derived in (a) when $a = 0$.

2–24 A hypothetical substance has an isothermal compressibility $\kappa = a/v$ and an expansivity $\beta = 2bT/v$, where a and b are constants. (a) Show that the equation of state is given by $v - bT^2 + aP = \text{constant}$. (b) If at a pressure P_0 and temperature T_0, the specific volume is v_0, evaluate the constant.

2–25 A substance has an isothermal compressibility $\kappa = aT^3/P^2$ and an expansivity $\beta = bT^2/P$ where a and b are constants. Find the equation of state of the substance and the ratio, a/b.

2–26 From the equation of state given in Eq. (2–12) compute (a) the rate at which the length of a rod changes with temperature when the tension is kept constant; (b) the rate at which the length changes with tension when the temperature is constant; (c) the change dT in temperature that is necessary to keep the length constant when there is a small change $d\mathscr{F}$ in the tension. Assume Young's modulus is independent of temperature.

2–27 A railroad track is laid without expansion joints in a desert where day and night temperatures differ by $\Delta T = 50$ K. The cross-sectional area of a rail is $A = 3.6 \times 10^{-3}$ m^3, the stretch modulus Y is 20×10^{10} N m^{-2}, and the coefficient of linear expansion $\alpha = 8 \times 10^{-6}\ (\text{K})^{-1}$. (a) If the length of the track is kept constant, what is the difference in the tension in the rails between day and night? (b) If the tension is zero when the temperature is a minimum, what is it when the temperature is a maximum? (c) If the track is 15,000 m long, and is free to expand, what is the change in its length between day and night? (d) What partial derivatives must be evaluated to answer the preceding questions?

2–28 Find the critical constants P_c, v_c, and T_c in terms of a, b, and R for a van der Waals gas.

2–29 Using the critical constants listed in Table 2–3, compute the value of b in the van der Waals equation for CO_2, (a) from v_c, and (b) from T_c and P_c.

2–30 Show that the critical constants of a substance obeying the Dieterici* equation of state, $P(v - b)\exp(a/vRT) = RT$, are

$$P_c = a/4e^2b^2, \qquad v_c = 2b, \qquad T_c = a/4Rb.$$

(b) Compare the ratio P_cv_c/RT_c for a Dieterici gas with the experimental values in Table 2–4.

2–31 Derive Eq. (2–38).

2–32 (a) Making use of the cyclic relation Eq. (2–41), find the expansivity β of a substance obeying the Dieterici equation of state given in Problem 2–30. (b) At high temperatures and large specific volumes all gases approximate ideal gases. Verify that for large values of T and v, the Dieterici equation and the expression for β derived in (a) both go over to the corresponding equations for an ideal gas.

2–33 Find $(\partial P/\partial T)_v$ for gases obeying (a) the van der Waals equation of state, (b) the approximate equation of state of Problem 2–23, and (c) the Dieterici equation of state (Problem 2–30).

2–34 From the equation of state of a paramagnetic material, show that the cyclic partial derivatives $(\partial M/\partial \mathcal{H})_T$, $(\partial \mathcal{H}/\partial T)_M$, and $(\partial T/\partial M)_{\mathcal{H}}$, satisfy Eq. (2–44).

2–35 (a) Use the fact that dv is an exact differential and the definitions of β and κ to prove that

$$\left(\frac{\partial \beta}{\partial P}\right)_T = -\left(\frac{\partial \kappa}{\partial T}\right)_P.$$

(b) From Fig. 2–16, obtain a linear equation that gives approximately the relation between κ and T for copper, at a constant pressure of 1 atm, and at $T = 1000$ K. (c) Compute the change of the expansivity of copper with pressure, at constant temperature. (d) Find the expansivity of copper at 1000 K and 1 atm, and compute the fractional change in volume of the copper when the pressure is isothermally increased to 1000 atm. Assume that $(\partial \beta/\partial P)_T$ is independent of pressure.

2–36 Use the relation of the previous problem to show that the data given in Problems 2–24 and 2–25 are consistent.

2–37 Show that the magnetic moment, M, of a paramagnetic material is a state function by showing that dM is an exact differential.

---

* Conrad H. Dieterici, German physicist (1858–1929).

3

The first law of thermodynamics

3–1 INTRODUCTION

The work-energy principle, in mechanics, is a consequence of Newton's laws of motion. It states that the work of the resultant force on a particle is equal to the change in kinetic energy of the particle. If a force is conservative, the work of this force can be set equal to the change in potential energy of the particle, and the work of all forces *exclusive* of this force is equal to the sum of the changes in kinetic and potential energy of the particle. The same statements apply to a rigid body. (For simplicity, assume that the lines of action of all forces pass through the center of mass so that rotational motion need not be considered.)

Work can also be done in processes in which there is no change in either the kinetic or potential energy of a system. Thus work is done when a gas is compressed or expanded, or when an electrolytic cell is charged or discharged, or when a paramagnetic rod is magnetized or demagnetized, even though the gas, or the cell, or the rod remains at rest at the same elevation. The science of thermodynamics is largely (but not exclusively) concerned with processes of this sort.

In mechanics, the work $d'W$ of a force $\mathbf{F}$ when its point of application is displaced a distance $d\mathbf{s}$ is defined as $F \cos \theta \, ds$, where θ is the angle between the vectors $\mathbf{F}$ and $d\mathbf{s}$. If $\mathbf{F}$ and $d\mathbf{s}$ are in the same direction, $\theta = 0°$, $\cos \theta = 1$, and work equals $F \, ds$. In thermodynamics, and for reasons that will be explained shortly, it is customary to reverse this sign convention and define the work as $d'W = -F \cos \theta \, ds$. Then when $\mathbf{F}$ and $d\mathbf{s}$ are in opposite directions, $\theta = 180°$, $\cos \theta = -1$, and the work is $+F \, ds$. The reason for using $d'W$ rather than dW will be explained in Section 3–4.

When a thermodynamic system undergoes a process, the work in the process can always be traced back ultimately to the work of some force. However, it is convenient to express the work in terms of the thermodynamic properties of the system and we begin by considering the work in a volume change.

3–2 WORK IN A VOLUME CHANGE

The full line in Fig. 3–1 represents the boundary of a system of volume V and arbitrary shape, acted on by a uniform external hydrostatic pressure P_e. Suppose the system expands against this pressure to the shape shown by the dotted outline. The external force acting on an element of the boundary surface of area dA is $dF_e = P_e \, dA$. When the element moves outward through a distance ds, the force and displacement are in opposite directions and the work of the force is $dF_e \, ds = P_e \, dA \, ds$. When all surface elements are included, the work $d'W$ in the process is found by integrating the product $P_e \, dA \, ds$ over the entire surface:

$$d'W = P_e \int dA \, ds.$$

The integral equals the volume between the two boundaries, or the increase dV in the volume of the system. Therefore

$$d'W = P_e \, dV. \tag{3–1}$$

Thus when a system expands against an external pressure, dV is positive, the work is positive, and we say that work is done *by* the system. When a system is compressed, dV is negative, the work is negative, and we say that work is done *on* the system. When the science of thermodynamics was first being developed, a quantity of primary interest was the work done *by* a system in a process in which steam in a cylinder expanded against a piston. It was convenient to consider the work in such a process as positive, which is the reason for reversing the usual sign convention as described above. Some texts in thermodynamics retain the sign convention of mechanics and hence express the work in a volume change as $d'W = -P_e\,dV$. Then positive work corresponds to work done *on* a system, and negative work to work done *by* a system. In this book, however, we shall retain the sign convention customarily used in thermodynamics, in which the work done *by* a system is positive.

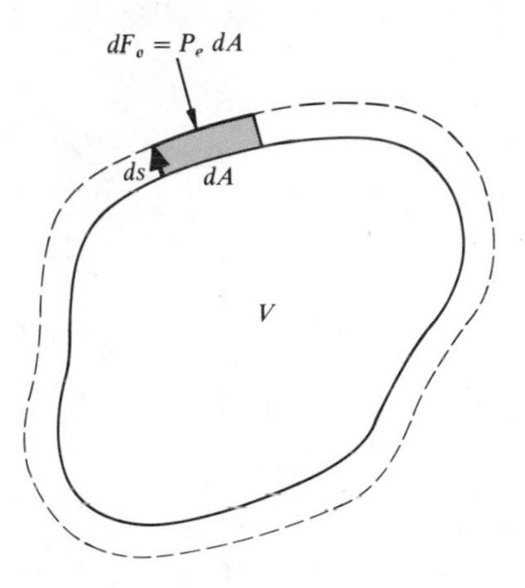

Fig. 3–1 The work done by a system expanding against an external force is given by $P_e\,dA\,ds$.

The MKS unit of pressure is 1 newton per square meter (1 N m^{-2}) and the unit of volume is 1 cubic meter (1 m^3). The unit of work is therefore 1 newton-meter (1 N m) or 1 joule (1 J).

The work of the *external* forces exerted on the boundary of a system is often spoken of as *external work*. The external work in a volume change is given by Eq. (3–1) whatever the nature of a process. If a process is *reversible*, the system is essentially in mechanical equilibrium at all times and the external pressure P_e equals the pressure P exerted against the boundary by the system. Hence in a *reversible* process we can replace P_e with P and write

$$d'W = P\,dV. \tag{3–2}$$

In a finite reversible process in which the volume changes from V_a to V_b, the total work W is

$$W = \int_{V_a}^{V_b} P\,dV. \tag{3–3}$$

When the nature of a process is specified, P can be expressed as a function of V through the equation of state of the system and the integral can be evaluated.

The relation between the pressure and volume of a system, in any reversible process, can be represented by a curve in the P-V plane. The work in a small volume change dV is represented graphically by the area $P\,dV$ of a narrow vertical strip such as that shown shaded in Fig. 3–2. The total work W in a finite process is proportional to the area between the curve representing the process and the horizontal axis, bounded by vertical lines as V_a and V_b. The work is positive if the process proceeds in the direction shown, from state a to state b. If the process proceeds in the opposite direction, the work is negative.

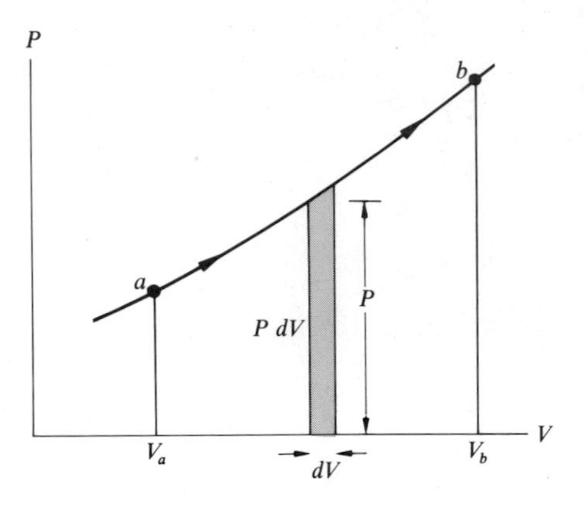

Fig. 3–2 The shaded area represents the work in a small volume change.

We next evaluate $\int P\,dV$ for a few reversible processes.

The work in any *isochoric* process is evidently zero since in such a process $V =$ constant.

In an *isobaric* process the pressure is constant and

$$W = P\int_{V_a}^{V_b} dV = P(V_b - V_a). \tag{3–4}$$

The work is represented by the area of the shaded rectangle in Fig. 3–3(a) of base $V_b - V_a$ and of height P.

If P is not constant, it must be expressed as a function of V through the equation of state. If the system is an ideal gas,

$$P = nRT/V.$$

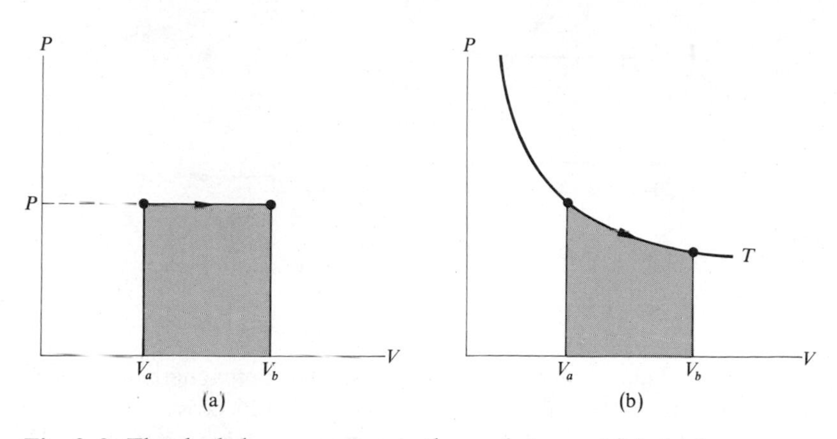

Fig. 3–3 The shaded area represents the work in an (a) isobaric process, (b) isothermal process.

For the special case of an *isothermal* process, T is constant and

$$W = nRT\int_{V_a}^{V_b} \frac{dV}{V} = nRT \ln \frac{V_b}{V_a}. \tag{3–5}$$

The work is represented by the shaded area in Fig. 3–3(b). If $V_b > V_a$, the process is an expansion, ln (V_b/V_a) is positive, and the work is positive. If $V_b < V_a$, the process is a compression, ln (V_b/V_a) is negative, and the work is negative.

It is left as a problem to calculate the work in an isothermal change in volume of a Van der Waals gas.

3–3 OTHER FORMS OF WORK

Figure 3–4 represents a wire under tension. The left end of the wire is fixed and an external stretching force $\mathscr{F}_e$ is exerted on the right end. When the wire is stretched a small additional amount $ds = dL$, $\mathscr{F}_e$ and dL are in the same direction and the work of the force $\mathscr{F}_e$ is $d'W = -\mathscr{F}_e\,dL$. If the process is reversible, the external force $\mathscr{F}_e$ equals the tension $\mathscr{F}$ in the wire and

$$d'W = -\mathscr{F}\,dL. \tag{3–6}$$

If dL is positive, dW is negative and work is done *on* the wire. If the wire is allowed to shorten, dL is negative, $d'W$ is positive, and work is done *by* the wire. The MKS unit of tension is 1 newton (1 N) and the unit of length is 1 meter (1 m).

One of the most important applications of thermodynamics is to the study of the behavior of paramagnetic substances at extremely low temperatures. This question will be considered more fully in Section 8–8, and for the present we consider only the expression for the work in a process in which the magnetic state of the substance is changed. The system is to consist of a long slender rod in an

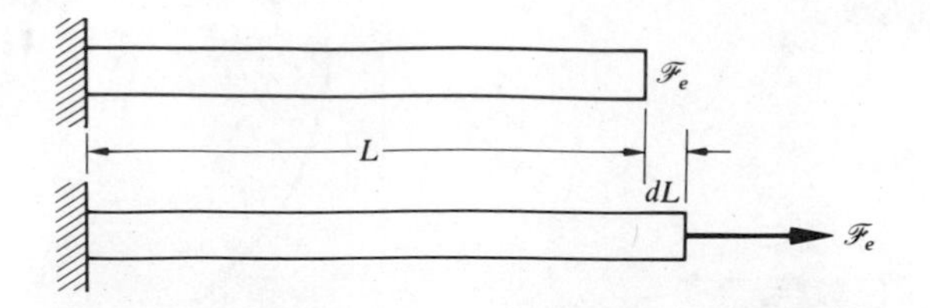

Fig. 3–4 The work done on a wire in increasing its length dL is $\mathscr{F}_e\, dL$.

external magnetic field parallel to its length, so that demagnetizing effects can be neglected. Let L represent the length of the rod and A its cross-sectional area, and suppose it to be wound uniformly with a magnetizing winding of negligible resistance, having N turns and carrying a current I. Let B represent the flux density in the rod and $\Phi = BA$ the total flux. When the current in the windings is increased by dI in a time dt, the flux changes by $d\Phi$ and the induced emf in the winding is

$$\mathscr{E} = -N\frac{d\Phi}{dt} = -NA\frac{dB}{dt}.$$

The power input $\mathscr{P}$ to the system is given by $\mathscr{P} = \mathscr{E}I$, and the work $d'W$ in time dt is

$$d'W = \mathscr{P}\, dt = \mathscr{E}I\, dt.$$

The magnetic intensity $\mathscr{H}$ produced by the current I in the winding is

$$\mathscr{H} = \frac{NI}{L};$$

and eliminating I, we get

$$d'W = V\mathscr{H}\, dB, \tag{3–7}$$

where $V = AL$ is the volume of the rod.

If $\mathscr{M}$ is the magnetization in the rod, or the magnetic moment per unit volume, the flux density B is

$$B = \mu_0(\mathscr{H} + \mathscr{M}).$$

When this expression for B is inserted in Eq. (3–7), we have

$$d'W = -\mu_0 V\mathscr{H}\, d\mathscr{H} - \mu_0 V\mathscr{H}\, d\mathscr{M}. \tag{3–8}$$

The first term on the right is the work that would be required to increase the field in a vacuum, if the rod were not present, since in such a case $\mathscr{M}$ and $d\mathscr{M}$ would be zero. The second term is therefore the work associated with the change in magnetization of the rod.

The magnetic moment M of a specimen of volume V is $M = V\mathscr{M}$, but to avoid the appearance of the magnetic constant $\mu_0 = 4\pi \times 10^{-7}$ henry m^{-1} (H m^{-1})* in our equations, let us define the magnetic moment as

$$\text{M} = \mu_0 V\mathscr{M}. \tag{3–9}$$

Then the work of magnetization, exclusive of the vacuum work, is simply

$$d'W = -\mathscr{H}\,d\text{M}. \tag{3–10}$$

The MKS unit of $\mathscr{H}$ is 1 ampere per meter (1 A m^{-1})†. The unit of magnetization $\mathscr{M}$ is also 1 A m^{-1}. Therefore the unit of magnetic moment defined in Eq. (3–9) is $4\pi \times 10^{-7}$ henry ampere meter ($4\pi \times 10^{-7}$ H A m).

Similar reasoning leads to the result that when the electric intensity E in a dielectric slab is changed, the work is

$$d'W = -E\,dP, \tag{3–11}$$

where P is the dipole moment of the slab, equal to the product of its polarization (dipole moment per unit volume) and its volume V.

The MKS unit of E is 1 volt per meter (1 V m^{-1})‡ and the unit of polarization is 1 coulomb per meter squared (1 C m^{-2}).§ The unit of dipole moment P is 1 coulomb meter (1 C m) and again the unit of work is 1 volt coulomb = 1 J.

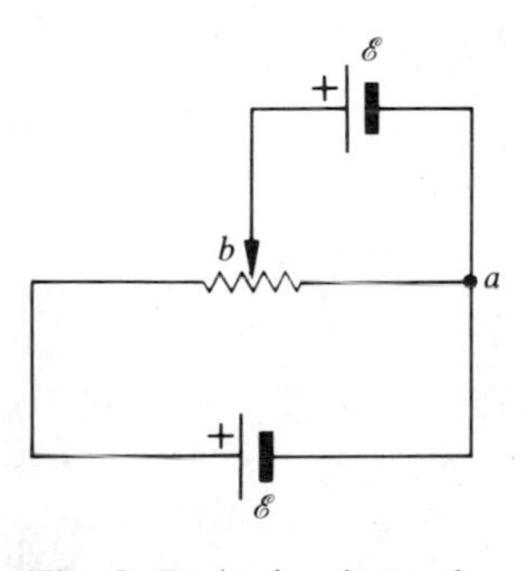

Fig. 3–5 A circuit to do work reversibly on an electrolytic cell of emf $\mathscr{E}$.

Consider next an electrolytic cell of emf $\mathscr{E}$ and of negligible internal resistance Let the terminals of the cell be connected respectively to one end a of a resistor, and to a sliding contact b on the resistor, as in Fig. 3–5. The resistor is connected across a second cell of emf $\mathscr{E}'$, greater than $\mathscr{E}$.

---

* Joseph Henry, American physicist (1797–1878).
† André M. Ampere, French physicist (1775–1836).
‡ Count Alessandro Volta, Italian physicist (1745–1827).
§ Charles A. de Coulomb, French engineer (1736–1806).

If the position of the sliding contact is adjusted so that the potential difference V_{ab}, due to the current in the resistor, is exactly equal to $\mathscr{E}$, the current in the cell will be zero. If V_{ab} is infinitesimally greater than $\mathscr{E}$, there will be a current in the cell from right to left, and if V_{ab} is infinitesimally less than $\mathscr{E}$, there will be a current in the cell in the opposite direction. Since the direction of the current in the cell can be reversed by an infinitesimal change in V_{ab}, the process taking place in the cell is reversible in the thermodynamic sense. If, in addition, the reacting substances in the cell are properly chosen, the direction of the chemical reaction within the cell will be reversed when the current reverses, and we speak of such a cell as a *reversible cell.*

The power $\mathscr{P}$ supplied to or by the cell is given by $\mathscr{P} = \mathscr{E}I$, where I is the current in the cell. The work in a short time interval dt is

$$d'W = \mathscr{P}\,dt = \mathscr{E}I\,dt.$$

In Chapter 2, we defined a quantity Z whose change dZ is the quantity of charge $I\,dt$ flowing past a point in the cell in time dt. To agree with the thermodynamic sign convention, we must write

$$d'W = -\mathscr{E}\,dZ. \tag{3–12}$$

If Z increases, as it does when the cell is being "charged," dZ is positive, dW is negative, and work is done *on* the cell.

The MKS unit of $\mathscr{E}$ is 1 volt (1 V) and the unit of Z is 1 coulomb (1 C). The unit of W is therefore 1 joule (1 J).

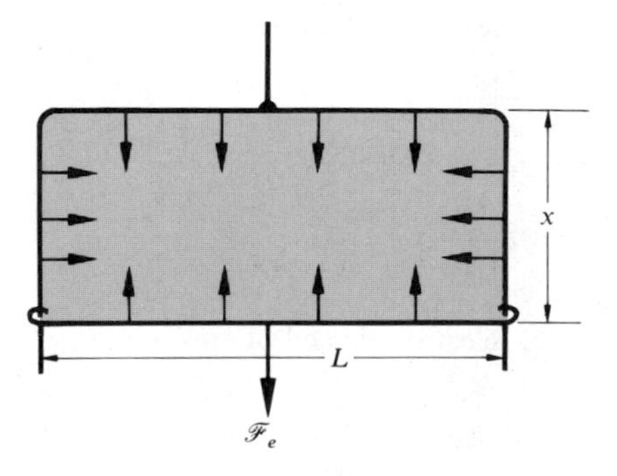

Fig. 3–6 Surface tension forces exerted at the boundary of a thin film.

As a final example, we calculate the work when the area of a surface film is changed. Figure 3–6 represents a common method for demonstrating the phenomenon of surface tension. A soap film is formed on a U-shaped frame having a sliding crossbar. Both surfaces of the film exert inward forces on the boundary of the film, and the crossbar is kept in equilibrium by an external force $\mathscr{F}_e$. The surface tension σ of the film is defined as the inward force exerted by one of the film surfaces, per unit length of boundary. Hence if L is the length of the crossbar, the total upward force on it is $2\sigma L$ (the film has two surfaces) and hence $\mathscr{F}_e = 2\sigma L$. When the crossbar is moved down a distance dx, the work of the force $\mathscr{F}_e$ is

$$d'W = -\mathscr{F}_e\,dx = -2\sigma L\,dx,$$

where the negative sign enters because $\mathscr{F}_e$ and dx are in the same direction. The total surface area of the film is $A = 2Lx$, so

$$dA = 2L\,dx$$

and hence

$$d'W = -\sigma\,dA. \tag{3–13}$$

The unit of σ is 1 newton per meter (1 N m^{-1}) and the unit of A is 1 square meter (1 m^2) so that the unit of work is 1 N m = 1 J.

3–4 WORK DEPENDS ON THE PATH

Suppose that a PVT system is taken from an initial equilibrium state a to a final equilibrium state b by two different *reversible* processes, represented by the two paths I and II in Fig. 3–7. The expression for the work W in either process is

$$W = \int_a^b d'W = \int_{V_a}^{V_b} P\,dV.$$

Although the work along either path is given by the integral of $P\,dV$, the pressure P is a different function of V along the two paths and hence the work is different also. The work in process I corresponds to the total shaded area under path I; the work in process II corresponds to the lightly shaded area under path II. Hence in contrast to the volume change $V_b - V_a$ between states a and b, which is the same for all paths between the states, the work W depends on the path and not simply on its endpoints. Therefore, as explained in Section 2–10, the quantity $d'W$ is an *inexact* differential and the work W is not a *property* of the system. Work is a *path function*, not a *point* function like V, and the work in a process cannot be set equal to the difference between the values of some property of a system in the end states of a process. Thus we use the symbol $d'W$ to emphasize that the work of an infinitesimal process is an inexact differential.

If the system in Fig. 3–7 is taken from state a to state b along path I and then returned from state b to state a along path II, the system performs a *cyclic* process. The positive work along path I is greater than the negative work along path II. The *net* work in the cycle is then positive, or work is done *by* the system, and the net work is represented by the area bounded by the closed path. If the cycle is traversed in the opposite sense, that is, first from a to b along path II and back from b to a along path I, the net work is negative and work is done *on* the system. In either case, the magnitude of the net work W is

$$W = \oint d'W = \oint P\,dV. \tag{3–14}$$

This is in contrast to the integral of an *exact* differential around a closed path, which always equals zero, as was shown in Section 2–10.

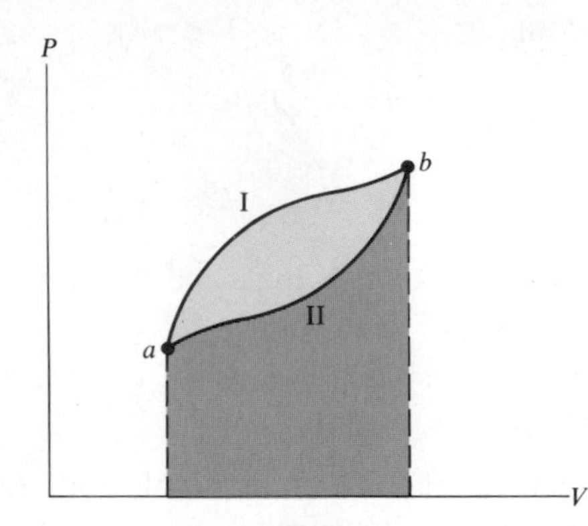

Fig. 3–7 Work depends upon the path.

3–5 CONFIGURATION WORK AND DISSIPATIVE WORK

In all of the examples in the preceding sections, the work in a reversible process is given by the product of some intensive variable (P, $\mathscr{H}$, $\mathscr{E}$, σ) and the change in some extensive variable (V, M, Z, A). Let Y represent any such intensive variable and X the corresponding extensive variable. In the most general case, where more than one pair of variables may be involved,

$$d'W = Y_1\,dX_1 + Y_2\,dX_2 + \cdots = \sum Y\,dX, \tag{3–15}$$

with the understanding that each product is to be taken with the proper algebraic sign: $P\,dV$, $-\mathscr{H}\,dM$, etc. The extensive variables X_1, X_2, etc., are said to determine the *configuration* of the system and the work $\sum Y\,dX$ is called *configuration work*.

It is possible that the configuration of a system can change *without* the performance of work. In Fig. 3–8, a vessel is divided into two parts by a diaphragm. The space above the diaphragm is evacuated and that below the diaphragm contains a gas. If the diaphragm is punctured, the gas expands into the evacuated region and fills the entire vessel. The end state would be the same if the diaphragm were a very light piston, originally fastened in place and then released. The process is known as a *free expansion.*

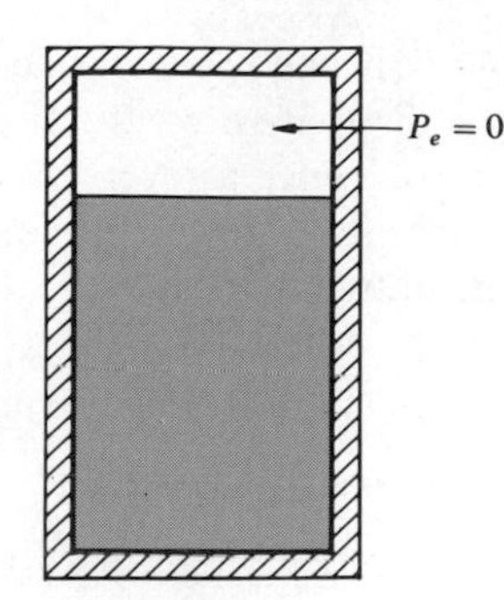

Fig. 3–8 In a free expansion of a gas, the configuration work is zero since P_e is zero.

Since the space above the diaphragm is evacuated, the external pressure P_e on the diaphragm is zero. The work in a free expansion is therefore

$$W = \int P_e \, dV = 0,$$

and the work is zero even though the volume of the gas increases.

Suppose that a stirrer is immersed in a fluid, the stirrer and fluid together being considered a system. The stirrer is attached to a shaft projecting through the wall of the container and an external torque is exerted on the outer end of the shaft. Regardless of the direction of rotation of the shaft, the external torque is always in the same direction as the angular displacement of the shaft and the work of the external torque is always *negative*, that is, work is always done *on* the composite system of fluid and stirrer. We speak of the work as *stirring work* or, more generally, as *dissipative work*.

Another common example of dissipative work is the work needed to maintain an electric current I in a resistor of resistance R. Electrical work of magnitude $\int I^2R \, dt$ must be done *on* the resistor, regardless of the direction of the current.

Unlike configuration work, the dissipative work in a process cannot be expressed in terms of a change in some property of a system on which the work is done. There is a close connection between dissipative work and a flow of heat, as we shall see later.

Any process in which dissipative work is done is necessarily *irreversible*. Work is done *on* a system when a stirrer in a fluid is rotated, but a small change in the external torque rotating the stirrer will not result in work being done *by* the system. Similarly, a small change in the terminal voltage of a source sending a current through a resistor will not result in work being done *by* the resistor.

In the general case, both configuration work and dissipative work may be done in a process. The *total* work in the process is defined as the algebraic sum of the configuration work and the dissipative work. If a process is to be *reversible*, the dissipative work must be zero. Since a reversible process is necessarily quasistatic, then to specify that a process is reversible implies (a) that the process is quasistatic and (b) that the dissipative work is zero. In a reversible process, then, the total work equals the configuration work.

3–6 THE FIRST LAW OF THERMODYNAMICS

There are many different processes by which a system can be taken from one equilibrium state to another, and in general the work done by the system is different in different processes. Out of all possible processes between two given states, let us select those that are *adiabatic*. That is, the system is enclosed by an adiabatic boundary and its temperature is independent of that of the surroundings. The boundary need not be rigid, so that configuration work can be done on or by the system. We assume also that dissipative work may be done on the system, and that there is no change in the kinetic and potential energies of the system.

Even though we consider only adiabatic processes, many such processes are possible between a given pair of states. A few of these are shown in Fig. 3–9. The system, initially in state a, first performs an adiabatic *free* expansion (represented by the cross-hatched line) from a to c. No configuration work is done in this process, and we assume there is no dissipative work. The system next performs a reversible adiabatic expansion to state b. In this process, the configuration work is represented by the shaded area under the curve cb, and since the dissipative work is zero in any reversible process, this shaded area represents the *total* work in the process a-c-b.

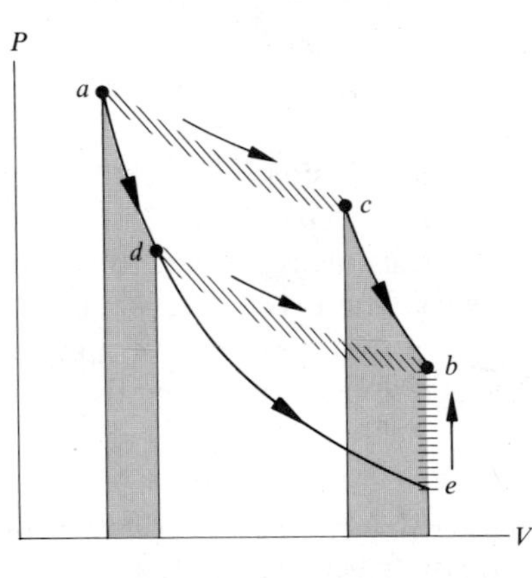

Fig. 3–9 The same amount of work is done in all adiabatic processes between the same pair of equilibrium states.

In a second process, starting again at state a, the system first performs a reversible adiabatic expansion to state d, this state being so chosen that a subsequent *free* expansion (again in the absence of any dissipative work) will terminate at state b. The total work in process a-d-b is then represented by the shaded area under the curve ad.

Although the two processes are very different, it is an experimental fact that the work, represented by the two shaded areas, is the same in both.

In a third possible process, the reversible adiabatic expansion starting at a is continued beyond point d to point e, at which the configuration (in this case, the volume) is the same as in state b. Then adiabatic dissipative work at constant configuration is done on the system (for example, a stirrer is rotated within the system) until it again reaches state b. (The dissipative work is not represented by an area in the diagram.)

The total work done by the system in the process a-e-b equals the configuration work done by the system in process a-e, represented by the area under the curve ae, minus the dissipative work done on the system in process e-b. It is found that this total work is the same as that in the first two processes, and it follows that the work *by* the system in the reversible expansion from d to e is equal to the work *on* the system in the dissipative process e-b.

It should not be inferred that experiments such as those illustrated in Fig. 3–9 have been carried out with high precision for all possible adiabatic processes between all possible pairs of equilibrium states. Nevertheless, the entire structure of thermodynamics is consistent with the conclusion that whatever the nature of the process,

the total work is the same in all adiabatic processes between any two equilibrium states having the same kinetic and potential energy.

The preceding statement is called the *first law of thermodynamics*. Processes in which the kinetic and potential energies in the end states are *not* the same are discussed in Section 3–13.

3–7 INTERNAL ENERGY

The total work W_{ad} in any adiabatic process is the sum of the works $d'W_{\text{ad}}$ in each stage of the process:

$$W_{\text{ad}} = \int_a^b d'W_{\text{ad}}.$$

Although in general the differential $d'W$ is inexact, and the work W has different values for different paths, the differential $d'W_{\text{ad}}$ is exact in the sense that the work is the same along *all adiabatic* paths between a given pair of states having the same kinetic and potential energies. It is therefore possible to define a *property* of a system, represented by U, such that the difference between its values in states a

and b is equal to the total work done by the system along *any* adiabatic path from a to b. We call this property the *internal energy* of the system.

The value of the internal energy (apart from an arbitrary constant which does not affect the values of *differences* in internal energy) depends only on the state of the system, and hence dU is an *exact* differential. It is conventional to define dU as the negative of the adiabatic work $d'W_{\text{ad}}$ done *by* a system, or as equal to the adiabatic work done *on* the system. Thus,

$$dU = -d'W_{\text{ad}}.$$

For two states that differ by a finite amount,

$$\int_{U_a}^{U_b} dU = U_b - U_a = -\int_a^b d'W_{\text{ad}} = -W_{\text{ad}},$$

or

$$U_a - U_b = W_{\text{ad}}. \tag{3–16}$$

That is, the total work W_{ad} done *by* a system in any adiabatic process between two states a and b having the same kinetic and potential energies is equal to the *decrease* $(U_a - U_b)$ in the internal energy of the system. Thus a gas expanding against a piston, in an adiabatic process, can do work even though there is no change in its kinetic or potential energy; the work is done at the expense of the *internal energy* of the gas.

It is evident that the unit of internal energy is equal to the unit of work, and that in the MKS system the unit is 1 joule (1 J).

Note that no assumptions or statements need be made regarding the *nature* of internal energy, from a molecular point of view. We shall see later how the methods of kinetic theory and statistical thermodynamics make it possible to interpret the internal energy of a system in terms of the energies of the particles of which the system is composed. From the standpoint of thermodynamics it suffices to know that the property of internal energy *exists*, and to know how it is defined.

We shall show in Chapter 5 that not all states of a system can be reached from a given state by adiabatic processes. However, if state b cannot be reached from state a by an adiabatic process, it is always true that state a can be reached from state b by an infinite number of adiabatic processes, in all of which the work W_{ad} is the same. The adiabatic work then defines the internal energy differences $U_b - U_a$.

3–8 HEAT FLOW

The first law of thermodynamics makes it possible to define the internal energy U of a system as a property of the system whose change between two equilibrium states equals the negative of the total work in any adiabatic process between the states. We now consider processes between a given pair of equilibrium states that are *not* adiabatic. That is, the system is not thermally insulated from its surrounding but makes contact *via* a nonadiabatic boundary with one or more other systems whose temperature differs from that of the system under consideration. Under these circumstances we say that there is a *flow of heat Q* (for brevity, a *heat flow Q*) between the system and its surroundings.

The heat flow Q is defined quantitatively in terms of the work in a process as follows. The total work W in a *non*adiabatic process between a given pair of equilibrium states differs from one such process to another, and differs also from the work W_{ad} in any adiabatic process between the same pair of states. We define the heat flow Q *into* the system in any process as the difference between the work W and the adiabatic work W_{ad}:

$$Q \equiv W - W_{ad}. \tag{3–17}$$

The heat flow into a system, like the change in its internal energy, is thus defined wholly in terms of mechanical work, and the unit of Q is evidently 1 joule. The procedure we have followed seems very different from that of defining a unit of heat as the heat flow into 1 gram of water when its temperature is increased by 1 degree Celsius (the gram-calorie) or the heat flow into 1 pound mass of water when its temperature is increased by 1 degree Fahrenheit (the British thermal unit or Btu). The advantage of the method we have used is that the unit of heat is defined in absolute terms and does not involve the properties of a particular material. We shall return to this point in Section 3–10.

Depending on the nature of a process, the work W may be greater or less than the adiabatic work W_{ad} and hence the algebraic sign of Q may be positive or negative. If Q is positive, there is a heat flow *into* the system; if Q is negative there is a heat flow *out* of the system. The heat flow may be positive during some parts of a process and negative in others. Then Q equals the *net* heat flow into the system.

Since numerical values of temperature are assigned in such a way that heat flows by conduction *from* a higher *to* a lower temperature, it follows that if the temperature of the surroundings is greater than that of a system, there will be a heat flow *into* the system and Q is positive. If the temperature of the surroundings is lower than that of the system, there will be a heat flow *out* of the system and Q is negative.

A *reversible* change in temperature of a system, as discussed in Section 1–9, can now be described in terms of a flow of heat. If the temperature of a system differs only infinitesimally from that of the surroundings, the direction of the heat flow can be reversed by an infinitesimal change in temperature of the system, and the heat flow is *reversible*.

If a process is adiabatic, the work W becomes simply the adiabatic work W_{ad} and from Eq. (3–17) the heat flow Q is zero. This justifies a statement made in Section 1–5, namely, that an adiabatic boundary can be described as one across which there is no flow of heat even if there is a difference in temperature between the surfaces of the boundary. An adiabatic boundary is an ideal heat insulator.

Since by definition the adiabatic work done by a system in a process from an initial equilibrium state a to a final equilibrium state b is equal to the decrease in internal energy of the system, $U_a - U_b$, Eq. (3–17) can be written

$$U_b - U_a = Q - W. \tag{3–18}$$

The difference $U_b - U_a$ is the *increase* in internal energy, and Eq. (3–18) states that *the increase in internal energy of a system, in any process in which there is no change in the kinetic and potential energies of the system, equals the net heat flow Q into the system minus the total work W done by the system.*

Had we used the sign convention of mechanics, in which the work of a force is defined as $F \cos \theta \, ds$ instead of $-F \cos \theta \, ds$, the sign of W would be reversed and we would have, instead of Eq. (3–18),

$$U_b - U_a = Q + W.$$

That is, Q is positive when there is a heat flow *into* the system and W is positive when work is done *on* the system. The increase in internal energy is then equal to the *sum* of the heat flow *into* the system and the work done *on* the system. This is a more logical sign convention and it is used by some authors.

If the heat flow and the work are both very small, the change in internal energy is very small also and Eq. (3–18) becomes

$$dU = d'Q - d'W. \tag{3–19}$$

Equation (3–18), or its differential form, Eq. (3–19), is commonly referred to as the analytical formulation of the first law of thermodynamics (and we shall continue to refer to it as such); but, in fact, these equations are nothing more than the *definitions* of Q or $d'Q$ and do not constitute a physical law. The true significance of the first law lies in the statement that the work is the same in all adiabatic processes between any two equilibrium states having the same kinetic and potential energy.

There is no restriction on the nature of the process to which Eqs. (3–18) and (3–19) refer; the process may be reversible or irreversible. If it *is* reversible, the only work is configuration work, and (for a PVT system) we can replace $d'W$ with $P\,dV$. Hence in a reversible process,

$$dU = d'Q - P\,dV. \tag{3–20}$$

More generally, for a system of any nature in a reversible process,

$$dU = d'Q - \sum Y\,dX. \tag{3–21}$$

3–9 HEAT FLOW DEPENDS ON THE PATH

Equations (3–18) and (3–19) can be written

$$Q = (U_b - U_a) + W,$$
$$d'Q = dU + d'W.$$

For a given pair of initial and final states, the values of $(U_b - U_a)$, or of dU, are the same for all processes between the states. However, as we have seen, the quantities W or $d'W$ are different for different processes and as a consequence the heat flows Q or $d'Q$ are different also. Thus $d'Q$, like $d'W$, is an *inexact* differential and Q is not a *property* of a system. Heat, like work, is a *path* function, not a *point* function, and it has a meaning only in connection with a *process*. The net heat flow Q into a system in any process between states a and b is the sum of the $d'Q$'s in each stage of the process, and we can write

$$Q = \int_a^b d'Q.$$

However, as with the work W in a process, we cannot set the integral equal to the difference between the values of some property of the system in the initial and final states. Thus suppose we were to arbitrarily pick some reference state of a system and assign a value Q_0 to the "heat in the system" in this reference state. The "heat" in some second state would then equal the "heat" Q_0, plus the heat flow Q into the system in a process from the reference state to the second state. But the heat flow is different for different processes between the states and it is impossible to assign any definite value to the "heat" in the second state.

If a process is cyclic, its end states coincide; there is no change in the internal energy; and from Eq. (3–18), $Q = W$. In such a process, the net heat flow Q into the system equals the net work W done by the system. But since the net work W is not necessarily zero, the net heat flow Q is not necessarily zero, and all we can say is that

$$\oint d'Q = Q.$$

This is analogous to the corresponding expression for the work W in a cyclic process and is in contrast to the integral of an exact differential around a closed path, which is always zero.

3–10 THE MECHANICAL EQUIVALENT OF HEAT

Suppose that dissipative work W_d is done on a system, in an adiabatic process at constant configuration. This will be the case, for example, if work is done on a friction device immersed in a fluid kept at constant volume and thermally insulated. The heat flow Q in the process is zero, the configuration work is zero, and the dissipative work is the total work. Then if U_a and U_b are respectively the initial and final values of the internal energy of the system, and since work done *on* a system is inherently negative, we can write

$$U_b - U_a = |W_\text{d}|. \tag{3–22}$$

That is, the increase in internal energy of the system equals the magnitude of the dissipative work done on the system.

On the other hand, in a process in which the configuration work and dissipative work are both zero, but in which there is a heat flow Q into the system, the change in internal energy is

$$U_b - U_a = Q. \tag{3–23}$$

If Eqs. (3–22) and (3–23) refer to the same pair of end states, the heat flow Q in the second process equals the dissipative work in the first. From the standpoint of the *system*, it is a matter of indifference whether the internal energy is increased by the performance of dissipative work, or by an inflow of heat from the surroundings.

These two processes illustrate what is meant by the common but imprecise statement that in a dissipative process, "work is converted to heat." All one can really say is that *the change in internal energy of a system*, in a dissipative process, is the same *as if* there had been a heat flow Q into the system, equal in magnitude to the dissipative work.

As another special case, suppose that dissipative work W_d is done on a system at constant configuration, and at the same time there is a heat flow Q *out* of the system, equal in magnitude to W_d. The internal energy of the system then remains constant. This will be the case if a resistor carrying a current is kept at constant temperature by a stream of cooling water. A heat flow out of the resistor into the cooling water is equal in magnitude to the dissipative work done on the resistor, and it is customary to say in this case also that "work is converted to heat."

For many years, the quantity of heat flowing into a system was expressed in terms of calories, or British thermal units, 1 calorie being defined as the heat flow into 1 gram of water in a process in which its temperature was increased by 1 Celsius degree, and 1 Btu as the heat flow into 1 pound-mass of water when its temperature was increased by 1 Fahrenheit degree. Careful measurements showed that these quantities of heat varied slightly with the particular location of the one-degree interval, for example, whether it was from 0°C to 1°C, or from 50°C to 51°C. To avoid confusion, the *15-degree calorie* was defined as the heat flow into 1 gram of water when its temperature was increased from 14.5°C to 15.5°C.

If the same rise in temperature is produced by the performance of dissipative work, the best experimental measurements find that 4.1858 joules are required, a value that is referred to as the *mechanical equivalent of heat*. We can then say,

$$1 \text{ 15-degree calorie} = 4.1858 \text{ joules}. \tag{3–24}$$

This relation between the joule and the 15-degree calorie is necessarily subject to some experimental uncertainty. For this reason, and also so as not to base the definition of the calorie on the properties of some particular material (i.e., water), an international commission has agreed to *define* the New International Steam Table calorie (the IT calorie) by the equation

$$1 \text{ IT calorie} = \tfrac{1}{860} \text{ watt hour} = \tfrac{3600}{860} \text{ joules (exactly)}.$$

Then to five significant figures,

$$1 \text{ IT calorie} = 4.1860 \text{ joules}. \tag{3–25}$$

The apparently arbitrary figure of 860 was chosen so that the IT calorie would agree closely with the experimental value of the 15-degree calorie.

Since the relations between the joule and the foot-pound, between the gram and the pound-mass, and between the Celsius and Fahrenheit degrees are also matters of definition and not subject to experimental uncertainty, the British thermal unit is also defined exactly in terms of the joule. To five significant figures,

$$1 \text{ Btu} = 778.28 \text{ foot-pounds}. \tag{3–26}$$

This definition of the calorie and the Btu as exact multiples of the joule has the effect of making these units obsolete; and in current experimental physics, quantities of heat are customarily expressed in joules. However, the calorie and the Btu are so deeply embedded in the scientific and engineering literature that in all probability it will be many years before their use disappears entirely.

For many years it was thought that heat was a substance contained in material. The first conclusive evidence that it was not was given by Count Rumford* who observed the temperature rise of the chips produced while boring cannons. He concluded that heat flow into the chips was caused by the work of boring. The earliest precision measurements of the mechanical equivalent of heat were made by Joule, who measured the mechanical dissipative work done on a system of paddle wheels immersed in a tank of water and calculated, from the known mass of water and its measured rise in temperature, the quantity of heat that would have to flow into the water to produce the same change in internal energy. The experiments were performed in a period from 1840 to 1878, and although Joule expressed his results in English units, they are equivalent to the remarkably precise value of

$$1 \text{ calorie} = 4.19 \text{ joules}.$$

(The energy unit, 1 joule, was not introduced or named until after Joule's death, and the standardized 15-degree calorie had not been agreed on at the time of Joule's work.)

However, the true significance of Joule's work went far beyond a mere determination of the mechanical equivalent of heat. By means of experiments like those above, and others of a similar nature, Joule demonstrated conclusively that there was in fact a direct proportion between "work" and "heat," and he succeeded in

* Benjamin Thompson, Count Rumford, American physicist (1753–1814).

dispelling the belief, current at that time, that "heat" was an invisible, weightless fluid known as "caloric." It may be said that Joule not only determined the value of the mechanical equivalent of heat but provided the experimental proof that such a quantity actually existed.

3–11 HEAT CAPACITY

Provided no changes of phase take place in a process, and except in certain special cases, the temperature of a system changes when there is a heat flow into the system. The *mean heat capacity* $\overline{C}$ of a system, in a given process, is defined as the ratio of the heat flow Q into the system, to the corresponding change in temperature, ΔT:

$$\overline{C} = \frac{Q}{\Delta T}. \tag{3–27}$$

The term "capacity" is not well chosen because it implies that a system has a definite "capacity" for holding so much heat and no more, like the "capacity" of a bucket for water. A better term, following the usage in electricity, would be "heat capacitance" or "thermal capacitance."

The *true* heat capacity at any temperature is defined as the limit approached by $\overline{C}$ as ΔT approaches zero:

$$C = \lim_{\Delta T \to 0} \frac{Q}{\Delta T} = \frac{d'Q}{dT}. \tag{3–28}$$

The MKS unit of C is 1 joule per kelvin (1 J K^{-1}).

Note carefully that the ratio $d'Q/dT$ *cannot* be interpreted as the *derivative* of Q with respect to T, since Q is not a property of the system and is not a function of T. The notation $d'Q$ simply means "a small flow of heat," and dT is the corresponding change in temperature.

A process is not *completely* defined by the temperature difference between its end states; and for a given temperature change dT the heat flow $d'Q$ may be positive, negative, or zero, depending on the nature of the process. The heat capacity of a system therefore depends both on the nature of the system and on the particular process the system may undergo, and for a given system it may have any value between $-\infty$ and $+\infty$.

The heat capacity in a process in which a system is subjected to a constant external hydrostatic pressure is called the *heat capacity at constant pressure* and is represented by C_P. The value of C_P, for a given system, depends both on the pressure and on the temperature. If a system is kept at constant volume while heat is supplied to it, the corresponding heat capacity is called the *heat capacity at constant volume* and is represented by C_V. Because of the large stresses set up when a solid or liquid is heated and not allowed to expand, direct experimental determinations of C_V for a solid or liquid are difficult and C_P is the quantity generally measured. However, as we shall show later, if C_P is known, the heat capacity for any other process can be calculated if, in addition, we know the equation of state of the system.

To measure the heat capacity of a system experimentally, we must measure the heat $d'Q$ flowing into the system in a process, and measure the corresponding change in temperature dT. The most precise method of measuring the heat flowing into a system is to insert a resistor into the system, or surround it with a coil of resistance wire, and measure the electrical dissipative work $d'W = \int I^2R\,dt$ done on the resistor. As we have shown, if the state of the resistor does not change, the heat flow $d'Q$ out of the resistor and into the system is equal in magnitude to the electrical work $d'W$. In such an experiment, the temperature of the resistor increases along with that of the system so that its internal energy does not remain constant and the heat flowing out of it into the system is not exactly equal to the electrical work. The difference, however, can be made negligibly small or a correction can be made for it. A correction must also be made for the heat flow between the system and the surroundings.

The concept of heat capacity applies to a given *system*. The *specific heat capacity*, or the *heat capacity per unit mass* or *per mole*, is characteristic of the *material* of which the system is composed and is represented by c_P or c_v. The MKS unit of specific heat capacity is 1 joule per kelvin, per kilogram (1 J kg^{-1} K^{-1}) or 1 joule per kelvin, per kilomole (1 J kilomole^{-1} K^{-1}).

Figure 3–10 shows the variation with temperature of the molal specific heat capacities c_P and c_v for copper, at a constant pressure of 1 atm. At low temperatures the two are nearly equal, and near absolute zero both drop rapidly to zero. (Compare with the graph of expansivity in Fig. 2–16.) This behavior is characteristic of most solids, although the temperature at which the sharp drop occurs varies widely from one substance to another. At high temperatures, c_P continues to increase while c_v becomes nearly constant and equal to about 25×10^3 J kilomole^{-1} K^{-1}. It is found that this same value of c_v is approached by many solids at high temperatures and it is called the *Dulong* and Petit*† value, after the men who first discovered this fact.

Although there seems to be little connection between the heat capacity of solids and the properties of gases at low pressure, it will be recalled that the gas constant R equals 8.31×10^3 J kilomole^{-1} K^{-1}, and 25×10^3 J kilomole^{-1} K^{-1} is almost exactly three times this; that is, the specific heat capacity at constant volume is nearly equal to $3R$ at high temperatures. We shall show in Section 9–8 that on theoretical grounds a value of $3R$ is to be expected for c_v for solids at high temperatures.

---

* Pierre L. Dulong, French chemist (1785–1838).

† Alexis T. Petit, French physicist (1791–1820).

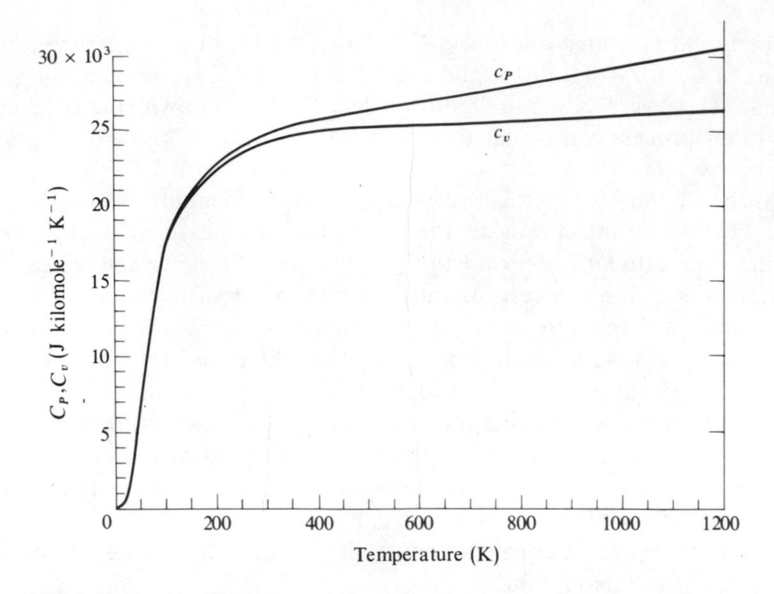

Fig. 3–10 Graphs of c_v and c_P for copper as functions of temperature at a constant pressure of 1 atm.

Figure 3–11 shows the change with pressure of c_P and c_v for mercury, at constant temperature. The pressure variation is relatively much smaller than the variation with temperature.

Some values of c_P and c_v for gases, also expressed in terms of R, are given in Table 9–1 for temperatures near room temperature. It will be noted that for monatomic gases $c_P/R \approx 5/2 = 2.50$, $c_v/R \approx 3/2 = 1.50$, and for diatomic gases, $c_P/R \approx 7/2 = 3.50$, $c_v/R \approx 5/2 = 2.50$.

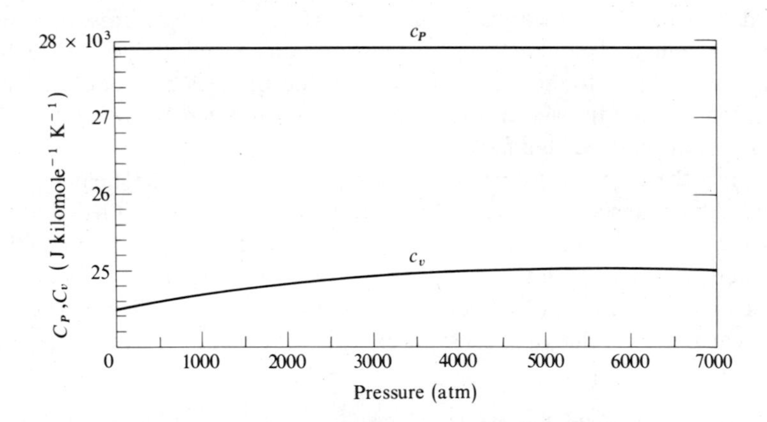

Fig. 3–11 Graphs of c_v and c_P for mercury as functions of pressure at a constant temperature of 0°C.

The total quantity of heat flowing into a system, in any process, is given by

$$Q = \int d'Q = \int_{T_1}^{T_s} C\,dT = n\int_{T_1}^{T_2} c\,dT, \tag{3–29}$$

where C is the heat capacity appropriate to the process and c is the corresponding molal value. Over a temperature interval in which C can be considered constant,

$$Q = C(T_2 - T_1) = nc(T_2 - T_1). \tag{3–30}$$

The larger the heat capacity of a system, the smaller its change in temperature for a given flow of heat, and by making the heat capacity very large indeed, the temperature change can be made as small as we please. A system of very large heat capacity is referred to as a *heat reservoir*, with the implication that the heat flow into or out of it can be as large as we please, without any change in the temperature of the reservoir. Thus any reversible process carried out by a system in contact with a heat reservoir is isothermal.

Heat capacities corresponding to C_P and C_V can be defined for systems other than PVT systems. Thus in a process in which the magnetic field intensity $\mathscr{H}$ is constant, a magnetic system has a heat capacity $C_{\mathscr{H}}$. If the magnetic moment M is constant, the corresponding heat capacity is C_M. For a polymer or stretched wire, the heat capacities are those at constant tension, $C_{\mathscr{F}}$, and at constant length, C_L.

3–12 HEATS OF TRANSFORMATION. ENTHALPY

In Section 2–5, the changes of phase of a pure substance were described but no reference was made to the work or heat accompanying these changes. We now consider this question.

Consider a portion of an isothermal process in either the solid-liquid, liquid-vapor, or solid-vapor region, and let the process proceed in such a direction that a mass m is converted from solid to liquid, liquid to vapor, or solid to vapor. The system then absorbs heat, and the *heat of transformation l* is defined as the ratio of the heat absorbed to the mass m undergoing the change of phase. (One can also define the molal heat of transformation as the ratio of the heat absorbed to the number of moles n undergoing a change.) The unit of heat of transformation is 1 J kg^{-1} or 1 J $kilomole^{-1}$.

Changes of phase are always associated with changes in volume, so that work is always done on or by a system in a phase change (except at the critical point, where the specific volumes of liquid and vapor are equal). If the change takes place at constant temperature, the pressure is constant also and the specific work done by the system is therefore

$$w = P(v_2 - v_1),$$

where v_2 and v_1 are the final and initial specific volumes. Then from the first law, the change in specific internal energy is

$$u_2 - u_1 = l - P(v_2 - v_1).$$

This equation can be written

$$l = (u_2 + Pv_2) - (u_1 + Pv_1).$$

The sum $(u + Pv)$ occurs frequently in thermodynamics. Since u, P, and v are all properties of a system, the sum is a property also and is called the specific *enthalpy* (accent on the second syllable) and is denoted by h:

$$h = u + Pv, \tag{3–31}$$

and the unit of h is also 1 joule per kilogram or 1 joule per kilomole.

Therefore,

$$l = h_2 - h_1. \tag{3–32}$$

The heat of transformation in any change of phase is equal to the difference between the enthalpies of the system in the two phases. We shall show later that this is a special case of the general property of enthalpy that the heat flow in any reversible isobaric process is equal to the change in enthalpy.

We shall use the notation l_{12}, l_{23}, l_{13} to represent heats of transformation from solid to liquid, liquid to vapor, and solid to vapor. These are called respectively the heats of *fusion*, *vaporization*, and *sublimation*. Particular properties of the solid, liquid, and vapor phases will be distinguished by one, two, or three primes respectively. The order of the numbers of primes follows the order of the phases of a substance as the temperature is increased.

As an example, consider the change in phase from liquid water to water vapor at a temperature of 100°C. The heat of vaporization at this temperature is

$$l_{23} = h''' - h'' = 22.6 \times 10^5 \text{ J kg}^{-1}.$$

The vapor pressure P at this temperature is 1 atm or 1.01×10^5 N m^{-2}, and the specific volumes of vapor and liquid are $v''' = 1.8$ m^3 kg^{-1} and $v'' = 10^{-3}$ m^3 kg^{-1}. The work in the phase change is then

$$w = P(v''' - v'') = 1.7 \times 10^5 \text{ J kg}^{-1}.$$

The change in specific internal energy is

$$u''' - u'' = l_{23} - w = 20.9 \times 10^5 \text{ J kg}^{-1}.$$

Thus about 92% of the heat of transformation is accounted for by the increase in internal energy, and about 8% by the work that must be done to push back the atmosphere to make room for the vapor.

Figure 3–12 is a graph of the heat of vaporization of water as a function of temperature. It decreases with increasing temperature and becomes zero at the critical temperature where the properties of liquid and vapor become identical.

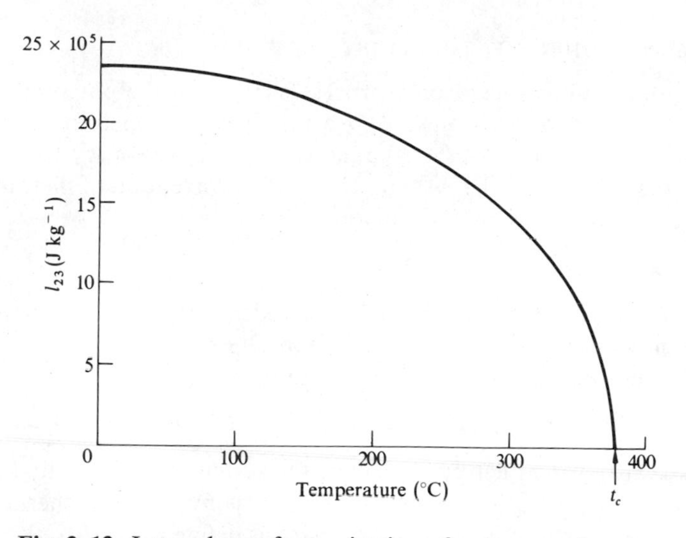

Fig. 3–12 Latent heat of vaporization of water as a function of temperature. The latent heat becomes zero at the critical temperature $t_c = 374°C$.

Since enthalpy h is a state function, its value depends only on the state of the system. If a system performs a cyclic process, the initial and final enthalpies are equal and the net enthalpy change in the process is zero. This makes it possible to derive a simple relation between the three heats of transformation at the triple point.

Consider a cyclic process performed around the triple point and close enough to it so that the only changes in enthalpy occur during phase transitions. Let the substance, initially in the solid phase, be first transformed to the vapor phase, then to the liquid phase, and finally returned to its initial state in the solid phase. (See Fig. 2–10.) There is a heat flow into the system in the first process and the increase in specific enthalpy is $\Delta h_1 = l_{13}$. In the second and third processes there is a heat flow out of the system, and the corresponding changes in enthalpy are $\Delta h_2 = -l_{23}$ and $\Delta h_3 = -l_{12}$. Then since

$$\Delta h_1 + \Delta h_2 + \Delta h_3 = 0,$$

it follows that

$$l_{13} - l_{23} - l_{12} = 0,$$

or

$$l_{13} = l_{23} + l_{12}. \tag{3–33}$$

That is, the heat of sublimation, at the triple point, equals the sum of the heat of vaporization and the heat of fusion.

3–13 GENERAL FORM OF THE FIRST LAW

Up to now we have considered only processes in which the potential and kinetic energies of a system remained constant. We now relax this constraint. In mechanics, the *work-energy theorem* states that the increase in kinetic energy ΔE_k of a system equals the work W done on the system. In the sign convention of thermodynamics, where work done *by* a system is positive, we have

$$\Delta E_k = -W.$$

More generally, the *internal* energy of a system, as well as its kinetic energy, can change in a process, and can change as a result of a flow of heat into the system as well as by the performance of work. Then in general,

$$\Delta U + \Delta E_k = Q - W.$$

If conservative forces act on a system, the system has a potential energy and the work of the conservative forces (in the sign convention of thermodynamics) equals the change in potential energy ΔE_p. Let us define a quantity W^* as the total work W, minus the work W_c of any conservative forces:

$$W^* = W - W_c \quad \text{or} \quad W = W^* + W_c.$$

Then

$$\Delta U + \Delta E_k = Q - W^* - W_c.$$

Now replace the "work" term W_c with the change in potential energy ΔE_p and transfer this term to the "energy" side of the equation. This gives

$$\Delta U + \Delta E_k + \Delta E_p = Q - W^*.$$

We now define the *total* energy E of the system as the sum of its internal energy, its kinetic energy, and its potential energy:

$$E = U + E_k + E_p.$$

Therefore

$$\Delta E = \Delta U + \Delta E_k + \Delta E_p;$$

and finally, if E_b and E_a represent the final and initial values of the total energy in a process,

$$\Delta E = E_b - E_a = Q - W^*. \tag{3–34}$$

If the heat flow and the work are both small,

$$dE = d'Q - d'W^*. \tag{3–35}$$

If the kinetic and potential energies are constant, $\Delta E = \Delta U$ and $W^* = W$, so Eqs. (3–34) and (3–35) reduce to

$$U_b - U_a = Q - W,$$
$$dU = d'Q - d'W.$$

Equations (3–34) and (3–35) are often referred to as the general form of the first law of thermodynamics, but they are better described as generalizations of the work-energy theorem of mechanics. That is, the principles of thermodynamics generalize this theorem by including the internal energy U of a system as well as its kinetic and potential energies, and by including the heat Q flowing into the system as well as the work W^*. Thus the change in the total energy ΔE of a system equals the new flow of heat Q into the system, minus the work W^* done by the system, exclusive of the work of any conservative forces.

If a system is completely isolated, that is, if it is enclosed in a rigid adiabatic boundary and is acted on only by conservative forces, the heat Q and the work W^* are both zero. Then $\Delta E = 0$ and the total energy of the system remains constant. This is the generalized form of the principle of *conservation of energy:* **the total energy of an isolated system is constant.** In the special case in which the kinetic and potential energies are constant, as for a system at rest in the laboratory, the internal energy U is constant.

Since Eqs. (3–34) and (3–35) do not apply to an isolated system, they should not be referred to as expressing the principle of conservation of energy.

3–14 ENERGY EQUATION OF STEADY FLOW

As a first illustration of the application of the general form of the first law, consider the apparatus shown schematically in Fig. 3–13. The large rectangle represents a device through which there is a flow of fluid. No restrictions are placed on the nature of the device, and we assume only that a *steady state* exists, that is, the state of the fluid at any point does not change with time. The fluid enters at an elevation z_1, with a velocity $\mathscr{V}_1$ and at a pressure P_1, and it leaves at an elevation z_2 with a velocity $\mathscr{V}_2$ and at a pressure P_2. During the time in which a mass m passes through the device, there is a heat flow Q into the fluid, and mechanical work W_{sh} (the so-called *shaft work*) is done by the fluid.

Let us imagine that at a certain instant pistons are inserted in the pipes through which the fluid enters and leaves, and that these are moved along the pipes with the velocities $\mathscr{V}_1$ and $\mathscr{V}_2$. The distances moved by the pistons during a time interval in which the mass m enters and leaves are respectively x_1 and x_2. The arrows $\mathscr{F}_1$ and $\mathscr{F}_2$ represent the forces exerted on the pistons by the adjacent fluid.

The work done by the forces $\mathscr{F}_1$ and $\mathscr{F}_2$ is

$$\mathscr{F}_2 x_2 - \mathscr{F}_1 x_1 = P_2 A_2 x_2 - P_1 A_1 x_1 = P_2 V_2 - P_1 V_1,$$

where V_1 and V_2 are respectively the volumes occupied by the mass m on entering and leaving.

The gravitational force on the mass m is mg, where g is the local acceleration of gravity, and the work of this force when a mass m is lifted from elevation z_1 to elevation z_2 is

$$W_c = mg(z_2 - z_1).$$

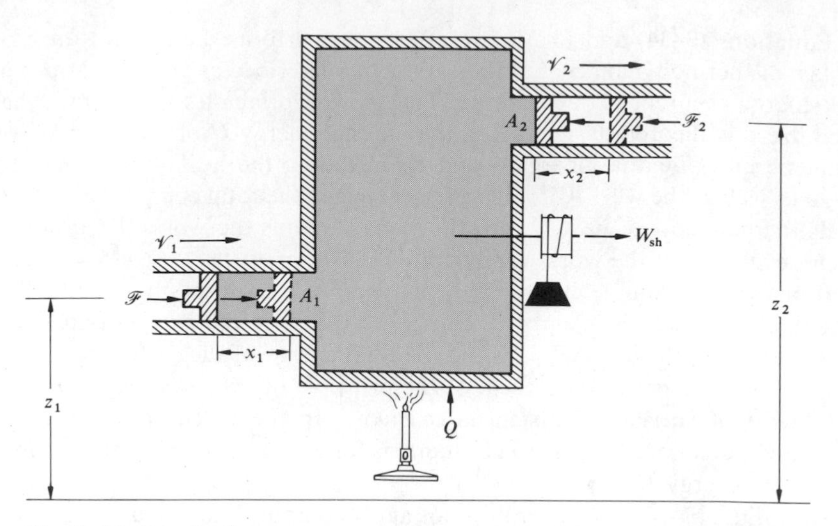

Fig. 3–13 Steady flow process.

The total work W, including the shaft work, is

$$W = W_{sh} + P_2V_2 - P_1V_1 + mg(z_2 - z_1).$$

The work W^*, or the total work minus the work W_c of the conservative gravitational force, is

$$W^* = W_{sh} + P_2V_2 - P_1V_1.$$

The increase in internal energy of the mass m is

$$\Delta U = m(u_2 - u_1),$$

where u_1 and u_2 are the respective specific internal energies.

The increase in kinetic energy is

$$\Delta E_k = \tfrac{1}{2}m(\mathscr{V}_2^2 - \mathscr{V}_1^2),$$

and the increase in potential energy is

$$\Delta E_p = mg(z_2 - z_1) = W_c.$$

We then have from Eq. (3–34)

$$m(u_2 - u_1) + \tfrac{1}{2}m(\mathscr{V}_2^2 - \mathscr{V}_1^2) + mg(z_2 - z_1) = Q - W_{sh} - P_2V_2 + P_1V_1. \tag{3–36}$$

Let v_1 and v_2 be the specific volumes of the fluid on entering and leaving, and let q and w_{sh} represent the heat flow and shaft work, per unit mass. Then

$$V_2 = mv_2, \quad V_1 = mv_1, \quad Q = mq, \quad W_{sh} = mw_{sh}.$$

After inserting these expressions in Eq. (3–36) canceling m, and rearranging terms, we have

$$(u_2 + Pv_2 + \tfrac{1}{2}\mathscr{V}_2^2 + gz_2) - (u_1 + Pv_1 + \tfrac{1}{2}\mathscr{V}_1^2 + gz_1) = q - w_{\text{sh}}.$$

Substituting the specific enthalpy h for $u + Pv$, Eq. (3–36) can be written

$$(h_2 + \tfrac{1}{2}\mathscr{V}_2^2 + gz_2) - (h_1 + \tfrac{1}{2}\mathscr{V}_1^2 + gz_1) = q - w_{\text{sh}}. \tag{3–37}$$

This is the *energy equation for steady flow*. We now apply it to some special cases.

The turbine The temperature in a steam turbine is higher than that of its surroundings but the flow of fluid through it is so rapid that only a relatively small quantity of heat is lost per unit mass of steam and we can set $q = 0$. The shaft work is of course not zero, but differences in elevation between inlet and outlet can usually be neglected. With these approximations, Eq. (3–37) becomes

$$-w_{\text{sh}} = (h_2 - h_1) + \tfrac{1}{2}(\mathscr{V}_2^2 - \mathscr{V}_1^2). \tag{3–38}$$

The shaft work obtained from the turbine, per unit mass of steam, depends on the enthalpy difference between inlet and outlet, and on the difference between the squares of the inlet and exhaust velocities.

Fig. 3–14 Flow through a nozzle.

Flow through a nozzle The steam entering a turbine comes from a boiler where its velocity is small, and before entering the turbine it is given a high velocity by flowing through a nozzle. Figure 3–14 shows a nozzle in which steam enters at a velocity $\mathscr{V}_1$ and leaves at a velocity $\mathscr{V}_2$. The shaft work is zero, the heat flow is small and can be neglected, and differences in elevation are small. Hence for a nozzle

$$\mathscr{V}_2^2 = \mathscr{V}_1^2 + 2(h_1 - h_2). \tag{3–39}$$

Bernoulli's* equation Consider the flow of an incompressible fluid along a pipe of varying cross section and elevation. No shaft work is done, and we assume the flow to be adiabatic and frictionless. Then

$$h_1 + \tfrac{1}{2}\mathscr{V}_1^2 + gz_1 = h_2 + \tfrac{1}{2}\mathscr{V}_2^2 + gz_2 = \text{constant},$$

or, writing out the expression for the enthalpy,

$$u + Pv + \tfrac{1}{2}\mathscr{V}^2 + gz = \text{constant}.$$

* Daniel Bernoulli, Swiss mathematician (1700–1782).

The change in internal energy of a system in any process equals the heat flow into the system minus the sum of the configuration work and the dissipative work. For a rigid body or an incompressible fluid, the configuration work is necessarily zero since the volume is constant. If the dissipative work and the heat flow are both zero, as in this case, the internal energy is constant. Therefore

$$Pv + \tfrac{1}{2}\mathscr{V}^2 + gz = \text{constant},$$

and replacing v by $1/\rho$, where ρ is the density, we have

$$P + \tfrac{1}{2}\rho\mathscr{V}^2 + \rho gz = \text{constant}. \qquad (3\text{–}40)$$

This is Bernoulli's equation for the steady flow of an incompressible frictionless fluid.

PROBLEMS

3–1 Compute the work done against atmospheric pressure when 10 kg of water is converted to steam occupying 16.7 m^3.

3–2 Steam at a constant pressure of 30 atm is admitted to the cylinder of a steam engine. The length of the stroke is 0.5 m and the diameter of the cylinder is 0.4 m. How much work in joules is done by the steam per stroke?

3–3 An ideal gas originally at a temperature T_1 and pressure P_1 is compressed reversibly against a piston to a volume equal to one-half of its original volume. The temperature of the gas is varied during the compression so that at each instant the relation $P = AV$ is satisfied, where A is a constant. (a) Draw a diagram of the process in the P-V plane. (b) Find the work done on the gas, in terms of n, R, and T_1.

3–4 Compute the work done by the expanding air in the left side of the U-tube in Problem 2–4. Assume the process to be reversible and isothermal.

3–5 Compute the work of the expanding gas in the left side of the U-tube in Problem 2–5. The process is reversible and isothermal. Explain why the work is not merely that required to raise the center of gravity of the mercury.

3–6 An ideal gas, and a block of copper, have equal volumes of 0.5 m^3 at 300 K and atmospheric pressure. The pressure on both is increased reversibly and isothermally to 5 atm. (a) Explain with the aid of a P-V diagram why the work is not the same in the two processes. (b) In which process is the work done greater? (c) Find the work done on each if the compressibility of the copper is 0.7×10^{-6} atm^{-1}. (d) Calculate the change in volume in each case.

3–7 (a) Derive the general expression for the work per kilomole of a van der Waals gas in expanding reversibly and at a constant temperature T from a specific volume v_1 to a specific volume v_2. (b) Using the constants in Table 2–1, find the work done when 2 kilomoles of steam expand from a volume of 30 m^3 to a volume of 60 m^3 at a temperature of 100°C. (c) Find the work of an ideal gas in the same expansion.

3–8 (a) Show that the work done in an arbitrary process on a gas can be expressed as

$$d'W = PV\beta\, dT - PV\kappa\, dP.$$

(b) Find the work of an ideal gas in the arbitrary process.

3–9 (a) Derive an equation similar to that in Problem 3–8 for the work $d'W$ when the temperature of a stretched wire changes by dT and the tension changes by $d\mathscr{F}$. (b) Find the expression for the work when the temperature is changed and the tension is held constant. What is the algebraic sign of W if the temperature increases? (c) Find the expression for the work when the tension is changed isothermally. What is the algebraic sign of W if the tension decreases?

3–10 (a) Derive an equation similar to that in Problem 3–8 for the work $d'W$ when the temperature of a paramagnetic salt changes by dT and the applied magnetic intensity changes by $d\mathscr{H}$. (b) Find the expression for the work when the temperature is changed and the magnetic intensity is held constant. What is the algebraic sign of W when the temperature rises? What is doing work in this process? (c) Find the expression for the work when the magnetic intensity is increased isothermally. What is the algebraic sign of W when the intensity is decreased?

3–11 Calculate the work necessary to reversibly and isothermally double the magnetization in a slender cylindrical paramagnetic rod which fills the volume V of a coaxial cylindrical solenoid of N turns having no resistance. Assume that the magnetic intensity is uniform inside the solenoid and neglect end effects. How does the problem change if the resistance of the coil must be considered?

3–12 Show that $d'W = -E\,dP$ by calculating the work necessary to charge a parallel plate capacitor containing a dielectric.

3–13 Calculate the work necessary to slowly increase the volume of a spherical rubber balloon by 20 percent. The initial radius of the balloon is 20 cm and the surface tension of a thin rubber sheet can be considered to be 3×10^4 N m^{-1}.

3–14 A volume of 10 m^3 contains 8 kg of oxygen at a temperature of 300 K. Find the work necessary to decrease the volume to 5 m^3, (a) at a constant pressure and (b) at constant temperature. (c) What is the temperature at the end of the process in (a)? (d) What is the pressure at the end of the process in (b)? (e) Show both processes in the P-V plane.

3–15 On a P-V diagram starting from an initial state P_0V_0 plot an adiabatic expansion to $2V_0$, an isothermal expansion to $2V_0$ and an isobaric expansion to $2V_0$. (a) Use this graph to determine in which process the least work is done by the system. (b) If, instead, the substance was compressed to $V_0/2$, in which process would the least work be done? (c) Plot the processes of parts (a) and (b) on a P-T diagram starting from P_0T_0. Indicate expansions and compressions and be careful to show relative positions at the endpoints of each process.

3–16 The temperature of an ideal gas at an initial pressure P_1 and volume V_1 is increased at constant volume until the pressure is doubled. The gas is then expanded isothermally until the pressure drops to its original value, where it is compressed at constant pressure until the volume returns to its initial value. (a) Sketch these processes in the P-V plane and in the P-T plane. (b) Compute the work in each process and the net work done in the cycle if $n = 2$ kilomoles, $P_1 = 2$ atm and $V_1 = 4$ m^3.

3–17 (a) Calculate the work done by one kilomole of an ideal gas in reversibly traversing the cycle shown in Fig. 3–15 ten times. (b) Indicate the direction of traversal around the cycle if the net work is positive.

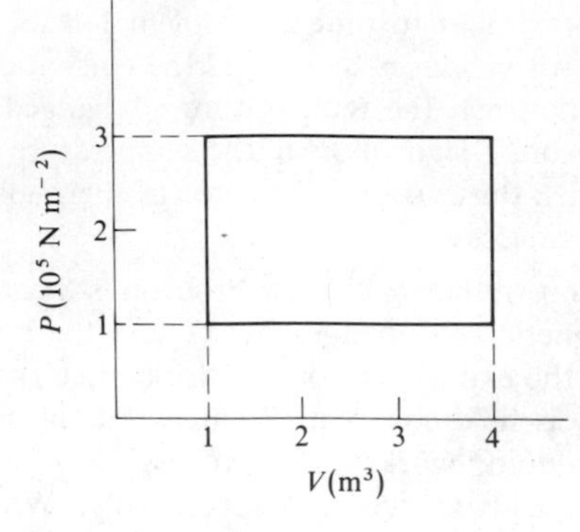

Figure 3–15

3–18 (a) Calculate the work done on 1 cm³ of a magnetic material in reversibly traversing the cycle shown in Fig. 3–16. (b) Indicate the direction in which the cycle must be traversed if the net work is positive.

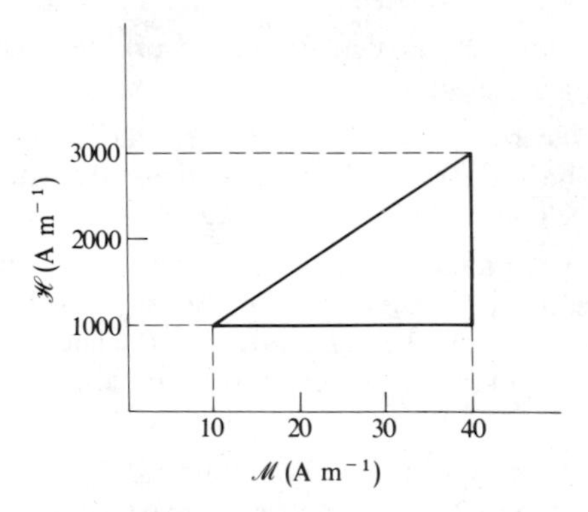

Figure 3–16

3–19 Calculate the work necessary to isothermally and reversibly remove a paramagnetic slender rod from a close fitting coaxial solenoid of zero resistance while the magnetic intensity $\mathscr{H}$ remains constant. Assume that the rod obeys Curie's law.

3–20 Consider only adiabatic processes which transform a system from state *a* to state *d* as shown in Fig. 3–17. The two curves *a-c-e* and *b-d-f* are reversible adiabatic processes. The processes indicated with cross-hatches are not reversible. (a) Prove that the total work done along paths *a-b-d*, *a-c-d*, *a-c-e-f-d* is the same. (b) Show that the configuration work along *a-b* = *c-d* = *e-f* = 0. (c) Show that the dissipative work along path *c-d* is greater than that along path *a-b* and less than that along path *e-f*.

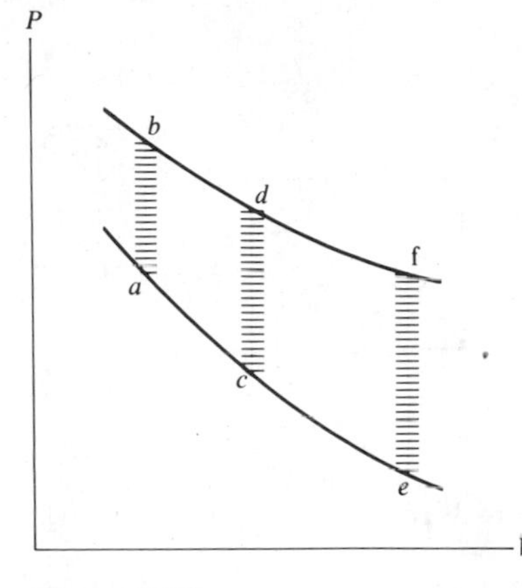

Figure 3–17

3–21 Make a sketch of the changes of internal energy as the volume of the system of the previous problem changes during the processes shown in Fig. 3–17.

3–22 Calculate the change in internal energy of a fluid in an adiabatic container when a current of 10 *A* is passed for 70 s through a 4-Ω resistor in contact with the fluid.

3–23 A gas explosion takes place inside a well-insulated balloon. As a result, the balloon expands 10 percent in volume. Does the internal energy of the balloon increase, decrease, or stay the same; or is there enough information given to determine the change in internal energy? Explain your answer.

3–24 A mixture of hydrogen and oxygen is enclosed in a rigid insulating container and exploded by a spark. The temperature and pressure both increase. Neglect the small amount of energy provided by the spark itself. (a) Has there been a flow of heat into the system? (b) Has any work been done by the system? (c) Has there been any change in internal energy U of the system?

3–25 The water in a rigid, insulated cylindrical tank is set in rotation and left to itself. It is eventually brought to rest by viscous forces. The tank and water constitute the system. (a) Is any work done during the process in which the water is brought to rest? (b) Is there a flow of heat? (c) Is there any change in the internal energy U?

3–26 When a system is taken from state a to state b, in Fig. 3–18 along the path a-c-b, 80 J of heat flow into the system, and the system does 30 J of work. (a) How much heat flows into the system along path a-d-b, if the work done is 10 J? (b) The system is returned from state b to state a along the curved path. The work done on the system is 20 J. Does the system absorb or liberate heat and how much? (c) If $U_a = 0$ and $U_d = 40$ J, find the heat absorbed in the processes a-d and d-b.

3–27 Compressing the system represented in Fig. 3–19 along the adiabatic path a-c requires1000 J. Compressing the system along b-c requires 1500 J but 600 J of heat flow out of the system. (a) Calculate the work done, the heat absorbed, and the internal energy change of the system in each process and in the total cycle a-b-c-a. (b) Sketch this cycle on a P-V diagram. (c) What are the limitations on the values that could be specified for process b-c given that 1000 J are required to compress the system along a-c.

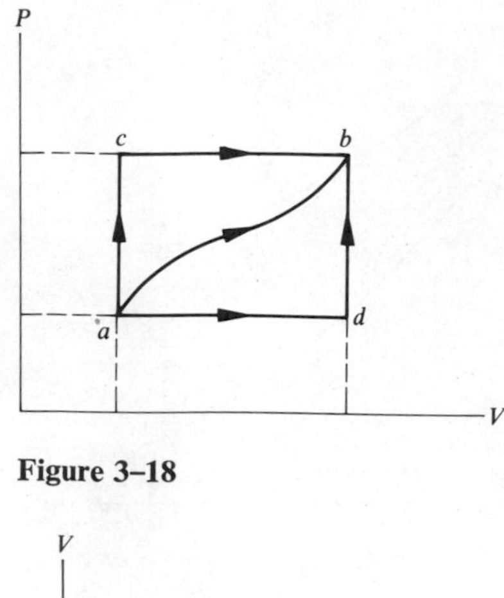

Figure 3–18

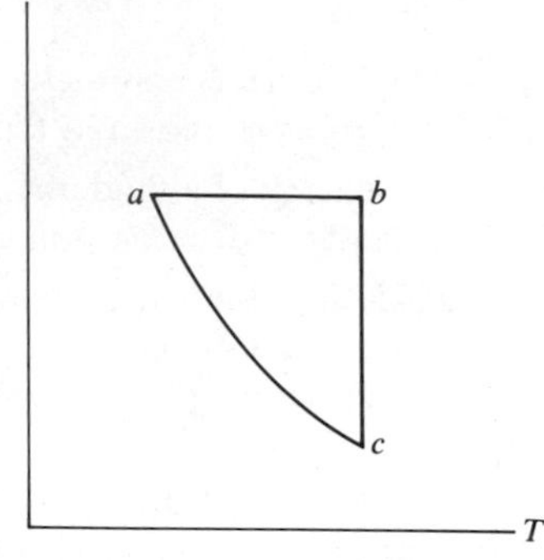

Figure 3–19

3–28 The molal specific heat capacity c_P of most substances (except at very low temperatures) can be satisfactorily expressed by the empirical formula

$$c_P = a + 2bT - cT^{-2},$$

where a, b, and c are constants and T is the Kelvin temperature. (a) In terms of a, b, and c, find the heat required to raise the temperature of n moles of the substance at constant pressure from T_1 to T_2. (b) Find the mean specific heat capacity between T_1 and T_2. (c) For magnesium, the numerical values of the constants are $a = 25.7 \times 10^3$, $b = 3.13$, $c = 3.27 \times 10^8$, when c_P is in J kilomole^{-1} K^{-1}. Find the true specific heat capacity of magnesium at 300 K, and the mean specific heat capacity between 300 K and 500 K.

3–29 The specific heat capacity c_v of solids at low temperature is given by the equation

$$c_v = A\left(\frac{T}{\theta}\right)^3,$$

a relation known as the Debye T^3 law. The quantity A is a constant equal to 19.4×10^5 J kilomole^{-1} K^{-1} and θ is the "Debye* temperature," equal to 320 K for NaCl. What is the

* Peter J. W. Debye, Dutch chemist (1884–1966)

molal specific heat capacity at constant volume of NaCl (a) at 10 K, (b) at 50 K? (c) How much heat is required to raise the temperature of 2 kilomoles of NaCl from 10 K to 50 K, at constant volume? (d) What is the mean specific heat capacity at constant volume over this temperature range?

3–30 Use Fig. 3–10 to estimate the energy necessary to heat one gram of copper from 300 to 600 K (a) at constant volume, (b) at constant pressure. (c) Determine the change in internal energy of the copper in each case. (d) Why is c_P larger than c_v?

3–31 Electrical energy is supplied to a thermally insulated resistor at the constant rate of $\mathscr{P}$ watts*, and the temperature T of the resistor is measured as a function of time t. (a) Derive an expression for the heat capacity of the resistor in terms of the slope of the temperature-time graph. (b) By means of a heating coil, heat is supplied at a constant rate of 31.2 watts to a block of cadmium of mass 0.5 kg. The temperature is recorded at certain intervals as follows:

t(s)	0	15	45	105	165	225	285	345	405	465	525
T(K)	34	45	57	80	100	118	137	155	172	191	208

Construct a graph of T versus t, and measure the slopes at a sufficient number of points to plot a graph of the molal specific heat capacity of cadmium, at constant pressure, as a function of temperature. The atomic weight of cadmium is 112.

3–32 A fictional metal of atomic weight 27 has a density of 3000 kg m^{-3}. The heat of fusion is 4×10^5 J kg^{-1} at the melting point (900 K), and at the boiling point (1300 K) the heat of vaporization is 1.20×10^7 J kg^{-1}. For the solid, c_P can be given by $750 + 0.5\,T$ in J kg^{-1} K^{-1} and in the liquid c_P is 1200 J kg^{-1} K^{-1} independent of temperature. (a) Draw a curve of temperature versus time as 10 g of this metal are heated at a constant rate of 1 W from 300 to 1200 K. (b) Determine the amount of heat necessary to cause this temperature change.

3–33 (a) Calculate the heat of sublimation of the metal sample of the previous problem assuming that the heats of vaporization and fusion are independent of temperature and pressure. (b) Calculate the change of internal energy of the metal sample upon melting. (c) Calculate the change of internal energy of the metal sample upon vaporizing. Justify the approximations which must be made.

3–34 Use physical arguments to show that for a system consisting of two phases in equilibrium the specific heat capacity at constant pressure and the coefficient of thermal expansion are infinite.

3–35 Consider a system consisting of a cylinder containing 0.2 kilomoles of an ideal gas and fitted with a massless piston of area 0.5 m^2. The force of friction between the piston and the cylinder walls is 10 N. The gas is initially at a pressure of 1 atm and the system is to be maintained at 300 K. The volume of the system is slowly decreased 10 percent by an external force. (a) Compute the work done on the system by the external force. (b) Compute

* James Watt, Scottish engineer (1736–1819).

the configurational work done on the system. (c) Compute the dissipative work done on the system. (d) How do the above answers change if the piston has a mass of 1 kg and the piston is displaced vertically?

3–36 A steam turbine receives a steam flow of 5000 kg hr^{-1} and its power output is 500 kilowatts. Neglect any heat loss from the turbine. Find the change in specific enthalpy of the steam flowing through the turbine, (a) if entrance and exist are at the same elevation, and entrance and exit velocities are negligible, (b) if the entrance velocity is 60 m s^{-1}, the exit velocity is 360 m s^{-1}, and the inlet pipe is 3 m above the exhaust.

4

Some consequences of the first law

4–1 THE ENERGY EQUATION

The specific internal energy u of a pure substance, in a state of thermodynamic equilibrium, is a function only of the state of the substance and is a *property* of the substance. We shall restrict the discussion for the present to systems whose state can be described by the properties P, v, and T.

The equation which expresses the internal energy of a substance as a function of the variables defining the state of the substance is called its *energy equation.* Like the equation of state, the energy equation is different for different substances. The equation of state and the energy equation together completely determine all properties of a substance. The energy equation cannot be derived from the equation of state but must be determined independently.

Since the variables P, v, and T are related through the equation of state, the values of any two of them suffice to determine the state. Hence the internal energy can be expressed as a function of any pair of these variables. Each of these equations defines a surface called the *energy surface*, in a rectangular coordinate system in which u is plotted on one axis while the other two axes may be P and v, P and T, or T and v.

As was explained in Chapter 2, in connection with the P-v-T surface of a substance, an energy surface can also be described in terms of the partial derivatives of u, at any point, or the slopes of lines in the surface in two mutually perpendicular directions. If the equation of the energy surface is known, the slopes can be found by partial differentiation. Conversely, if the slopes or partial derivatives are known or have been measured experimentally, in principle the equation of the surface can be found, to within a constant, by integration.

4–2 T AND v INDEPENDENT

We begin by considering u as a function of T and v. Then as explained in Chapter 2, the difference in internal energy du between two equilibrium states in which the temperature and volume differ by dT and dv is

$$du = \left(\frac{\partial u}{\partial T}\right)_v dT + \left(\frac{\partial u}{\partial v}\right)_T dv. \tag{4–1}$$

The partial derivatives are the slopes of isothermal and isochoric lines on a surface in which u is plotted as a function of T and v.

We shall show in a later chapter that, making use of the *second* law of thermodynamics, the partial derivative $(\partial u/\partial v)_T$ can be calculated from the equation of state. This is not true of the derivative $(\partial u/\partial T)_v$, which must be measured experimentally and whose physical significance we now derive. To do this, we make use of the first law for a reversible process,

$$d'q = du + P\,dv. \tag{4–2}$$

When the expression for du from Eq. (4–1) is inserted in this equation, we obtain

$$d'q = \left(\frac{\partial u}{\partial T}\right)_v dT + \left[\left(\frac{\partial u}{\partial v}\right)_T + P\right] dv. \tag{4–3}$$

In the special case of a process at constant volume, $dv = 0$ and $d'q = c_v\,dT$. Then in such a process,

$$c_v\,dT_v = \left(\frac{\partial u}{\partial T}\right)_v dT_v,$$

and hence

$$\left(\frac{\partial u}{\partial T}\right)_v = c_v. \tag{4–4}$$

Thus the geometrical significance of c_v is the slope of an isochoric line on a u-T-v surface, and experimental measurements of c_v determine this slope at any point. This is analogous to the fact that the slope of an isobaric line on a P-v-T surface, $(\partial v/\partial T)_P$ is equal to the expansivity β multiplied by the volume v. Then just as this partial derivative can be replaced in any equation by βv, so can the derivative $(\partial u/\partial T)_v$ be replaced with c_v. Equation (4–3) can therefore be written for *any* reversible process as

$$d'q = c_v\,dT + \left[\left(\frac{\partial u}{\partial v}\right)_T + P\right] dv. \tag{4–5}$$

In a process at constant pressure, $d'q = c_P\,dT$ and

$$c_P\,dT_P = c_v\,dT_P + \left[\left(\frac{\partial u}{\partial v}\right)_T + P\right] dv_P.$$

Dividing through by dT_P and replacing dv_P/dT_P with $(\partial v/\partial T)_P$, we get

$$c_P - c_v = \left[\left(\frac{\partial u}{\partial v}\right)_T + P\right]\left(\frac{\partial v}{\partial T}\right)_P. \tag{4–6}$$

It should be noted that this equation does not refer to a *process* between two equilibrium states. It is simply a general relation that must hold between quantities that are all properties of a system in any one equilibrium state. Since all of the quantities on the right can be calculated from the equation of state, we can find c_v if c_P has been measured experimentally.

For a process at constant temperature, $dT = 0$, and Eq. (4–5) becomes

$$d'q_T = \left[\left(\frac{\partial u}{\partial v}\right)_T + P\right] dv_T = \left(\frac{\partial u}{\partial v}\right)_T dv_T + P\,dv_T. \tag{4–7}$$

This equation merely states that the heat supplied to a system in a reversible isothermal process equals the sum of the work done by the system and the increase in its internal energy. Note that it serves no purpose to define a specific heat capacity at constant *temperature*, c_T by the equation $d'q_T = c_T\,dT$, because $d'q_T$

is not zero while $dT = 0$. Hence $c_T = \pm\infty$, since $d'q_T$ can be positive or negative. In other words, a system behaves in an isothermal process as if it had an infinite heat capacity, since any amount of heat can flow into or out of it without producing a change in temperature.

Finally, we consider a reversible *adiabatic* process, in which $d'q = 0$. The changes in the properties of the system in such a process will be designated by the subscript s, the reason being that the specific entropy s (see Section 5–3) remains constant in such a process. Equation (4–5) becomes

$$c_v\left(\frac{\partial T}{\partial v}\right)_s = -\left[\left(\frac{\partial u}{\partial v}\right)_T + P\right]. \tag{4–8}$$

4–3 *T* AND *P* INDEPENDENT

The enthalpy h of a pure substance, like its internal energy u, is a property of the substance that depends on the state only and can be expressed as a function of any two of the variables P, v, and T. Each of these relations defines an enthalpy surface in a rectangular coordinate system in which h is plotted along one axis while the other two axes are P and v, P and T, or T and v. Equations in which the temperature T and pressure P are considered independent can be derived most directly by considering the h-T-P surface.

The enthalpy difference between two neighboring states is

$$dh = \left(\frac{\partial h}{\partial T}\right)_P dT + \left(\frac{\partial h}{\partial P}\right)_T dP. \tag{4–9}$$

We shall show later that the derivative $(\partial h/\partial P)_T$ can be calculated from the equation of state. To evaluate $(\partial h/\partial T)_P$, we start with the definition of enthalpy for a PvT system:

$$h = u + Pv.$$

For any two states that differ by dv and dP,

$$dh = du + P\,dv + v\,dP,$$

and when this is combined with the first law,

$$d'q = du + P\,dv,$$

we obtain

$$d'q = dh - v\,dP. \tag{4–10}$$

When the expression for dh from Eq. (4–9) is inserted in this equation, we have

$$d'q = \left(\frac{\partial h}{\partial T}\right)_P dT + \left[\left(\frac{\partial h}{\partial P}\right)_T - v\right] dP, \tag{4–11}$$

which is the analogue of Eq. (4–3).

In a process at constant pressure, $dP = 0$ and $d'q = c_P\,dT$. Therefore

$$\left(\frac{\partial h}{\partial T}\right)_P = c_P, \tag{4–12}$$

and the slope of an isobaric line on the h-T-P surface equals the specific heat capacity at constant pressure. Comparison with Eq. (4–4) shows that the enthalpy h plays the same role in processes at constant pressure as does the internal energy u in processes at constant volume. The derivative $(\partial h/\partial T)_P$ can therefore be replaced with c_P in any equation in which it occurs and Eq. (4–11) can be written for any reversible process,

$$d'q = c_P\,dT + \left[\left(\frac{\partial h}{\partial P}\right)_T - v\right] dP, \tag{4–13}$$

which is the analogue of Eq. (4–5).

In a process at constant volume, $d'q = c_v\,dT$ and

$$c_P - c_v = -\left[\left(\frac{\partial h}{\partial P}\right)_T - v\right]\left(\frac{\partial P}{\partial T}\right)_v, \tag{4–14}$$

which is the analogue of Eq. (4–6).

If the temperature is constant,

$$d'q_T = \left[\left(\frac{\partial h}{\partial P}\right)_T - v\right] dP_T. \tag{4–15}$$

In an adiabatic process, $d'q = 0$ and

$$c_P\left(\frac{\partial T}{\partial P}\right)_s = -\left[\left(\frac{\partial h}{\partial P}\right)_T - v\right]. \tag{4–16}$$

4–4 P AND v INDEPENDENT

Equations corresponding to those derived in Sections 4–2 and 4–3, but in terms of P and v as independent variables, can be derived as follows. The energy difference between two neighboring equilibrium states in which the pressure and volume differ by dP and dv is

$$du = \left(\frac{\partial u}{\partial P}\right)_v dP + \left(\frac{\partial u}{\partial v}\right)_P dv. \tag{4–17}$$

However, the partial derivatives $(\partial u/\partial P)_v$ and $(\partial u/\partial v)_P$ do not involve any properties other than those already introduced. To show this, we return to the expression for du in terms of dT and dv, namely,

$$du = \left(\frac{\partial u}{\partial T}\right)_v dT + \left(\frac{\partial u}{\partial v}\right)_T dv.$$

Then since

$$dT = \left(\frac{\partial T}{\partial P}\right)_v dP + \left(\frac{\partial T}{\partial v}\right)_P dv,$$

we can eliminate dT between these equations and obtain

$$du = \left[\left(\frac{\partial u}{\partial T}\right)_v\left(\frac{\partial T}{\partial P}\right)_v\right] dP + \left[\left(\frac{\partial u}{\partial T}\right)_v\left(\frac{\partial T}{\partial v}\right)_P + \left(\frac{\partial u}{\partial v}\right)_T\right] dv.$$

Comparison with Eq. (4–17) shows that

$$\left(\frac{\partial u}{\partial P}\right)_v = \left(\frac{\partial u}{\partial T}\right)_v\left(\frac{\partial T}{\partial P}\right)_v, \tag{4–18}$$

and

$$\left(\frac{\partial u}{\partial v}\right)_P = \left(\frac{\partial u}{\partial T}\right)_v\left(\frac{\partial T}{\partial v}\right)_P + \left(\frac{\partial u}{\partial v}\right)_T. \tag{4–19}$$

The partial derivatives on the right sides of these equations have already been introduced in the preceding sections.

It is left as a problem to obtain expressions corresponding to Eqs. (4–18) and (4–19) for the partial derivatives of h with respect to P and v.

Later on, we shall encounter other properties in addition to u and h that can be expressed as functions of P, v, and T. For any such property w, and any three variables x, y, and z, the general forms of Eqs. (4–18) and (4–19) are

$$\left(\frac{\partial w}{\partial x}\right)_y = \left(\frac{\partial w}{\partial z}\right)_y\left(\frac{\partial z}{\partial x}\right)_y, \tag{4–20}$$

$$\left(\frac{\partial w}{\partial x}\right)_y = \left(\frac{\partial w}{\partial z}\right)_x\left(\frac{\partial z}{\partial x}\right)_y + \left(\frac{\partial w}{\partial x}\right)_z. \tag{4–21}$$

The first of these equations is simply the chain rule for partial derivatives, in which one of the variables is constant.

It is left as a problem to show that

$$\left(\frac{\partial u}{\partial P}\right)_v = c_v\left(\frac{\partial T}{\partial P}\right)_v, \tag{4–22}$$

$$\left(\frac{\partial h}{\partial v}\right)_P = c_P\left(\frac{\partial T}{\partial v}\right)_P, \tag{4–23}$$

$$d'q_T = c_P\left(\frac{\partial T}{\partial v}\right)_P dv_T + c_v\left(\frac{\partial T}{\partial P}\right)_v dP_T, \tag{4–24}$$

and

$$c_v\left(\frac{\partial P}{\partial v}\right)_s = c_P\left(\frac{\partial P}{\partial v}\right)_T. \tag{4–25}$$

4–5 THE GAY-LUSSAC–JOULE EXPERIMENT AND THE JOULE-THOMSON EXPERIMENT

It was mentioned in the preceding sections that on the basis of the second law of thermodynamics, the partial derivatives $(\partial u/\partial v)_T$ and $(\partial h/\partial P)_T$, which describe the way in which the internal energy of a substance varies with volume and in which the enthalpy varies with pressure, at constant temperature, can be calculated from the equation of state of the substance. We now describe how they can also be determined experimentally, for a gaseous system. Since there are no instruments that measure internal energy and enthalpy directly, we first express these derivatives in terms of measurable properties. Making use of Eq. (2–44), we can write

$$\left(\frac{\partial u}{\partial v}\right)_T \left(\frac{\partial v}{\partial T}\right)_u \left(\frac{\partial T}{\partial u}\right)_v = -1.$$

Therefore

$$\left(\frac{\partial u}{\partial v}\right)_T = -c_v\left(\frac{\partial T}{\partial v}\right)_u, \tag{4–26}$$

and the desired partial derivative can be found from a measurement of the rate of change of temperature with volume, in a process at constant internal energy.

In the same way, we find that

$$\left(\frac{\partial h}{\partial P}\right)_T = -c_P\left(\frac{\partial T}{\partial P}\right)_h, \tag{4–27}$$

and the partial derivative can be found from a measurement of the rate of change of temperature with pressure, for states at the same enthalpy.

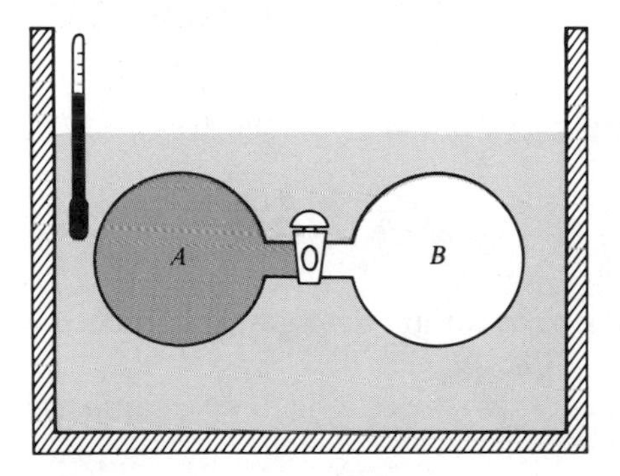

Fig. 4–1 Principle of the Gay-Lussac–Joule experiment.

The earliest attempts to determine the dependence of the internal energy of a gas on its volume were made by Gay-Lussac* and later by Joule, at about the middle of the last century. The apparatus used is shown in principle in Fig. 4-1. Vessel A, containing a sample of the gas to be investigated, is connected to an *evacuated* vessel B by a tube in which there is a stopcock, initially closed. The vessels are immersed in a tank of water of known mass, whose temperature can be measured by a thermometer. Heat losses from the tank to its surroundings will be assumed negligible, or will be allowed for.

* Joseph L. Gay-Lussac, French chemist (1778–1850).

The entire system is first allowed to come to thermal equilibrium and the thermometer reading is noted. The stopcock is then opened and the gas performs a *free* expansion into the evacuated vessel. The work W in this expansion is zero. Eventually, the system comes to a new equilibrium state in which the pressure is the same in both vessels. If the temperature of the gas changes in the free expansion, there will be a flow of heat between the gas and the water bath and the reading of the thermometer in the water will change.

Both Gay-Lussac and Joule found that the temperature change of the water bath, if any, was too small to be detected. The difficulty is that the heat capacity of the bath is so large that a small heat flow into or out of it produces only a very small change in temperature. Similar experiments have been performed more recently with modified apparatus, but the experimental techniques are difficult and the results are not of great precision. All experiments show, however, that the temperature change of the gas itself, even if there were no heat flow to the surroundings, is not large; and hence we postulate as an additional property of an *ideal* gas that its temperature change in a free expansion is zero. There is then no heat flow from the gas to the surroundings and both Q and W are zero. Therefore the internal energy is constant, and for an ideal gas,

$$\left(\frac{\partial T}{\partial v}\right)_u = 0 \text{ (ideal gas).} \tag{4–28}$$

The partial derivative above is called the *Joule coefficient* and is represented by η:

$$\eta \equiv \left(\frac{\partial T}{\partial v}\right)_u. \tag{4–29}$$

Although it is equal to zero for an ideal gas, the Joule coefficient of a real gas is not zero.

It follows from Eq. (4–26), since c_v is finite, that *for an ideal gas*

$$\left(\frac{\partial u}{\partial v}\right)_T = 0. \tag{4–30}$$

That is, the specific internal energy of an ideal gas is independent of the volume and is a function of temperature only. For an ideal gas, the partial derivative $(\partial u/\partial T)_v$ is a *total* derivative and

$$c_v = \frac{du}{dT}, \qquad du = c_v\, dT. \tag{4–31}$$

The energy equation of an ideal gas can now be found by integration. We have

$$\int_{u_0}^{u} du = u - u_0 = \int_{T_0}^{T} c_v\, dT,$$

where u_0 is the internal energy at some reference temperature T_0. If c_v can be considered constant,

$$u = u_0 + c_v(T - T_0). \tag{4–32}$$

The energy surface of an ideal gas (of constant c_v) is shown in Fig. 4–2, plotted as a function of T and v. At constant temperature, the internal energy is constant, independent of the volume. At constant volume, the internal energy increases linearly with temperature.

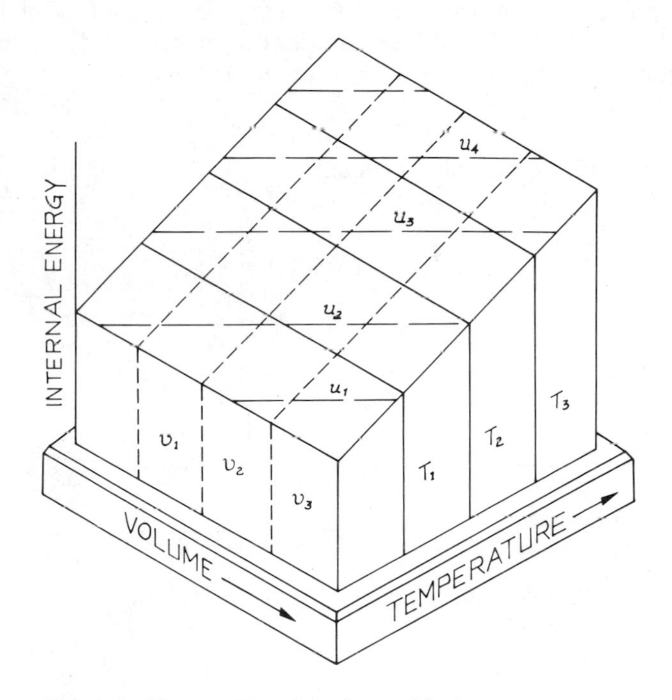

Fig. 4–2 The u-v-T surface for an ideal gas.

Because of the difficulty of measuring precisely the extremely small temperature changes in a free expansion, Joule and Thomson (who later became Lord Kelvin) devised another experiment in which the temperature change of an expanding gas would not be masked by the relatively large heat capacity of its surroundings. Many gases have been carefully investigated in this way. Not only do the results provide information about intermolecular forces but they can be used to reduce gas thermometer temperatures to thermodynamic temperatures without the necessity of extrapolation to zero pressure. The temperature drop produced in the process is utilized in some of the methods for liquefying gases.

The apparatus used by Joule and Thomson is shown schematically in Fig. 4–3. A continuous stream of gas at a pressure P_1 and a temperature T_1 is forced through a porous plug in a tube, from which it emerges at a lower pressure P_2 and a temperature T_2. The device is thermally insulated, and after it has operated for a time long enough for the steady state to become established, the only heat flow from the gas stream is the small flow through the insulation. That is, in the steady state, no heat flows from the gas to *change* the temperature of the walls, and the large heat capacity of the walls does not mask the temperature change of the gas, which is practically what it would be were the system truly an isolated one.

The process is then one of *steady flow*, in which the heat flow Q and the shaft W_{sh} are both zero, and in which there is no change in elevation. The initial and final velocities are both small and their squares can be neglected. Then from the energy equation of steady flow, Eq. (3-38), we have

$$h_1 = h_2,$$

and the initial and final enthalpies are equal.

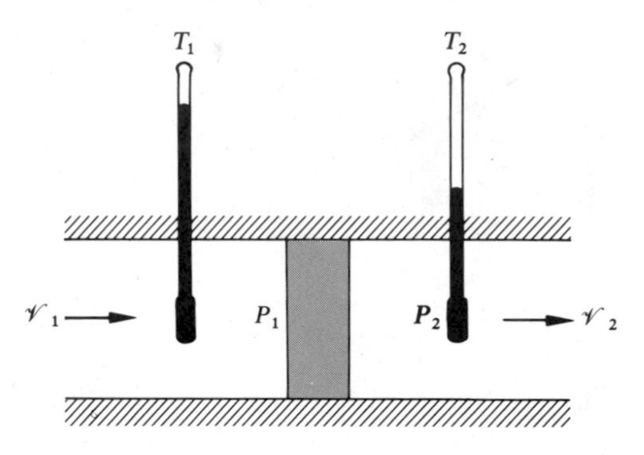

Fig. 4-3 Principle of the Joule-Thomson experiment.

Suppose that a series of measurements are made on the same gas, keeping the initial pressure P_1 and the temperature T_1 the same but varying the pumping rate so that the pressure P_2 on the downstream side of the plug is made to take on a series of values P_2, P_3, etc. Let the temperatures T_2, T_3, etc. be measured in each experiment. (Note that once the pressure on the downstream side is fixed, nothing can be done about the temperature. The properties of the gas determine what the temperature will be.) The corresponding pairs of values of P_2 and T_2, P_3 and T_3, etc., will determine a number of points in a pressure-temperature diagram as in Fig. 4-4(a). Since $h_1 = h_2 = h_3$, etc., the enthalpy is the same at all of these points and a smooth curve drawn through the points is a curve of *constant enthalpy*. Note carefully that this curve does *not* represent the *process* executed by the gas in passing through the plug, since the process is not quasistatic and the gas does not pass through a series of equilibrium states. The final pressure and temperature must be measured at a sufficient distance from the plug for local nonuniformities in the stream to die out, and the gas passes by a nonquasistatic process from one point on the curve to another.

By performing other series of experiments, again keeping the initial pressure and temperature the same in each series but varying them from one series to another, a family of curves corresponding to different values of h can be obtained. Such a family is shown in Fig. 4–4(b), which is typical of all real gases. If the initial temperature is not too great, the curves pass through a maximum called the *inversion point*. The locus of the inversion points is the *inversion curve*.

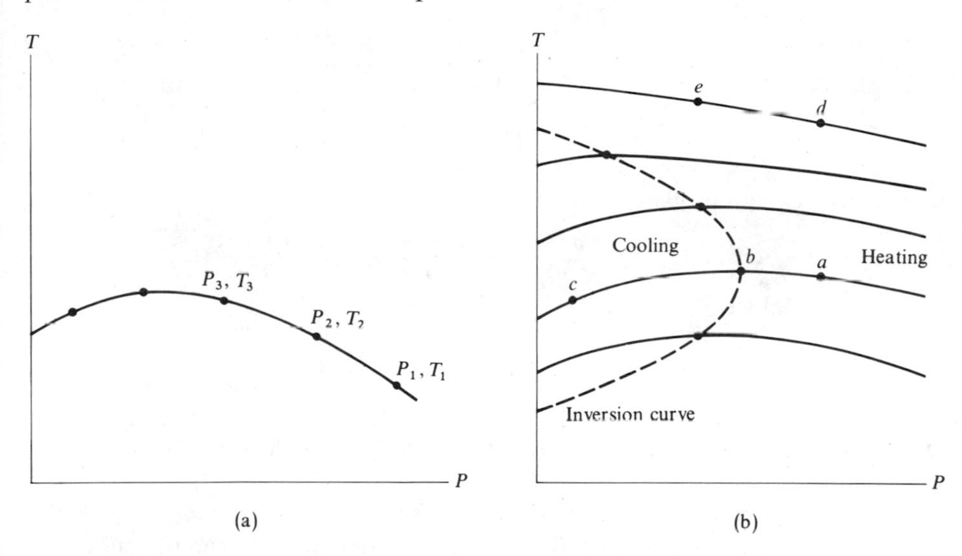

Fig. 4–4 (a) Points of equal enthalpy. (b) Isenthalpic curves and the inversion curve.

When the Joule-Thomson expansion is to be used in the liquefaction of gases, it is evident that the initial temperature and pressure, and the final pressure, must be so chosen that the temperature decreases during the process. This is possible only if the pressure and temperature lie on a curve having a maximum. Thus a drop in temperature would be produced by an expansion from point a or b to point c, but a temperature *rise* would result in an expansion from d to e.

The slope of an isenthalpic curve at any point is the partial derivative, $(\partial T/\partial P)_h$. It is called the *Joule-Thomson* (or the *Joule-Kelvin*) *coefficient* and is represented by μ.

$$\mu \equiv \left(\frac{\partial T}{\partial P}\right)_h. \tag{4–33}$$

At low pressures and high temperatures, where the properties of real gases approach those of an ideal gas, the isenthalpic curves become nearly horizontal and their slope approaches zero. We therefore postulate that an ideal gas shows no temperature change when forced through a porous plug. Hence for such a gas $\mu = 0$, and from Eq. (4–27),

$$\left(\frac{\partial h}{\partial P}\right)_T = 0 \text{ (ideal gas).} \tag{4–34}$$

We shall return in Section 6–10 to a further discussion of the Joule-Thomson experiment, after it has been shown how μ can be calculated from the equation of state.

Since for an ideal gas,

$$\left(\frac{\partial u}{\partial v}\right)_T = \left(\frac{\partial h}{\partial P}\right)_T = 0,$$

Eqs. (4–6) and (4–14) become

$$c_P - c_v = P\left(\frac{\partial v}{\partial T}\right)_P = v\left(\frac{\partial P}{\partial T}\right)_v;$$

and from the equation of state, $Pv = RT$,

$$P\left(\frac{\partial v}{\partial T}\right)_P = v\left(\frac{\partial P}{\partial T}\right)_v = R.$$

Thus for an ideal gas,

$$c_P - c_v = R. \tag{4–35}$$

Table 9–1 gives experimental values of $(c_p - c_v)/R$ for a number of real gases at temperatures near room temperature. This ratio, exactly unity for an ideal gas at all temperatures, is seen to differ from unity by less than 1 percent for nearly all of the gases listed.

If h_0 is the specific enthalpy of an ideal gas in a reference state in which the internal energy is u_0 and the temperature is T_0, it follows that if c_P can be considered constant, the enthalpy equation of an ideal gas is

$$h = h_0 + c_P(T - T_0), \tag{4–36}$$

which is the analogue of Eq. (4–30).

4–6 REVERSIBLE ADIABATIC PROCESSES

We have from Eq. (4–25), for any substance in a reversible adiabatic process,

$$\left(\frac{\partial P}{\partial v}\right)_s = \frac{c_P}{c_v}\left(\frac{\partial P}{\partial v}\right)_T.$$

For an ideal gas,

$$\left(\frac{\partial P}{\partial v}\right)_T = -\frac{P}{v}.$$

Let us represent the ratio c_P/c_v by γ:

$$\gamma \equiv \frac{c_P}{c_v}. \tag{4–37}$$

Replacing $(\partial P/\partial v)_s$ by dP_s/dv_s, and omitting the subscript s for simplicity, we have for an ideal gas,

$$\frac{dP}{P} + \gamma \frac{dv}{v} = 0.$$

In an interval in which γ can be considered constant, this integrates to

$$\ln P + \gamma \ln v = \ln K,$$

or

$$Pv^{\gamma} = K, \tag{4-38}$$

where K is an integration constant. That is, when an ideal gas for which γ is constant performs a reversible adiabatic process, the quantity Pv^{γ} has the same value at all points of the process.

Since the gas necessarily obeys its equation of state in *any* reversible process, the relations between T and P, or between T and v, can be found from the equation above by eliminating v or P between it and the equation of state. They can also be found by integrating Eq. (4–8) and Eq. (4–16). The results are

$$TP^{(1-\gamma)/\gamma} = \text{constant}, \tag{4-39}$$

$$Tv^{\gamma-1} = \text{constant}. \tag{4-40}$$

It was stated in Section 3–11 that the value of c_p for monatomic gases is very nearly equal to $5R/2$ and that for diatomic gases is nearly equal to $7R/2$. Since the difference $c_P - c_v$ is equal to R for an ideal gas and is very nearly equal to R for all gases, we can write for a monatomic gas

$$\gamma = \frac{c_P}{c_v} = \frac{c_P}{c_P - R} = \frac{5R/2}{(5R/2) - R} = \frac{5}{3} = 1.67;$$

for a diatomic gas,

$$\gamma = \frac{7R/2}{(7R/2) - R} = 1.40.$$

Table 9–1 includes the experimental values of γ for a number of common gases.

The curves representing adiabatic processes are shown on the ideal gas P-v-T surface in Fig. 4–5(a), and their projections on the P-v plane in Fig. 4–5(b).

The adiabatic curves projected onto the P-v plane have at every point a somewhat steeper slope than the isotherms. The temperature of an ideal gas increases in a reversible adiabatic compression, as will be seen from an examination of Fig. 4–5(a) or from Eqs. (4–39) or (4–40). This increase in temperature may be very large and it is utilized in the Diesel type of internal combustion engine, where, on the compression stroke, air is compressed in the cylinders to about 1/15 of its volume at atmospheric pressure. The air temperature at the completion of the compression stroke is so high that fuel oil injected into the air burns without the necessity of a spark to initiate the combustion process.

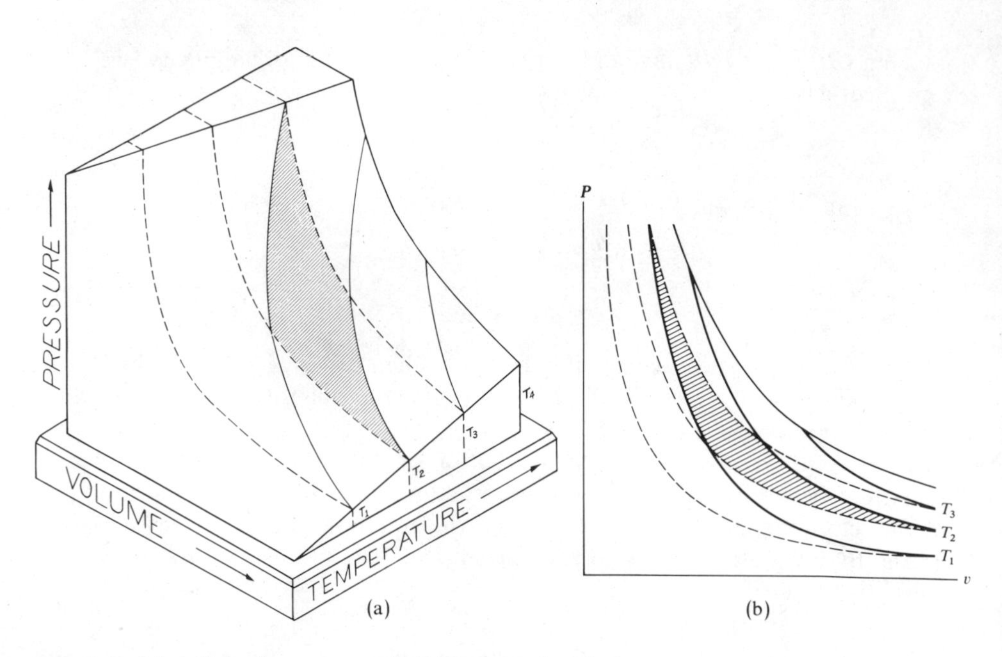

Fig. 4–5 (a) Adiabatic processes (full lines) on the ideal gas *P-v-T* surface. (b) Projection of the adiabatic processes in (a) onto the *P-v* plane. The shaded area is a Carnot cycle (see Section 4–7).

The specific work in a reversible adiabatic expansion of an ideal gas is

$$w = \int_{v_1}^{v_2} P\,dv = K\int_{v_1}^{v_2} v^{-\gamma}\,dv$$

$$= \frac{1}{1-\gamma}\,[Kv^{1-\gamma}]_{v_1}^{v_2}, \tag{4–41}$$

where K is the integration constant in Eq. (4–36). But to state that $Pv^{\gamma} = \text{const} = K$ means that

$$P_1v_1^{\gamma} = P_2v_2^{\gamma} = K.$$

Hence when inserting the upper limit in Eq. (4–39) we let $K = P_2v_2^{\gamma}$, while at the lower limit we let $K = P_1v_1^{\gamma}$. Then

$$w = \frac{1}{1-\gamma}\,(P_2v_2 - P_1v_1). \tag{4–42}$$

The work can also be found by realizing that since there is no heat flow into or out of a system in an adiabatic process, the work is done wholly at the expense of the internal energy of the system. Hence

$$w = u_1 - u_2,$$

and for an ideal gas for which c_v is constant,

$$w = c_v(T_1 - T_2). \tag{4–43}$$

4–7 THE CARNOT CYCLE

In 1824, Carnot* first introduced into the theory of thermodynamics a simple cyclic process now known as a *Carnot cycle*. Carnot was primarily interested in improving the efficiencies of steam engines, but instead of concerning himself with mechanical details he concentrated on an understanding of the basic physical principles on which the efficiency depended. It may be said that the work of Carnot laid the foundation of the science of thermodynamics. Although actual engines have been constructed which carry a system through essentially the same sequence of processes as in a Carnot cycle, the chief utility of the cycle is as an aid in thermodynamic reasoning. In this section we shall describe the Carnot cycle and in the next section will consider its relation to the efficiency of an engine.

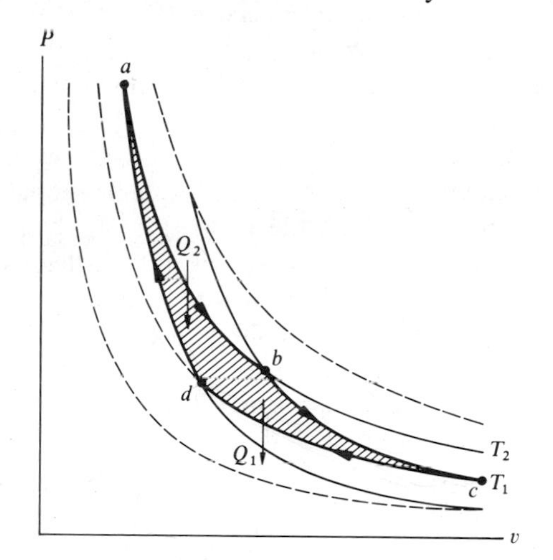

Fig. 4–6 The Carnot cycle.

A Carnot cycle can be carried out with a system of any nature. It may be a solid, liquid, or gas, or a surface film, or a paramagnetic substance. The system may even undergo a change of phase during the cycle. A Carnot cycle for an ideal gas is represented by the shaded area on the *P-v-T* surface of Fig. 4–5(a), and its projection onto the *P-v* plane is shown in Fig. 4–5(b) and again in Fig. 4–6.

Starting at state *a*, the system at a temperature T_2 is brought in contact with a heat reservoir at this temperature and performs a *reversible isothermal* process that takes it to state *b*. For an ideal gas, this process is an *expansion*. For a paramagnetic material, it would be an increase in the magnetic moment *M*, etc. In this process there is a heat flow Q_2 into the system and work W_2 is done by the system.

At state *b*, the system is thermally insulated and performs a *reversible adiabatic* process to state *c*. In this process the temperature decreases to a lower value

* N. L. Sadi Carnot, French engineer (1796–1832).

T_1. The heat flow into the system is zero and additional work W' is done by the system.

The system is next brought in contact with a heat reservoir at temperature T_1 and performs a *reversible isothermal* process to state d. There is a heat flow Q_1 *out* of the system and work W_1 is done *on* the system.

State d must be chosen so that a final *reversible adiabatic* process will return the system to its initial state a. The heat flow is zero in this process and work W'' is done on the system.

The significant features of any Carnot cycle are therefore (a) the entire heat flow *into* the system takes place at single higher temperature T_2; (b) the entire heat flow *out* of the system takes place at a single lower temperature T_1; (c) the system, often referred to as the *working substance*, is carried through a *cyclic* process; and (d) all processes are *reversible*. We can say in general that any cyclic process bounded by two reversible isothermals and two reversible adiabatics constitutes a Carnot cycle.

Although the magnitudes of the heat flows and quantities of work are arbitrary (they depend on the actual changes in volume, magnetic moment, etc.), it is found that the *ratio* Q_2/Q_1 depends only on the temperatures T_2 and T_1. To calculate this ratio, it is necessary to know the equation of state of the system, and its energy equation. (It is necessary to know these at this stage of our development of the principles of thermodynamics. We shall show in Section 5–2 that for two given temperatures T_2 and T_1 the ratio T_2/T_1 has the same value for *all* working substances.) Let us therefore assume that the system is an ideal gas.

Since the internal energy of an ideal gas is a function of its temperature only, the internal energy is constant in the isothermal process *a-b* and the heat flow Q_2 into the system in this process is equal to the work W_2. Hence from Eq. (3–5),

$$Q_2 = W_2 = nRT_2 \ln \frac{V_b}{V_a}, \tag{4–44}$$

where V_b and V_a are the volumes in states b and a, respectively. Similarly, the magnitude of the heatflow Q_1 equals the work W_1 and

$$Q_1 = W_1 = nRT_1 \ln \frac{V_c}{V_d}. \tag{4–45}$$

But states b and c lie on the same adiabatic, and hence from Eq. (4–40),

$$T_2 V_b^{\gamma-1} = T_1 V_c^{\gamma-1}.$$

Similarly, since states a and b lie on the same adiabatic,

$$T_2 V_a^{\gamma-1} = T_1 V_d^{\gamma-1}$$

When the first of these equations is divided by the second, we find that

$$\frac{V_b}{V_a} = \frac{V_c}{V_d}. \tag{4–46}$$

It follows from Eqs. (4–44), (4–45), and (4–46) that

$$\frac{Q_2}{Q_1} = \frac{T_2}{T_1}. \tag{4–47}$$

Thus *for an ideal gas*, the ratio Q_2/Q_1 depends only on the temperatures T_2 and T_1.

4–8 THE HEAT ENGINE AND THE REFRIGERATOR

A system carried through a Carnot cycle is the prototype of all cyclic *heat engines*. The feature that is common to all such devices is that they receive an input of heat at one or more higher temperatures, do mechanical work on their surroundings, and reject heat at some lower temperature.

When any working substance is carried through a cyclic process, there is no change in its internal energy in any complete cycle and from the first law the net flow of heat Q into the substance, in any complete cycle, is equal to the work W done by the engine, per cycle. Thus if Q_2 and Q_1 are the absolute magnitudes of the heatflows into and out of the working substance, per cycle, the net heat flow Q per cycle is

$$Q = Q_2 - Q_1.$$

The net work W per cycle is therefore

$$W = Q = Q_2 - Q_1. \tag{4–48}$$

The *thermal efficiency* η of a heat engine is defined as the ratio of the work output W to the heat *input* Q_2:

$$\eta \equiv \frac{W}{Q_2} = \frac{Q_2 - Q_1}{Q_2}. \tag{4–49}$$

The work output is "what you get," the heat input is "what you pay for." Of course, the rejected heat Q_1 is in a sense a part of the "output" of the engine, but ordinarily this is wasted (as in the hot exhaust gases of an automobile engine, or as a contribution to the "thermal pollution" of the surroundings) and has no economic value. If the rejected heat were included as a part of its output, the thermal efficiency of every heat engine would be 100%. The definition of thermal efficiency as work output divided by heat input applies to every type of heat engine and is not restricted to a Carnot engine.

If the working substance is an ideal gas, then for a Carnot cycle we have shown that

$$\frac{Q_1}{Q_2} = \frac{T_1}{T_2}.$$

The thermal efficiency is then

$$\eta = \frac{Q_2 - Q_1}{Q_2} = 1 - \frac{Q_1}{Q_2} = 1 - \frac{T_1}{T_2}, \tag{4–50}$$

or

$$\eta = \frac{T_2 - T_1}{T_2}. \tag{4–51}$$

The thermal efficiency therefore depends only on the temperatures T_2 and T_1. We shall show in Section 5–2 that the thermal efficiency of any Carnot cycle is given by the expression above, whatever the nature of the working substance.

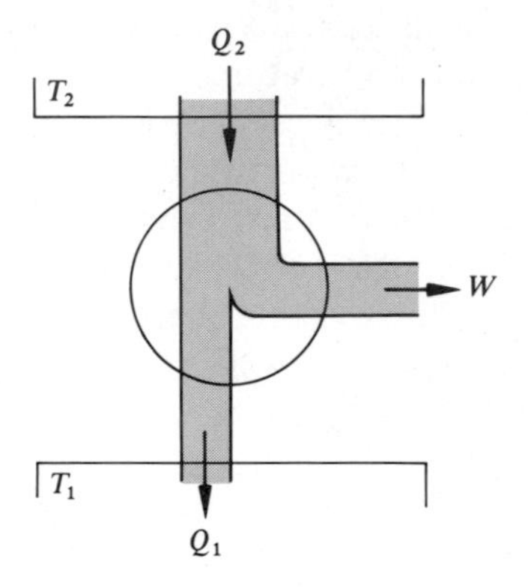

Fig. 4–7 Schematic flow diagram of a heat engine.

It is helpful to represent the operation of any heat engine by a schematic flow diagram like that in Fig. 4–7. The width of the "pipeline" from the high temperature reservoir is proportional to the heat Q_2, the width of the line to the low temperature reservoir is proportional to Q_1, and the width of the line leading out from the side of the engine is proportional to the work output W. The circle is merely a schematic way of indicating the engine. The goal of an engine designer is to make the work output pipeline as large as possible, and the rejected heat pipeline as small as possible, for a given incoming pipeline from the high temperature reservoir.

We may mention that Carnot would not have constructed his flow diagram in the same way as that in Fig. 4–7. In Carnot's time, it was believed that "heat" was some sort of indestructible fluid, in which case the pipelines Q_2 and Q_1 would have the same width. How then could there be any pipeline W? It was thought that work W could be abstracted from a "downhill" flow of heat in the same way that work can be obtained from a flow of water through a turbine, from a higher to a lower elevation. The quantities of water flowing into and out of the turbine are equal, and the mechanical work is done at the expense of the decrease in potential energy of the water. But in spite of his erroneous ideas as to the nature of heat, Carnot did obtain the correct expression for the efficiency of a Carnot engine.

If the Carnot cycle in Fig. 4–6 is traversed in a counterclockwise rather than a clockwise direction, the directions of all arrows in Figs. 4–6 and 4–7 are reversed, and since all processes in the cycle are reversible (in the thermodynamic sense), there is no change in the *magnitudes* of Q_2, Q_1, and W. Heat Q_1 is now removed from the low-temperature reservoir, work W is done *on* the system, and heat Q_2 equal to $W + Q_1$ is delivered to the high-temperature reservoir. We now have a *Carnot refrigerator* or a *heat pump*, rather than a Carnot engine. That is, heat is pumped out of a system at low temperature (the interior of a household refrigerator, for example, or out of the atmosphere or the ground in the case of a heat pump used for house heating), mechanical work is done (by the motor driving the refrigerator), and heat equal to the sum of the mechanical work and the heat removed from the low-temperature reservoir is liberated at a higher temperature.

The useful result of operating a *refrigerator* is the heat Q_1 removed from the low-temperature reservoir; this is "what you get." What you have to pay for is the work input, W. The greater the ratio of what you get to what you pay for, the better the refrigerator. A refrigerator is therefore rated by its *coefficient of performance*, c, defined as the ratio of Q_1 to W. Again making use of Eq. (4–48), we can write

$$c \equiv \frac{Q_1}{W} = \frac{Q_1}{Q_2 - Q_1}. \tag{4–52}$$

The coefficient of performance of a refrigerator, unlike the thermal efficiency of a heat engine, can be much larger than 100%.

The definition above of coefficient of performance applies to any refrigerator, whether or not it operates in a Carnot cycle. For a Carnot refrigerator, $Q_2/Q_1 = T_2/T_1$ and

$$c = \frac{T_1}{T_2 - T_1}. \tag{4–53}$$

PROBLEMS

4–1 The specific internal energy of a van der Waals gas is given by

$$u = c_v T - \frac{a}{v} + \text{constant}.$$

(a) Sketch a u-T-v surface assuming c_v is a constant. (b) Show that for a van der Waals gas,

$$c_P - c_v = R\frac{1}{1 - \dfrac{2a(v-b)^2}{RTv^3}}.$$

4–2 The equation of state of a certain gas is $(P + b)v = RT$ and its specific internal energy is given by $u = aT + bv + u_0$. (a) Find c_v. (b) Show that $c_P - c_v = R$. (c) Using Eq. (4–8) show that $Tv^{R/c_v} = \text{constant}$.

4–3 The specific internal energy of a substance can be given by

$$u - u_0 = 3T^2 + 2v,$$

in an appropriate set of units. (a) Sketch a u-T-v diagram for this substance. (b) Compute the change in temperature of the substance if 5 units of heat are added while the volume of the substance is held constant. Show this process on the u-T-v diagram. (c) Can the change in temperature of the substance during an adiabatic decrease in volume of 20% be determined from the information given? If so, compute it. If not, state what additional information must be supplied.

4–4 At temperatures above 500 K, the value of c_P for copper can be approximated by a linear relation of the form $c_P = a + bT$. (a) Find as accurately as you can from Fig. 3–10 the values of a and b. (b) Compute the change in the specific enthalpy of copper at a pressure of 1 atm when the temperature is increased from 500 to 1200 K.

4–5 Show that $\left(\dfrac{\partial h}{\partial P}\right)_T = -c_P\left(\dfrac{\partial T}{\partial P}\right)_h$.

4–6 Show that $\left(\dfrac{\partial u}{\partial T}\right)_P = c_P - P\beta v$.

4–7 Compare the magnitudes of the terms c_P and $P\beta v$ in the previous problem (a) for copper at 600 K and 1 atm, and (b) for an ideal gas for which $c_P = 5R/2$. (c) When heat is supplied to an ideal gas in an isobaric process, what fraction goes into an increase in internal energy? (d) When heat is supplied to copper in an isobaric process, what fraction goes into an increase in internal energy?

4–8 (a) Show that the specific enthalpy of the gas of Problem 4–2 can be written as $h = (a + R)T + \text{constant}$. (b) Find c_P. (c) Using Eq. (4–16) show that $T(P + b)^{-R/c_P} = \text{constant}$. (d) Show that $(\partial h/\partial v)_P = c_P T/v$.

4–9 Derive expressions analogous to Eq. (4–18) and Eq. (4–19) for h as a function of P and v.

4–10 Complete the derivations of Eqs. (4–22) to (4–25).

4–11 An ideal gas for which $c_v = 5R/2$ is taken from point a to point b in Fig. 4–8 along the three paths a-c-b, a-d-b, and a-b. Let $P_2 = 2P_1$, and $v_2 = 2v_1$. (a) Compute the heat supplied to the gas, per mole, in each of the three processes. Express the answer in terms of R and T_1. (b) Compute the molal specific heat capacity of the gas, in terms of R, for the process a-b.

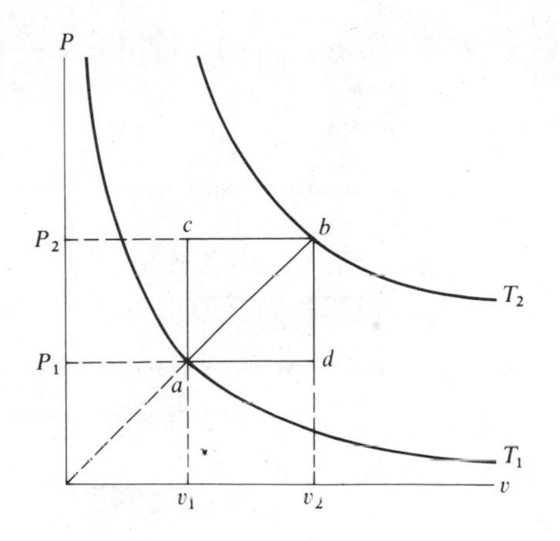

Figure 4–8

4–12 For a van der Waals gas obeying the energy equation of Problem 4–1 show that

$$\left(\frac{\partial T}{\partial v}\right)_s = \frac{\gamma}{v\kappa}\left(\frac{\partial T}{\partial P}\right)_s.$$

4–13 For a paramagnetic substance obeying Curie's law the internal energy is a function of T only. Show that (a) $d'Q = C_M\, dT - \mathscr{H}\, dM$; (b) $d'Q = C_{\mathscr{H}}\, dT - M\, d\mathscr{H}$; and (c) $C_{\mathscr{H}} - C_M = M\mathscr{H}/T$.

4–14 For a one-dimensional system show that (a) $C_L = \left(\frac{\partial U}{\partial T}\right)_L$; (b) $C_{\mathscr{F}} = \left(\frac{\partial H}{\partial T}\right)_{\mathscr{F}}$; and (c) $C_L\left(\frac{\partial \mathscr{F}}{\partial L}\right)_S = C_{\mathscr{F}}\left(\frac{\partial \mathscr{F}}{\partial L}\right)_T$.

4–15 For an ideal gas show that (a) $\left(\frac{\partial u}{\partial P}\right)_T = 0$, and (b) $\left(\frac{\partial T}{\partial P}\right)_u = 0$.

4–16 Suppose one of the vessels in the Joule apparatus of Fig. 4–1 contains n_A moles of a van der Waals gas and the other contains n_B moles, both at an initial temperature T_1. The volume of each vessel is V. Find the expression for the change in temperature when the stopcock is opened and the system is allowed to come to a new equilibrium state. Neglect any flow of heat to the vessels. Verify your solution for the cases when $n_B = 0$, using Eq. (4–26), and when $n_A = n_B$. Assume the energy equation of Problem 4–1.

4–17 (a) Show that for an ideal gas $h - h_0 = c_P(T - T_0)$ and (b) sketch an h-P-T surface for an ideal gas.

4–18 Assume the energy equation given in Problem 4–1. (a) Find the expression for the Joule coefficient η for a van der Waals gas. (b) Find the expression for the enthalpy of a van der Waals gas, as a function of v and T. (c) Find the expression for the Joule-Thomson coefficient μ for a van der Waals gas. (d) Show that the expressions in (a) and (c) reduce to those for an ideal gas if $a = b = 0$. [*Hint:* See Problem 2–22.]

4–19 Show that (a) $\left(\frac{\partial h}{\partial P}\right)_T = -\mu c_P$, (b) $\left(\frac{\partial h}{\partial T}\right)_v = c_P\left[1 - \frac{\beta\mu}{\kappa}\right]$, (c) $\left(\frac{\partial h}{\partial v}\right)_T = \frac{\mu c_P}{v\kappa}$, (d) $\left(\frac{\partial T}{\partial v}\right)_h = \frac{\mu}{(v\kappa - \mu v\beta)}$.

4–20 For an ideal gas, show that in a reversible adiabatic process (a) $TP^{(\gamma-1)/\gamma}$ = constant, and (b) $Tv^{(\gamma-1)}$ = constant.

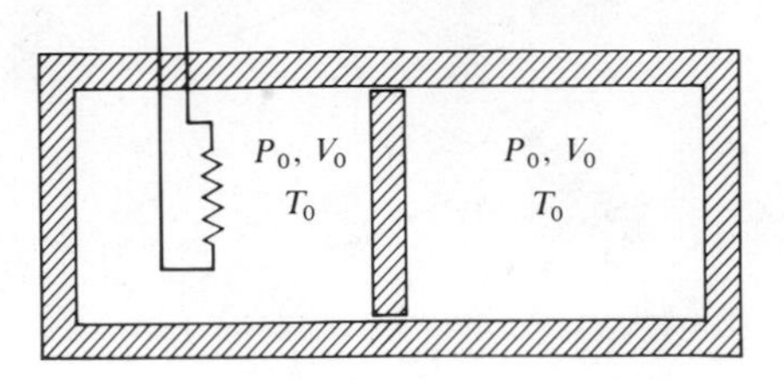

Figure 4–9

4–21 Figure 4–9 represents a cylinder with thermally insulated walls containing a movable frictionless thermally insulated piston. On each side of the piston are n moles of an ideal gas. The initial pressure P_0, volume V_0, and temperature T_0 are the same on both sides of the piston. The value of γ for the gas is 1.50, and c_v is independent of temperature. By means of a heating coil in the gas on the left side of the piston, heat is supplied slowly to the gas on this side. It expands and compresses the gas on the right side until its pressure has increased to 27 $P_0/8$. In terms of n, c_v, and T_0, (a) how much work is done on the gas on the right side? (b) what is the final temperature of the gas on the right? (c) what is the final temperature of the gas on the left? (d) how much heat flows into the gas on the left?

4–22 In the compression stroke of a Diesel engine, air is compressed from atmospheric pressure and room temperature to about 1/15 of its original volume. Find the final temperature, assuming a reversible adiabatic compression. (Take $\gamma_{\text{air}} = 1.4$.)

4–23 (a) Show that the work done on an ideal gas to compress it isothermally is greater than that necessary to compress the gas adiabatically if the pressure change is the same in the two processes, and (b) that the isothermal work is less than the adiabatic work if the volume change is the same in the two processes. As a numerical example, take the initial pressure and volume to be 10^6 N m^{-2} and 0.5 m^3 kilomole^{-1}, and take γ to be 5/3. Compute the work necessary to change the value of the appropriate variable by a factor of 2. (c) Plot these processes on a P-V diagram and explain physically why the isothermal work should be greater than the adiabatic work in part (a) and why it should be less in part (b).

4–24 An ideal gas for which $c_v = 3R/2$ occupies a volume of 4 m^3 at a pressure of 8 atm and a temperature of 400 K. The gas expands to a final pressure of 1 atm. Compute the final volume and temperature, the work done, the heat absorbed, and the change in internal energy, for each of the following processes: (a) a reversible, isothermal expansion; (b) a reversible adiabatic expansion; and (c) an expansion into a vacuum.

4–25 One mole of an ideal gas is taken from $P = 1$ atm and $T = 273$ K to $P = 0.5$ atm and $T = 546$ K by a reversible isothermal process followed by a reversible isobaric process. It is returned to its initial state by a reversible isochoric process followed by a reversible adiabatic process. Assume that $c_v = (3/2)R$. (a) Draw this cycle on a P-V

diagram. (b) For each process and for the whole cycle, find the change in T, V, P, W, Q, U, and H. A tabular arrangement of the results will be useful. (c) Draw this cycle on a V-T diagram and on a U-V diagram.

4–26 (a) Use Eq. (4–8) to derive for a van der Waals gas the equations corresponding to Eqs. (4–38) and (4–40). (b) Compute the work in a reversible adiabatic expansion by direct evaluation of $\int P\,dv$ and by use of the energy equation of Problem 4–1.

4–27 The equation of state for radiant energy in equilibrium with the temperature of the walls of a cavity of volume V is $P = aT^4/3$. The energy equation is $U = aT^4V$. (a) Show that the heat supplied in an isothermal doubling of the volume of the cavity is $4aT^4V/3$. (b) Use Eq. (4–3) to show that in an adiabatic process VT^3 is a constant.

4–28 Sketch a Carnot cycle for an ideal gas on a (a) u-v diagram, (b) u-T diagram, (c) u-h diagram, (d) P-T diagram.

4–29 Sketch qualitatively a Carnot cycle (a) in the V-T plane for an ideal gas; (b) in the P-V plane for a liquid in equilibrium with its vapor; (c) in the $\mathscr{E}$-Z plane for a reversible electrolytic cell whose emf is a function of T alone and assuming that reversible adiabatics have a constant positive slope.

4–30 A Carnot engine is operated between two heat reservoirs at temperatures of 400 K and 300 K. (a) If the engine receives 1200 Cal from the reservoir at 400 K in each cycle, how many Calories does it reject to the reservoir at 300 K? (b) If the engine is operated as a refrigerator (i.e., in reverse) and receives 1200 Cal from the reservoir at 300 K, how many Calories does it deliver to the reservoir at 400 K? (c) How much work is done by the engine in each case?

4–31 (a) Show that for Carnot engines operating between the same high temperature reservoirs and different low temperature reservoirs, the engine operating over the largest temperature difference has the greatest efficiency. (b) Is the more effective way to increase the efficiency of a Carnot engine to increase the temperature of the hotter reservoir, keeping the temperature of the colder reservoir constant, or vice versa? (c) Repeat parts (a) and (b) to find the optimum coefficient of performance for a Carnot refrigerator.

4–32 Derive a relationship between the efficiency of a Carnot engine and the coefficient of performance of the same engine when operated as a refrigerator. Is a Carnot engine whose efficiency is very high particularly suited as a refrigerator? Give reasons for your answer.

4–33 An ideal gas for which $c_v = 3R/2$ is the working substance of a Carnot engine. During the isothermal expansion the volume doubles. The ratio of the final volume to the initial volume in the adiabatic expansion is 5.7. The work output of the engine is 9×10^5 J in each cycle. Compute the temperature of the reservoirs between which the engine operates.

4–34 Calculate the efficiency and the coefficient of performance of the cycles shown in (a) Problem 3–26, and (b) Problem 3–27.

4–35 An electrolytic cell is used as the working substance of a Carnot cycle. In the appropriate temperature range the equation of state for the cell is $\mathscr{E} = \mathscr{E}_0 - \alpha(T - T_0)$, where $\alpha > 0$ and $T > T_0$. The energy equation is

$$U - U_0 = \left(\mathscr{E} - T\frac{d\mathscr{E}}{dT}\right)Z + C_Z(T - T_0)$$

where C_Z is the heat capacity at constant Z which is assumed to be a constant and Z is the charge which flows through the cell. (a) Sketch the Carnot cycle on an $\mathscr{E} - Z$ diagram and indicate the direction in which the cycle operates as an engine. (b) Use the expression for the efficiency of a Carnot cycle to show that charge transferred in the isothermal processes must have the same magnitude.

4–36 A building is to be cooled by a Carnot engine operated in reverse (a Carnot refrigerator). The outside temperature is 35°C (95°F) and the temperature inside the building is 20°C (68°F). (a) If the engine is driven by a 12×10^3 watt electric motor, how much heat is removed from the building per hour? (b) The motor is supplied with electricity generated in a power plant which consists of a Carnot engine operating between reservoirs at temperatures of 500°C and 35°C. Electricity (transmitted over a 5 ohm line), is received at 220 volts. The motors which operate the refrigerator and the generator at the power plant each have an efficiency of 90%. Determine the number of units of refrigeration obtained per unit of heat supplied. (c) How much heat must be supplied per hour at the power plant? (d) How much heat is rejected per hour from the power plant?

4–37 Refrigerator cycles have been developed for heating buildings. Heat is absorbed from the earth by a fluid circulating in buried pipes and heat is delivered at a higher temperature to the interior of the building. If a Carnot refrigerator were available for use in this way, operating between an outside temperature of 0°C and an interior temperature of 20°C, how many kilowatt-hours of heat would be supplied to the building for every kilowatt-hour of electrical energy needed to operate the refrigerator?

4–38 The temperature of a household refrigerator is 5°C and the temperature of the room in which it is located is 20°C. The heat flowing from the warmer room every 24 hours is about 3×10^6 J (enough to melt about 20 lb of ice) and this heat must be pumped out again if the refrigerator is to be kept cold. If the refrigerator is 60% as efficient as a Carnot engine operating between reservoirs having temperatures of 5°C and 20°C, how much power in watts would be required to operate it? Compare the daily cost at 3 cents per kilowatt-hour with the cost of 20 lb of ice (about 75 cents).

4–39 An approximate equation of state for a gas is $P(v - b) = RT$, where b is a constant. The specific internal energy of a gas obeying this equation of state is $u = c_v T + \text{constant}$. (a) Show that the specific heat at constant pressure of this gas is equal to $c_v + R$. (b) Show that the equation of a reversible adiabatic process is $P(v - b)^\gamma = \text{constant}$. (c) Show that the efficiency of a Carnot cycle using this gas as the working substance is the same as that for an ideal gas, assuming $(\partial u/\partial v)_T = 0$.

5

Entropy and the second law of thermodynamics

5–1 THE SECOND LAW OF THERMODYNAMICS

Figure 5–1 shows three different systems, each enclosed in a rigid adiabatic boundary. In part (a), a body at a temperature T_1 makes thermal contact with a large heat reservoir at a higher temperature T_2. In part (b), a rotating flywheel drives a generator that sends a current through a resistor immersed in a heat reservoir. In part (c), a gas is confined to the left portion of the container by a diaphragm. The remainder of the container is evacuated. We know from experience that in part (a) there will be a heat flow from the reservoir into the body and that, eventually, the body will come to the same temperature T_2 as the reservoir. (The heat capacity of a reservoir is so large that its temperature is not changed appreciably by a flow of heat into or out of it.) In part (b) the flywheel will eventually be brought to rest. Dissipative work will be done on the resistor and there will be a heat flow out of it into the reservoir, equal in magnitude to the original kinetic energy of the flywheel. If the diaphragm in part (c) is punctured, the gas will perform a free expansion into the evacuated region and will come to a new equilibrium state at a larger volume and a lower pressure. In each of these processes, the total energy of the system, including any kinetic energy of the flywheel in part (b), remains constant.

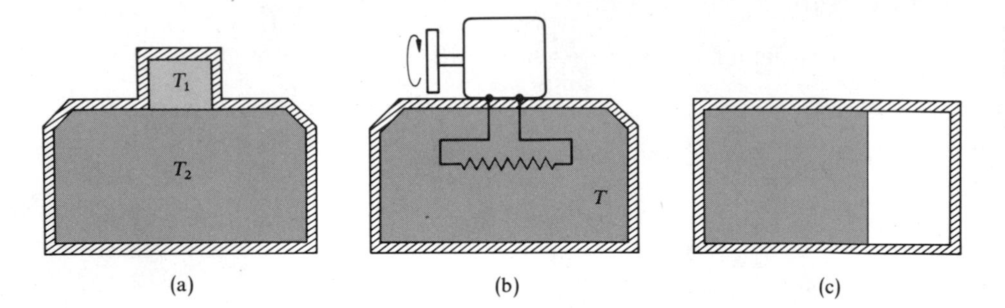

Fig. 5–1 In part (a) there is a reversible heat flow between a body at temperature T_1 and a large heat reservoir at a higher temperature T_2. In (b), a rotating flywheel drives a generator which sends a current through a resistor in a heat reservoir. In (c), a gas in the left portion of the container performs a free expansion into the evacuated region when the diaphragm is punctured.

Now suppose we start with the three systems at the end states of the above processes and imagine the processes to take place in the reversed direction. In the first example, the body originally at the same temperature as the reservoir would spontaneously cool down until its original temperature was restored. In the second, there would be a heat flow out of the reservoir into the resistor, which would send a current through the generator (now serving as a motor), and the flywheel would be set in rotation with its original kinetic energy. In the third, the gas would compress itself back into its original volume.

Everyone realizes that these reversed processes do not happen. But why not? The total energy in each case would remain constant in the reversed process as it did in the original, and there would be no violation of the principle of conservation of energy. There must be some other natural principle, in addition to the first law and not derivable from it, which determines the *direction* in which a natural process will take place. This principle is contained in the *second law of thermodynamics.* The second law, like the first, is a generalization from experience and it asserts that certain processes, of which the three considered above are examples, are essentially one-way processes and will proceed in one direction only.

The three impossible, reversed processes were chosen as examples because they appear at first sight to differ widely from one another. In the first, a composite system originally at a uniform temperature would separate spontaneously into two portions at different temperatures. In the second, there would be a flow of heat out of a reservoir and an equivalent amount of kinetic energy would appear. In the third, the volume of an isolated sample of gas would decrease and its pressure would increase. Many other illustrations could be given. In the field of chemistry, for example, oxygen and hydrogen gas in the proper proportions can be enclosed in a vessel and a chemical reaction can be initiated by a spark. If the enclosure has rigid adiabatic walls the internal energy of the system remains constant. After the reaction has taken place, the system consists of water vapor at a high temperature and pressure, but the water vapor will not spontaneously dissociate into hydrogen and oxygen at a lower temperature and pressure.

Can we find some feature which all of these dissimilar impossible processes have in common? Given two states of an isolated system, in both of which the energy is the same, can we find a criterion that determines which is a possible initial state and which is a possible final state of a process taking place in the system? What are the conditions under which no process at all can occur, and in which a system is in equilibrium? These questions could be answered if there existed some property of a system, that is, some function of the state of a system, which has a different value at the beginning and at the end of a possible process. This function cannot be the energy, since that is constant. A function having the desired property can be found, however. It was devised by Clausius* and is called the *entropy* of the system. Like the energy, it is a function of the state of the system only and, as we shall prove, it either remains constant or increases in any possible process taking place in an isolated system. In terms of entropy, the second law can be stated:

Processes in which the entropy of an isolated system would decrease do not occur: or in every process taking place in an isolated system the entropy of the system either increases or remains constant.

Furthermore, if an isolated system is in such a state that its entropy is a maximum, any change from that state would necessarily involve a decrease in entropy

* Rudolph J. E. Clausius, German physicist (1822–1888).

and hence will not happen. Therefore the necessary condition for the equilibrium of an isolated system is that its entropy shall be a maximum.

Note carefully that the statements above apply to isolated systems only. It is quite possible for the entropy of a nonisolated system to decrease in an actual process, but it will always be found that the entropy of other systems with which the first interacts increases by at least an equal amount.

The second law has been stated here without defining entropy. In the next sections the concept of entropy is developed by using first the properties of the Carnot cycle and then by calculating entropy changes during reversible and irreversible processes. After a discussion of the physical significance of entropy production, equivalent alternative statements of the second law are presented.

5–2 THERMODYNAMIC TEMPERATURE

Before proceeding to the development of the concept of entropy, we shall use the Carnot cycle to define the thermodynamic temperature. In Chapter 1, we introduced the symbol T to represent temperature on the ideal gas thermometer scale, with the promise that it would later be shown to equal the thermodynamic temperature. Let us therefore return to the symbol θ, as used in Chapter 1, to designate an *empirical* temperature defined in terms of an arbitrary thermometric property X, such as the resistance R of a platinum resistance thermometer or the pressure P of a constant-volume hydrogen thermometer.

The Carnot cycle for a $PV\theta$ system is shown in the θ-V plane in Fig. 5–2. The shape of the adiabatics varies, of course, from one substance to another. Let us first carry out the cycle *a-b-c-d-a*. In the process *a-b* there is a heat flow Q_2 into the system from a reservoir at a temperature θ_2, and in the process *c-d* there is a smaller heat flow Q_1 out of the system into a reservoir at a temperature θ_1. The heat flows are zero in the adiabatic processes *b-c* and *d-a*. Since the system is returned to its initial state at point a, there is no change in its internal energy; and from the first law, since $\Delta U = 0$, the work W in the cycle is

$$W = |Q_2| - |Q_1|.$$

This is the only condition imposed on Q_2 and Q_1 by the first law: the work W in the cycle equals the difference between the absolute magnitudes of Q_2 and Q_1.

In Section 5–1 the second law was stated in terms of the entropy of a system, but since we have not as yet defined this property we must begin with a consequence of the second law that does not involve the entropy concept. Thus our starting point will be the assertion that **for any two temperatures θ_2 and θ_1, the ratio of the magnitudes of Q_2 and Q_1 in a Carnot cycle has the same value for *all* systems, whatever their nature.** That is, the ratio $|Q_2|/|Q_1|$ is a function only of the temperatures θ_2 and θ_1:

$$\frac{|Q_2|}{|Q_1|} = f(\theta_2, \theta_1). \tag{5–1}$$

The form of the function f depends on the particular empirical temperature scale on which θ_2 and θ_1 are measured, but it does not depend on the nature of the system performing the cycle.

It should not be inferred that the quantities of heat absorbed and liberated in a Carnot cycle have been measured experimentally for all possible systems and all possible pairs of temperatures. The justification of the preceding assertion lies in the correctness of all conclusions that can be drawn from it.

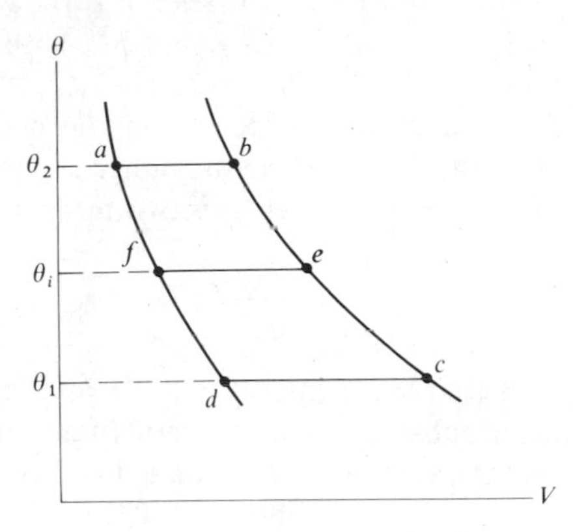

Fig. 5–2 Carnot cycles represented in the θ–V plane. Curves *a-f-d* and *b-e-c* are reversible adiabatics.

The function $f(\theta_2, \theta_1)$ has a very special form. To show this, suppose we first carry out the cycle *a-b-e-f-a* in Fig. 5–2 in which the isothermal process *e-f* is at some temperature θ_i, intermediate between θ_1 and θ_2. Let Q_2 be the heat absorbed at temperature θ_2 and Q_i the heat rejected at temperature θ_i. Then

$$\frac{|Q_2|}{|Q_i|} = f(\theta_2, \theta_i). \tag{5–2}$$

Now carry out the cycle *f-e-c-d-f*, between temperatures θ_i and θ_1, and let the heat Q_i absorbed in this cycle, in the process *f-e*, equal the heat rejected in the first cycle in the process *e-f*. Then if Q_1 is the heat rejected at the temperature θ_1,

$$\frac{|Q_i|}{|Q_1|} = f(\theta_i, \theta_1). \tag{5–3}$$

When Eqs. (5–2) and (5–3) are multiplied, we get

$$\frac{|Q_2|}{|Q_i|} \cdot \frac{|Q_i|}{|Q_1|} = \frac{|Q_2|}{|Q_1|} = f(\theta_2, \theta_i) \cdot f(\theta_i, \theta_1),$$

and hence from Eq. (5–1),

$$f(\theta_2, \theta_1) = f(\theta_2, \theta_i) \cdot f(\theta_i, \theta_1).$$

Since the left side is a function only of θ_2 and θ_1, this must be true of the right side also. The form of the function f must therefore be such that the product on the right does not contain θ_i and this is possible only if

$$f(\theta_2, \theta_i) = \frac{\phi(\theta_2)}{\phi(\theta_i)}, \qquad f(\theta_i, \theta_1) = \frac{\phi(\theta_i)}{\phi(\theta_1)}.$$

That is, although $f(\theta_2, \theta_i)$ is a function of *both* θ_2 and θ_i, and $f(\theta_i, \theta_1)$ is a function of *both* θ_i and θ_1, the function f must have the special form such that it is equal to the ratio of two functions ϕ, where $\phi(\theta_2)$, $\phi(\theta_i)$, and $\phi(\theta_1)$ are functions only of the *single* empirical temperatures θ_2, θ_i, and θ_1, respectively.

Again, the form of the function ϕ depends on the choice of the empirical temperature scale but not on the nature of the substance carried through the Carnot cycle. Then for a cycle carried out between any two temperatures θ_2 and θ_1,

$$\frac{|Q_2|}{|Q_1|} = \frac{\phi(\theta_2)}{\phi(\theta_1)}. \tag{5–4}$$

It was proposed by Kelvin that since the ratio $\phi(\theta_2)/\phi(\theta_1)$ is independent of the properties of any particular substance, the thermodynamic temperature T corresponding to the empirical temperature θ could be defined by the equation

$$T = A\phi(\theta), \tag{5–5}$$

where A is an arbitrary constant.

Then

$$\frac{|Q_2|}{|Q_1|} = \frac{T_2}{T_1}, \tag{5–6}$$

and the ratio of two thermodynamic temperatures is equal to the ratio of the quantities of heat absorbed and liberated when *any system whatever* is carried through a Carnot cycle between reservoirs at these temperatures. In particular, if one reservoir is at the triple point temperature T_3 and the other is at some arbitrary temperature T, and if Q_3 and Q are the corresponding heat flows,

$$\frac{|Q|}{|Q_3|} = \frac{T}{T_3}$$

and

$$T = T_3 \frac{|Q|}{|Q_3|}. \tag{5–7}$$

If the numerical value of 273.16 is assigned to T_3, the corresponding unit of T is 1 Kelvin.

In principle, then, a thermodynamic temperature can be determined by carrying out a Carnot cycle and measuring the heat flows Q and Q_3, which take the place of some arbitrary thermometric property X.

Note that the form of the function $\phi(\theta)$ need not be known to determine T experimentally, but we shall show in Section 6–11 how this function can be determined in terms of the properties of the thermometric substance used to define the empirical temperature θ.

Since the absolute values of the heat flows are necessarily positive, it follows from Eq. (5–6) that the thermodynamic or Kelvin temperature is necessarily positive also. This is equivalent to stating that there is an *absolute zero* of thermodynamic temperature, and that the thermodynamic temperature cannot be negative.*

In Section 4–7, we analyzed a Carnot cycle for the special case of an ideal gas. Although the results were expressed in terms of thermodynamic temperature T, this temperature had not at that point been defined, and, strictly speaking, we should have used the gas temperature θ defined by Eq. (1–4). Then if we define an ideal gas as one whose equation of state is

$$Pv = R\theta,$$

and for which

$$\left(\frac{\partial u}{\partial v}\right)_\theta = 0,$$

the analysis in Section 4–7 would lead to the result that

$$\frac{\theta_2}{\theta_1} = \frac{|Q_2|}{|Q_1|}.$$

It follows, then, that the ratio of two ideal gas thermometer temperatures is equal to the ratio of the corresponding thermodynamic temperatures. This justifies our replacing θ with T in earlier chapters.

5–3 ENTROPY

In the preceding section, the relation between the temperatures T_2 and T_1, and the heat flows Q_2 and Q_1 in a Carnot cycle, were exprcssed in terms of the absolute values $|Q_2|$ and $|Q_1|$. However, since Q_2 is a heat flow *into* the system and Q_1 is a heat flow *out* of the system, the heat flows have opposite signs; and hence for a Carnot cycle, we should write

$$\frac{T_2}{T_1} = -\frac{Q_2}{Q_1},$$

or

$$\frac{Q_1}{T_1} + \frac{Q_2}{T_2} = 0.$$

* However, see Section 13–5.

Now consider any arbitrary reversible cyclic process such as that represented by the closed curve in Fig. 5–3. The net result of such a process can be approximated as closely as desired by performing a large number of small Carnot cycles, all traversed in the same sense. Those adiabatic portions of the cycles which coincide are traversed twice in opposite directions and will cancel. The outstanding result consists of the heavy zig-zag line. As the cycles are made smaller, there is a more complete cancellation of the adiabatic portions but the isothermal portions remain outstanding.

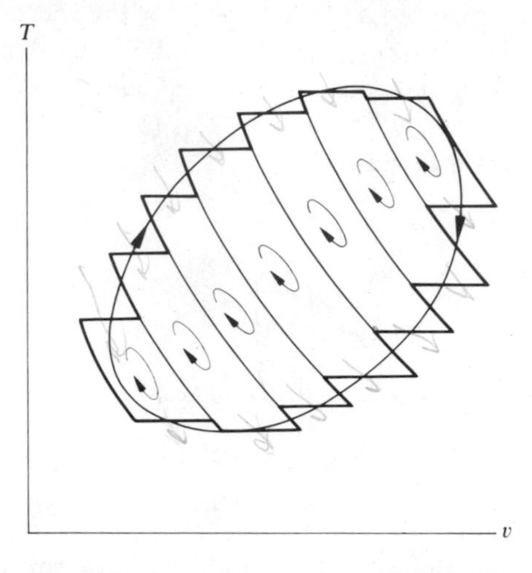

Fig. 5–3 Any arbitrary reversible cyclic process can be approximated by a number of small Carnot cycles.

If one of the small cycles is carried out between temperatures T_2 and T_1, and ΔQ_2 and ΔQ_1 are the corresponding heat flows, then for that cycle,

$$\frac{\Delta Q_1}{T_1} + \frac{\Delta Q_2}{T_2} = 0,$$

and when all such terms are added, for all cycles, we have

$$\sum \frac{\Delta Q_{\mathrm{r}}}{T} = 0.$$

The subscript "r" serves as a reminder that the result above applied to *reversible* cycles only.

In the limit, as the cycles are made narrower, the zig-zag processes correspond more and more closely to the original cyclic process. The sum can then be replaced by an integral and we can write for the original process,

$$\oint \frac{d'Q_r}{T} = 0. \tag{5–8}$$

That is, if the heatflow $d'Q_r$ into the system at any point is divided by the temperature T of the system at this point, and these quotients are summed over the entire cycle, the sum equals zero. At some points of the cycle $d'Q_r$ is positive and at others, negative. The temperature T is positive always*. The negative contributions to the integral just cancel the positive contributions.

Since the integral of any exact differential such as dV or dU around a closed path is zero, we see from Eq. (5–8), that although $d'Q_r$ is not an exact differential, the ratio $d'Q_r/T$ *is* an exact differential. It is therefore possible to define a *property* S of a system whose value depends only on the state of the system and whose differential dS is

$$dS \equiv \frac{d'Q_r}{T}. \tag{5–9}$$

Then in any cyclic process,

$$\oint dS = 0. \tag{5–10}$$

Another property of an exact differential is that its integral between any two equilibrium states is the same for all paths between the states. Hence for *any* path between states a and b,

$$\int_a^b dS = S_b - S_a. \tag{5–11}$$

The property S is called the *entropy* of the system. The MKS unit of entropy is evidently 1 joule per kelvin (1 J K^{-1}). Entropy is an extensive property, and we define the *specific entropy* s as the entropy per mole or per unit mass:

$$s = \frac{S}{n}, \quad \text{or} \quad s = \frac{S}{m}.$$

Equations (5–9) or (5–11) define only *differences* in entropy. We shall see later in Section 7–7 that it is possible to assign an absolute value to the entropy of certain systems; but on the basis of the equations above, the entropy of a system is determined only to within some arbitrary constant.

* However, see Section 13–5.

5–4 CALCULATIONS OF ENTROPY CHANGES IN REVERSIBLE PROCESSES

In any adiabatic process, $d'Q = 0$, and hence in any *reversible adiabatic* process,

$$d'Q_r = 0 \quad \text{and} \quad dS = 0.$$

The entropy of a system is therefore constant in any reversible adiabatic process, and such a process can be described as *isentropic*. This explains the use of the subscript s in earlier chapters to designate a reversible adiabatic process.

In a *reversible isothermal* process, the temperature T is constant and may be taken outside the integral sign. The change in entropy of a system in a finite reversible isothermal process is therefore,

$$S_b - S_a = \int_a^b \frac{d'Q_r}{T} = \frac{1}{T}\int_a^b d'Q_r = \frac{Q_r}{T}. \tag{5–12}$$

To carry out such a process, the system is brought in contact with a heat reservoir at a temperature infinitesimally greater (or less) than that of the system. In the first case there is a heat flow *into* the system, Q_r is positive, $S_b > S_a$, and the entropy of the system *increases*. In the second case there is a heat flow *out* of the system, Q_r is negative, and the entropy of the system *decreases*.

A common example of a reversible isothermal process is a change in phase at constant pressure during which the temperature remains constant also. The heat flow into the system, per unit mass or per mole, equals the heat of transformation l, and the change in (specific) entropy is simply

$$s_2 - s_1 = l/T. \tag{5–13}$$

For example, the latent heat of transformation from liquid water to water vapor at atmospheric pressure and at the corresponding temperature of (approximately) 373 K is $l_{23} = 22.6 \times 10^5$ J kg^{-1}. The specific entropy of the vapor therefore exceeds that of the liquid by

$$s''' - s'' = \frac{l_{23}}{T} = \frac{22.6 \times 10^5 \text{ J kg}^{-1}}{373 \text{ K}} = 6060 \text{ J kg}^{-1}\text{ K}^{-1}.$$

In most processes a reversible flow of heat into or out of a system is accompanied by a change in temperature, and calculation of the corresponding entropy change requires an evaluation of the integral

$$\int \frac{d'Q_r}{T}.$$

If the process takes place at constant volume, for example, and if changes in phase are excluded, the heat flow per unit mass or per mole equals $c_v\, dT$ and

$$(s_2 - s_1)_v = \int_{T_1}^{T_2} c_v \frac{dT}{T}. \tag{5–14}$$

If the process is at constant pressure, the heat flow equals $c_P\,dT$ and

$$(s_2 - s_1)_P = \int_{T_1}^{T_2} c_P \frac{dT}{T}. \tag{5–15}$$

To evaluate these integrals for a given system, we must know c_v or c_P as functions of T. In a temperature interval in which the specific heat capacities can be considered constant,

$$(s_2 - s_1)_v = c_v \ln(T_2/T_1), \tag{5–16}$$

$$(s_2 - s_1)_P = c_P \ln(T_2/T_1). \tag{5–17}$$

To raise the temperature from T_1 to T_2 reversibly, we require a large number of heat reservoirs having temperatures $T_1 + dT$, $T_1 + 2\,dT, \ldots, T_2 - dT$, T_2. The system at temperature T_1 is brought in contact with the reservoir at temperature $T_1 + dT$ and kept in contact with this reservoir until thermal equilibrium is reached. The system, now at temperature $T_1 + dT$, is then brought into contact with the reservoir at temperature $T_1 + 2\,dT$, etc., until the system reaches the temperature T_2.

For example, the value of c_P for liquid water, in the temperature interval from T_1 = 273 K (0°C) to T_2 = 373 K (100°C) is 4.18×10^3 J kg^{-1} K^{-1} (assumed constant). The specific entropy of liquid water at 373 K therefore exceeds that at 273 K by

$$(s_2 - s_1)_P = c_P \ln \frac{T_2}{T_1} = 4.18 \times 10^3 \text{ J kg}^{-1}\text{ K}^{-1} \times \ln \frac{373}{273} = 1310 \text{ J kg}^{-1}\text{ K}^{-1}.$$

In every process in which there is a *reversible* flow of heat between a system and its surroundings, the temperatures of system and surroundings are essentially equal; and the heat flow into the surroundings, at every point, is equal in magnitude and opposite in sign to the heat flow into the system. Hence the entropy change of the *surroundings* is equal in magnitude and opposite in sign to that of the *system*, and the net entropy change of system plus surroundings is zero. (In an isothermal process, the surroundings consist of a single reservoir. In a process in which the temperature of the system changes, the surroundings consist of all those reservoirs at different temperatures that exchanged heat with the system.) Since systems and surroundings together constitute a *universe*, we can say that the entropy of the universe remains constant in every change in state in which there is only a reversible heat flow into (or out of) a system.

If the boundary of the original system is enlarged so as to include the reservoirs with which the system exchanges heat, all heat flows take place *within* this composite system. There are no heat flows across the enlarged boundary and the process is adiabatic for the composite system. Hence we can also say that any *reversible* heat flows within a composite system enclosed by an adiabatic boundary produce no net change in the entropy of the composite system.

5–5 TEMPERATURE-ENTROPY DIAGRAMS

Since entropy is a property of a system, its value in any equilibrium state of the system (apart from an arbitrary constant) can be expressed in terms of the variables specifying the state of the system. Thus for a *PVT* system, the entropy can be expressed as a function of *P* and *V*, *P* and *T*, or *T* and *V*. Then, just as with the internal energy, we can consider the entropy as one of the variables specifying the state of the system, and we can specify the state in terms of the entropy *S* and one other variable. If the temperature *T* is selected as the other variable, every state of a system corresponds to a point in a *T-S* diagram and every reversible process corresponds to a curve in this diagram.

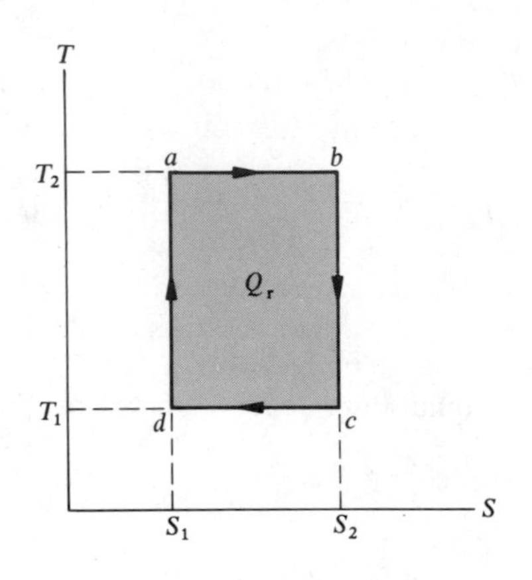

Fig. 5–4 The temperature-entropy diagram of a Carnot cycle.

A Carnot cycle has an especially simple form in such a diagram, since it is bounded by two isotherms, along which *T* is constant, and two reversible adiabatics, along which *S* is constant. Thus Fig. 5–4 represents the Carnot cycle *a-b-c-d-a* of Fig. 5–2.

The area under the curve representing any reversible process in a *T-S* diagram is

$$\int_a^b T\,dS = \int_a^b d'Q_r = Q_r,$$

so the area under such a curve represents the *heat flow* in the same way that the area under a curve in a *P-V* diagram represents *work*. The area enclosed by the graph of a reversible *cyclic* process corresponds to the *net* heat flow into a system in the process.

5–6 ENTROPY CHANGES IN IRREVERSIBLE PROCESSES

The change in entropy of a system is defined by Eq. (5–9) for a *reversible* process only; but since the entropy of a system depends only on the state of the system, the entropy difference between two given equilibrium states is the same regardless of the nature of a process by which the system may be taken from one state to the other. We can, therefore, find the change in entropy of a system in an *irreversible* process by devising some *reversible* process (any reversible process will do) between the end states of the irreversible process.

Consider first the process in Fig. 5–1(a) in which the temperature of a body is increased from T_1 to T_2 by bringing it in contact with a *single* reservoir at a temperature T_2, instead of using a series of reservoirs at temperatures between T_1 and T_2. The process is irreversible since there is a finite temperature difference between the body and the reservoir during the process, and the direction of the heat flow cannot be reversed by an infinitesimal change in temperature. The initial and final states of the body are the same, however, whether the temperature is changed reversibly or irreversibly, so the change in entropy of the body is the same in either process. Then, from Eq. (5–17), if the process is at constant pressure and the heat capacity C_P of the body can be considered constant, the entropy change of the body is

$$\Delta S_{\text{body}} = C_P \ln \frac{T_2}{T_1}.$$

Since $T_2 > T_1$, there is a heat flow *into* the body, $\ln (T_2/T_1)$ is positive, and the entropy of the body increases.

How does the entropy of the *reservoir* change in the process? The reservoir temperature remains constant at the value T_2; hence its change in entropy is the same as in a reversible *isothermal* process in which the heat flow into it is equal in magnitude to that in the irreversible process. Again assuming C_P to be constant, the heat flow into the body is

$$Q = C_P(T_2 - T_1).$$

The heat flow into the *reservoir* is the negative of this, and the change in entropy of the reservoir is

$$\Delta S_{\text{reservoir}} = -\frac{Q}{T_2} = -C_P \frac{T_2 - T_1}{T_2}.$$

Since $T_2 > T_1$, there is a heat flow *out* of the reservoir, the fraction $(T_2 - T_1)/T_2$ is positive, the entropy change of the reservoir is negative, and its entropy decreases.

The total change in entropy of the composite system, body plus reservoir, is

$$\Delta S = \Delta S_{\text{body}} + \Delta S_{\text{reservoir}} = C_P \left| \ln \frac{T_2}{T_1} - \frac{T_2 - T_1}{T_2} \right|.$$

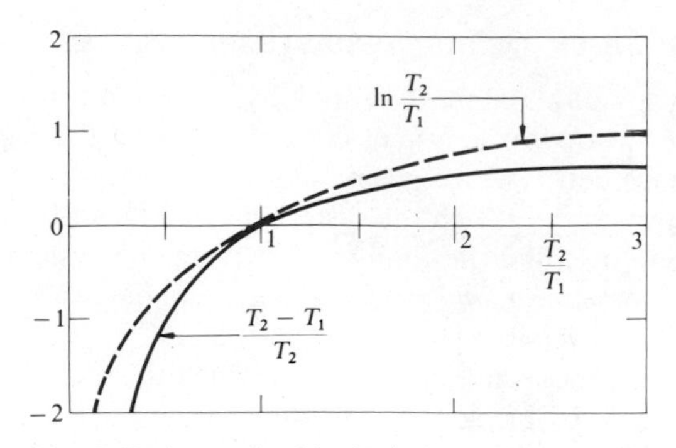

Fig. 5–5 A graph of ln (T_2/T_1) and $(T_2 - T_1)/T_1$ as a function of T_2/T_1.

Figure 5–5 shows graphs of ln (T_2/T_1) and of $(T_2 - T_1)/T_2$, as functions of the ratio T_2/T_1. It will be seen that when $T_2 > T_1$, or when $T_2/T_1 > 1$, the quantities ln (T_2/T_1) and $(T_2 - T_1)/T_2$ are both positive, but the former is greater than the latter. The increase in entropy of the body is then greater than the decrease in entropy of the reservoir, and the entropy of the universe (body plus reservoir) increases in the irreversible process.

As an example, suppose that the temperature of liquid water is increased from 273 K to 373 K by bringing it in contact with a heat reservoir at a temperature of 373 K. We have shown in the preceding example that the increase in specific entropy of the water in this process is 1310 J kg^{-1} K^{-1}. The heat flow into the water, per kilogram, equal to the heat flow out of the reservoir, is

$$\begin{aligned} q &= c_P(T_2 - T_1) \\ &= 4.18 \times 10^3 \text{ J kg}^{-1} \text{ K}^{-1} (373 \text{ K} - 273 \text{ K}) \\ &= 418 \times 10^3 \text{ J kg}^{-1}. \end{aligned}$$

The decrease in entropy of the reservoir is

$$\Delta S = -\frac{q}{T_2} = -\frac{418 \times 10^3 \text{ J kg}^{-1}}{373 \text{ K}} = -1120 \text{ J kg}^{-1} \text{ K}^{-1},$$

and the increase in entropy of the water is greater than the decrease in entropy of the reservoir.

If the body is initially at a higher temperature than the reservoir, heat flows out of the body into the reservoir. The entropy of the body decreases and that of the reservoir increases. We leave it as a problem to show that in this irreversible process the entropy of the universe also increases. Hence *the entropy of the universe always increases in a process during which heat flows across a finite temperature difference.*

Consider next the process in part (b) of Fig. 5–1 in which a rotating flywheel drives a generator which sends a current through a resistor in a heat reservoir.

The temperature of the resistor remains constant. Therefore, if the resistor alone is considered the system, none of the properties of the system change and there is no change in the entropy of the system. We assume that the temperature of the resistor during the process differs only slightly from that of the reservoir, so the heat flow between resistor and reservoir is reversible; and if Q is the magnitude of the heat flow, the entropy of the *reservoir* increases by Q/T. This is also the entropy increase of the composite system, resistor plus reservoir, and again there is an increase in entropy of the universe.

There appears at first sight to be a discrepancy here. If the entropy of the *reservoir* increases as a result of a reversible flow of heat into it, why does not the entropy of the *resistor* decrease by an equal amount, since there is an equal heat flow out of it? Nevertheless, the entropy of the resistor does not change since there is no change in its state. We can take two points of view. One is that since the entropy of the resistor does not change, the performance of dissipative work on it results in an increase in its entropy, even in the absence of a heat flow into it. The same can be said of dissipative work of any form, such as that done in stirring a viscous fluid. Thus the entropy increase of the resistor as a result of the performance of dissipative work on it just balances the entropy decrease due to the heat flow out of it.

The second point of view, as has been stated earlier, is that the performance of dissipative work on a system is equivalent to a flow of heat into the system, equal in magnitude to the dissipative work. Then the *net* heat flow into the resistor is zero, and there is no change in its entropy; the only heat flow that need be considered is that into the reservoir.

If we choose to consider resistor and reservoir together as a single composite system, there is no heat flow into it from its surroundings, but dissipative work is done on it with a corresponding increase in entropy.

Finally, in the irreversible free expansion of a gas in part (c) of Fig. 5–1, there are no heat flows within the system and there is no dissipative work. The same final state of the gas can be reached, however, by a *reversible* expansion. In such an expansion some external work will be done; and since the internal energy of the gas is constant, there will be a reversible heat flow into it, equal in magnitude to this work. The entropy of the gas would therefore increase in this reversible process and there will be the same increase in entropy as in the original free expansion.

5–7 THE PRINCIPLE OF INCREASE OF ENTROPY

In all of the irreversible processes described in the preceding section, it was found that the entropy of the Universe increased. This is found to be the case in any irreversible process that may be analyzed, and we conclude that it is true for *all* irreversible processes. This conclusion is known as the *principle of increase of entropy* and is considered as a part of the second law of thermodynamics: *The entropy of the Universe increases in every irreversible process.* If all systems that interact in a process are enclosed in a rigid adiabatic boundary, they form a completely isolated system and constitute their own universe. Hence we can also say that the entropy of a completely isolated system increases in every irreversible process taking place within the system. Since, as discussed in Section 5–4, the entropy remains constant in a *reversible* process within an isolated system, we have justified the statement of the second law in Section 5–1 namely, that **in every process taking place in an isolated system, the entropy of the system either increases or remains constant.**

We can now gain a further insight into the concepts of reversible and irreversible processes. Consider again the first example in Section 5–1 in which a body at a temperature T_1 eventually comes to thermal equilibrium with a reservoir at a different temperature T_2. This process is irreversible in the sense in which we originally defined the term; that is, the direction of the heat flow between the body and the reservoir cannot be reversed by an infinitesimal change in the temperature of either. This is not to say, however, that the original state of the composite system cannot be *restored*. For example, we can bring the body back to its original temperature, in a reversible process, by making use of a series of auxiliary reservoirs at temperatures between T_1 and T_2; and the original state of the reservoir can be restored by a reversible flow of heat into or out of it to an auxiliary reservoir at an infinitesimally different temperature. In these reversible processes, the decrease in entropy of the original composite system is equal in magnitude and opposite in sign to its increase in the original irreversible process, so there is no outstanding change in its entropy, but the entropy increase of the auxiliary reservoirs is the *same* as that of the composite system in the first process. Hence the original entropy increase has simply been passed on to the auxiliary reservoirs. If the state of the composite system is restored by an *irreversible* process, the entropy increase of the auxiliary reservoirs is even greater than the entropy increase in the original process. Hence, although a system can be restored to its original state after an irreversible process, the entropy increase associated with the process can never be wiped out. At best, it can be passed on from one system to another. This is the true significance of the term, *irreversible*. The state of the Universe can never be completely restored.

In mechanics, one of the reasons that justifies the introduction of the concepts of energy, momentum, and angular momentum is that they obey a *conservation principle*. Entropy is *not* conserved, however, except in reversible processes, and this unfamiliar property, or lack of property, of the entropy function is one reason why an aura of mystery usually surrounds the concept of entropy. When hot and cold water are mixed, the heat flow out of the hot water equals the heat flow into the cold water and energy is conserved. But the increase in entropy of the cold water is larger than the decrease in entropy of the hot water, and the total entropy of the system is greater at the end of the process than it was at the beginning. Where did this additional entropy come from? The answer is that it was *created* in the mixing process. Furthermore, once entropy has been created, it can never be destroyed. The Universe must forever bear this additional burden of entropy (a statement that implies the assumption, which may be questionable, that the Universe constitutes an isolated, closed system). "Energy can neither be created nor destroyed," says the first law of thermodynamics. "Entropy cannot be destroyed," says the second law, "but it can be created."

The preceding discussion relates to the thermodynamic definition of the entropy concept. The methods of *statistics*, to be discussed in later chapters, will give additional insight into the entropy concept.

In Section 3–7, the difference in internal energy between two states of a system was defined as equal to the negative of the work in any adiabatic process between the states. It was mentioned at that time that not all states of a system could be reached from a given initial state by an adiabatic process, but that whenever a final state b could not be reached from an initial state a by an adiabatic process, state a could always be reached from state b by such a process. We can now understand why this should be the case.

Only those states having the *same* entropy as the initial state can be reached from this state by a *reversible* adiabatic process, along which the entropy is constant. To reach any *arbitrary* state one must also make use of an *irreversible* adiabatic process, such as a free expansion or a stirring process, as shown in Fig. 5–1. But in the irreversible process the entropy always *increases* and never decreases. Hence the only states that can be reached from a given initial state by adiabatic processes are those in which the entropy is equal to, or greater than, that in the initial state.

However, if the entropy in some arbitrary state is *less* than that in the initial state, the entropy in the initial state is necessarily *greater* than that in the arbitrary state, and the (original) initial state can always be reached from the arbitrary state by an adiabatic process.

In a process in which two bodies at different temperatures are brought in contact and come to thermal equilibrium, the net change in *energy* of the system is zero, since the heat flow out of one body equals the heat flow into the other. In what significant way have things changed? Who cares whether or not the entropy of the system has increased?

The mechanical engineer is concerned, among other things, with heat engines, whose energy input is a flow of heat from a reservoir and whose *useful* output is mechanical work. At the end of the process above, we have a single system all at one temperature, while at the start we had two systems at different temperatures. These systems could have been utilized as the reservoirs of a heat engine, withdrawing heat from one, rejecting heat to the other, and diverting a part of the heat to produce mechanical work. Once the entire system has come to the same temperature, this opportunity no longer exists. Thus any irreversible process in a heat engine, with an associated increase in entropy, reduces the amount of mechanical work that can be abstracted from a given amount of heat flowing out of the high temperature reservoir. What has been "lost" in the irreversible process is not *energy*, but *opportunity*—the opportunity to convert to mechanical work a part of the internal energy of a system at a temperature higher than that of its surroundings.

The physical chemist is concerned not so much with the *magnitude* of the entropy increase in an irreversible process as with the fact that a process can take place in an isolated system only if the entropy of the system increases. Will two substances react chemically or will they not? If the reaction would result in a decrease in entropy, the reaction is impossible. However, while the entropy might decrease if the reaction were to take place at one temperature and pressure, it is possible that it could increase at other values of temperature and pressure. Hence a knowledge of the entropies of substances as functions of temperature and pressure is all-important in determining the possibilities of chemical reactions.

5–8 THE CLAUSIUS AND KELVIN-PLANCK STATEMENTS OF THE SECOND LAW

We have chosen to consider the second law as a statement regarding possible entropy changes during arbitrary processes. Entropy was defined in terms of heat flows into and out of a Carnot cycle. Two other statements are often taken as the starting point for defining entropy both of which lead, of course, to the same end result but by a somewhat more lengthy argument. The *Clausius statement* of the second law is:

No process is possible whose *sole* result is a heat flow out of one system at a given temperature and a heat flow of the same magnitude into a second system at a higher temperature.

The Clausius statement seems at first to be a trivial and obvious assertion, since heat can flow by *conduction* only from a higher to a lower temperature. However, the mechanism of heat conduction is used to define what is meant by "higher" and "lower" temperatures; numerical values are assigned to temperature such that heat flows by *conduction* from a higher to a lower temperature. But the Clausius statement goes further and asserts that *no process whatever* is possible whose *sole* result would conflict with the statement.

The Clausius statement can be seen to be a direct consequence of the principle of increase of entropy. Suppose that the *sole* result of a process were a heat flow Q out of system A at a temperature T_1, and a heat flow of equal magnitude into a system B at a higher temperature T_2. Such a process would not violate the first law, since the work in the process would be zero and the increase in internal energy of B would equal the decrease in internal energy of A. The entropy changes of the systems would be

$$\Delta S_A = -\frac{|Q|}{T_1}, \qquad \Delta S_B = \frac{|Q|}{T_2}.$$

But $T_1 < T_2$, so $|\Delta S_A| > |\Delta S_B|$ and the net result would be a *decrease* in the entropy of the universe.

It might appear at first sight that the outstanding result of operating a refrigerator would contradict the Clausius statement. Suppose for example, that a Carnot refrigerator is operated between a reservoir at a temperature T_1 and a second reservoir at a higher temperature T_2. In each cycle, there is a heat flow Q_1 out of the reservoir at the lower temperature T_1 and a heat flow Q_2 into the reservoir at the higher temperature T_2. The magnitudes of the heat flows are not equal, however, since $Q_2/Q_1 = T_2/T_1$, and $T_2 > T_1$. Thus, although there is a *transfer* of heat from a lower to a higher temperature, the heat flow out of one reservoir is not equal to the heat flow into the other; and the heat flows are not the *sole* result of the process because work, equal in magnitude to $|Q_2| - |Q_1|$, must be done in order to carry out the cycle.

The *Kelvin-Planck statement* of the second law is:

No process is possible whose *sole* result is a heat flow Q out of a reservoir at a single temperature, and the performance of work W equal in magnitude to Q.

Such a process, if it took place, would not violate the first law, but the principle of increase of entropy forbids such a process because the entropy of the reservoir would *decrease* by an amount $|Q|/T$, with no compensating increase in the entropy of any other system. In the operation of any heat engine, there is a heat flow out of a high-temperature reservoir and work is done, but this is not the *sole* result of the process because some heat is always rejected to a reservoir at a lower temperature.

The Clausius statement of the second law can be used to show that there is an upper limit to the thermal efficiency of any heat engine, and to the coefficient of performance of a refrigerator. Thus let the circle in Fig. 5–6(a) represent a Carnot engine operating between two reservoirs at temperatures T_2 and T_1, taking in heat $|Q_2|$ from the reservoir at the higher temperature T_2, rejecting heat $|Q_1|$ to the reservoir at the lower temperature T_1, and doing work $W = |Q_2| - |Q_1|$. The thermal efficiency $\eta = W/|Q_2|$ is about 50%. The rectangle at the right of the diagram represents an assumed engine having a higher thermal efficiency than the Carnot engine (about 75%). Let primed symbols refer to the assumed higher efficiency engine. We assume that the engines are constructed so that each delivers the same mechanical work and hence $W' = W$. The thermal efficiency of the assumed engine is

$$\eta' = \frac{W'}{|Q_2'|} = \frac{W}{|Q_2'|}.$$

Since we assume that $\eta' > \eta$, it follows that $|Q_1'| < |Q_2|$. The assumed engine therefore takes in a smaller quantity of heat from the high-temperature reservoir than does the Carnot engine. It also rejects a smaller quantity of heat to the low-temperature reservoir, since the work, or the difference between the heats absorbed and rejected, is the same for both engines.

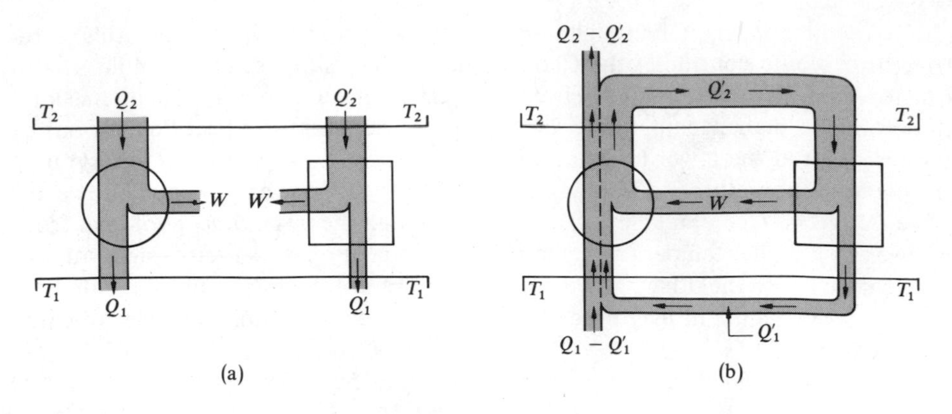

Fig. 5–6 In part (a), the circle represents a Carnot engine and the rectangle an assumed engine having a higher thermal efficiency. If the assumed engine were to drive the Carnot engine in reverse as a refrigerator, as in part (b), the result would violate the Clausius statement of the second law.

Because the Carnot engine is reversible (in the thermodynamic sense) it can also be operated as a refrigerator with no change in the magnitudes of W, $|Q_2|$, and $|Q_1|$. Hence let the assumed engine be connected to the Carnot engine as in Fig. 5–6(b). The system will run itself because the work output of the assumed engine is equal to the work required to operate the Carnot refrigerator. The assumed engine withdraws heat $|Q_2'|$ from the high-temperature reservoir, while the Carnot refrigerator delivers a larger quantity of heat $|Q_2|$ to this reservoir. Also, the assumed engine rejects heat $|Q_1'|$ to the low-temperature reservoir while the Carnot refrigerator withdraws from this reservoir a larger quantity of heat $|Q_1|$.

It should be evident from the diagram that a part of the heat delivered to the high-temperature reservoir can be diverted to provide the heat input to the assumed engine, and that the heat delivered to the low-temperature reservoir will provide a part of the heat removed from this reservoir by the Carnot refrigerator.

The *sole* result of operating the composite system is then a transfer of heat from the low- to the high-temperature reservoir, represented in Fig. 5–6(b) by the width of the "pipeline" at the left side of the diagram, in violation of the Clausius statement of the second law. It follows that the assumed engine cannot exist and that *no engine operating between two reservoirs at given temperatures can have a higher thermal efficiency than a Carnot engine operating between the same pair of reservoirs.*

The same reasoning as that above shows that no refrigerator can have a higher coefficient of performance than a Carnot refrigerator, for two reservoirs at given temperatures.

The statement of the second law in terms of entropy as stated in Section 5–1 was used directly to verify the Clausius and Kelvin-Planck statements of the second law. The Kelvin-Planck statement can be used to show that the ratios of heat flows in and out of a Carnot cycle depend only on the temperatures of the reservoirs between which the cycle operates. (See Problem 5–33.) This property of the Carnot cycle was used to define entropy and thermodynamic temperature.

PROBLEMS

5–1 Suppose a temperature scale is defined in terms of a substance A such that the efficiency of a Carnot engine operating between the boiling and melting points of this substance (at a pressure of 1 atm) is exactly 50%. One degree on this new scale is equal to two degrees on the Fahrenheit scale and there are 75 A-degrees between the melting and boiling points of the substance. Determine the melting- and boiling-point temperatures of the substance on the Kelvin scale.

5–2 Analyze a Carnot cycle for the special case of an ideal paramagnet to show that the ratio of two empirical temperatures defined by Curie's law, $\theta_i = C_C\mathscr{H}/M_i$, is equal to the ratio of the corresponding thermodynamic temperatures. The internal energy of an ideal paramagnet depends on T alone; and during an adiabatic process $\mathscr{H}/\theta_i$ remains constant.

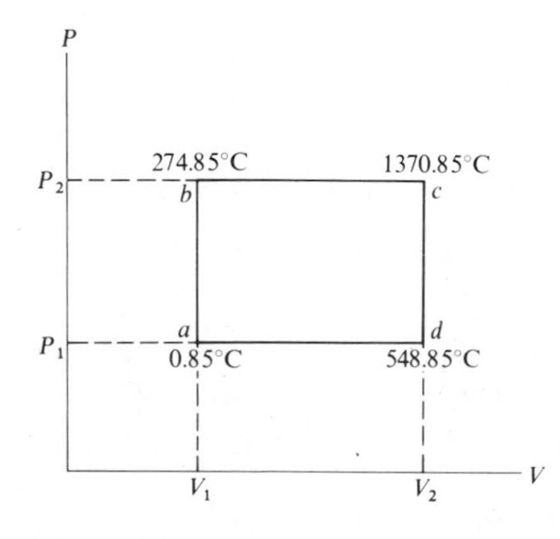

Figure 5–7

5–3 Find the change in entropy of the system during the following processes: (a) 1 kg of ice at 0°C and 1 atm pressure melts at the same temperature and pressure. The latent heat of fusion is 3.34×10^5 J kg^{-1}. (b) 1 kg of steam at 100°C and one atm pressure condenses to water at the same temperature and pressure. The latent heat of vaporization is 2.26×10^6 J kg^{-1}.

5–4 A system is taken reversibly around the cycle *a-b-c-d-a* shown in Fig. 5 7. The temperatures t are given in degrees Celsius. Assume that the heat capacities are independent of temperature and $C_V = 8$ J K^{-1} and $C_P = 10$ J K^{-1}. (a) Calculate the heat flow $\int d'Q$ into the system in each portion of the cycle. According to the first law, what is the significance of the sum of these heat flows? (b) If $V_1 = 9 \times 10^{-3}$ m^3 and $V_2 = 20 \times 10^{-3}$ m^3, calculate the pressure difference $(P_2 - P_1)$. (c) Calculate the value of $\int \frac{d'Q}{T}$ along each portion of the cycle. According to the second law, what is the significance of the value of the sum of these integrals? (d) Suppose that a temperature T' were defined as the Celsius temperature plus some value other than 273.15. Would it then be true that $\oint \frac{d'Q}{T'} = 0$? Explain.

5–5 A 50-ohm resistor carrying a constant current of 1 A is kept at a constant temperature of 27°C by a stream of cooling water. In a time interval of 1 s, (a) what is the change in entropy of the resistor? (b) what is the change in entropy of the universe?

5–6 A Carnot engine operates on 1 kg of methane, which we shall consider to be an ideal gas. The ratio of the specific heat capacities γ is 1.35. If the ratio of the maximum volume to the minimum volume is 4 and the cycle efficiency is 25%, find the entropy increase of the methane during the isothermal expansion.

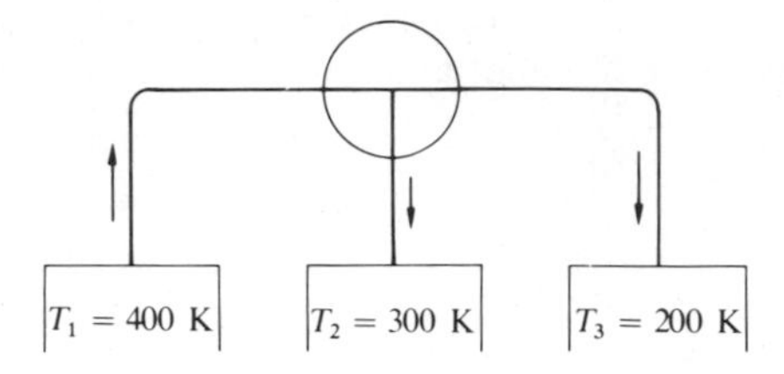

Figure 5–8

5–7 The circle in Fig. 5–8 represents a reversible engine. During some integral number of complete cycles the engine absorbs 1200 J from the reservoir at 400 K and performs 200 J of mechanical work. (a) Find the quantities of heat exchanged with the other reservoirs, and state whether the reservoir gives up or absorbs heat. (b) Find the change in entropy of each reservoir. (c) What is the change in entropy of the universe?

5–8 One kilogram of water is heated reversibly by an electric heating coil from 20°C to 80°C. Compute the change in entropy of (a) the water, (b) the universe. (Assume that the specific heat capacity of water is a constant.)

5–9 A thermally insulated 50-ohm resistor carries a current of 1 A for 1 s. The initial temperature of the resistor is 10°C, its mass is 5 g, and its specific heat capacity is 850 J kg^{-1} K^{-1}. (a) What is the change in entropy of the resistor? (b) What is the change in entropy of the universe?

5–10 The value of c_P for a certain substance can be represented by $c_P = a + bT$. (a) Find the heat absorbed and the increase in entropy of a mass m of the substance when its temperature is increased at constant pressure from T_1 to T_2. (b) Using this equation and Fig. 3–10, find the increase in the molal specific entropy of copper, when the temperature is increased at constant pressure from 500 K to 1200 K.

5–11 A body of finite mass is originally at a temperature T_2, which is higher than that of a heat reservoir at a temperature T_1. An engine operates in infinitesimal cycles between the body and the reservoir until it lowers the temperature of the body from T_2 to T_1. In this process there is a heat flow Q out of the body. Prove that the maximum work obtainable from the engine is $Q + T_1(S_1 - S_2)$, where $S_1 - S_2$ is the decrease in entropy of the body.

5–12 On a single *T-S* diagram, sketch curves for the following reversible processes for an ideal gas starting from the same initial state: (a) an isothermal expansion, (b) an adiabatic expansion, (c) an isochoric expansion, and (d) an isochoric process in which heat is added.

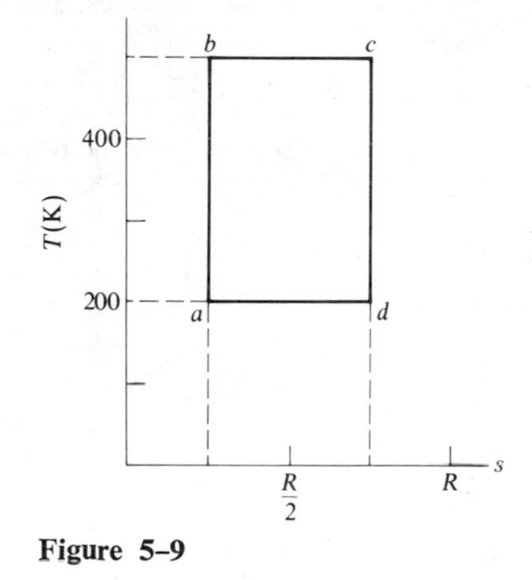

Figure 5–9

5–13 A system is taken reversibly around the cycle *a-b-c-d-a* shown on the *T-S* diagram of Fig. 5–9. (a) Does the cycle *a-b-c-d-a* operate as an engine or a refrigerator? (b) Calculate the heat transferred in each process. (c) Find the efficiency of this cycle operating as an engine graphically as well as by direct calculation. (d) What is the coefficient of performance of this cycle operating as a refrigerator?

5–14 Show that if a body at temperature T_1 is brought in contact with a heat reservoir at temperature $T_2 < T_1$, the entropy of the universe increases. Assume that the heat capacity of the body is constant.

5–15 Suppose the heat capacity of the body discussed in Section 5–6 is 10 J K^{-1} and $T_1 = 200$ K. Calculate the changes in entropy of the body and of the reservoir if (a) $T_2 = 400$ K, (b) $T_2 = 600$ K, (c) $T_2 = 100$ K. (d) Show that in each case the entropy of the universe increases.

5–16 (a) One kilogram of water at 0°C is brought into contact with a large heat reservoir at 100°C. When the water has reached 100°C, what has been the change in entropy of the water, of the heat reservoir, and of the universe? (b) If the water had been heated from 0°C to 100°C by first bringing it into contact with a reservoir at 50°C and then with a reservoir at 100°C, what would have been the change in entropy of the universe? (c) Explain how the water might be heated from 0°C to 100°C with no change in the entropy of the universe.

5–17 Liquid water having a mass of 10 kg and a temperature of 20°C is mixed with 2 kg of ice at a temperature of −5°C at 1 atm pressure until equilibrium is reached. Compute the final temperature and the change in entropy of the system. [c_P(water) $= 4.18 \times 10^3$ J kg^{-1} K^{-1}; c_P(ice) $= 2.09 \times 10^3$ J kg^{-1} K^{-1}; and $l_{12} = 3.34 \times 10^5$ J kg^{-1}.]

5–18 Construct a reversible process to show explicitly that the entropy increases during a free expansion of an ideal gas.

5–19 What are the difficulties in showing explicitly that the entropy of an ideal gas must increase during an irreversible adiabatic compression.

5–20 Two identical finite systems of constant heat capacity C_P are initially at temperatures T_1 and T_2 where $T_2 > T_1$. (a) These systems are used as the reservoirs of a Carnot engine which does an infinitesimal amount of work $d'W$ in each cycle. Show that the final equilibrium temperature of the reservoirs is $(T_1T_2)^{1/2}$. (b) Show that the final temperature of the systems if they are brought in contact in a rigid adiabatic enclosure is $(T_1 + T_2)/2$. (c) Which final temperature is greater? (d) Show that the total amount of work done by the Carnot engine in part (a) is $C_P(T_2^{1/2} - T_1^{1/2})^2$. (e) Show that the total work available in the process of part (b) is zero.

5–21 A mass m of a liquid at a temperature T_1 is mixed with an equal mass of the same liquid at a temperature T_2. The system is thermally insulated. Show that the entropy change of the universe is

$$2mc_P \ln \frac{(T_1 + T_2)/2}{\sqrt{T_1T_2}},$$

and prove that this is necessarily positive.

5–22 One mole of monatomic ideal gas initially at temperature T_i expands adiabatically against a massless piston until its volume doubles. The expansion is not necessarily quasistatic or reversible. It can be said, however, that the work done, the internal energy change, and the entropy change of the system, and the entropy change of the universe must fall within certain limits. Evaluate the limits for these quantities and describe the process associated with each limit.

5–23 When there is a heat flow out of a system during a reversible isothermal process, the entropy of the system decreases. Why does this not violate the second law?

5–24 Show that $(\partial s/\partial T)_x > 0$ for all processes where x is an arbitrary intensive or extensive property of the system.

5–25 Use Fig. 5–10 to show that whenever a system is taken around a closed cycle, the sum of the heat flow Q_i divided by the reservoir temperture T_i for each process is less than or equal to zero; i.e.,

$$\sum \frac{Q_i}{T_i} \leq 0. \qquad (5\text{–}18)$$

This is the *Clausius inequality*. [*Hint:* Arrange for $Q_{1A} = Q_1$ and $Q_2 = Q_{2B}$ and use the Kelvin-Planck statement of the second law.]

5–26 (a) In the operation of a refrigerator, there is a heat flow out of one reservoir at a lower temperature and a heat flow into a second reservoir at a higher temperature. Explain why this process does not contradict the Clausius statement of the second law. (b) In the operation of a heat engine, there is a heat flow Q out of a reservoir, and mechanical work W is done. Explain why this process does not violate the Kelvin-Planck statement of the second law.

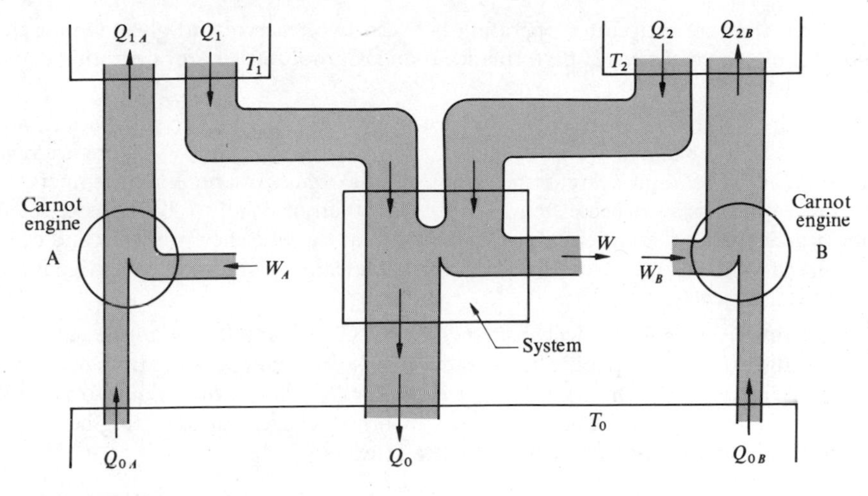

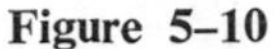

Figure 5–10

5–27 An inventor claims to have developed an engine that takes in 10^7 J at a temperature of 400 K, rejects 4×10^6 J at a temperature of 200 K, and delivers 3.6×10^6 J of mechanical work. Would you advise investing money to put this engine on the market? How would you describe this engine?

5–28 Show that if the Kelvin-Planck statement of the second law were not true, a violation of the Clausius statement would be possible.

5–29 Show that if the Clausius statement of the second law were not true, a violation of the Kelvin-Planck statement would be possible.

5–30 Assume that a certain engine has a greater efficiency than a Carnot engine operating between the same pair of reservoirs, and that in each cycle both engines reject the same quantity of heat to the low-temperature reservoir. Show that the Kelvin-Planck statement of the second law would be violated in a process in which the assumed engine drove the Carnot engine in the reversed direction as a refrigerator.

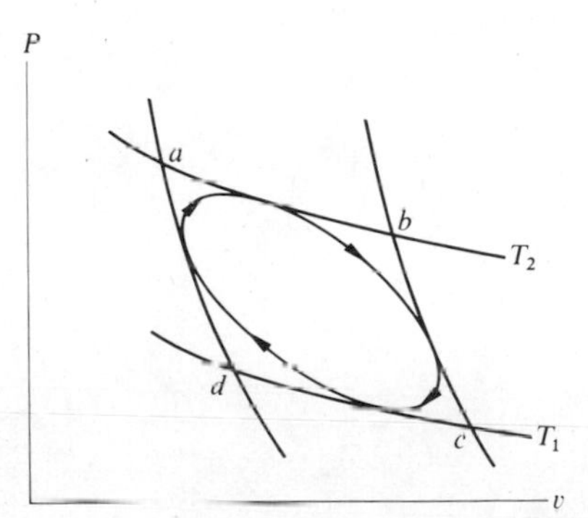

Figure 5–11

5–31 Show that no refrigerator operating between two reservoirs at given temperatures can have a higher coefficient of performance than a Carnot refrigerator operating between the same two reservoirs.

5–32 In Fig. 5–11, *abcd* represents a Carnot cycle, bounded by two adiabatics and by two isotherms at the temperatures T_1 and T_2, where $T_2 > T_1$. The oval figure is a reversible cycle for which T_2 and T_1 are, respectively, the maximum and minimum temperatures. In this cycle, heat is absorbed at temperatures less than or equal to T_2 and is rejected at temperatures greater than or equal to T_1. Prove that the efficiency of the second cycle is less than that of the Carnot cycle. [*Hint:* Approximate the second cycle by a large number of small Carnot cycles.]

5–33 Starting from either the Kelvin-Planck or the Clausius statement of the second law, show that the ratio $|Q_2|/|Q_1|$ must be the same for all Carnot cycles operating between the same pair of reservoirs. [*Hint:* Arrange for a heat flow Q from a Carnot engine to a reservoir in n cycles and have the same heat flow into a Carnot refrigerator operating between the same reservoirs in m cycles where n and m are integers.]

6

Combined first and second laws

6–1 INTRODUCTION

We now combine the first and second laws to obtain several important thermodynamic relations. The analytical formulation of the first law of thermodynamics, in differential form, is

$$d'Q = dU + d'W. \tag{6–1}$$

The second law states that for a reversible process between two equilibrium states,

$$d'Q_r = T\,dS. \tag{6–2}$$

Also, the work in a reversible process, for a *PVT* system, is

$$d'W = P\,dV. \tag{6–3}$$

It follows that in any infinitesimal reversible process, for a *PVT* system,

$$T\,dS = dU + P\,dV. \tag{6–4}$$

Equation (6–4) is one formulation of the *combined first and second laws* for a *PVT* system. For other systems, such as a stretched wire or a surface film, the appropriate expression for the work replaces the term $P\,dV$.

Although Eqs. (6–2) and (6–3) are true only for a reversible process, it is important to realize that Eq. (6–4) is not restricted to a *process* at all, since it simply expresses a relation between the *properties* of a system and the differences between the values of these properties, in two neighboring equilibrium states. That is, although we made use of a reversible process to derive the relation between dS, dU, and dV, once we have determined what this relation is it must be true for any pair of neighboring equilibrium states, whatever the nature of a process between the states, or even if no process at all takes place between them.

Suppose a system undergoes an *irreversible* process between two equilibrium states. Then both Eqs. (6–1) and (6–4) can be applied to the process, since the former is correct for any process, reversible or not, and the latter is correct for any two equilibrium states. However, if the process is irreversible, the term $T\,dS$ in Eq. (6–4) *cannot* be identified with the term $d'Q$ in Eq. (6–1), and the term $P\,dV$ in Eq. (6–4) *cannot* be identified with the term $d'W$ in Eq. (6–1). As an example, consider an irreversible process in which adiabatic stirring work $d'W$ is done on a system kept at constant volume. The entropy of the system increases so $T\,dS \neq 0$, but $d'Q = 0$ because the process is adiabatic. Also, $P\,dV = 0$ because the process is at constant volume, while $d'W \neq 0$.

A large number of thermodynamic relations can now be derived by selecting T and v, T and P, or P and v as independent variables. Furthermore, since the state of a pure substance can be defined by *any* two of its properties, the partial derivative of any one property with respect to any other, with any one of those remaining held constant, has a physical meaning, and it is obviously out of the question to attempt to tabulate all possible relations between all of these derivatives. However, every partial derivative can be expressed in terms of the coefficient of volume expansion $\beta = (1/v)(\partial v/\partial T)_P$, the isothermal compressibility $\kappa = -(1/v)(\partial v/\partial P)_T$, and c_P, together with the properties P, v, and T themselves, so that no physical properties of a substance other than those already discussed need be measured. A derivative is said to be in *standard form* when it is expressed in terms of the quantities above.

Once the partial derivatives have been evaluated, the results can be collected in a systematic way devised by P. W. Bridgman*, so that when a particular derivative is needed, it is not necessary to calculate it from first principles. The procedure is explained in Appendix A.

We next demonstrate the general method by which the derivatives are evaluated, and work out a few relations that will be needed later.

6–2 T AND v INDEPENDENT

Let us write our equations in terms of specific quantities, so that the results are independent of the mass of any particular system and refer only to the *material* of which the system is composed. From the combined first and second laws, we have

$$ds = \frac{1}{T}(du + P\,dv),$$

and considering u as a function of T and v,

$$du = \left(\frac{\partial u}{\partial T}\right)_v dT + \left(\frac{\partial u}{\partial v}\right)_T dv. \tag{6–5}$$

Therefore

$$ds = \frac{1}{T}\left(\frac{\partial u}{\partial T}\right)_v dT + \frac{1}{T}\left[\left(\frac{\partial u}{\partial v}\right)_T + P\right] dv.$$

But we can also write,

$$ds = \left(\frac{\partial s}{\partial T}\right)_v dT + \left(\frac{\partial s}{\partial v}\right)_T dv. \tag{6–6}$$

Note that one could *not* carry out a corresponding procedure on the basis of the first law alone, which states that

$$d'q = du + d'w.$$

One *cannot* write

$$d'q = \left(\frac{\partial q}{\partial T}\right)_v dT + \left(\frac{\partial q}{\partial v}\right)_T dv,$$

because q is not a function of T and v, and $d'q$ is *not* an exact differential. It is only because ds *is* an exact differential that we can express it in terms of dT and dv.

* Percy W. Bridgman, American physicist (1882–1961).

Since dT and dv are independent, their coefficients in the preceding equations must be equal. Therefore

$$\left(\frac{\partial s}{\partial T}\right)_v = \frac{1}{T}\left(\frac{\partial u}{\partial T}\right)_v, \tag{6–7}$$

$$\left(\frac{\partial s}{\partial v}\right)_T = \frac{1}{T}\left[\left(\frac{\partial u}{\partial v}\right)_T + P\right]. \tag{6–8}$$

Furthermore, as shown in Section 2–10, the *second* derivatives of s and u with respect to T and v (the "mixed" second-order partial derivatives) are independent of the order of differentiation. Thus:

$$\left[\frac{\partial}{\partial v}\left(\frac{\partial s}{\partial T}\right)_v\right]_T = \left[\frac{\partial}{\partial T}\left(\frac{\partial s}{\partial v}\right)_T\right]_v = \frac{\partial^2 s}{\partial v\,\partial T} = \frac{\partial^2 s}{\partial T\,\partial v}.$$

Hence from Eqs. (6–7) and (6–8), differentiating the first partially with respect to v and the second with respect to T, we obtain

$$\frac{1}{T}\frac{\partial^2 u}{\partial v\,\partial T} = \frac{1}{T}\left[\frac{\partial^2 u}{\partial T\,\partial v} + \left(\frac{\partial P}{\partial T}\right)_v\right] - \frac{1}{T^2}\left[\left(\frac{\partial u}{\partial v}\right)_T + P\right],$$

which simplifies to

$$\left(\frac{\partial u}{\partial v}\right)_T = T\left(\frac{\partial P}{\partial T}\right)_v - P = \frac{T\beta}{\kappa} - P. \tag{6–9}$$

The dependence of internal energy on volume, at constant temperature, can therefore be calculated from the equation of state, or from the values of β, κ, T, and P.

Since $(\partial u/\partial T)_v = c_v$, Eq. (6–5) may now be written:

$$du = c_v\,dT + \left[T\left(\frac{\partial P}{\partial T}\right)_v - P\right] dv. \tag{6–10}$$

Hill and Lounasmaa have measured the specific heat capacity at constant volume and the pressure of liquid He^4 as a function of temperature between 3 and 20 K and for a range of densities.* The data for c_v and P are shown on Figs. 6–1(a) and 6–1(b), plotted as a function of a reduced density ρ_r which is the ratio of the actual density of He^4 to its density at the critical point, taken by them to be 68.8 kg m^{-3}. The molal specific volume is, then, $0.0582/\rho_r$ m^3 kilomole^{-1}.

For example, at a temperature of 6 K and a pressure of 19.7 atm, $\rho_r = 2.2$, thus $v = 2.64 \times 10^{-2}$ m^3 kilomole^{-1}. The isothermal compressibility of He^4 at 6 K and 19.7 atm can be found to be 9.42×10^{-8} m^2 N^{-1} by measuring the slope of the 6 K isotherm at 19.7 atm and dividing by $\rho_r = 2.2$. The value of the expansivity $\beta = 5.35 \times 10^{-2}$ K^{-1} is calculated by dividing the fractional change of the reduced density along the 19.7 atm isobar as the temperature is varied by ± 1 K and dividing by the temperature change.

* R. W. Hill and O. V. Lounasmaa, *Philosophical Transactions of the Royal Society of London*, **252A,** (1960): 357. Actually $(\partial P/\partial T)_v$ was also directly measured, making it possible to calculate all the thermodynamic properties of He^4 except c_P to an accuracy of 1% by direct numerical integration of the data. Data used by permission.

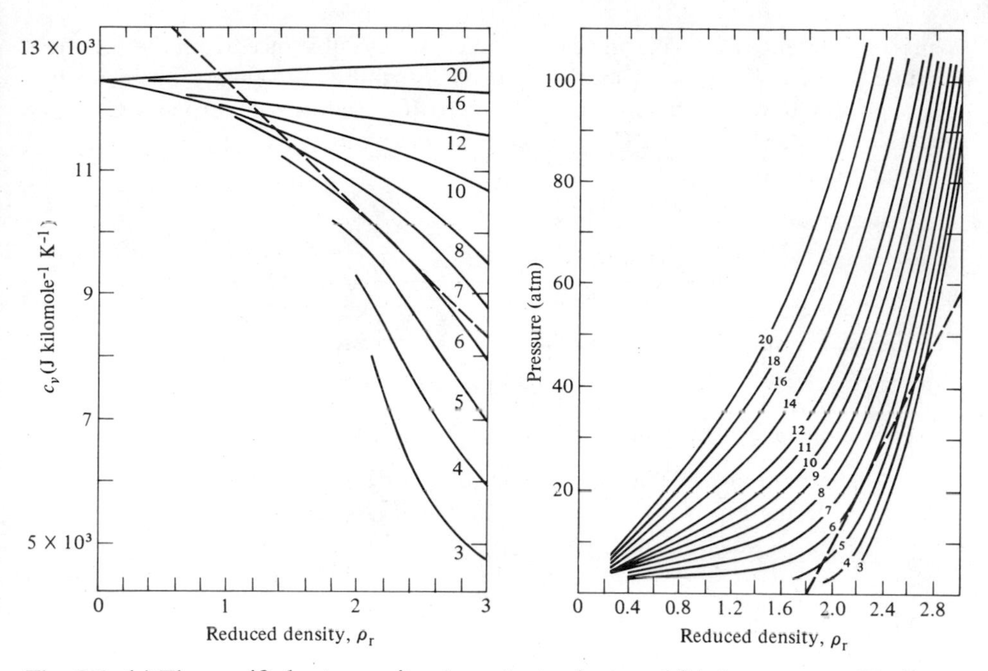

Fig. 6–1 (a) The specific heat capacity at constant volume and (b) the pressure of He^4 as a function of reduced density at temperatures between 3 and 20 K. Each curve is marked with the temperature in kelvins. The reduced density ρ_r is the ratio of the actual density of He^4 to 68.8 kg m^{-3}. The dashed lines are the tangents to the 6 K isotherm at $\rho_r = 2.2$. The experiments were performed by Hill and Lounasmaa. (These figures are reprinted by permission from O. V. Lounasmaa's article, "The Thermodynamic Properties of Fluid Helium, *Philosophical Transactions of the Royal Society of London* 252A (1960): 357 (Figs. 4 and 7).)

These data can be used to calculate $(\partial u/\partial v)_T$, by Eq. (6–9):

$$\left(\frac{\partial u}{\partial v}\right)_T = \frac{T\beta}{\kappa} - P = \frac{(6)(5.35 \times 10^{-2})}{9.42 \times 10^{-8}} - 19.7(1.01 \times 10^5) = 1.42 \times 10^6 \text{ J m}^{-3}.$$

By using values of $(\partial u/\partial v)_T$ and c_v, determined at various temperatures and densities, Eq. (6–5) can be integrated numerically to obtain values of the change in internal energy.

In Section 4–2, using the first law alone, we derived the equation

$$c_P - c_v = \left[\left(\frac{\partial u}{\partial v}\right)_T + P\right]\left(\frac{\partial v}{\partial T}\right)_P. \tag{6–11}$$

Making use of Eq. (6–9), we see that

$$c_P - c_v = T\left(\frac{\partial P}{\partial T}\right)_v\left(\frac{\partial v}{\partial T}\right)_P = \frac{\beta^2 T v}{\kappa}. \tag{6–12}$$

Thus the difference $c_P - c_v$ can be calculated for any substance, from the equation of state or from β and κ. The quantities, T, v, and κ are always positive, and although β may be positive, negative, or zero (for water, it is zero at 4°C and is negative between 0°C and 4°C), β^2 is always positive or zero. It follows that c_P is never smaller than c_v.

Using the data for He^4 given above,

$$c_P - c_v = \frac{(5.35 \times 10^{-2})^2(6)(2.64 \times 10^{-2})}{9.42 \times 10^{-8}} = 4810 \text{ J kilomole}^{-1}\text{ K}^{-1}.$$

Since c_v is measured to be 9950 J kilomole^{-1} K^{-1} at 6 K and $\rho_r = 2.2$,

$$c_P = 14{,}760 \text{ J kilomole}^{-1}\text{ K}^{-1}.$$

Even at these low temperatures $(c_P - c_v)/c_v = 48$ percent.

Let us now return to the expressions for $(\partial s/\partial T)_v$ and $(\partial s/\partial v)_T$ in Eqs. (6–7) and (6–8). Using Eq. (6–9) and the fact that $(\partial u/\partial T)_v = c_v$,

$$\left(\frac{\partial s}{\partial T}\right)_v = \frac{c_v}{T} \tag{6–13}$$

and

$$\left(\frac{\partial s}{\partial v}\right)_T = \left(\frac{\partial P}{\partial T}\right)_v. \tag{6–14}$$

Therefore from Eq. (6–6),

$$ds = \frac{c_v}{T}\,dT + \left(\frac{\partial P}{\partial T}\right)_v dv,$$

or

$$T\,ds = c_v\,dT + T\left(\frac{\partial P}{\partial T}\right)_v dv. \tag{6–15}$$

For liquid He^4 at 6 K and 19.7 atm,

$$\left(\frac{\partial s}{\partial T}\right)_v = \frac{9950}{6} = 1.66 \times 10^3 \text{ J kilomole}^{-1}\text{ K}^{-2},$$

and

$$\left(\frac{\partial s}{\partial v}\right)_T = \frac{5.35 \times 10^{-2}}{9.42 \times 10^{-8}} = 5.68 \times 10^5 \text{ J K}^{-1}\text{ m}^{-3}.$$

Using the values of these quantities determined at various temperatures and densities, Eq. (6–6) or Eq. (6–15) can be numerically integrated to yield values of the entropy as a function of temperature and volume.

Finally equating the mixed partial second derivatives of s with respect to v and T, we get

$$\left(\frac{\partial c_v}{\partial v}\right)_T = T\left(\frac{\partial^2 P}{\partial T^2}\right)_v. \tag{6–16}$$

For any substance for which the pressure is a linear function of temperature at constant volume, $(\partial^2 P/\partial T^2)_v = 0$ and c_v is independent of volume, although it may be dependent on temperature.

The value for $(\partial c_v/\partial v)_T$ for He^4 is calculated by measuring the slope of the 6 K isotherm on Fig. 6–1(a) at $\rho_r = 2.2$. The slope, $(\partial c_v/\partial \rho_r)_T$, is related to $(\partial c_v/\partial v)_T$ by

$$\left(\frac{\partial c_v}{\partial v}\right)_T = \left(\frac{\partial c_v}{\partial \rho_r}\right)_T \left(\frac{\partial \rho_r}{\partial v}\right)_T = -\left(\frac{\partial c_v}{\partial \rho_r}\right)_T \frac{\rho_r^2}{0.0582} = 1.7 \times 10^5 \text{ J K}^{-1} \text{ m}^{-3}.$$

The value for $(\partial^2 P/\partial T^2)_v$ for He^4 is estimated by calculating values for the change in pressure as the temperature is changed by 1 K, keeping ρ_r constant at 2.2, and measuring the slope of the curve obtained by plotting these values of $\Delta P/\Delta T$ versus T. This process yields a value of $T(\partial^2 P/\partial T^2)$, which is close to 1.7×10^5 J K^{-1} m^{-3}.

6–3 *T* AND *P* INDEPENDENT

In terms of the enthalpy $h = u + Pv$, the combined first and second laws can be written,

$$ds = \frac{1}{T}(dh - v\,dP),$$

and considering h as a function of T and P,

$$dh = \left(\frac{\partial h}{\partial T}\right)_P dT + \left(\frac{\partial h}{\partial P}\right)_T dP. \tag{6–17}$$

Therefore

$$ds = \frac{1}{T}\left(\frac{\partial h}{\partial T}\right)_P dT + \frac{1}{T}\left[\left(\frac{\partial h}{\partial P}\right)_T - v\right] dP.$$

But

$$ds = \left(\frac{\partial s}{\partial T}\right)_P dT + \left(\frac{\partial s}{\partial P}\right)_T dP, \tag{6–18}$$

and hence

$$\left(\frac{\partial s}{\partial T}\right)_P = \frac{1}{T}\left(\frac{\partial h}{\partial T}\right)_P, \tag{6–19}$$

$$\left(\frac{\partial s}{\partial P}\right)_T = \frac{1}{T}\left[\left(\frac{\partial h}{\partial P}\right)_T - v\right]. \tag{6–20}$$

Equating the mixed second-order partial derivatives of s, we find that

$$\left(\frac{\partial h}{\partial P}\right)_T = -T\left(\frac{\partial v}{\partial T}\right)_P + v = -\beta v T + v, \tag{6–21}$$

which is the analogue of Eq. (6–9). The dependence of enthalpy on pressure, at constant temperature, can therefore be calculated from the equation of state, or from β, v, and T.

Since $(\partial h/\partial T)_P = c_P$, Eq. (6–17) can be written,

$$dh = c_P\, dT - \left[T\left(\frac{\partial v}{\partial T}\right)_P - v\right] dP. \tag{6–22}$$

Using Eq. (6–21) and the fact that $(\partial h/\partial T)_P = c_P$, the partial derivatives of s with respect to T and P are

$$\left(\frac{\partial s}{\partial T}\right)_P = \frac{c_P}{T}, \tag{6–23}$$

$$\left(\frac{\partial s}{\partial P}\right)_T = -\left(\frac{\partial v}{\partial T}\right)_P. \tag{6–24}$$

Hence

$$T\, dS = c_P\, dT - T\left(\frac{\partial v}{\partial T}\right)_P dP, \tag{6–25}$$

and

$$\left(\frac{\partial c_P}{\partial P}\right)_T = -T\left(\frac{\partial^2 v}{\partial T^2}\right)_P. \tag{6–26}$$

Continuing with our example of liquid He^4 at 6 K and 19.7 atm

$$\left(\frac{\partial h}{\partial P}\right)_T = (2.64 \times 10^{-2})[-(5.35 \times 10^{-2})(6) + 1] = 1.79 \times 10^{-2} \text{ m}^3 \text{ kilomole}^{-1}.$$

Similarly

$$\left(\frac{\partial s}{\partial T}\right)_P = \frac{14760}{6} = 2460 \text{ J kilomole K}^{-2},$$

and

$$\left(\frac{\partial s}{\partial P}\right)_T = -(5.35 \times 10^{-2})(2.64 \times 10^{-2}) = -14.1 \times 10^{-4} \text{ m}^3 \text{ kilomole}^{-1} \text{ K}^{-1}.$$

6–4 *P* AND *v* INDEPENDENT

It is left as an exercise to show that if P and v are considered independent, we can write

$$\left(\frac{\partial s}{\partial P}\right)_v = \frac{c_v}{T}\left(\frac{\partial T}{\partial P}\right)_v = \frac{c_v}{T}\frac{\kappa}{\beta} \tag{6–27}$$

$$\left(\frac{\partial s}{\partial v}\right)_P = \frac{c_P}{T}\left(\frac{\partial T}{\partial v}\right)_P = \frac{c_P}{Tv\beta} \tag{6–28}$$

$$T\, ds = c_P\left(\frac{\partial T}{\partial v}\right)_P dv + c_v\left(\frac{\partial T}{\partial P}\right)_v dP. \tag{6–29}$$

For liquid He^4,

$$\left(\frac{\partial s}{\partial P}\right)_v = 2.92 \times 10^{-3}\ \text{m}^3\ \text{kilomole}^{-1}\ \text{K}^{-1},$$

and

$$\left(\frac{\partial s}{\partial v}\right)_P = 1.74 \times 10^6\ \text{J K}^{-1}\ \text{m}^{-3}.$$

6–5 THE $T\,ds$ EQUATIONS

The three expressions for $T\,ds$ derived in the preceding sections are collected below:

$$T\,ds = c_v\,dT + T\left(\frac{\partial P}{\partial T}\right)_v dv, \tag{6–30}$$

$$T\,ds = c_P\,dT - T\left(\frac{\partial v}{\partial T}\right)_P dP, \tag{6–31}$$

$$T\,ds = c_P\left(\frac{\partial T}{\partial v}\right)_P dv + c_v\left(\frac{\partial T}{\partial P}\right)_v dP. \tag{6–32}$$

These are called the "$T\,ds$" equations. They enable one to compute the heat flow $d'q_r = T\,ds$ in a reversible process; and when divided through by T, they express ds in terms of each pair of variables. They also provide relations between pairs of variables in a reversible adiabatic process in which s is a constant, and $ds = 0$.

The increase in temperature of a solid or liquid when it is compressed adiabatically can be found from the first $T\,ds$ equation. In terms of β and κ, we have

$$T\,ds = 0 = c_v\,dT_s + \frac{\beta T}{\kappa}\,dv_s,$$

$$dT_s = -\frac{\beta T}{\kappa c_v}\,dv_s. \tag{6–33}$$

If the volume is decreased, dv_s is negative and dT_s is positive when β is positive, but is negative when β is negative. Thus while ordinarily the temperature of a solid or liquid increases when the volume is decreased adiabatically, the temperature of water between 0°C and 4°C *decreases* in an adiabatic compression.

If the increase in pressure, rather than the decrease in volume, is specified, the temperature change can be found from the second $T\,ds$ equation:

$$T\,ds = 0 = c_P\,dT_s - \beta v T\,dP_s,$$

$$dT_s = \frac{\beta v T}{c_P}\,dP_s. \tag{6–34}$$

If β is positive, the temperature increases when pressure is applied. Hence if it is desired to keep the temperature constant, there must be a heat flow out of the system. This heat flow can also be found from the second $T\,ds$ equation, setting $dT = 0$ and $T\,ds = d'q_T$. Thus

$$d'q_T = -\beta v T\,dP_T. \tag{6–35}$$

Comparison of Eqs. (6–34) and (6–35) shows that for a given change in pressure the heat flow in an isothermal process equals the temperature rise in an adiabatic process, multiplied by the specific heat capacity at constant pressure.

Consider an adiabatic compression of 10^{-3} kilomole of liquid He^4 which decreases the volume by 1%. Assume that for He^4, β, T, κ, c_v and c_P remain essentially constant during the compression. Then by Eq. (6–33)

$$dT_s = -\frac{(5.35 \times 10^{-2})(6)(2.64 \times 10^{-5})}{(9.42 \times 10^{-8})(9.95 \times 10^{3})}(-.01) = 9 \times 10^{-5}\ \mathrm{K}.$$

Similarly if the pressure on 10^{-3} kilomole of He^4 is increased by 1%, by Eq. (6–34)

$$dT_s = \frac{(5.35 \times 10^{-2})(2.64 \times 10^{-5})(6)(19.7)(1.01 \times 10^{5})(.01)}{1.48 \times 10^{4}} = 1.1 \times 10^{-5}\ \mathrm{K}.$$

Helium is a rather soft solid, for which β is large and κ is small. Even so, the temperature changes during adiabatic processes are very small. For gases the temperature changes during an adiabatic process can become significant.

The heat which must flow out of the same sample of He^4 in order to keep the temperature constant during an isothermal process for the same change in volume is

$$d'q_T = T\left(\frac{\partial P}{\partial T}\right)_v dv_T = -\frac{(6)(5.35 \times 10^{-2})(2.64 \times 10^{-5})(.01)}{9.42 \times 10^{-8}} = -0.9\ \mathrm{J\ kilomole^{-1}}.$$

For an isothermal increase in pressure,

$$\begin{aligned} d'q_T &= -(5.35 + 10^{-2})(2.64 \times 10^{-5})(6)(19.7)(1.01 \times 10^{5})(.01) \\ &= -0.17\ \mathrm{J\ kilomole^{-1}}. \end{aligned}$$

The pressure needed to decrease the volume of a substance adiabatically is found from the third $T\,ds$ equation:

$$T\,ds = 0 = \frac{\kappa c_v}{\beta}\,dP_s + \frac{c_P}{\beta v}\,dv_s,$$

and hence

$$-\frac{1}{v}\left(\frac{\partial v}{\partial P}\right)_s = \kappa\,\frac{c_v}{c_P}. \tag{6–36}$$

It will be recalled that the compressibility κ is the *isothermal* compressibility, defined by the equation

$$\kappa = -\frac{1}{v}\left(\frac{\partial v}{\partial P}\right)_T.$$

The left side of Eq. (6–36) defines the *adiabatic* compressibility, which we shall write as κ_s. (To be consistent, the isothermal compressibility should have been written κ_T; we will continue to use κ, however.) Denoting the ratio c_P/c_v by γ, Eq. (6–36) becomes

$$\kappa_s = \frac{\kappa}{\gamma}. \tag{6–37}$$

Since c_P is always greater than (or equal to) c_v, γ is always greater than (or equal to) unity even for a solid or liquid, and the adiabatic compressibility is always less than (or equal to) the isothermal compressibility. This is natural, because an increase in pressure causes a rise in temperature (except when $\beta = 0$) and the expansion resulting from this temperature rise offsets to some extent the contraction brought about by the pressure. Thus for a given pressure increase dP, the volume change dv is less in an adiabatic than in an isothermal compression and the compressibility is therefore smaller.

When a sound wave passes through a substance, the compressions and rarefactions are adiabatic rather than isothermal. The velocity of a compressional wave, it will be recalled, equals the square root of the reciprocal of the product of density and compressibility, and the adiabatic rather than the isothermal compressibility should be used. Conversely, the adiabatic compressibility can be determined from a measurement of the velocity of a compressional wave and such measurements provide the most precise method of determining the ratio c_P/c_v.

For our example of liquid He^4, $\gamma = 14760/9950 = 1.48$ and $\rho = 4/2.64 \times 10^{-2} = 162$ kg m^{-3}. Therefore the velocity of sound is given by

$$v = \left[\frac{1.48}{162(9.43 \times 10)^{-8}}\right]^{1/2} = 3.11 \times 10^3 \text{ m s}^{-1}.$$

This is about 10% lower than an extrapolation of sound velocity data taken at 20 atm below 4.5 K would yield.

6–6 PROPERTIES OF A PURE SUBSTANCE

The general relations derived in the preceding sections can be used to compute the entropy and enthalpy of a pure substance from its directly measurable properties, namely, the *P-v-T* data and the specific heat capacity at constant pressure c_P. Since temperature and pressure are the quantities most readily controlled experimentally, these are the variables usually selected. We have, from the second $T\,ds$ equation, Eq. (6–31),

$$ds = \frac{c_P}{T}\,dT - \left(\frac{\partial v}{\partial T}\right)_P dP,$$

and from Eq. (6–22),

$$dh = c_P\,dT + \left[v - T\left(\frac{\partial v}{\partial T}\right)_P\right] dP.$$

Let s_0 and h_0 represent the entropy and enthalpy in an arbitrary reference state P_0, v_0, and T_0. Then

$$s = \int_{T_0}^{T} \frac{c_P}{T}\, dT - \int_{P_0}^{P} \left(\frac{\partial v}{\partial T}\right)_P dP + s_0, \tag{6–38}$$

and

$$h = \int_{T_0}^{T} c_P\, dT + \int_{P_0}^{P} \left[v - T\left(\frac{\partial v}{\partial T}\right)_P \right] dP + h_0. \tag{6–39}$$

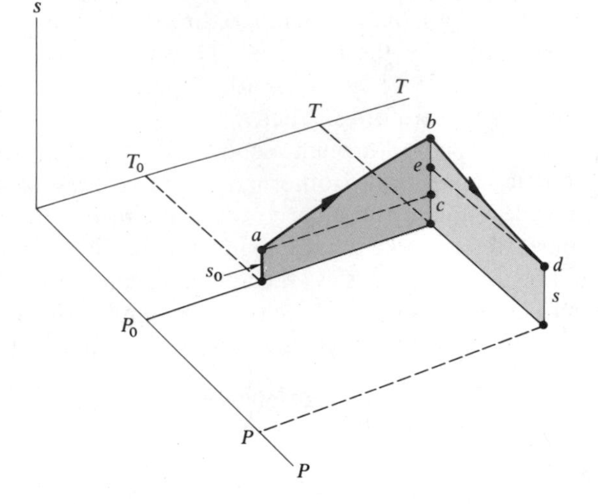

Fig. 6–2 Integration paths used in evaluation of entropy.

Since s and h are properties of a system, the difference between their values in any two equilibrium states depends only on the states and not on the process by which the system is taken from the first state to the second. Let us therefore evaluate the first integrals in each of the preceding equations at the constant pressure P_0, and the second integrals at a constant temperature T. The paths of integration are illustrated in Fig. 6–2. The vertical height of point a above the P-T plane represents the entropy s_0 at the reference pressure P_0 and the reference temperature T_0. Curve ab is the first integration path, at the constant pressure P_0. The first integral in Eq. (6–38) is represented by the length of the line segment bc. Curve bd is the second integration path, at the constant temperature T, and the second integral is represented by the length of the line segment be. The vertical height of point d above the P-T plane represents the entropy s at the pressure P and temperature T. The change in entropy of the system as it is taken from state a to state d is just the difference in the vertical heights of a and d above the P-T plane. In practice, other integration paths are often used because they simplify the treatment of experimental data.

In evaluating the first integral, we must use the specific heat capacity at the reference pressure P_0, or c_{P_0}. This, of course, must be expressed as a function of temperature. The coefficient of dP in the second integral must be expressed as a function of P, at the constant temperature T.

Experimental data on c_P are often available only at a pressure P different from the reference pressure P_0. Equation (6–26) can then be used to compute c_{P_0} from c_P and the P-v-T data. Integrating Eq. (6–26) at the constant temperature T, we get

$$c_{P_0} = c_P + T\int_{P_0}^{P}\left(\frac{\partial^2 v}{\partial T^2}\right)_P dP. \tag{6–40}$$

Thus the entropy and enthalpy of a system can be determined from a knowledge of its equation of state and of its specific heat capacity as a function of temperature, both of which can be measured experimentally.

6–7 PROPERTIES OF AN IDEAL GAS

The integrals in Eqs. (6–38), (6–39), and (6–40) are readily evaluated for an ideal gas. We have

$$v = RT/P, \qquad (\partial v/\partial T)_P = R/P, \qquad (\partial^2 v/\partial T^2)_P = 0.$$

Hence, from Eq. (6–40) the value of c_P is the same at all pressures, and c_P is a function of temperature only. The entropy and enthalpy are then

$$s = \int_{T_0}^{T}\frac{c_P}{T}\,dT - R\ln\frac{P}{P_0} + s_0, \tag{6–41}$$

$$h = \int_{T_0}^{T} c_P\,dT + h_0. \tag{6–42}$$

Over a temperature range in which c_P can be considered constant, these simplify further to

$$s = c_P\ln\frac{T}{T_0} - R\ln\frac{P}{P_0} + s_0, \tag{6–43}$$

$$h = c_P(T - T_0) + h_0. \tag{6–44}$$

The quantities s_0 and h_0 are *arbitrary* values that may be assigned to s and h in the reference state T_0, P_0.

The entropy as a function of temperature and volume, or of pressure and volume, can now be obtained from the equation of state, or by integration of the first and third $T\,ds$ equations. We give the results only for a range of variables in which the specific heat capacities can be considered constant:

$$s = c_v \ln \frac{T}{T_0} + R \ln \frac{v}{v_0} + s_0, \tag{6-45}$$

$$s = c_v \ln \frac{P}{P_0} + c_P \ln \frac{v}{v_0} + s_0. \tag{6-46}$$

The internal energy u, as a function of T and P, is

$$\begin{aligned} u &= h - Pv \\ &= \int_{T_0}^{T} c_P\,dT + h_0 - RT. \end{aligned}$$

Since for an ideal gas, $c_P = c_v + R$, this can be written

$$u = \int_{T_0}^{T} c_v\,dT + u_0, \tag{6-47}$$

where u_0 is the internal energy in the reference state. This equation could have been obtained more simply by the direct integration of Eq. (6-10). The method above was used to illustrate how u can be obtained from h and the equation of state. Since for an ideal gas, c_v, (like c_P), is a function of temperature only, the internal energy is a function of temperature only. If c_v can be considered constant, then

$$u = c_v(T - T_0) + u_0. \tag{6-48}$$

To find the equation of a reversible adiabatic process, we can set s = constant in any expression for the entropy. Thus from Eq. (6-46),

$$\begin{aligned} c_v \ln P + c_P \ln v &= \text{constant} \\ \ln P^{c_v} + \ln v^{c_P} &= \text{constant} \\ Pv^{c_P/c_v} &= \text{constant}, \end{aligned}$$

a familiar result.

The heat absorbed in a reversible process can be found from any of the $T\,ds$ equations, setting $T\,ds = d'q$. Thus in a reversible isothermal process, from the first $T\,ds$ equation,

$$d'q_T = P\,dv_T.$$

6-8 PROPERTIES OF A VAN DER WAALS GAS

We next make the same calculations as in the preceding section, but for a van der Waals gas. These serve to illustrate how the properties of a real gas can be found if its equation of state and if its specific heat capacity are known. A van der Waals gas has been selected because of its relatively simple equation of state,

$$\left(P + \frac{a}{v^2}\right)(v - b) = RT.$$

The expressions for the properties of a van der Waals gas are simpler if T and v, rather than T and P, are selected as variables. From the first $T\,ds$ equation,

$$ds = \frac{c_v}{T}\,dT + \left(\frac{\partial P}{\partial T}\right)_v dv.$$

From Eq. (6–16),

$$\left(\frac{\partial c_v}{\partial v}\right)_T = T\left(\frac{\partial^2 P}{\partial T^2}\right)_v = 0, \tag{6–49}$$

since P is a linear function of T. That is, c_v is a function of temperature only and does not vary with the volume at constant temperature.

From the equation of state,

$$\left(\frac{\partial P}{\partial T}\right)_v = \frac{R}{v - b}.$$

Then if s_0 is the entropy in a reference state P_0, v_0, T_0, we have

$$s = \int_{T_0}^{T} \frac{c_v}{T}\,dT + R\ln\left(\frac{v - b}{v_0 - b}\right) + s_0.$$

If c_v can be considered constant,

$$s = c_v \ln\frac{T}{T_0} + R\ln\left(\frac{v - b}{v_0 - b}\right) + s_0. \tag{6–50}$$

The internal energy is obtained from Eq. (6–10),

$$du = c_v\,dT + \left[T\left(\frac{\partial P}{\partial T}\right)_v - P\right] dv$$

$$= c_v\,dT + \frac{a}{v^2}\,dv.$$

If u_0 is the energy in the reference state,

$$u = \int_{T_0}^{T} c_v\,dT - a\left(\frac{1}{v} - \frac{1}{v_0}\right) + u_0,$$

and if c_v is constant,

$$u = c_v(T - T_0) - a\left(\frac{1}{v} - \frac{1}{v_0}\right) + u_0. \tag{6–51}$$

The internal energy of a van der Waals gas therefore depends on its specific volume as well as on its temperature. Note that only the van der Waals constant a appears in the energy equation. The reason is that this constant is a measure of the force of attraction between the molecules, or of their mutual potential energy, which changes as the specific volume changes and the intermolecular separation increases or decreases. The constant b is proportional to the volume occupied by the molecules themselves and does not affect the internal energy. It does, however, enter into the expression for the entropy because the entropy of a gas depends on the volume throughout which its molecules are dispersed, and the fact that the molecules themselves occupy some space makes the available volume less than the volume of the container.

The difference between the specific heat capacities, from Eq. (6–12), is

$$c_P - c_v = \frac{\beta^2 T v}{\kappa} = R \frac{1}{1 - \dfrac{2a(v-b)^2}{RTv^3}}.$$

The second term in the denominator is a small correction term, so in this term we can approximate $(v - b)$ by v, and assume that $Pv = RT$. Then, approximately,

$$c_P - c_v \approx R\left(1 + \frac{2aP}{R^2T^2}\right). \tag{6–52}$$

The constant a for carbon dioxide is 366×10^3 J m^3 kilomole^{-2}; and at a pressure of 1 bar $= 10^5$ N m^{-2} and a temperature of 300 K,

$$\frac{2aP}{R^2T^2} \approx 10^{-2},$$

so that within 1 percent, $c_P - c_v = R$.

The relation between T and v, in a reversible adiabatic process, is obtained by setting $s =$ constant. If we assume $c_v =$ constant, then from Eq. (6–50),

$$c_v \ln T + R \ln (v - b) = \text{constant},$$

or

$$T(v - b)^{R/c_v} = \text{constant}. \tag{6–53}$$

The heat absorbed in a reversible isothermal process, from the first $T\,ds$ equation, is

$$d'q_T = RT \frac{dv}{v - b}.$$

Since the change in internal energy is

$$du_T = a \frac{dv}{v^2},$$

the work $d'w$, from the first law, is

$$d'w_T = d'q_T - du_T = \left(\frac{RT}{v-b} - \frac{a}{v^2}\right) dv = P\,dv;$$

and in a finite process,

$$w_T = RT \ln \frac{v_2 - b}{v_1 - b} + a\left(\frac{1}{v_2} - \frac{1}{v_1}\right). \tag{6–54}$$

6–9 PROPERTIES OF A LIQUID OR SOLID UNDER HYDROSTATIC PRESSURE

The expressions for the properties of a liquid or solid under hydrostatic pressure can be obtained by introducing β, κ, and c_P in the general equations as functions of T and P, T and v, or P and v. We shall, however, consider only the special case in which β and κ can be assumed constant.

Let us first obtain the equation of state of a solid or liquid under hydrostatic pressure. We have

$$dv = \left(\frac{\partial v}{\partial T}\right)_P dT + \left(\frac{\partial v}{\partial P}\right)_T dP = \beta v\,dT - \kappa\,v\,dP.$$

Therefore

$$v = v_0 + \int_{T_0}^{T} \beta v\,dT - \int_{P_0}^{P} \kappa v\,dP,$$

where v_0 is the specific volume at the temperature T_0 and the pressure P_0. The first integral is evaluated at the pressure P_0 and the second at the temperature T. Because of the small values of β and κ for liquids and solids, the specific volume v will change only very slightly, even with large changes in T and P. Hence only a small error will be made if we assume v to be constant in the integrals and equal to v_0. Then if β and κ are constant also, we have the approximate equation of state

$$v = v_0[1 + \beta(T - T_0) - \kappa(P - P_0)]. \tag{6–55}$$

The entropy as a function of T and P can be found from the second $T\,ds$ equation:

$$s = \int_{T_0}^{T} \frac{c_P}{T}\,dT - \int_{P_0}^{P} \left(\frac{\partial v}{\partial T}\right)_P dP + s_0. \tag{6–56}$$

Following the procedure described in Section 6–6 and Fig. 6–2, we evaluate the first integral at the pressure P_0 (so that $c_P = c_{P_0}$) and the second at the temperature T. If c_P has been measured at atmospheric pressure P, then from Eq. (6–40)

$$c_{P_0} = c_P + \int_{P_0}^{P} \left(\frac{\partial^2 v}{\partial T^2}\right)_P dP.$$

From the approximate equation of state, given in Eq. (6–55),

$$\left(\frac{\partial v}{\partial T}\right)_P = \beta v_0, \qquad \left(\frac{\partial^2 v}{\partial T^2}\right)_P = 0.$$

Hence, to within the approximation that β can be considered constant, we can assume that c_{P_0} is equal to its value c_P at atmospheric pressure, and can be taken outside the integral sign in Eq. (6–56).

Replacing $(\partial v/\partial T)_P$ in Eq. (6–56) by the constant βv_0, which can also be taken outside the integral sign, we have the approximate expression for the entropy:

$$s = c_P \ln \frac{T}{T_0} - \beta v_0(P - P_0) + s_0. \tag{6–57}$$

The enthalpy can be calculated from Eq. (6–39), replacing $(\partial v/\partial T)_P$ by βv_0. The difference $c_P - c_v$ is

$$c_P - c_v = \frac{\beta^2 T v}{\kappa}.$$

For copper at 1000 K,

$$\beta \simeq 6 \times 10^{-5}\ \text{K}^{-1}, \qquad v \simeq 7.2 \times 10^{-3}\ \text{m}^3\ \text{kilomole}^{-1},$$
$$\kappa \simeq 10 \times 10^{-12}\ \text{m}^2\ \text{N}^{-1},$$

and hence

$$c_P - c_v \simeq 4300\ \text{J kilomole}^{-1}\ \text{K}^{-1}$$

which equals $0.52R$ and is in good agreement with the graphs of c_P and c_v in Fig. 3–10. At lower temperatures, both β and T are smaller, and below about 350 K, c_P and c_v are practically equal.

6–10 THE JOULE AND JOULE-THOMSON EXPERIMENTS

The experiments of Gay-Lussac and Joule, and of Joule and Thomson, were described in Section 4–5 where, on the basis of the first law alone, we derived the equations

$$\eta \equiv \left(\frac{\partial T}{\partial v}\right)_u = -\frac{1}{c_v}\left(\frac{\partial u}{\partial v}\right)_T,$$

$$\mu \equiv \left(\frac{\partial T}{\partial P}\right)_h = -\frac{1}{c_P}\left(\frac{\partial h}{\partial P}\right)_T.$$

We have now shown from the combined first and second laws that the quantities $(\partial u/\partial v)_T$ and $(\partial h/\partial P)_T$ can be calculated from the equation of state of a system through Eqs. (6–9) and (6–21):

$$\left(\frac{\partial u}{\partial v}\right)_T = T\left(\frac{\partial P}{\partial T}\right)_v - P,$$

$$\left(\frac{\partial h}{\partial P}\right)_T = -T\left(\frac{\partial v}{\partial T}\right)_P + v.$$

For a van der Waals gas,

$$\left(\frac{\partial u}{\partial v}\right)_T = \frac{a}{v^2},$$

$$\left(\frac{\partial h}{\partial P}\right)_T = \frac{RTv^3b - 2av(v-b)^2}{RTv^3 - 2a(v-b)^2}.$$

Hence in a Joule expansion of a van der Waals gas,

$$\eta = \left(\frac{\partial T}{\partial v}\right)_u = -\frac{a}{c_v v^2},$$

and in a finite change in volume (dropping the subscript u for simplicity)

$$T_2 - T_1 = \frac{a}{c_v}\left(\frac{1}{v_2} - \frac{1}{v_1}\right). \tag{6–58}$$

Thus for a given change in specific volume, the expected temperature change is proportional to the van der Waals constant a, which is a measure of the attractive force between the molecules. For an ideal gas, $a = 0$ and the temperature change is zero. Because v_2 is necessarily larger than v_1, T_2 is less than T_1 for all real gases.

In a Joule-Thomson expansion of a van der Waals gas

$$\mu = \left(\frac{\partial T}{\partial P}\right)_h = -\frac{1}{c_P}\frac{RTv^3b - 2av(v-b)^2}{RTv^3 - 2a(v-b)^2}. \tag{6–59}$$

The *inversion curve* in Fig. 4–4(b) is the locus of points at which $(\partial T/\partial P)_h = 0$, and the temperature at such a point is the *inversion temperature*, T_i. Hence, setting $(\partial T/\partial P)_h = 0$ in Eq. (6–59), we obtain the equation of the inversion curve of a van der Waals gas,

$$T_i = \frac{2a(v-b)^2}{Rv^2b}. \tag{6–60}$$

The relation between T_i and the corresponding pressure P_i is obtained by eliminating v between this equation and the equation of state. The resulting curve has the same general shape as those observed for real gases, although the numerical agreement is not close.

When the Joule-Thomson effect is to be used in the liquefaction of gases, the gas must first be cooled below its maximum inversion temperature, which occurs when the pressure is small and the specific volume is large. We can then approximate $(v - b)$, in Eq. (6–60), by v, and for a van der Waals gas,

$$T_i(\max) = \frac{2a}{Rb}. \tag{6–61}$$

Reference to Table 2–1 will show that the values of b (which is measure of molecular size) are nearly the same for all gases, so that maximum value of T_i for a van der Waals gas is very nearly proportional to a. Table 6–1 list values of $2a/Rb$ for carbon dioxide, hydrogen, and helium; and for comparison, the observed values of T_i are also given. The agreement is surprisingly good. In order to be cooled in a Joule-Thomson expansion, hydrogen must be precooled to about 200 K, which is usually done with the aid of liquid nitrogen. Helium must be cooled to about 40 K and that can be accomplished with liquid hydrogen or by allowing the helium to do adiabatic work.

Table 6–1 Calculated and observed values of the maximum inversion temperature

Gas	a (J m^3 kilomole^{-2})	b (m^3 kilomole^{-1})	$2a/Rb$	T_i (max)
CO_2	366×10^3	.0429	2040 K	~1500 K
H_2	24.8	.0266	224 K	200 K
He	3.44	.0234	35 K	~40 K

6–11 EMPIRICAL AND THERMODYNAMIC TEMPERATURE

In Section 5–2, thermodynamic temperature T was defined by the equation

$$T = A\phi(\theta), \tag{6–62}$$

where A is an arbitrary constant and $\phi(\theta)$ is a function of the empirical temperature θ as measured by a thermometer using any arbitrary thermometric property. The form of the function $\phi(\theta)$ need not be known, however, to determine the temperature T of a system, because it follows from the definition above that the ratio of two thermodynamic temperatures is equal to the ratio of the quantities of heat absorbed and rejected in a Carnot cycle. In principle, then, the thermodynamic temperature of a system can be determined by measuring these heat flows; and, in fact, this procedure is sometimes followed in experiments at very low temperatures.

We now show how the function $\phi(\theta)$ can be determined for any gas thermometer filled to a specified pressure P_3 at the triple point, so that T can be found from Eq. (6–62) without the necessity of extrapolating to zero pressure P_3 as in Fig. 1–4. We assume that the equation of state of the gas, and its energy equation, have been determined on the empirical temperature scale θ defined by the gas, so that P and U are known experimentally as functions of V and θ. We start with Eq. (6–9),

$$\left(\frac{\partial U}{\partial V}\right)_T = T\left(\frac{\partial P}{\partial T}\right)_V - P.$$

Because T is a function of θ only, constant T implies constant θ and $(\partial\theta/\partial T)_V = d\theta/dT$. Therefore we can write,

$$\left(\frac{\partial U}{\partial V}\right)_\theta = T\left(\frac{\partial P}{\partial \theta}\right)_V \frac{d\theta}{dT} - P,$$

or

$$\frac{dT}{T} = \frac{(\partial P/\partial\theta)_V}{P + (\partial U/\partial V)_\theta}\, d\theta. \tag{6–63}$$

Since the left side of this equation is a function of T only, the right side must be a function of θ only. If we represent the coefficient of $d\theta$ by $g(\theta)$,

$$g(\theta) \equiv \frac{(\partial P/\partial\theta)_V}{P + (\partial U/\partial V)_\theta},$$

then

$$\frac{dT}{T} = g(\theta)\, d\theta;$$

and

$$\ln T = \int g(\theta)\, d\theta + \ln A',$$

$$T = A' \exp\left[\int g(\theta)\, d\theta\right], \tag{6–64}$$

where A' is an integration constant. Comparison with Eq. (6–62) shows that the function $\phi(\theta)$ is

$$\phi(\theta) = \exp\left[\int g(\theta)\, d\theta\right], \tag{6–65}$$

if $A = A'$. Since $g(\theta)$ can be found experimentally, the thermodynamic temperature T, corresponding to any empirical temperature θ, can be calculated from Eq. (6–64).

As an example, suppose the gas is a "Boyle's law" gas, for which we have found by experiment that the product PV is constant at constant temperature. We choose the product PV as the thermometric property X and define the empirical temperature θ as

$$\theta = \theta_3 \frac{PV}{(PV)_3}, \tag{6–66}$$

where $(PV)_3$ is the value of the product PV at the triple point and θ_3 is the arbitrary value assigned to θ at the triple point. Then

$$P = \frac{(PV)_3}{\theta_3}\frac{\theta}{V}$$

and

$$\left(\frac{\partial P}{\partial\theta}\right)_V = \frac{(PV)_3}{\theta_3 V}.$$

If, in addition, we have found from the Joule experiment that the internal energy of the gas is independent of its volume and is a function of temperature only,

$$\left(\frac{\partial U}{\partial V}\right)_\theta = 0$$

and

$$g(\theta) = \frac{(PV)_3}{PV\theta_3} = \frac{1}{\theta}.$$

Then

$$\int g(\theta)\, d\theta = \int \frac{d\theta}{\theta} = \ln \theta,$$

$$\phi(\theta) = \exp\left[\int g(\theta)\, d\theta\right] = \exp(\ln \theta) = \theta,$$

and finally

$$T = A\theta.$$

In this case, the function $\phi(\theta)$ equals θ and the thermodynamic temperature T is directly proportional to the empirical temperature θ. But a gas which obeys Boyle's law and whose internal energy is a function of temperature only is an ideal gas, and the empirical temperature θ is the ideal gas temperature. This is in agreement with the result obtained earlier when we analyzed a Carnot cycle carried out by an ideal gas.

It may be noted that if the only condition imposed on the gas is that it obeys Boyle's law, the empirical temperature defined by Eq. (6–66) is not directly proportional to the thermodynamic temperature. Only if in addition $(\partial U/\partial V)_\theta = 0$ will $g(\theta)$ reduce to $1/\theta$.

6–12 MULTIVARIABLE SYSTEMS. CARATHÉODORY PRINCIPLE

Thus far, we have considered only systems whose state can be defined by the values of *two* independent variables such as the pressure P and the temperature T. The volume V is then determined by the equation of state, and the internal energy U by the energy equation. For generality, let X represent the extensive variable corresponding to the volume V, and Y the associated intensive variable corresponding to the pressure P. The work $d'W$ in an infinitesimal reversible process is then $Y\,dX$ and the first law states that in such a process

$$d'Q_r = dU + d'W = dU + Y\,dX. \tag{6–67}$$

If we choose U and X as the independent variables specifying the state of the system, then from the equation of state and the energy equation we can find Y as a function of U and X and Eq. (6–67) expresses the inexact differential $d'Q_r$ in terms of U and X and their differentials.

It is shown in textbooks of mathematics that any equation expressing an inexact differential in terms of *two* independent variables and their differentials *always* has an *integrating denominator*, and when the equation is divided through by this denominator, the left side becomes an *exact* differential. But we have shown that $d'Q_r/T$ is the exact differential dS, so that in this case the integrating denominator is the thermodynamic temperature T and

$$\frac{d'Q_r}{T} = dS = \frac{1}{T}\,dU + \frac{Y}{T}\,dX,$$

or

$$T\,dS = dU + Y\,dX. \tag{6–68}$$

Now consider the more general case of a *multivariable* system, for which the values of *more* than two independent variables are necessary to specify the state. It will suffice to consider a 3-variable system (that is, three *independent* variables). An example is a paramagnetic gas in an external magnetic field $\mathscr{H}$, whose state can be specified by its volume V, its magnetic moment M, and its temperature T. The work $d'W$ in a reversible process undergone by such a system is

$$d'W = P\,dV - \mathscr{H}\,dM. \tag{6–69}$$

Let X_1 and X_2 represent the two extensive variables (corresponding to V and $-M$) and Y_1 and Y_2 the associated intensive variables (corresponding to P and $\mathscr{H}$). Then in general

$$d'W = Y_1\,dX_1 + Y_2\,dX_2;$$

and from the first law,

$$d'Q_r = dU + d'W = dU + Y_1\,dX_1 + Y_2\,dX_2. \tag{6–70}$$

If we choose U, X_1, and X_2 as the independent variables specifying the state of the system, this equation expresses the inexact differential $d'Q_r$ in terms of *three* independent variables and their differentials. Unlike the corresponding Eq. (6–67) for a 2-variable system, an equation such as Eq. (6–70), expressing an inexact differential in terms of the differentials of *three* (or more) independent variables, does not *necessarily* have an integrating denominator, although it *may* have one, and indeed does have one if the variables are those defining a thermodynamic system.

To show that this is true, we return to the assertion in Section 5–2 that when *any system whatever* is carried throughout a Carnot cycle, the ratio $|Q_2|/|Q_1|$ has the same value, for the same pair of reservoir temperatures. Hence regardless of the complexity of a system, we can still define thermodynamic temperature by the equation

$$\frac{|Q_2|}{|Q_1|} = \frac{T_2}{T_1},$$

and by exactly the same reasoning as in Section 5–3, the entropy change of a multi-variable system can be defined as

$$dS = \frac{d'Q_r}{T}.$$

Hence when Eq. (6–70) is divided through by T, the left side becomes the exact differential dS and the thermodynamic temperature T is an integrating denominator for $d'Q_r$, regardless of the complexity of a system. Equation (6–70) can therefore be written

$$\frac{d'Q_r}{T} = dS = \frac{1}{T}[dU + Y_1\,dX_1 + Y_2\,dX_2],$$

or

$$T\,dS = dU + Y_1\,dX_1 + Y_2\,dX_2. \tag{6–71}$$

Since the entropy S is a property of any system, it can be considered a function of any three of the variables specifying the state of a 3-variable system. Thus if we consider X_1, X_2, and the temperature T as independent variables, the *entropy equation* of a system is

$$S = S(T, X_1, X_2).$$

If S is constant, the preceding equation is the equation of a *surface* in a three-dimensional T-X_1-X_2 space. That is, all isentropic processes carried out by the system, and for which S has some constant value, say S_1, lie on a single surface in a T-X_1-X_2 diagram. All processes for which S has a constant value S_2 lie on a second surface, and so on. These isentropic *surfaces* are generalizations of the isentropic *curves* for a 2-variable system. Similarly, all *isothermal* processes at a given temperature lie on a single surface which, in a T-X_1-X_2 diagram, is a plane perpendicular to the temperature axis. In general, for a system defined by m independent variables, where $m > 3$, isothermal and isentropic processes lie on hypersurfaces of $(m - 1)$ dimensions, in an m-dimensional hyperspace.

It is of interest to consider the geometrical representation, in a T-X_1-X_2 diagram, of the possible Carnot cycles that can be carried out by a 3-variable system. Figure 6-3 shows portions of two isothermal surfaces at temperatures T_2 and T_1, and of two isentropic surfaces at entropies S_2 and S_1, where $S_2 > S_1$.

Suppose we start a Carnot cycle at a point at which $T = T_2$ and $S = S_1$. Then any curve in the plane $T = T_2$, from the intersection of this plane with the surface $S = S_1$ to its intersection with the surface $S = S_2$, is an isothermal process at temperature T_2 in which the entropy increases from S_1 to S_2. The process might start at any one of the points a_1, a_2, a_3, etc., and terminate at any one of the points b_1, b_2, b_3, etc. Even a process such as a_1-a_3-b_1-b_3 satisfies the conditions. (Any process represented by the line of intersection of an isothermal and an isentropic surface, such as processes a_1-a_3 and b_1-b_3, has the interesting property of being *both isothermal and isentropic*.) Thus in contrast to a 2-variable system, for which only *one* isothermal process between entropies S_1 and S_2 is possible at a given temperature, there are in a 3-variable system (or in a multivariable system) an infinite number of such processes.

The next step in the cycle could consist of any curve on the isentropic surface $S = S_2$, from any point such as b_1, b_2, b_3, etc., to any point such as c_1, c_2, c_3, etc. The cycle is completed by any process in the plane $T = T_1$ to the surface $S = S_1$, and a final process in this surface to the starting point.

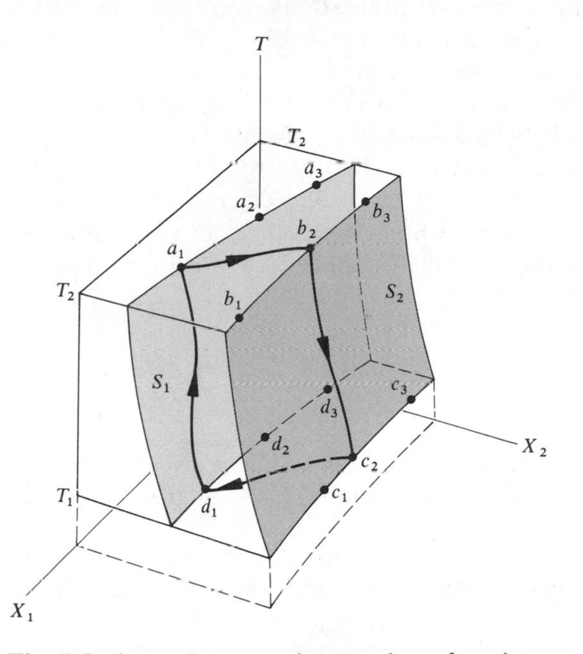

Fig. 6–3 Any process such as a_1-b_2-c_2-d_1-a_1 is a Carnot cycle for a 3-variable system.

Note that the heat flow Q is the same in all reversible isothermal processes at a given temperature between the isentropic surfaces S_1 and S_2, since in any such process $Q = T(S_2 - S_1)$.

When any one of the cyclic processes described above is represented in the T-S plane, it has exactly the same form as that for a 2-variable system, namely, as shown in Fig. 5–4, a rectangle with sides parallel to the T- and S-axes.

We have pointed out earlier that the only states of a 2-variable system that can be reached from a given state by an adiabatic process are those for which the entropy is equal to or greater than that in the initial state. All *adiabatically accessible* states then either lie on the isentropic *curve* through the given state, or lie on the same side of that curve. The same is true for a 3-variable system, except that the accessible states either lie on the isentropic *surface* through the given state, or lie on the same side of that surface, namely, that side for which the entropy is greater. States for which the entropy is less than that in the initial state lie on the other side of the surface and are *adiabatically inaccessible* from the given state.

Carathéodory* took the property of *adiabatic inaccessibility* as the starting point of the formulation of the second law. The Carathéodory principle asserts that **in the immediate vicinity of every equilibrium state of a thermodynamic system, there are other states that cannot be reached from the given state by an adiabatic process.** Carathéodory was then able to show, by a lengthy mathematical argument, that if this is the case, an expression like Eq. (6–70), in three (or more) independent variables, necessarily *does* have an integrating denominator. The mathematics is not easy to follow and we shall not pursue the matter further.

Starting with the Carathéodory principle, one can deduce the existence of thermodynamic temperature and the entropy function. We have reversed the argument and, by starting with a statement regarding the quantities of heat absorbed and liberated in a Carnot cycle, together with the principle of increase of entropy, have shown that the Carathéodory principle is a necessary consequence.

PROBLEMS

6–1 Express $(\partial u/\partial P)_T$ in standard form by (a) the method used to obtain Eq. (6–9) and (b) the method devised by Bridgman. (c) Find $(\partial u/\partial P)_T$ for an ideal gas.

6–2 (a) Find the difference $c_P - c_v$ for mercury at a temperature of 0°C and a pressure of 1 atm taking the values of β and κ from Fig. 2–17. The density of mercury is 13.6×10^3 kg m^{-3} and the atomic weight is 200.6. (b) Determine the ratio $(c_P - c_v)/3R$.

6–3 The equation of state of a certain gas is $(P + b)v = RT$. (a) Find $c_P - c_v$. (b) Find the entropy change in an isothermal process. (c) Show that c_v is independent of v.

6–4 The energy equation of a substance is given by $u = aT^2v$, where a is a constant. (a) What information can be deduced about the entropy of the substance? (b) What are the limitations on the equation of state of the substance? (c) What other measurements must be made to determine the entropy and the equation of state?

6–5 The equation of state of a substance is given as $(P + b)v = RT$. What information can be deduced about the entropy, the internal energy, and the enthalpy of the substance? What other experimental measurement(s) must be made to determine all of the properties of the substance?

6–6 A substance has the properties that $(\partial u/\partial v)_T = 0$ and $(\partial h/\partial P)_T = 0$. (a) Show that the equation of state must be $T = APv$ where A is a constant. (b) What additional information is necessary to specify the entropy of the substance?

6–7 Express $(\partial h/\partial v)_T$ in standard form by (a) the method used to derive Eq. (6–21) and (b) by the method devised by Bridgman. (c) Find the value of $(\partial h/\partial v)_T$ for an ideal gas.

* Constantin Carathéodory, Greek mathematician (1873–1950).

6–8 Show that $T\left(\frac{\partial s}{\partial T}\right)_h = \frac{c_P}{1 - \beta T}$.

6–9 Show that $\left(\frac{\partial T}{\partial P}\right)_h - \left(\frac{\partial T}{\partial P}\right)_s = -\frac{v}{c_P}$.

6–10 Derive (a) Eq. (6–21), (b) Eq. (6–27), (c) Eq. (6–28), and (d) Eq. (6–29).

6–11 Derive Eq. (6–27) by the Bridgman method.

6–12 Derive Eq. (6–12), the relation for $c_P - c_v$, from the $T\,ds$ equations.

6–13 Show that the difference between the isothermal and adiabatic compressibilities is

$$\kappa - \kappa_S = \frac{T\beta^2 v}{c_P}.$$

6–14 Show that $(\partial h/\partial v)_s = \gamma/\kappa$.

6–15 Can the equation of state and c_P as a function of T be determined for a substance if $s(P, T)$ and $h(P, T)$ are known? If not, what additional information is needed?

6–16 Hill and Lounasmaa state that all the thermodynamic properties of liquid helium can be calculated in the temperature range 3 to 20 K and up to 100 atm pressure from their measurements of c_v, $(\partial P/\partial T)_v$ and P as a function of T for various densities of helium. (a) Show that they are correct by deriving expressions for u, s, and h in terms of the experimentally determined quantities. (b) Which of the measurements are not absolutely necessary in order to completely specify all the properties of He^4 in the temperature and pressure range given? Explain.

6–17 Use the data of Figs. 6–1(a) and 6–1(b) to calculate the change of entropy of 10^{-3} kilomoles of He^4 as its temperature and reduced density are changed from 6 K and 2.2 to 12 K and 2.6.

6–18 (a) Derive Eqs. (6–45) and (6–46). (b) Derive expressions for $h(T, v)$ and $h(P, v)$ for an ideal gas.

6–19 Assume that c_P for an ideal gas is given by $c_P = a + bT$, where a and b are constants. (a) What is the expression for c_v for this gas? (b) Use the expression for c_P in Eqs. (6–41) and (6–42) to obtain expressions for the specific entropy and enthalpy of an ideal gas in terms of the values in some reference state. (c) Derive an expression for the internal energy of an ideal gas.

6–20 One kilomole of an ideal gas undergoes a throttling process from a pressure of 4 atm to 1 atm. The initial temperature of the gas is 50°C. (a) How much work could have been done by the ideal gas had it undergone a reversible process to the same final state at constant temperature? (b) How much does the entropy of the universe increase as a result of the throttling process?

6–21 Show that the specific enthalpy of a van der Waals gas is given by $c_vT - 2a/v - RTv/(v - b) + \text{constant}$.

6–22 The pressure on a block of copper at a temperature of 0°C is increased isothermally and reversibly from 1 atm to 1000 atm. Assume that β, κ, and ρ are constant and equal respectively to $5 \times 10^{-5}\ \text{K}^{-1}$, $8 \times 10^{-12}\ \text{N}^{-1}\,\text{m}^2$, and $8.9 \times 10^3\ \text{kg m}^{-3}$. Calculate (a) the work done on the copper per kilogram, and (b) the heat evolved. (c) How do you account for the fact that the heat evolved is greater than the work done? (d) What would be the rise in temperature of the copper, if the compression were adiabatic rather than isothermal? Explain approximations made.

6–23 For a solid whose equation of state is given by Eq. (6–55) and for which c_P and c_v are independent of T, show that the specific internal energy and specific enthalpy are given by

$$u = c_v(T - T_0) + \left[\left(2\beta T_0 + \frac{v}{v_0} - 1\right)\frac{1}{2\kappa} - P_0\right](v - v_0) + u_0$$

and

$$h = c_P(T - T_0) + v_0(P - P_0)\left[1 - \beta T_0 - \frac{\kappa}{2}(P - P_0)\right] + h_0.$$

6–24 Figures 2–16, 2–17, 3–10 and 3–11 give data on copper and mercury. Are these data sufficient to determine all of the properties of copper and mercury between 500 and 1000 K? If so, determine expressions for the entropy and enthalpy. If not, specify the information needed.

6–25 The table below gives the volume of 1 g of water at a number of temperatures at a pressure of 1 atm.

t(°C)	V(cm^3)	t(°C)	V(cm^3)
0	1.00013	20	1.00177
2	1.00003	50	1.01207
4	1.00000	75	1.02576
6	1.00003	100	1.04343
10	1.00027		

Estimate as closely as you can the temperature change when the pressure on water in a hydraulic press is increased reversibly and adiabatically from a pressure of 1 atm to a pressure of 1000 atm, when the initial temperature is (a) 2°C, (b) 4°C, (c) 50°C. Make any reasonable assumptions or approximations, but state what they are.

6–26 The isothermal compressibility of water is 50×10^{-6} atm^{-1} and $c_P = 4.18 \times 10^3$ J kg^{-1} K^{-1}. Other properties of water are given in the previous problem. Calculate the work done as the pressure on 1 g of water in a hydraulic press is increased reversibly from 1 atm to 10,000 atm (a) isothermally, (b) adiabatically. (c) Calculate the heat evolved in the isothermal process.

6–27 Sketch a Carnot cycle in the h-s plane for (a) an ideal gas, (b) a van der Waals gas, (c) a solid. Make reasonable approximations but state what they are. (See Problem 6–21 and 6–23 for expressions for the specific enthalpy.)

6–28 Compute η and μ for a gas whose equation of state is given by (a) $P(v - b) = RT$ and (b) $(P + b)v = RT$, where b is a constant. Assume that c_v and c_P are constants.

6–29 Assuming that helium obeys the van der Waals equation of state, determine the change in temperature when one kilomole of helium gas undergoes a Joule expansion at 20 K to atmospheric pressure. The initial volume of the helium is 0.12 m^3. (See Tables 2–1 and 9–1 for data.) Describe approximations.

6–30 Carbon dioxide at an initial pressure of 100 atm and a temperature of 300 K undergoes an adiabatic free expansion in which the final volume is ten times the original volume.

Find the change in temperature and the increase in specific entropy, assuming that CO_2 is (a) an ideal gas, (b) a van der Waals gas. (Use Tables 2–1 and 9–1 and make any other assumptions that seem reasonable.)

6–31 Beginning with the van der Waals equation of state, derive Eqs. (6–59) and (6–60).

6–32 Assuming that helium is a van der Waals gas, calculate the pressure so that the inversion temperature of helium is 20 K. (See Table 6–1 for data.)

6–33 The helium gas of Problem 6–29 undergoes a throttling process. Calculate the Joule-Thomson coefficient at (a) 20 K and (b) 150 K. (c) For each process calculate the change of the temperature of the helium if the final pressure is 1 atm, assuming μ is independent of P and T.

6–34 Calculate the maximum inversion temperature of helium.

6–35 Show that if P and θ are chosen as independent variables, the relation between thermodynamic temperature T and empirical temperature θ on the scale of any gas thermometer is

$$\frac{dT}{T} = \frac{(\partial v/\partial \theta)_P}{v - (\partial h/\partial P)_\theta}\, d\theta.$$

6–36 (a) Show that on the empirical temperature scale θ of any gas thermometer,

$$\frac{dT}{T} = \frac{(\partial P/\partial \theta)_v}{P - \eta c_v}\, d\theta = \frac{(\partial v/\partial \theta)_P}{v + \mu c_P}\, d\theta,$$

where η and μ are, respectively, the Joule and Joule-Thomson coefficients of the gas. (b) Show also that

$$\frac{dT}{T} = \frac{(\partial P/\partial \theta)_v}{(c_P - c_v)(\partial \theta/\partial v)_P}\, d\theta.$$

6–37 For a paramagnetic substance, the specific work in a reversible process is $-\mathscr{H}\, dm$. (a) Consider the state of the substance to be defined by the magnetic moment per unit volume m and some empirical temperature θ. Show that

$$\frac{dT}{T} = \frac{(\partial \mathscr{H}/\partial \theta)_m}{\mathscr{H} - (\partial u/\partial m)_\theta}\, d\theta.$$

(b) It is found experimentally that over a range of variables which is not too great, the ratio $(\mathscr{H}/m)$ is constant at constant temperature. (This corresponds to the property of a "Boyles' law" gas that PV is constant at constant temperature.) Choose the ratio $(\mathscr{H}/m)$ as the thermometric property X, and define an empirical temperature θ in the usual way. Show that the thermodynamic temperature T is directly proportional to θ only if the internal energy u is independent of m at constant temperature.

6–38 (a) On a T-V-M diagram sketch two surfaces of constant entropy for an ideal gas obeying Curie's law. (b) Using the two surfaces of part (a) together with two isothermal surfaces, sketch two possible Carnot cycles for this system. (c) Derive the relation between M and V for processes which are both isothermal and isentropic. Sketch the process in the V-M plane.

6–39 On Fig. 6–4, the states a and b lie on a line of constant x_1 and x_2. (a) Show that both a and b cannot be reached by isentropic processes from the state i by proving that the cycle i-a-b-i violates the Kelvin-Planck statement of the second law.

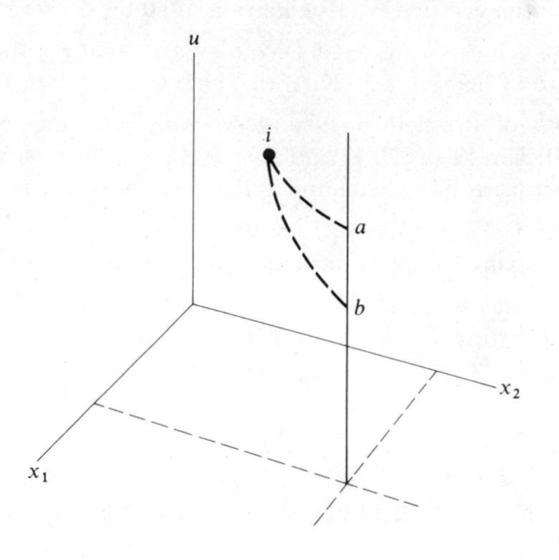

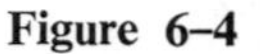
Figure 6–4

7

Thermodynamic potentials

7–1 THE HELMHOLTZ FUNCTION AND THE GIBBS FUNCTION

In addition to the internal energy and the entropy of a system, several other useful quantities can be defined that are combinations of these and the state variables. One such quantity, already introduced, is the enthalpy, H, defined for a PVT system as

$$H = U + PV. \tag{7-1}$$

There are two other important quantities, the Helmholtz* function F and the Gibbs† function G, which are now defined.

From the first law, when a system performs any process, reversible or irreversible, between two equilibrium states, the work W in the process is

$$W = (U_1 - U_2) + Q;$$

that is, the work is provided in part by the system, whose internal energy decreases by $(U_1 - U_2)$, and in part by the heat reservoirs with which the system is in contact and out of which there is a heat flow of magnitude Q.

We now derive expressions for the maximum amount of work that can be obtained when a system undergoes a process between two equilibrium states, for the special case in which the only heat flow is from a single reservoir at a temperature T and the initial and final states are at this same temperature. From the principle of the increase of entropy, the sum of the increase in entropy of the system, $(S_2 - S_1)$, and that of reservoir, ΔS_R, is equal to or greater than zero:

$$(S_2 - S_1) + \Delta S_R \geq 0.$$

The entropy change of the reservoir is

$$S_R = -\frac{Q}{T}.$$

Hence

$$(S_2 - S_1) - \frac{Q}{T} \geq 0.$$

and

$$T(S_2 - S_1) \geq Q.$$

Therefore from the first law,

$$W_T \leq (U_1 - U_2) - T(S_1 - S_2). \tag{7-2}$$

Let us define a property of the system called its *Helmholtz function* F, by the equation

$$F \equiv U - TS. \tag{7-3}$$

---

* Herman L. F. Helmholtz, German physicist (1821–1894).
† Josiah Willard Gibbs, American physicist (1839–1903).

Then for two equilibrium states at the same temperature T,

$$(F_1 - F_2) = (U_1 - U_2) - T(S_1 - S_2),$$

and from Eq. (7–2),

$$W_T \leq (F_1 - F_2). \tag{7–4}$$

That is, the decrease in the Helmholtz function of a system sets an upper limit to the work in *any* process between two equilibrium states at the same temperature, during which there is a heat flow into the system from a single reservoir at this temperature. If the process is *reversible*, the total entropy of system plus reservoir is constant, $Q = T(S_2 - S_1)$, and

$$W_T = (U_1 - U_2) - T(S_1 - S_2) = (F_1 - F_2). \tag{7–5}$$

The equality sign then holds in Eq. (7–4) and the work is a *maximum*. If the process is irreversible, the work is less than this maximum.

Because its decrease equals the maximum energy that can be "freed" in a process and made available for work, the quantity F is sometimes called the *free energy* of a system. However, since the same term is also applied to another property to be defined shortly, we shall use the term "Helmholtz function" to avoid confusion. Note, however, that although the decrease in the Helmholtz function of a system equals the maximum work that can be obtained under the conditions above, the energy converted to work is provided only in part by the system, the remainder coming from heat withdrawn from a heat reservoir.

Equation (7–2) is perfectly general and applies to a system of any nature. The process may be a change of state, or a change of phase, or a chemical reaction. In general, the work in a differential process will be given by $P\,dV$ plus a sum of terms such as $-\mathscr{E}\,dZ$ or $-\mathscr{H}\,dM$, but for simplicity we assume only one additional term which will be represented by $Y\,dX$. The total work in any finite process is then the sum of the "$P\,dV$" work and the "$Y\,dX$" work. Let us now represent the former by W' and the latter by A. The work in any process is then $W' + A$ and Eq. (7–4) becomes

$$W'_T + A_T \leq (F_1 - F_2). \tag{7–6}$$

In a process at constant volume, the "$P\,dV$" work $W' = 0$ and in such a process,

$$A_{T,V} \leq (F_1 - F_2). \tag{7–7}$$

The decrease in the Helmholtz function therefore sets an upper limit to the "non-$P\,dV$" work in a process at constant *temperature* and *volume*. If the process is reversible, this work equals the decrease in the Helmholtz function. If both V and X are constant, then $A = 0$ and

$$0 \leq (F_1 - F_2)$$

or

$$F_2 \leq F_1. \tag{7–8}$$

That is, in a process at constant volume for which $A = 0$ and T is constant, the Helmholtz function can only decrease or, in the limit, remain constant. Conversely, such a process is possible only if $F_2 \leq F_1$.

Consider next a process at a constant external pressure P. The work W' in such a process is $P(V_2 - V_1)$, and from Eq. (7–6),

$$A_{T,P} \leq (F_1 - F_2) + P(V_1 - V_2)$$

$$A_{T,P} \leq (U_1 - U_2) - T(S_1 - S_2) + P(V_1 - V_2).$$

Let us define a function G called the *Gibbs function* by the equation

$$G \equiv F + PV = H - TS = U - TS + PV. \tag{7–9}$$

Then for two states at the same temperature T and pressure P,

$$G_1 - G_2 = (U_1 - U_2) - T(S_1 - S_2) + P(V_1 - V_2),$$

and

$$A_{T,P} \leq (G_1 - G_2). \tag{7–10}$$

The decrease in the Gibbs function therefore sets an upper limit to the "non-$P\,dV$" work in any process between two equilibrium states at the same *temperature* and *pressure*. If the process is reversible, this work is equal to the decrease in the Gibbs function. Because its decrease in such a process equals the maximum energy that can be "freed" and made available for "non-$P\,dV$" work, the Gibbs function has also been called the *free energy* of a system, but as stated earlier, we shall use the term "Gibbs function" to avoid confusion with the Helmholtz function.

If the variable X is constant in a process, or if the only work is "$P\,dV$" work, then $A = 0$ and

$$G_2 \leq G_1. \tag{7–11}$$

That is, in such a process the Gibbs function either remains constant or decreases. Conversely, such a process is possible only if G_2 is equal to or less than G_1.

In Sections 6–7 and 6–8, we derived expressions for the specific enthalpy and entropy of an ideal gas and of a van der Waals gas. Making use of Eqs. (6–41) and (6–42), the specific Gibbs function $g = u - Ts + Pv = h - Ts$ for an ideal gas, selecting T and P as independent variables, is

$$g = \int_{T_0}^{T} c_P\,dT - T\int_{T_0}^{T} c_P \frac{dT}{T} + RT \ln \frac{P}{P_0} + h_0 - s_0 T. \tag{7–12}$$

If c_P can be considered constant,

$$g = c_P(T - T_0) - c_P T \ln \frac{T}{T_0} + RT \ln \frac{P}{P_0} - s_0(T - T_0) + g_0, \tag{7–13}$$

which can be written more compactly as

$$g = RT(\ln P + \phi), \tag{7–14}$$

where

$$RT\phi = c_P(T - T_0) - c_P T \ln \frac{T}{T_0} - RT \ln P_0 - s_0(T - T_0) + g_0. \tag{7–15}$$

Note that ϕ is a function of T only.

We see that while s, u, and h are indeterminate to within arbitrary *constants* s_0, u_0, and h_0, the Gibbs function is indeterminate to within an arbitrary *linear function of the temperature*, $h_0 - s_0 T$.

It is left as a problem to show that the specific Helmholtz function $f = u - Ts$ for an ideal gas, selecting T and v as the independent variables, is

$$f = c_v(T - T_0) - c_v T \ln \frac{T}{T_0} - RT \ln \frac{v}{v_0} - s_0(T - T_0) + f_0. \tag{7–16}$$

For a van der Waals gas

$$f = c_v(T - T_0) - c_v T \ln \frac{T}{T_0} - a\left(\frac{1}{v} - \frac{1}{v_0}\right) - RT \ln\left(\frac{v - b}{v_0 - b}\right) - s_0(T - T_0) + f_0 \tag{7–17}$$

which is seen to reduce to the ideal gas expression when $a = b = 0$.

7–2 THERMODYNAMIC POTENTIALS

The differences between the values of the Helmholtz and Gibbs functions in two neighboring equilibrium states of a *closed** PVT system are

$$dF = dU - T\,dS - S\,dT, \tag{7–18}$$

$$dG = dU - T\,dS - S\,dT + P\,dV + V\,dP. \tag{7–19}$$

Since

$$dU = T\,dS - P\,dV, \tag{7–20}$$

we can eliminate dU between Eqs. (7–18) and (7–19), obtaining

$$dF = -S\,dT - P\,dV, \tag{7–21}$$

$$dG = -S\,dT + V\,dP. \tag{7–22}$$

Also, from the definition of enthalpy,

$$dH = T\,dS + V\,dP. \tag{7–23}$$

* No *matter* crosses the boundary of a closed system.

The coefficients of the differentials on the right sides of the four preceding equations can be identified with the partial derivatives of the variable on the left side. For example, considering U as a function of S and V, we have

$$dU = \left(\frac{\partial U}{\partial S}\right)_V dS + \left(\frac{\partial U}{\partial V}\right)_S dV. \tag{7–24}$$

Comparison with Eq. (7–20) shows that $(\partial U/\partial S)_V = T$ and $(\partial U/\partial V)_S = -P$. Similar relations can be written for dF, dG, and dH. It follows that

$$\left(\frac{\partial U}{\partial S}\right)_V = T, \qquad \left(\frac{\partial U}{\partial V}\right)_S = -P, \tag{7–25}$$

$$\left(\frac{\partial F}{\partial T}\right)_V = -S, \qquad \left(\frac{\partial F}{\partial V}\right)_T = -P, \tag{7–26}$$

$$\left(\frac{\partial G}{\partial T}\right)_P = -S, \qquad \left(\frac{\partial G}{\partial P}\right)_T = V, \tag{7–27}$$

$$\left(\frac{\partial H}{\partial S}\right)_P = T, \qquad \left(\frac{\partial H}{\partial P}\right)_S = V. \tag{7–28}$$

It will be recalled that the intensity $\mathbf{E}$ of an electrostatic field is, at every point, equal to the negative of the gradient of the potential ϕ at that point. Thus the components of $\mathbf{E}$ are

$$E_x = -\left(\frac{\partial \phi}{\partial x}\right), \qquad E_y = -\left(\frac{\partial \phi}{\partial y}\right), \qquad E_z = -\left(\frac{\partial \phi}{\partial z}\right).$$

Because the properties P, V, T, and S can be expressed in a similar way in terms of the partial derivatives of U, F, G, and H, these quantities can be described as *thermodynamic potentials*, although the term is more commonly applied to F and G only. But to avoid confusion as to which of these is meant by the term "thermodynamic potential," we shall refer to F simply as the Helmholtz function, and to G as the Gibbs function.

Although there are mnemonic aids to remembering Eqs. (7–20) to (7–23), there is a certain useful symmetry to these equations which can also be used to remember them. The differential of each thermodynamic potential is expressed in terms of the differentials of the "characteristic variables" for that potential; S and V for the potential U; T and V for the potential F; T and P for the potential G; and S and P for the potential H. Furthermore, dS and dP always appear with the plus sign and dT and dV always appear with the minus sign. Also, each term in the expressions for the differentials must have the dimensions of energy.

It was pointed out earlier that the properties of a substance are not completely specified by its equation of state alone, but that in addition we must know the energy equation of the substance. Suppose, however, that the expression for any thermodynamic potential is known in terms of its characteristic variables. That is, suppose

U is known as a function of S and V, or F is known as a function of T and V, or G is known as a function of T and P, or that H is known as a function of S and P. If so, then *all* thermodynamic properties can be obtained by differentiation of the thermodynamic potential, and the equation for the thermodynamic potential in terms of its characteristic variables is known as the *characteristic equation* of the substance.

For example, suppose that the Helmholtz function F is known as a function of T and V. Then from the second of Eqs. (7–26) we can calculate P as a function of V and T, which is the equation of state of the substance. The entropy S can be found from the first of these equations, and from the definition of F we then have the energy equation. Thus

$$P = -\left(\frac{\partial F}{\partial V}\right)_T,$$

$$S = -\left(\frac{\partial F}{\partial T}\right)_V,$$

$$U = F + TS = F - T\left(\frac{\partial F}{\partial T}\right)_V. \tag{7–29}$$

In the same way, if G is known as a function of T and P, then

$$V = \left(\frac{\partial G}{\partial P}\right)_T,$$

$$S = -\left(\frac{\partial G}{\partial T}\right)_P,$$

$$H = G + TS = G - T\left(\frac{\partial G}{\partial T}\right)_P. \tag{7–30}$$

Equations (7–29) and (7–30) are known as the *Gibbs-Helmholtz* equations.

All of the preceding equations can be written for systems other than PVT systems. Suppose, for example, that the system is a wire in tension for which the work in a differential reversible process is $-\mathscr{F}\,dL$. Then considering the Helmholtz function $F = U - TS$ as a function of T and L, we would have

$$\left(\frac{\partial F}{\partial L}\right)_T = \mathscr{F}.$$

The Gibbs function for the wire is defined as

$$G = U - TS - \mathscr{F}L,$$

where the product $\mathscr{F}L$ is preceded by a minus sign because the work dW equals $-\mathscr{F}\,dL$. Then

$$\left(\frac{\partial G}{\partial \mathscr{F}}\right)_T = -L.$$

The preceding equations are the analogues of the second of Eqs. (7–26) and (7–27).

We now consider a multivariable closed system, but limit the discussion to one whose state can be described by its temperature T, two extensive variables X_1 and X_2, and the corresponding intensive variables Y_1 and Y_2. The work in a differential reversible process is

$$d'W = Y_1\,dX_1 + Y_2\,dX_2,$$

and the combined first and second laws take the form

$$dU = T\,dS - Y_1\,dX_1 - Y_2\,dX_2. \tag{7-31}$$

Because the system has two equations of state, the equilibrium state of the system can be considered a function of T, and either of the two extensive variables X_1 and X_2, or the two intensive variables Y_1 and Y_2, or one extensive variable X_1 and the other intensive variable Y_2. We could equally well let Y_1 and X_2 represent these variables.

We first consider the state of the system to be expressed as a function of T, X_1, and X_2. The Helmholtz function F is defined, as for a system described by two independent variables, as

$$F = U - TS,$$

so that

$$dF = dU - T\,dS - S\,dT,$$

and eliminating dU between this equation and Eq. (7-31), we have

$$dF = -S\,dT - Y_1\,dX_1 - Y_2\,dX_2.$$

The coefficient of each differential on the right side of this equation is the corresponding partial derivative of F, with the other variables held constant. Thus

$$\left(\frac{\partial F}{\partial T}\right)_{X_1,X_2} = -S, \quad \left(\frac{\partial F}{\partial X_1}\right)_{T,X_2} = -Y_1, \quad \left(\frac{\partial F}{\partial X_2}\right)_{T,X_1} = -Y_2. \tag{7-32}$$

The Gibbs function of the system is defined as

$$G = U - TS + Y_1X_1 + Y_2X_2.$$

When the expression for dG is written out, and dU eliminated, making use of Eq. (7-31), the result is

$$dG = -S\,dT + X_1\,dY_1 + X_2\,dY_2.$$

It follows that

$$\left(\frac{\partial G}{\partial T}\right)_{Y_1,Y_2} = -S, \quad \left(\frac{\partial G}{\partial Y_1}\right)_{T,Y_2} = X_1, \quad \left(\frac{\partial G}{\partial Y_2}\right)_{T,Y_1} = X_2. \tag{7-33}$$

In the special case in which Y_2 is the intensity of a conservative force field (gravitational, electric, or magnetic), the system has a potential energy $E_p = Y_2X_2$, and its total energy E is

$$E = U + E_p = U + Y_2X_2.$$

We then define a new function F^* as

$$F^* \equiv E - TS = U - TS + Y_2X_2. \tag{7-34}$$

The function $F^* = E - TS$ can be considered a generalized Helmholtz function, corresponding to $F = U - TS$ for a system whose total energy equals its internal energy only. Proceeding in the same way as before, we find that

$$\left(\frac{\partial F^*}{\partial T}\right)_{X_1, Y_2} = -S; \quad \left(\frac{\partial F^*}{\partial X_1}\right)_{T, Y_2} = -Y_1; \quad \left(\frac{\partial F^*}{\partial Y_2}\right)_{T, X_1} = +X_2. \tag{7-35}$$

It is left as a problem to show that if X_1 and X_2 are selected as variables, we have the generalized Gibbs-Helmholtz equation,

$$U = F - T\left(\frac{\partial F}{\partial T}\right)_{X_1, X_2}. \tag{7-36}$$

The enthalpy H is defined as

$$H = U + Y_1 X_1 + Y_2 X_2,$$

and we find that

$$H = G - T\left(\frac{\partial G}{\partial T}\right)_{Y_1, Y_2}. \tag{7-37}$$

If Y is the intensity of a conservative force field,

$$E = F^* - T\left(\frac{\partial F^*}{\partial T}\right)_{X_1, Y_2}. \tag{7-38}$$

From the purely thermodynamic viewpoint, we are at liberty to consider either X_1 and X_2, Y_1 and Y_2, or X_1 and Y_2 as independent, in addition to T. We shall show later that the methods of statistics lead directly to expressions for F, G, or F^*, in terms of the parameters that determine the energy of the system. All other thermodynamic properties can be calculated when any one of these is known.

7-3 THE MAXWELL RELATIONS

A set of equations called the Maxwell† relations can be derived from the fact that the differentials of the thermodynamic potentials are exact. In Section 2-10 it was pointed out that if

$$dz = M(x, y)\, dx + N(x, y)\, dy,$$

dz is exact when

$$\left(\frac{\partial M}{\partial y}\right)_x = \left(\frac{\partial N}{\partial x}\right)_y. \tag{7-39}$$

† James Clerk Maxwell, Scottish physicist (1831–1879).

Applying Eq. (7–39) to Eq. (7–20) through (7–23) we have

$$\left(\frac{\partial T}{\partial V}\right)_S = -\left(\frac{\partial P}{\partial S}\right)_V, \tag{7–40}$$

$$\left(\frac{\partial S}{\partial V}\right)_T = \left(\frac{\partial P}{\partial T}\right)_V, \tag{7–41}$$

$$\left(\frac{\partial S}{\partial P}\right)_T = -\left(\frac{\partial V}{\partial T}\right)_P, \tag{7–42}$$

and

$$\left(\frac{\partial T}{\partial P}\right)_S = \left(\frac{\partial V}{\partial S}\right)_P. \tag{7–43}$$

These equations are useful because they provide expressions for the entropy change in terms of P, V, and T, and they are called the *Maxwell relations*. These equations can also be derived from the fact that the mixed partial derivatives of U, F, G, and H are independent of the order of differentiation.

Note that in each of the Maxwell relations the cross product of the differentials has the dimensions of energy. The independent variable in the denominator of one side of an equation is the constant on the other side. The sign can be argued from considering the physics of the process for a simple case. As an example, consider Eq. (7–41). During an isothermal expansion of an ideal gas, heat must be added to the gas to keep the temperature constant. Thus the right side of Eq. (7–41) has a value greater than zero. At constant volume, increasing the temperature of an ideal gas increases the pressure, and the left side of Eq. (7–41) also has a value greater than zero.

Maxwell relations can also be written for systems having equations of state which depend on thermodynamic properties other than P and V.

7–4 STABLE AND UNSTABLE EQUILIBRIUM

Thus far, it has been presumed that the "equilibrium state" of a system implies a state of *stable* equilibrium. In some circumstances, a system can persist for a long period of time in a state of *metastable* equilibrium, but eventually the system transforms spontaneously to a stable state. We now consider the necessary condition that a state shall be one of stable equilibrium.

Our earlier definitions of the properties of a substance were restricted to states of stable equilibrium only, and according to these definitions it is meaningless to speak of the entropy, Gibbs function etc., of a system in a metastable state. However, since a substance can remain in a metastable state for a long period of time, its directly measurable properties, such as pressure and temperature, can be determined in the same way as for a system in a completely stable state. We simply assume that the entropy, Gibbs function, etc., are related to the directly measurable properties in the same way as they are in a stable equilibrium state. The assumption is justified by the correctness of the conclusions drawn from it.

Figure 7–1 is a schematic diagram of the *P-V-T* surface representing the states of *stable* equilibrium of a pure substance. Suppose the substance is originally in the vapor phase at point *a* and the temperature is decreased at constant pressure. In the absence of condensation nuclei such as dust particles or ions, the temperature can be reduced considerably below that at point *b*, where the isobaric line intersects the saturation line, without the appearance of the liquid phase. The state of the vapor is then represented by point *c*, which lies *above* the *P-V-T* surface. If no condensation nuclei are present, it will remain in this state for a long period of time and is in metastable equilibrium. It is in mechanical and thermal equilibrium, but not in complete thermodynamic equilibrium. If a condensation nucleus is introduced, and if pressure and temperature are kept constant, the vapor transforms spontaneously to the liquid phase at point *f*. The vapor at point *c* is said to be *supercooled*.

A supercooled vapor can also be produced by the adiabatic expansion of a saturated vapor. In such a process, the volume increases and the pressure and temperature both decrease. If no condensation nuclei are present, the state of the

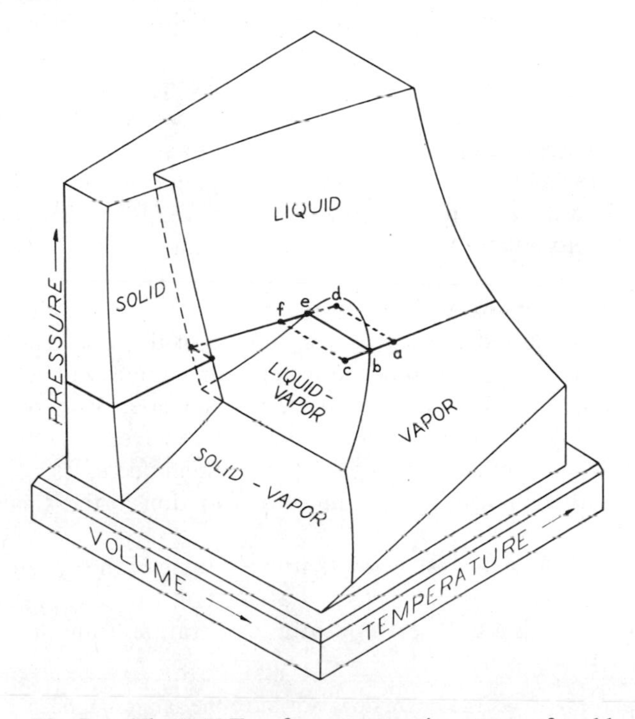

Fig. 7–1 The *P-V-T* surface representing states of stable equilibrium for a pure substance.

vapor again lies at some point above the equilibrium surface. This is the method used to obtain a supercooled vapor in the Wilson cloud chamber. When an ionizing particle passes through the chamber, the ions it forms serve as condensation nuclei and liquid droplets are formed along its path.

The temperature of a *liquid* can also be reduced below that at which it is in stable equilibrium with the solid, and the liquid is also described as *supercooled.* Thus if a molten metal in a crucible is cooled slowly, it may remain in the liquid phase at temperatures well below the normal freezing point. The converse does not seem to happen—as the temperature of a solid is increased, it starts to melt promptly at the normal melting point.

If the substance is originally in the liquid phase at point f in Fig. 7–1, and if the temperature is increased at constant pressure, the vapor phase may not form when point e is reached, and the liquid may be carried to the state represented by point d, which lies *below* the equilibrium surface. This is also a metastable state, and the liquid is said to be superheated.* A slight disturbance will initiate a spontaneous vaporization process, and if pressure and temperature are kept constant the system transforms to the vapor phase at point a.

In the *bubble chamber*, a superheated liquid (usually liquid hydrogen) is produced by an adiabatic reduction of pressure on a saturated liquid. Small bubbles of vapor are then formed on ions produced by an ionizing particle passing through the chamber.

We now consider the specific conditions that determine which of two possible states of a system is the stable state. If a system is completely isolated from its surroundings, a spontaneous process from one state to another can take place only if the entropy of the system increases, that is, if the entropy $(S_U)_2$ in the second state is greater than the entropy $(S_U)_1$ in the first state. The final state of stable equilibrium is therefore that in which the entropy is larger, that is, when $(S_U)_2 > (S_U)_1$.

Very often, however, we wish to compare two states of a system that is *not* completely isolated. Suppose first that the volume of the system is constant, so that the work in a process is zero, but the system is in contact with a heat reservoir at a temperature T, and we wish to compare two states at this temperature. By Eq. (7–8), under these conditions, a spontaneous process from one state to another can occur only if the Helmholtz function for the system decreases. The final state of equilibrium is that in which the Helmholtz function is the smaller, that is, $(F_{T,V})_2 < (F_{T,V})_1$.

Finally, let us remove the restriction that the volume of the system is constant, but assume that the system is subjected to a constant external pressure P. The system is in contact with a heat reservoir at a temperature T and its pressure is P

* The term "superheated" as used here does not have the same significance as when one speaks of "superheated steam" in a reciprocating steam engine or turbine. See Section 8–9.

in the initial and final states of a process. By Eq. (7-11) a spontaneous process can only occur under these conditions if the Gibbs function decreases. The state of stable equilibrium is that in which the Gibbs function is smaller, that is, $(G_{T,P})_2 < (G_{T,P})_1$.

As a corollary of the preceding conclusions, if a completely isolated system can exist in more than one state of stable equilibrium, the entropy S must be the same in all such states. If a system at constant volume and in contact with a single heat reservoir can exist in more than one state of stable equilibrium, the Helmholtz function F must be the same in all such states; and if a system, in contact with a single heat reservoir and in surroundings at constant pressure, can exist in more than one stable state, the Gibbs function G must be the same in all such states.

The preceding discussion referred to a system whose initial state was a *metastable* one. But we assumed it possible to assign values to the entropy, Helmholtz function, and so on to this state, even though strictly speaking these properties are defined only for states of *stable* equilibrium. From the definition of a state of stable equilibrium as one in which the properties of a system do not change with time, it is evident that no *spontaneous* process can take place from an initial state of stable equilibrium. Such processes can occur, however, if some of the *constraints* imposed on a system are changed. For example, suppose a system enclosed in a rigid adiabatic boundary consists of two parts at different temperatures, separated by an adiabatic wall. Each of the parts will come to a state of stable equilibrium, but they will be at different temperatures. The adiabatic wall separating them then constitutes a constraint that prevents the temperatures from equalizing.

As a second example, suppose that a system in contact with a reservoir at a temperature T is divided internally by a partition. Each portion of the system contains a gas, but the pressures on opposite sides of the partition are different. Both gases are in a state of equilibrium, and the partition constitutes a constraint that prevents the pressures from equalizing.

As a third example, suppose that on opposite sides of the partition in the preceding case there are two *different* gases, both at the same pressure. If the partition is removed, each gas will diffuse into the other until a homogeneous mixture results, and the partition constitutes a constraint that prevents this from happening.

If now the adiabatic wall in the first example is removed, or if the partition in the next two examples is removed, the state immediately following the removal of the constraint is no longer one of stable equilibrium, and a spontaneous process will take place until the system settles down to a new state of stable equilibrium. *During* the process, while the temperature, pressure, or composition of the gas mixture is not uniform, the system is in a nonequilibrium state. The entropy, Helmholtz function, etc., are undefined and no definite values can be assigned to them. However, if we compare the initial state of stable equilibrium, *before* the removal of the constraint, with the final equilibrium state *after* its removal, all of the results derived earlier in the section will apply. Thus in the first example, in which the system is completely isolated, the final entropy is greater than the initial entropy. In the second example, if the volume of the system is kept constant, the final value of the Helmholtz function is smaller than its initial value. In the third example, if the pressure is kept constant, the final value of the Gibbs function is less than its initial value.

7-5 PHASE TRANSITIONS

Suppose we have a system consisting of the liquid and vapor phases of a substance in equilibrium at a pressure P and a temperature T. In Fig. 7-2(a), the total specific volume of the system is v_1. The number of moles in the liquid phase is n''_1 and the number of moles in the vapor phase is n'''_1. The state of the system corresponds to point b_1 in Fig. 7-2(c). In Fig. 7-2(b), the total specific volume of the system is v_2, and the numbers of moles in the liquid and vapor phases are respectively n''_2 and n'''_2. The state of the system corresponds to point b_2 in Fig. 7-2(c).

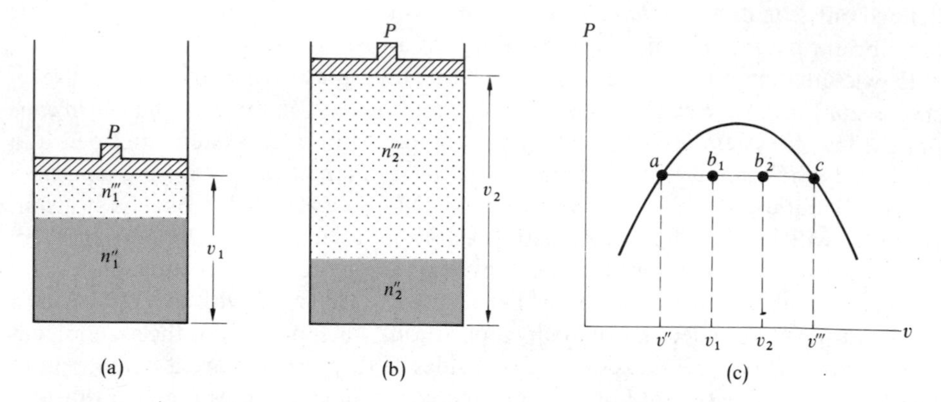

Fig. 7-2 The equilibrium between a liquid and its vapor at the two different molal volumes shown in (a) and (b) is represented on the portion of the P-v diagram in (c).

The states of the liquid and vapor portions of the system shown in Figs. 7-2(a) and 7-2(b) are represented in Fig. 7-2(c) by points a and c respectively, and the states differ only in the relative numbers of moles of liquid and vapor. If g'' and g''' are the specific Gibbs functions of the liquid and vapor phases, the Gibbs functions of the two states are, respectively,

$$G_1 = n''_1 g'' + n'''_1 g''',$$

$$G_2 = n''_2 g'' + n'''_2 g'''.$$

Since the total number of moles of the system is constant,

$$n''_1 + n'''_1 = n''_2 + n'''_2;$$

and since both states are stable,

$$G_1 = G_2.$$

It follows from these equations that

$$g'' = g''';\tag{7–44}$$

that is, *the specific Gibbs function has the same value in both phases.* The same result holds for any two phases in equilibrium. At the triple point, the specific Gibbs functions of all three phases are equal.

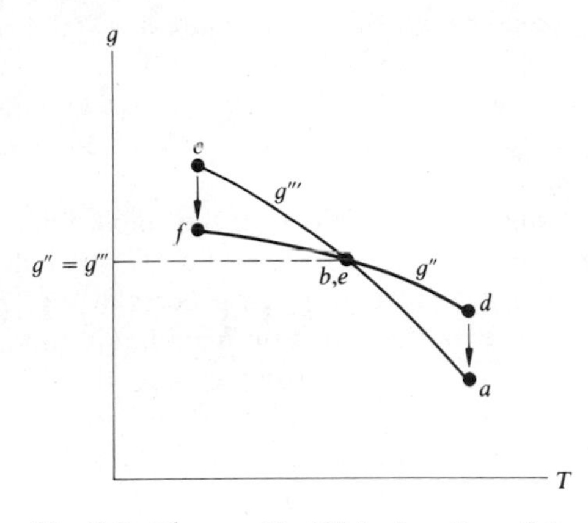

Fig. 7–3 The specific Gibbs function of the vapor and liquid in processes *a-b-c* and *d-e-f* of Fig. 7–1.

Let us now return to a consideration of the stable and metastable states illustrated in Fig. 7–1. Figure 7–3, which is lettered to correspond to Fig. 7–1, shows graphs of the specific Gibbs functions of the vapor and liquid in the processes *a-b-c* and *d-e-f* of Fig. 7–1. Since

$$\left(\frac{\partial g'''}{\partial T}\right)_P = -s''',$$

where s''' is the specific entropy of the vapor phase, the curve *abc* has a negative slope, of magnitude equal to the specific entropy s'''. Similarly, the curve *def* also has a negative slope, equal to the specific entropy s'' of the liquid. The difference between the entropies s''' and s'' equals the latent heat of transformation, l_{23}, divided by the temperature T:

$$s''' - s'' = \frac{l_{23}}{T}.$$

Since l_{23} is positive, $s''' > s''$ and the magnitude of the slope of the curve *abc* is greater than that of the curve *def*. The curves intersect at point b, e where $g'' = g'''$.

Points c and f represent two possible states of the system at the same temperature and pressure, but the Gibbs function in state c is greater than that in state f. We have shown that in a spontaneous process between two states at the same temperature and pressure, the Gibbs function must decrease. Hence a spontaneous transition from state c to state f is possible, while one from state f to state c is not. State f is therefore the state of *stable* equilibrium, while the equilibrium at state c is metastable.

Similarly, states d and a are at the same temperature and pressure, but the Gibbs function at d is greater than that at a. State a is stable and state d is metastable.

At points b and e, where the Gibbs functions are equal, the equilibrium is *neutral.* At this temperature and pressure the substance can exist indefinitely, in either phase, or in both.

If the substance in Fig. 7–1 is taken from the stable liquid state at point f to the stable vapor state at point a, in the process f-e-b-a which does not carry it into a metastable state, the curve representing the process in Fig. 7–3 consists only of the segments fe and ba. The phase transition from liquid to vapor, in the process e-b, is called a *first-order* transition because although the specific Gibbs function is itself continuous across the transition, its first derivative, equal to $-s''$ or $-s'''$ and represented by the slopes of the curves fe and ba, is discontinuous.

In principle there should also be phase transitions in which both the specific Gibbs function and its first derivative are continuous, but the second derivative changes discontinuously. In such transitions the latent heat of transformation is zero and the specific volume does not change for PvT systems. But, since

$$\left(\frac{\partial^2 g}{\partial T}\right)_P = -\left(\frac{\partial s}{\partial T}\right)_P = -\frac{c_P}{T}, \tag{7–45}$$

the value of c_P must be different in the two phases. Examples of such transitions would be the liquid-vapor transition at the critical point, the transition of a superconductor from the superconducting to the normal state in zero magnetic field, ferromagnetic to paramagnetic transitions in a simple model, order-disorder transformations, etc. Very careful experiments have been done on many systems, some to within one-millionth of a degree of the phase transition. It appears that the superconducting transition may be the only true second-order transition.

An example of a third type of transition, known as a *lambda-transition*, is that between the two liquid phases of He^4, ordinary liquid helium He I, and superfluid helium He II. This transition can take place at any point along the line separating these two liquid phases in Fig. 2–13. A graph of c_P versus T for the two phases has the general shape shown in Fig. 7–4, and the transition takes its name from the resemblance of this curve to the shape of the Greek letter λ. The value of c_P does uot change discontinuously, but its variation with temperature is different in the two phases.

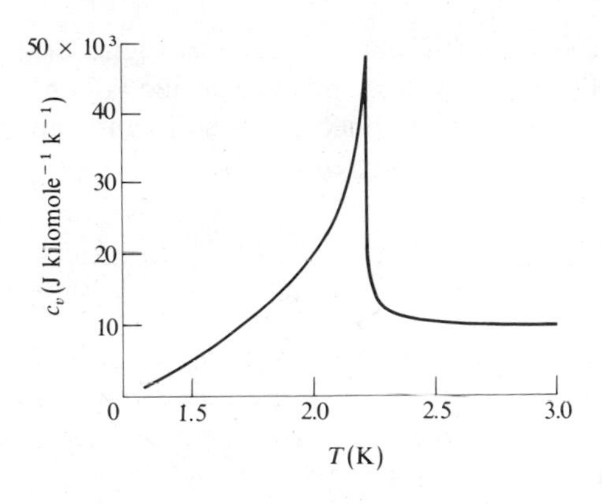

Fig. 7–4 The lambda transition for liquid He^4.

7–6 THE CLAUSIUS-CLAPEYRON EQUATION

The Clausius-Clapeyron* equation is an important relation describing how the pressure varies with temperature for a system consisting of two phases in equilibrium. Suppose a liquid and its vapor are in equilibrium at a pressure P and a temperature T, so that under these conditions $g'' = g'''$. At a temperature $T + dT$, the vapor pressure is $P + dP$ and the Gibbs functions are respectively $g'' + dg''$ and $g''' + dg'''$. But since the liquid and vapor are in equilibrium at the new temperature and pressure, it follows that the changes dg'' and dg''' are equal.

We have shown that

$$dg = -s\,dT + v\,dP.$$

The changes in temperature and pressure are the same for both phases, so

$$-s''\,dT + v''\,dP = -s'''\,dT + v'''\,dP,$$

or

$$(s''' - s'')\,dT = (v''' - v'')\,dP.$$

But the difference in specific entropies, $(s''' - s'')$, equals the heat of vaporization l_{23} divided by the temperature T, and hence

$$\left(\frac{\partial P}{\partial T}\right)_{23} = \frac{l_{23}}{T(v''' - v'')}, \tag{7–46}$$

which is the Clausius-Clapeyron equation for liquid-vapor equilibrium. Geometrically speaking, it expresses the *slope* of the equilibrium line between the

* Benoit-Pierre-Emile Clapeyron, French chemist (1799–1864).

liquid and vapor phases in a P-T diagram such as Fig. 2–8(a), in terms of the heat of transformation, the temperature, and the specific volumes of the phases.

When the same reasoning is applied to the solid and vapor, or solid and liquid phases, we obtain the corresponding equations

$$\left(\frac{dP}{dT}\right)_{13} = \frac{l_{13}}{T(v''' - v')}, \qquad \left(\frac{dP}{dT}\right)_{12} = \frac{l_{12}}{T(v'' - v')}. \tag{7–47}$$

Although the latent heat of any transformation varies with temperature, it is always positive (except for He^3 below 0.3 K), as is the temperature T. Also, the specific volume of the vapor phase is always greater than that of either the liquid or solid phase and the quantities $(v''' - v'')$ and $(v''' - v')$ are always positive. The slopes of the vapor pressure curves and sublimation pressure curves are therefore always positive. The specific volume of the solid phase, however, may be greater or less than that of the liquid phase, and so the slope of the solid-liquid equilibrium line may be either positive or negative. We can now understand more fully why the P-v-T surface for a substance like water, which expands on freezing, differs from that for a substance which contracts on freezing. (See Figs. 2–6 and 2–7). The term $(v'' - v')$ is negative for a substance that expands on freezing and is positive for a substance that contracts on freezing. Therefore the solid-liquid equilibrium surface, or its projection as a line in the P-T plane, slopes upward to the left for a substance like water that expands and upward to the right for a substance that contracts. Projections of the liquid-vapor and solid-vapor surfaces always have positive slopes.

An examination of Fig. 2–10 will show that Ice I (ordinary ice) is the only form of the solid phase with a specific volume greater than that of the liquid phase. Hence the equilibrium line between Ice I and liquid water is the only one that slopes upward to the left in a P-T diagram; all others slope upward to the right.

For changes in temperature and pressure that are not too great, the heats of transformation and the specific volumes can be considered constant, and the slope of an equilibrium line can be approximated by the ratio of the finite pressure and temperature changes, $\Delta P/\Delta T$. Thus the latent heat at any temperature can be found approximately from measurements of equilibrium pressures at two nearby temperatures, if the corresponding specific volumes are known. Conversely, if the equilibrium pressure and the latent heat are known at any one temperature, the pressure at a nearby temperature can be calculated. In calculations of this sort we usually assume that the vapor behaves like an ideal gas.

To integrate the Clausius-Clapeyron equation and obtain an expression for the pressure itself as a function of temperature, the heats of transformation and the specific volumes must be known as functions of temperature. This is an important problem in physical chemistry but we shall not pursue it further here except to mention that if variations in latent heat can be neglected, and if one of the phases is a vapor, and if the vapor is assumed to be an ideal gas, and if the specific volume of the liquid or solid is neglected in comparison with that of the vapor, the integration can be readily carried out. The resulting expression is

$$\left(\frac{dP}{dT}\right)_{23} = \frac{l_{23}}{T(RT/P)},$$

$$\frac{dP}{P} = \frac{l_{23}}{R}\frac{dT}{T^2},$$

$$\ln P = -\frac{l_{23}}{RT} + \ln \text{constant}. \tag{7–48}$$

The Clausius-Clapeyron equation can also be used to explain why the triple-point temperature of water, $T_3 = 273.16$ K, should be higher than the ice-point temperature $T_i = 273.15$ K. This appears puzzling at first, since at both temperatures ice and water are in equilibrium.

The triple-point temperature T_3 is defined as the temperature at which water vapor, liquid water, and ice are in equilibrium. At this temperature, the vapor pressure of water equals the sublimation pressure of ice and the pressure of the system equals this pressure, P_3, which has a value of 4.58 Torr. Water at its triple point is represented in Fig. 2–9(a).

The *ice point* is defined as the temperature at which pure ice and air-saturated water are in equilibrium under a total pressure of 1 atm. There is air in the space above the solid and liquid, as well as water vapor, and air is also dissolved in the water. The total pressure P is 1 atm and by definition the temperature is the ice-point temperature T_i. Thus the triple-point temperature and the ice-point temperature differ for two reasons; one is that the total pressure is different, and the other is that, at the ice point, the liquid phase is not pure water.

Let us first neglect any effect of the dissolved air and find the equilibrium temperature of ice and pure water when the pressure is increased from the triple point to a pressure of 1 atm. From Eq. (7–47), we have for the liquid-solid equilibrium,

$$dT = \frac{T(v'' - v')}{l_{12}}\, dP.$$

The changes in temperature and pressure are so small that we can assume that all terms in the coefficient of dP are constant. Let T_i' represent the equilibrium temperature of ice and pure water. Integrating the left side between T_3 and T_i', and the right side between P_3 and atmospheric pressure P, we have

$$T_i' - T_3 = \frac{T(v'' - v')}{l_{12}}(P - P_3).$$

To three significant figures, $T = 273$ K, $v' = 1.09 \times 10^{-3}$ m^3 kg^{-1}, $v'' = 1.00 \times 10^{-3}$ m^3 kg^{-1}, $l_{12} = 3.34 \times 10^5$ J kg^{-1}, and $P - P_3 = 1.01 \times 10^5$ N m^{-2}. Hence

$$T_i' - T_3 = -0.0075\text{K}.$$

That is, the ice-point temperature T_i' is 0.0075 K *below* the temperature of the triple point.

The effect of the dissolved air is to lower the temperature at which the liquid phase is in equilibrium with pure ice at atmospheric pressure by 0.0023 K below the equilibrium temperature for pure water. Hence the ice-point temperature T_i lies 0.0023 K below T'_i, or $0.0023 + 0.0075 = 0.0098$ K below the triple-point temperature T_3. In other words, the triple-point temperature is 0.0098 K or approximately 0.01 K above the temperature of the ice point. Then since a temperature of exactly 273.16 K is arbitrarily assigned to the triple point, the temperature of the ice point is approximately 273.15 K.

7–7 THE THIRD LAW OF THERMODYNAMICS

The principle known as the third law of thermodynamics governs the behavior of systems, which are in internal equilibrium, as their temperature approaches absolute zero. Its history goes back more than one hundred years, having its origin in attempts to find the property of a system that determines the direction in which a chemical reaction takes place; and, of equal importance, to find what determines when *no* reaction will take place and a system is in chemical equilibrium as well as in thermal and mechanical equilibrium.

A complete discussion of this problem would take us too far into the field of chemical thermodynamics, but the basic ideas are as follows. Suppose that a chemical reaction takes place in a container at constant pressure, and that the container makes contact with a reservoir at a temperature T. If the temperature of the system increases as a result of the reaction, there will be a heat flow to the reservoir until the temperature of the system is reduced to its original value T. For a process at constant pressure the heat flow to the reservoir is equal to the change of enthalpy of the system. If the subscripts *1* and *2* refer to the initial and final states of the system, before and after the reaction, then

$$\Delta H = H_2 - H_1 = -Q, \tag{7–49}$$

where $-Q$, the heat flow out of the system, is the *heat of reaction*. The components and products of the reaction will of course be different chemical substances. Thus if the reaction is

$$Ag + HCl \rightleftarrows AgCl + \tfrac{1}{2}H_2,$$

then H_1 is the enthalpy of the silver and hydrochloric acid and H_2 is the enthalpy of the silver chloride and hydrogen.

Before the second law of thermodynamics was well understood, it was assumed that all of the heat generated in a chemical process at constant pressure should be available to perform useful work. All spontaneous processes would proceed in a direction so that heat flows to the reservoir and the speed of the reaction would depend upon the heat of reaction. Many experiments were done by Thomsen* and by Berthelot†. They found some spontaneous processes which *absorb* heat during

* H. P. J. Julius Thomsen, Danish chemist (1826–1909).

† Pierre M. Berthelot, French chemist (1827–1907).

the reaction. Thus the heat of reaction cannot always be used to determine the direction in which a process takes place.

On the basis of the second law we have shown in Section 7–4 that a spontaneous process can occur in a system subjected to a constant pressure and in contact with reservoir at a temperature T if the Gibbs function, and not the enthalpy, decreases. The two are related by Eq. (7–30), the Gibbs-Helmholtz equation. The change in the Gibbs function can be related to the change in enthalpy by

$$G_2 - G_1 = H_2 - H_1 + T\left(\frac{\partial[G_2 - G_1]}{\partial T}\right)_P,$$

which can be rewritten as

$$\Delta G = \Delta H + T\left(\frac{\partial \Delta G}{\partial T}\right)_P. \tag{7–50}$$

Thus the change in enthalpy and the change in Gibbs function are equal only when $T(\partial \Delta G/\partial T)_P$ approaches zero.

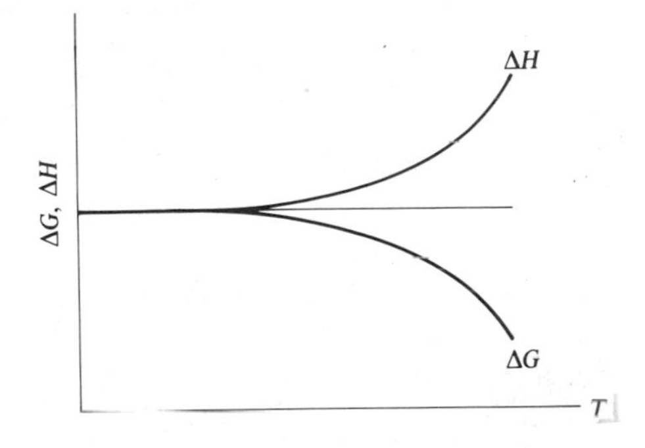

Fig. 7–5 The temperature dependence of the change in the Gibbs function and in the enthalpy for an isobaric process.

Nernst* noted from the results of the experiments by Thomsen and by Berthelot and careful experiments with galvanic cells, that in a reaction ΔG generally approached ΔH more closely as the temperature was reduced, even at quite high temperatures. In 1906, he therefore proposed as a general principle that as the temperature approached zero, not only did ΔG and ΔH approach equality, but their rates of change with temperature both approached zero. That is,

$$\lim_{T\to 0}\left(\frac{\partial\, \Delta G}{\partial T}\right)_P = 0, \qquad \lim_{T\to 0}\left(\frac{\partial\, \Delta H}{\partial T}\right)_P = 0. \tag{7–51}$$

In geometric terms this means that the graphs of ΔG and ΔH as a function of T both have the same horizontal tangent at $T = 0$ as shown in Fig. 7–5.

* Walter H. Nernst, German chemist (1864–1941).

The first of Eq. (7–51) can be written as

$$\lim_{T\to 0}\left(\frac{\partial(G_2 - G_1)}{\partial T}\right)_P = \lim_{T\to 0}\left[\left(\frac{\partial G_2}{\partial T}\right)_P - \left(\frac{\partial G_1}{\partial T}\right)_P\right] = 0.$$

But $(\partial G/\partial T)_P = -S$ so that

$$\lim_{T\to 0}(S_1 - S_2) = 0. \tag{7–52}$$

This is the Nernst heat theorem which states that:

in the neighborhood of absolute zero, all reactions in a liquid or solid in internal equilibrium take place with no change in entropy.

Planck, in 1911, made the further hypothesis that not only does the entropy *difference* vanish as $T \to 0$, but that:

the entropy of every solid or liquid substance in internal equilibrium at absolute zero is itself zero,

that is

$$\lim_{T\to 0} S = 0. \tag{7–53}$$

This is known as the third law of thermodynamics. Then if the reference temperature in the thermodynamic definition of entropy is taken at $T_0 = 0$, the arbitrary constant $S_0 = 0$, and the arbitrary linear function of the temperature appearing in the expressions for the Gibbs and Helmholtz functions for an ideal gas is zero.

If the substance is heated reversibly at constant volume or pressure from $T = 0$ to $T = T$, its entropy at a temperature T is

$$S(V, T) = \int_0^T C_V \frac{dT}{T}, \qquad S(P, T) = \int_0^T C_P \frac{dT}{T}. \tag{7–54}$$

Since the entropy at a temperature T must be finite, the integrals may not diverge; and C_V and C_P must approach zero as the temperature approaches zero:

$$\lim_{T\to 0} C_V = \lim_{T\to 0} C_P = 0. \tag{7–55}$$

We leave it as a problem to show, however, that $C_P/T = (\partial S/\partial T)_P$ may in fact diverge as T approaches 0 K (Problem 7–29).

The Nernst theorem implies that the change in entropy is zero in any process at 0 K. For example,

$$\lim_{T\to 0}\left(\frac{\partial S}{\partial P}\right)_T = \lim_{T\to 0}\left(\frac{\partial S}{\partial V}\right)_T = 0. \tag{7–56}$$

Using the Maxwell relations (Section 7–3), we obtain

$$\lim_{T\to 0}\left(\frac{\partial V}{\partial T}\right)_P = \lim_{T\to 0}\left(\frac{\partial P}{\partial T}\right)_V = 0; \tag{7–57}$$

and since V remains finite as $T \to 0$, we can also write

$$\lim_{T\to 0} \beta = 0. \tag{7–58}$$

Reference to Figs. 3–10 and 2–16, which show what is typical of all solids, will show in fact that the specific heat capacities and the expansivities do approach zero at $T \to 0$. The methods of statistics, as will be shown in Chapter 13, predict that at very low temperatures the specific heat capacities do approach zero. Statistical methods also lead to an expression for the entropy at absolute zero, and in certain systems the entropy does become zero in agreement with the Planck hypothesis.

The third law also implies that **it is impossible to reduce the temperature of a system to absolute zero in any finite number of operations,** as we shall see. The most efficient method for reaching absolute zero is to isolate the system from its surroundings and reduce its temperature below that of the surroundings in an adiabatic process in which the work is done by the system solely at the expense of its internal energy. Consider a reversible adiabatic process which takes a system in a state 1 to state 2 by a path which changes a property X and the temperature T of the system. It follows from Eq. (7–54) that

$$S_1(X_a, T_a) = \int_0^{T_a} \frac{C_{X_a}}{T}\, dT$$

and

$$S_2(X_b, T_b) = \int_0^{T_b} \frac{C_{X_b}}{T}\, dT.$$

In a reversible adiabatic process,

$$S_1(X_a, T_a) = S_2(X_b, T_b);$$

and therefore,

$$\int_0^{T_a} \frac{C_{X_a}}{T}\, dT = \int_0^{T_b} \frac{C_{X_b}}{T}\, dT. \tag{7–59}$$

If the process continues until $T_b = 0$, since each of the integrals converges,

$$\int_0^{T_a} \frac{C_{X_a}}{T}\, dT = 0.$$

However, C_{X_a} is greater than zero for T_a not equal to zero and Eq. (7–59) cannot be true. Therefore the absolute zero of temperature cannot be attained. This is sometimes called the *unattainability* statement of the third law. Mathematically the unattainability statement can be stated as

$$(\partial T/\partial X)_S = 0 \quad \text{at} \quad T = 0\,\text{K}. \tag{7–60}$$

Temperatures of 10^{-3} K have been reached in the laboratory. In fact the nuclei of copper have been cooled to almost 10^{-6} K but the poor thermal contact between the nuclear spin system and the lattice prevented the entire lattice from reaching such low temperatures.

PROBLEMS

7–1 Derive Eqs. (7–16) and (7–17).

7–2 Draw a careful sketch of a Carnot cycle of an ideal gas on a g-s diagram. Label each process and show the direction traversed if the cycle is that of a refrigerator. Assume that s is larger than c_P.

7–3 Show that if F is known as a function of V and T,

$$H = F - T\left(\frac{\partial F}{\partial T}\right)_V - V\left(\frac{\partial F}{\partial V}\right)_T$$

and

$$G = F - V\left(\frac{\partial F}{\partial V}\right)_T.$$

7–4 Use Eq. (7–16) to derive (a) the equation of state, (b) the energy equation, (c) the Gibbs function, and (d) the enthalpy of an ideal gas.

7–5 Derive the equation of state and the energy equation for a van der Waals gas from Eq. (7–17).

7–6 The specific Gibbs function of a gas is given by

$$g = RT \ln (P/P_0) - AP,$$

where A is a function of T only. (a) Derive expressions for the equation of state of the gas and its specific entropy. (b) Derive expressions for the other thermodynamic potentials. (c) Derive expressions for c_P and c_v. (d) Derive expressions for the isothermal compressibility and the expansivity. (e) Derive an expression for the Joule-Thomson coefficient.

7–7 The specific Gibbs function of a gas is given by

$$g = -RT \ln (v/v_0) + Bv,$$

where B is a function of T only. (a) Show explicitly that this form of the Gibbs function does not completely specify the properties of the gas. (b) What further information is necessary so that the properties of the gas can be completely specified?

7–8 Does the expression

$$f = RT \ln (v_0/v) + CT^2v,$$

where C is a positive constant, result in a reasonable specification of the properties of a gas at normal temperatures and pressures?

7–9 Derive Eqs. (7–36), (7–37), and (7–38).

7–10 Let us define a property of a system represented by Φ which is given by the equation

$$\Phi = S - \frac{U + PV}{T}.$$

Show that

$$V = -T\left(\frac{\partial \Phi}{\partial P}\right)_T,$$

$$U = T\left[T\left(\frac{\partial \Phi}{\partial T}\right)_P + P\left(\frac{\partial \Phi}{\partial P}\right)_T\right],$$

and

$$S = \Phi + T\left(\frac{\partial \Phi}{\partial T}\right)_P.$$

7–11 The work necessary to stretch a wire is given by Eq. (3–6). (a) Derive expressions for the differentials of the thermodynamic potentials. (b) Derive the four Maxwell relations for this system. (c) Derive the $T\,dS$ equations.

7–12 (a) Derive the thermodynamic potentials and their differentials for an $\mathscr{E}ZT$ system. (b) Derive the Maxwell relations and (c) the $T\,dS$ equations for the system.

7–13 The work $d'W$ in a reversible process undergone by a paramagnetic gas is given by Eq. (6–69). (a) Write expressions for dE, dU, dH, dF, dG, and dF^* for this system. (b) Use the expressions of part (a) to derive Maxwell relations for this system. (c) Write the $T\,dS$ equations for a paramagnetic gas.

7–14 Give an example of a change in the constraint imposed on a system which will cause its properties to change if the system is (a) completely isolated, (b) at constant temperature and pressure, (c) at constant temperature and volume.

7–15 Show that the internal energy of a system at constant entropy and volume must decrease in any spontaneous process.

7–16 If the Gibbs function of a system must decrease during any spontaneous processes in which the temperature and pressure remain constant, show that the entropy of an isolated system must increase during a spontaneous process. [*Hint:* Show that $(\Delta G)_{T,P}$ must increase for any process that includes a stage in which $(\Delta S)_U$ decreases.]

7–17 By the same method as used in the previous problem, show that if the Gibbs function of a system must decrease during any spontaneous process in which the temperature and pressure remain constant, (a) the Helmholtz function must also decrease in any spontaneous process at constant volume and temperature; and (b) the enthalpy must decrease in any spontaneous process at constant pressure and entropy.

7–18 What can be stated about the change of Gibbs function during a spontaneous process of a completely isolated system?

7–19 Sketch qualitative curves in a g-P and a g-T plane of the phases of a substance which sublimates rather than melts.

7–20 Sketch qualitative curves which represent the solid, liquid, and vapor phase of pure water in (a) the g-P plane at $T = -10°\text{C}$ and (b) the g-T plane at $P = 2$ atm so that the transitions from one phase to the other can be indicated.

7–21 Sketch graphs of g and its first and second derivatives as a function of T and P for (a) a first-order and (b) a second-order phase transition.

7–22 The specific Gibbs function of the solid phase and of the liquid phase of a substance are plotted in Fig. 7–6 as a function of temperature at a constant pressure of 10^5 N m^{-2}.

At higher pressures the curves of g versus T are parallel to those shown. The molal volume of the solid and of the liquid are respectively 0.018 and 0.020 m^3 kilomole^{-1}. (a) Sketch, approximately to scale, curves of g versus P for the solid and liquid phases. Justify your curves. (b) If one kilomole of the liquid is supercooled to 280 K and then transformed to solid isothermally and isobarically at 10^5 N m^{-2} calculate ΔG, ΔS, ΔH, ΔU, and ΔF for the system and ΔS for the universe.

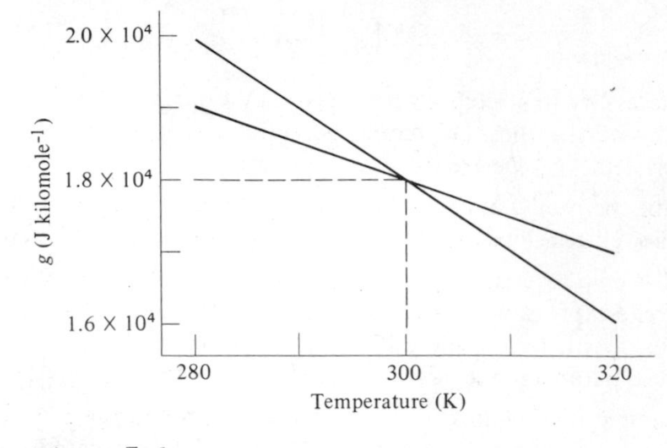

Figure 7–6

7–23 (a) Calculate the slope of the fusion curve of ice, in (N m^{-2} K^{-1}), at the normal melting point. The heat of fusion at this temperature is 3.34×10^5 J kg^{-1} and the change in specific volume on melting is -9.05×10^{-5} m^3 kg^{-1}. (b) Ice at $-2°$C and atmospheric pressure is compressed isothermally. Find the pressure at which the ice starts to melt. (c) Calculate $(\partial P/\partial T)_v$ for ice at $-2°$C. ($\beta = 15.7 \times 10^{-5}$ K^{-1} and $\kappa = 120 \times 10^{-12}$ m^2 N^{-1}). (d) Ice at $-2°$C and atmospheric pressure is kept in a container at constant volume, and the temperature is gradually increased. Find the temperature and pressure at which the ice starts to melt. Show this process and that in part (b) on a P-T diagram like the one in Fig. 2–9(a), and on a P-V-T surface like the one in Fig. 2–7. Assume that the fusion curve and the rate of change of pressure with temperature, at constant volume, are both linear.

7–24 Prove that in the P-V plane the slope of the sublimation curve at the triple point is greater than that of the vaporization curve at the same point.

7–25 The vapor pressure of a particular solid and of a liquid of the same material are given by $\ln P = 0.04 - 6/T$ and $\ln P = 0.03 - 4/T$ respectively, where P is given in atmospheres. (a) Find the temperature and pressure of the triple point of this material. (b) Find the values of the three heats of transformation at the triple point. State approximations.

7–26 An idealized diagram for the entropy of the solid phase and the entropy of the liquid phase of He^3 are shown in Fig. 7–7 plotted against temperature at the melting pressure. (He^3 does not liquefy at atmospheric pressure.) The molal volume of the liquid is greater than the molal volume of the solid by 10^{-3} m^3 kilomole^{-1} throughout the temperature range. (a) Draw a careful and detailed plot of the melting curve on a P-T diagram. The melting pressure at 0.3 K is 30 atm. (b) Discuss processes to freeze He^3 below 0.2 K.

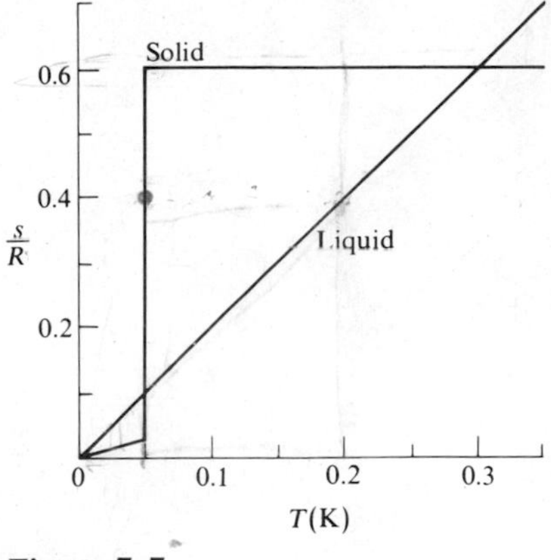

Figure 7–7

7–27 (a) Liquid He^3 at 0.2 K at a pressure just below the melting pressure is adiabatically compressed to a pressure just above the melting pressure. Use Fig. 7–7 to calculate the change in temperature of the He^3. Explain approximations. (b) How can this effect be used as a refrigerator at low temperatures?

7–28 In a second-order phase transition $s_i = s_f$ or $v_i = v_f$ at a particular temperature and pressure where f and i denote the final and initial phase. Show that in these cases the Clausius-Clapeyron equation can be written as

$$\frac{dP}{dT} = \frac{1}{Tv}\frac{c_{Pf} - c_{Pi}}{\beta_f - \beta_i} \qquad \text{or} \qquad \frac{dP}{dT} = \frac{\beta_f - \beta_i}{\kappa_f - \kappa_i},$$

respectively. [*Hint:* Begin with an appropriate $T\,dS$ relation.]

7–29 A low temperature physicist wishes to publish his experimental result that the heat capacity of a nonmagnetic dielectric material between 0.05 and 0.5 K varies as $AT^{1/2} + BT^3$. As editor of the journal, should you accept the paper for publication?

7–30 Show that the Planck statement of the third law can be derived from the unattainability statement.

7–31 The Planck statement of the third law states that one isentropic surface covers the $T = 0$ K plane. Derive Eq. (7–60) by showing that if this surface had a branch to higher temperatures, the specific heat capacity at constant X would have to be negative.

7–32 A polymer, held at constant tension shrinks as the temperature is increased. Sketch a curve of the length of the polymer as a function of temperature near 0 K and give reasons for all pertinent parts of your sketch.

7–33 (a) Show that Curie's law for an ideal paramagnet and the van der Waals equation of state cannot be valid near 0 K. (b) Show that there can be no first-order phase transition at 0 K.

8

Applications of thermodynamics to simple systems

8–1 CHEMICAL POTENTIAL

In this chapter the thermodynamic principles developed in the preceding chapters are applied to some simple systems. We begin by relaxing the constraint that the system be closed, and we investigate how the relationships developed are changed if mass enters or leaves the system or if mass is interchanged between parts of a system.

Suppose that a container of volume V is divided into two parts by a partition. On one side of the partition there are n_1 moles of an ideal gas and on the other side there are n_2 moles of a different ideal gas, both gases being at the same temperature T and pressure P.

The partition is now removed, each gas diffuses into the other, and a new equilibrium state is eventually attained in which both gases occupy the same total volume V. If the gases are ideal, there is no change in the temperature T or in the total pressure P. The final partial pressures of the gases are p_1 and p_2, and $p_1 + p_2 = P$.

The initial Gibbs function of the system is

$$G_i = n_1 g_{1i} + n_2 g_{2i},$$

where g_{1i} and g_{2i} are the initial values of the specific Gibbs function of the respective gases. From Eq. (7–14),

$$g_{1i} = RT(\ln P + \phi_1), \qquad g_{2i} = RT(\ln P + \phi_2),$$

where ϕ_1 and ϕ_2 are functions of temperature only.

The final value of the Gibbs function is

$$G_f = n_1 g_{1f} + n_2 g_{2f};$$

and since the final pressure of each gas is its partial pressure p,

$$g_{1f} = RT(\ln p_1 + \phi_1), \qquad g_{2f} = RT(\ln p_2 + \phi_2).$$

The quantities ϕ_1 and ϕ_2 have the same value in the initial and final states, since they are functions of temperature only.

The *mole fractions* x_1 and x_2 of each constituent, in the final state, are defined as

$$x_1 = \frac{n_1}{n_1 + n_2} = \frac{n_1}{n}, \qquad x_2 = \frac{n_2}{n_1 + n_2} = \frac{n_2}{n}, \tag{8–1}$$

where the total number of moles $n = n_1 + n_2$. Since both constituents are ideal gases, and both occupy the same volume V at the same temperature T,

$$n_1 = \frac{p_1 V}{RT}, \qquad n_2 = \frac{p_2 V}{RT}, \qquad n = \frac{PV}{RT};$$

and hence

$$x_1 = \frac{p_1}{P}, \qquad x_2 = \frac{p_2}{P}. \tag{8-2}$$

Then

$$\ln p_1 = \ln P + \ln x_1, \qquad \ln p_2 = \ln P + \ln x_2,$$

and

$$g_{1f} = RT(\ln P + \phi_1 + \ln x_1), \qquad g_{2f} = RT(\ln P + \phi_2 + \ln x_2).$$

The *chemical potential* μ of each gas in the mixture is defined as

$$\begin{aligned} \mu &\equiv RT(\ln P + \phi + \ln x) \\ &= RT(\ln p + \phi) \\ &= g + RT \ln x, \end{aligned} \tag{8-3}$$

where g is the specific Gibbs function at temperature T and *total* pressure P. The final Gibbs function of the system is therefore

$$G_f = n_1\mu_1 + n_2\mu_2.$$

The change in the Gibbs function in the mixing process is

$$\begin{aligned} G_f - G_i &= n_1(\mu_1 - g_1) + n_2(\mu_2 - g_2) \\ &= RT(n_1 \ln x_1 + n_2 \ln x_2). \end{aligned} \tag{8-4}$$

The expression in parenthesis is necessarily *negative*, since x_1 and x_2 are both fractions, less than 1; and hence the Gibbs function *decreases* in the irreversible mixing process, which we have shown is always the case in any such process at constant temperature and pressure.

As an example, consider a container of volume V divided into two parts by a partition. On the left side are 2 kilomoles of helium gas and on the right side is 1 kilomole of neon gas. Both gases have a temperature of 300 K and a pressure of 1 atm. After the partition is removed, the gases diffuse into each other and a new equilibrium state is reached. The mole fraction of each of the gases in the mixture is given by Eq. (8-1):

$$x_{He} = \frac{2}{2+1} = \frac{2}{3} \quad \text{and} \quad x_{Ne} = \frac{1}{2+1} = \frac{1}{3};$$

and their partial pressures are

$$p_{He} = 0.67 \text{ atm} \quad \text{and} \quad p_{Ne} = 0.33 \text{ atm}.$$

The chemical potential of each gas is

$$\mu_{He} = g_{He} + R(300) \ln 0.67; \qquad \mu_{Ne} = g_{Ne} + R(300) \ln 0.33,$$

where g_{He} and g_{Ne} are the specific Gibbs functions of the separated gas at the same temperature and pressure. The chemical potential of each constituent of the gas is a linear function of temperature and depends upon the natural logarithm of the mole fraction of that constituent in the gas.

The change in the Gibbs function in the mixing process is

$$\Delta G = G_f - G_i = RT(2 \ln 0.67 + 1 \ln 0.33),$$
$$= -5 \times 10^6 \text{ J}.$$

The change in entropy during the mixing process can be calculated from the first of Eq. (7–27):

$$\Delta S = -\left(\frac{\partial \Delta G}{\partial T}\right)_P = -R(n_1 \ln x_1 + n_2 \ln x_2),$$
$$= 2R,$$
$$= 16.6 \times 10^3 \text{ J K}^{-1}.$$

We have introduced the concept of chemical potential through the simple example of a mixture of two ideal gases. The concept has a much wider significance, however, and is basic to many problems in physical chemistry. It is applicable to solutions as well as gases, to substances that can react chemically, and to systems in which more than one phase is present. In the next section we prove that a system is in chemical equilibrium when the chemical potential of each constituent has the same value in each phase.

The general relation between μ and g, for any constituent in any phase, has the same form as Eq. (8–3):

$$\mu = g + RT \ln x,$$

where x is the mole fraction of the constituent:

$$x_j = \frac{n_j}{\sum n_j}.$$

If a phase consists of only one constituent, $x = 1$, $\ln x = 0$, and

$$\mu = g. \tag{8–5}$$

In this case, the chemical potential equals the specific Gibbs function.

The problem of liquid-vapor equilibrium discussed in Section 7–5 is an example. In this case there is only one constituent, $\mu = g$, and, as we have shown, the specific Gibbs functions g'' and g''' are equal in the state of stable equilibrium.

For a system consisting of a single pure substance, the concept of chemical potential can be arrived at in a different way. The combined first and second laws for a *closed PVT* system lead to the result that

$$dU = T\,dS - P\,dV.$$

Considering U as a function of S and V, we can also write

$$dU = \left(\frac{\partial U}{\partial S}\right)_V dS + \left(\frac{dU}{\partial V}\right)_S dV, \tag{8–6}$$

from which it follows that

$$\left(\frac{\partial U}{\partial S}\right)_V = T, \qquad \left(\frac{\partial U}{\partial V}\right)_S = -P. \tag{8–7}$$

The internal energy U is an extensive property and is proportional to the number of moles included in the system. It is implied in Eq. (8–6) that we are considering a *closed* system for which the number of moles n is constant. If, however, the system is *open*, so that we can add or remove material, the internal energy becomes a function of n as well as of S and V, and

$$dU = \left(\frac{\partial U}{\partial S}\right)_{V,n} dS + \left(\frac{\partial U}{\partial V}\right)_{S,n} dV + \left(\frac{\partial U}{\partial n}\right)_{S,V} dn. \tag{8–8}$$

For the special case in which $dn = 0$, this must reduce to Eq. (8–7), and hence

$$\left(\frac{\partial U}{\partial S}\right)_{V,n} = T, \qquad \left(\frac{\partial U}{\partial V}\right)_{S,n} = -P. \tag{8–9}$$

The additional subscript n on the partial derivatives simply makes explicit what is implied in Eqs. (8–7), namely, that in these equations n is assumed constant. The coefficient of dn in Eq. (8–8) is now defined as the chemical potential μ:

$$\mu \equiv \left(\frac{\partial U}{\partial n}\right)_{S,V}; \tag{8–10}$$

that is, the chemical potential is the change of internal energy per mole of substance added to the system in a process at constant S and V; and Eq. (8–8) can be written

$$dU = T\,dS - P\,dV + \mu\,dn. \tag{8–11}$$

This equation is the general form of the combined first and second laws for an *open PVT* system. More generally, if X represents any extensive variable corresponding to the volume V, and Y the intensive variable corresponding to the pressure P, the work in a differential reversible process is $Y\,dX$ and

$$dU = T\,dS - Y\,dX + \mu\,dn. \tag{8–12}$$

The chemical potential can be expressed in a number of different ways. If we write Eq. (8–12) as

$$dS = \frac{1}{T}\,dU + \frac{Y}{T}\,dX - \frac{\mu}{T}\,dn,$$

and consider S as a function of U, X, and n, it follows that the partial derivatives of S with respect to U, X, and n, respectively, the other two variables being held constant, are equal to the coefficients of the differentials dU, dX, and dn. Therefore

$$\mu = -T\left(\frac{\partial S}{\partial n}\right)_{U,X}. \tag{8–13}$$

The difference in the Helmholtz function $F = U - TS$, between two neighboring equilibrium states, is

$$dF = dU - T\,dS - S\,dT;$$

and when dU is eliminated between this equation and Eq. (8–12), we have for an open system,

$$dF = -S\,dT - Y\,dX + \mu\,dn,$$

from which it follows that

$$\mu = \left(\frac{\partial F}{\partial n}\right)_{T,X}. \tag{8–14}$$

In the same way, the difference in the Gibbs function $G = U - TS + YX$, for an open system, is

$$dG = -S\,dT + X\,dY + \mu\,dn \tag{8–15}$$

and

$$\mu = \left(\frac{\partial G}{\partial n}\right)_{T,Y}. \tag{8–16}$$

This equation is equivalent to the definition of μ for the special case discussed earlier in this section. For a single constituent, $G = ng$ and hence

$$\mu = \left(\frac{\partial G}{\partial n}\right)_{T,Y} = g.$$

In summary, we have the following expressions for the chemical potential:

$$\mu = -T\left(\frac{\partial S}{\partial n}\right)_{U,X} = \left(\frac{\partial F}{\partial n}\right)_{T,X} = \left(\frac{\partial G}{\partial n}\right)_{T,Y}.$$

8–2 PHASE EQUILIBRIUM AND THE PHASE RULE

The discussion of the previous section can be easily extended to the case of a phase composed of k constituents rather than just one. The internal energy of the phase is

$$U = U(S, V, n_1, n_2, \ldots, n_k), \tag{8–17}$$

where n_i is the number of moles of the ith constituent present in the phase. Equation (8–8) can be rewritten as

$$dU = \left(\frac{\partial U}{\partial S}\right)_{V,n} dS + \left(\frac{\partial U}{\partial V}\right)_{S,n} dV + \left(\frac{\partial U}{\partial n_1}\right)_{S,V,n'} dn_1 + \cdots + \left(\frac{\partial U}{\partial n_k}\right)_{S,V,n'} dn_k, \tag{8–18}$$

where the subscript n' signifies that the number of moles of all constituents is constant except for the constituent appearing in the derivative.

Equation (8–11) can be written as

$$dU = T\,dS - P\,dV + \mu_1\,dn_1 + \cdots + \mu_k\,dn_k, \tag{8–19}$$

where

$$\mu_i = \left(\frac{\partial U}{\partial n_i}\right)_{S,V,n'}, \text{ etc.} \tag{8–20}$$

The last equation defines the chemical potential of the ith constituent in the phase.

Similarly, the difference in the Gibbs function between two states at the same temperature and pressure for an open system of k constituents is

$$dG = dU - T\,dS + P\,dV.$$

Comparison with Eq. (8–19) yields

$$dG = \mu_1\,dn_1 + \cdots + \mu_k\,dn_k, \tag{8–21}$$

and

$$\mu_i = \left(\frac{\partial G}{\partial n_i}\right)_{P,T,n'}. \tag{8–22}$$

It now remains to be shown that the chemical potential of a constituent is not dependent on the size of the phase, but is specified by the relative composition, the pressure, and the temperature. Consider the phase to consist of two parts which are equal in every respect. If Δn_i moles of constituent i are added to each half of the phase without changing the pressure or the temperature of either half, the pressure and the temperature of the whole phase do not change and we can write for each half

$$\mu_i = \frac{\Delta G}{\Delta n_i}.$$

For the two halves, we get

$$\mu_i = \frac{2\,\Delta G}{2\,\Delta n_i} = \frac{\Delta G}{\Delta n_i}.$$

Hence the chemical potential μ is independent of the size of the phase.

Now assume that we have a phase at temperature T, pressure P, and Gibbs function G_0, and that we add mass which is at the same temperature and pressure. As a result of the above discussion, Eq. (8–21) can now be written

$$G - G_0 = \mu_1 n_1 + \cdots + \mu_k n_k. \tag{8–23}$$

Therefore we can also write

$$\begin{aligned} U &= TS - PV + \mu_1 n_1 + \cdots + \mu_k n_k + G_0, \\ H &= TS + \mu_1 n_1 + \cdots + \mu_k n_k + G_0, \\ F &= -PV + \mu_1 n_1 + \cdots + \mu_k n_k + G_0. \end{aligned} \tag{8–24}$$

It was shown in Section 7–5 that if two phases of a pure substance are in equilibrium at constant temperature and pressure, the specific Gibbs function has the same value in both phases. From this consideration we were able to derive the Clausius-Clapeyron equation. We now consider equilibrium in a system composed of more than one phase.

It is clear that only one gaseous phase can exist, since constituents added to this phase will diffuse until a homogeneous mixture is obtained. However, more than one liquid phase can exist because the immiscibility of certain liquids precludes the possibility of homogeneity. Generally speaking, mixtures of solids do not form a homogeneous mixture except in special circumstances. For example, a mixture of iron filings and sulfur, or the different types of ice, must be regarded as forming separate solid phases. On the other hand, some metal alloys may be considered to comprise a single solid phase.

Our previous observation that the specific Gibbs function has the same value in each phase for equilibrium between phases of a single constituent requires modification when more than one constituent is present in the system. We consider a closed system consisting of π phases and k constituents in equilibrium at constant temperature and pressure. As before, a constituent will be designated by a subscript $i = 1, 2, 3, \ldots, k$, and a phase by a superscript $(j) = 1, 2, 3, \ldots, \pi$. Thus the symbol $\mu_1^{(2)}$ means the chemical potential of constituent 1 in phase 2.

The Gibbs function of constituent i in phase j is the product of the chemical potential $\mu_i^{(j)}$ of that constituent in phase j, and the number of moles $n_i^{(j)}$ of the constituent in phase j. The total Gibbs function of phase j is the sum of all such products over all constituents, that is, it equals

$$\sum_{i=1}^{i=k} \mu_i^{(j)} n_i^{(j)}.$$

Finally, the total Gibbs function of the entire system is the sum of all such sums over all phases of the system, and can be written

$$G = \sum_{j=1}^{j=\pi} \sum_{i=1}^{i=k} \mu_i^{(j)} n_i^{(j)}. \tag{8–25}$$

We have shown in Section 7–1 that the necessary condition for stable equilibrium of a system at constant temperature and pressure is that the Gibbs function of the system shall be a *minimum*. That is, when we compare the equilibrium state with a second state at the same temperature and pressure, but differing slightly from the equilibrium state, the first variation in the Gibbs function is zero: $dG_{T,P} = 0$.

In the second state, the numbers of moles $n_i^{(j)}$ of each constituent in each phase are slightly different from their equilibrium values. Then since the chemical potentials are constant at constant temperature and pressure, we have from Eq. (8–25),

$$dG_{T,P} = \sum_{j=1}^{j=\pi} \sum_{i=1}^{i=k} \mu_i^{(j)} \, dn_i^{(j)} = 0, \tag{8–26}$$

where any $dn_i^{(j)}$ represents the small difference in the number of moles of constituent i in phase j. Writing out a few terms in the double sum, we have

$$\begin{aligned} &\mu_1^{(1)}\,dn_1^{(1)} + \mu_1^{(2)}\,dn_1^{(2)} + \cdots + \mu_1^{(\pi)}\,dn_1^{(\pi)} \\ &+ \mu_2^{(1)}\,dn_2^{(1)} + \mu_2^{(2)}\,dn_2^{(2)} + \cdots + \mu_2^{(\pi)}\,dn_2^{(\pi)} \\ &\quad\vdots \\ &+ \mu_k^{(1)}\,dn_k^{(1)} + \mu_k^{(2)}\,dn_k^{(2)} + \cdots + \mu_k^{(\pi)}\,dn_k^{(\pi)} = 0. \end{aligned} \tag{8–27}$$

If each of the differentials $dn_i^{(j)}$ in this formidable equation were *independent*, so that each could be given some arbitrary value, the equation could be satisfied only if the coefficient $\mu_i^{(j)}$ of each were zero. Thus although we might find a set of $\mu_i^{(j)}$'s such that the sum would be zero for some arbitrary choice of the $dn_i^{(j)}$'s, it would not be zero for a different arbitrary choice. However, the total amount of each constituent in all phases together must be constant, since none of the constituents is being created, destroyed, or transformed. A reduction of the amount of a constituent in one phase must result in an increase of the amount of that constituent in other phases. Thus the differentials $dn_i^{(j)}$ are not independent; but

$$\begin{aligned} dn_1^{(1)} + dn_1^{(2)} + \cdots + dn_1^{(\pi)} &= 0, \\ dn_2^{(1)} + dn_2^{(2)} + \cdots + dn_2^{(\pi)} &= 0, \\ &\vdots \\ dn_k^{(1)} + dn_k^{(2)} + \cdots + dn_k^{(\pi)} &= 0. \end{aligned} \tag{8–28}$$

The solution of Eq. (8–27) is constrained by the k conditions expressed by these *condition equations*.

To find this solution, the value of $dn_i^{(1)}$ obtained from each of Eqs. (8–28) is substituted into the corresponding line of Eq. (8–27). The first line of Eq. (8–27) becomes

$$-\mu_1^{(1)}(dn_1^{(2)} + dn_1^{(3)} + \cdots + dn_1^{(\pi)}) + \mu_1^{(2)}\,dn_1^{(2)} + \cdots + \mu_1^{(\pi)}\,dn_1^{(\pi)},$$

which can be rewritten as

$$(\mu_1^{(2)} - \mu_1^{(1)})\,dn_1^{(2)} + (\mu_1^{(3)} - \mu_1^{(1)})\,dn_1^{(3)} + \cdots + (\mu_1^{(\pi)} - \mu_1^{(1)})\,dn_1^{(\pi)}.$$

Similar expressions can be written for each line of Eq. (8–27); but now each of these remaining $dn_i^{(j)}$ (in which $j \neq 1$) is independent and can be varied arbitrarily. In order that Eq. (8–27) have a solution for all arbitrary variations of these $dn_i^{(j)}$, their coefficients must each be equal to zero. For the first line of Eq. (8–27), we obtain

$$\mu_1^{(2)} = \mu_1^{(1)},\ \mu_1^{(3)} = \mu_1^{(1)},\ \ldots,\ \mu_1^{(\pi)} = \mu_1^{(1)};$$

that is, the chemical potential of this constituent must have the same value in all phases. Continuing the procedure for each constituent gives the result that *the chemical potential of each constituent must have the same value in all phases*, that is,

$$\begin{aligned} \mu_1^{(1)} &= \mu_1^{(2)} = \cdots = \mu_1^{(\pi)}, \\ \mu_2^{(1)} &= \mu_2^{(2)} = \cdots = \mu_2^{(\pi)}, \\ &\vdots \\ \mu_k^{(1)} &= \mu_k^{(2)} = \cdots = \mu_k^{(\pi)}. \end{aligned} \tag{8–29}$$

If this is the case, we can omit the superscripts in the preceding equations and simply write μ_1, μ_2, etc., for the chemical potentials. The first line in Eq. (8–27) then becomes

$$\mu_1(dn_1^{(1)} + dn_1^{(2)} + \cdots dn_1^{(\pi)})$$

which from the first of the condition equations equals zero. The same is true for every other constituent and Eq. (8–27) is satisfied. It is not obvious that Eqs. (8–29) are *necessary* as well as *sufficient*. A proof of this will be found in Appendix B. Equations (8–29) are generalizations of the result derived earlier that when two or more phases of a *single* constituent are in equilibrium, the chemical potential has the same value in all phases.

Suppose the phases of a system are not in equilibrium. Then the molal Gibbs function of each constituent will not have the same value in each phase. For each constituent for which a difference in the molal Gibbs function exists, there will be a tendency, called the *escaping tendency*, to escape spontaneously from the phase in which its molal Gibbs function is higher to that phase in which the molal Gibbs function is lower, until equilibrium exists between the phases, i.e., until the molal Gibbs function has the same value in all phases. Conversely, the escaping tendency of any constituent is the same in all phases when the system is in equilibrium.

The *phase rule*, which was first derived by Gibbs, follows logically from the conclusions reached above. First we shall consider a heterogeneous system in which the constituents are present in all phases. Equations (8–29), which specify the conditions of phase equilibrium and hence will be called the equations of phase equilibrium, are $k(\pi - 1)$ in number. Now the composition of each phase containing k constituents is fixed if $k - 1$ constituents are known, since the sum of the mole fractions of each constituent in the phase must equal unity. Therefore, for π phases, there are a total of $\pi(k - 1)$ variables, in addition to temperature and pressure, which must be specified. There are, then, $\pi(k - 1) + 2$ variables altogether.

If the number of variables is equal to the number of equations, then whether or not we can actually solve the equations, the temperature, pressure, and composition of each phase are *determined*. The system is then called *nonvariant* and is said to have zero *variance*.

If the number of variables is one greater than the number of equations, an arbitrary value can be assigned to one of the variables and the remainder are completely determined. The system is then called *monovariant* and is said to have a variance of 1.

In general, the variance f is defined as the excess of the number of variables over the number of equations, and

$$f = [\pi(k - 1) + 2] - [k(\pi - 1)],$$

or

$$f = k - \pi + 2. \quad \text{(No chemical reactions)} \tag{8–30}$$

This equation is called the *Gibbs phase rule.*

As an example, consider liquid water in equilibrium with its vapor. There is only one constituent (H_2O) and $k = 1$. There are two phases, $\pi = 2$, and the number of equations of phase equilibrium is

$$k(\pi - 1) = 1.$$

This single equation states simply that, as we have previously shown, the chemical potential μ has the same value in both phases.

The number of variables is

$$\pi(k - 1) + 2 = 2.$$

These variables are the temperature T and pressure P, since in both phases the mole fraction of the single constituent must be 1. The variance f is therefore

$$f = k - \pi + 2 = 1,$$

which means that an arbitrary value can be assigned to *either* the temperature T or the pressure P, but not to *both.* (Of course, limitations are imposed on these arbitrary values since they must lie within a range in which liquid water and water vapor *can* exist in equilibrium.) Thus if we specify the temperature T, the pressure P will then be the vapor pressure of water at this temperature and it cannot be given some arbitrary value. If we make the pressure greater than the vapor pressure, keeping the temperature constant, all the vapor will condense to liquid as shown in the isotherm in Fig. 2–9. If we make the pressure less than the vapor pressure, all the liquid will evaporate.

At the *triple point* of water, all three phases are in equilibrium and $\pi = 3$. Then $k(\pi - 1) = 2$, and there are *two* equations of phase equilibrium stating that the chemical potential in any one phase is equal to its value in each of the other phases. The number of variables is $\pi(k - 1) + 2 = 2$, which is equal to the number of equations. The variance is

$$f = k - \pi + 2 = 0,$$

and the system is therefore *invariant.* We cannot assign an arbitrary value to *either* the temperature or the pressure. Once a system such as the triple point cell in Fig. 1–3 has been set up in *any* laboratory, its temperature is *necessarily* that of the triple point of water, and its pressure is the vapor pressure at this temperature. It is for this reason that the temperature of the triple point of water has been chosen as the single fixed point of the thermodynamic temperature scale; it can be reproduced precisely at any point and at any time. Of course, the triple point of any other pure substance would serve, but water was chosen because of its universal availability in a pure state.

It can be readily shown that if a constituent is absent from a phase, the number of variables and the number of equations are each reduced by one. Hence the original restriction that every constituent be present in every phase can be removed, and Eq. (8–30) remains valid.

If chemical reaction takes place within the system, the constituents are not completely independent. Let us suppose that the four constituents A, B, C, and D undergo the reaction

$$n_A A + n_B B \rightleftarrows n_C C + n_D D,$$

where the n's are the number of moles of the constituents. We now have an additional independent equation, so that the total number of independent equations is $k(\pi - 1) + 1$. The number of variables is $\pi(k - 1) + 2$, as before. Therefore the number of degrees of freedom is

$$f = (k - 1) - \pi + 2.$$

But it is possible to conceive of a system where a number of chemical reactions could take place, and accordingly we express the phase rule in the more general form

$$f = (k - r) - \pi + 2 \qquad \text{(with chemical reaction)}, \tag{8–31}$$

where r is the number of independent reversible chemical reactions.

8–3 DEPENDENCE OF VAPOR PRESSURE ON TOTAL PRESSURE

As an application of the concepts developed in the last two sections, we consider the dependence of the vapor pressure of a liquid on the total pressure. Figure 8–1(a) represents a liquid in equilibrium with its vapor. The total pressure in the system is the vapor pressure. An indifferent gas (that is, one that does not react chemically with the liquid or its vapor), as represented by open circles in Fig. 8–1(b), is pumped into the space above the liquid, thereby increasing the total pressure. The question is: Will the vapor pressure be changed when this is done at constant temperature?

We make use of the condition that the chemical potential of the original substance must have the same value in the liquid and gas phases. Since the liquid phase consists of a single constituent, the chemical potential in this phase equals the specific Gibbs function of the liquid:

$$\mu'' = g''.$$

The gas phase can be considered a mixture of ideal gases and we can use the results of Section 8–1:

$$\mu''' = RT(\ln p + \phi),$$

where μ''' is the chemical potential of the vapor and p is the vapor pressure.

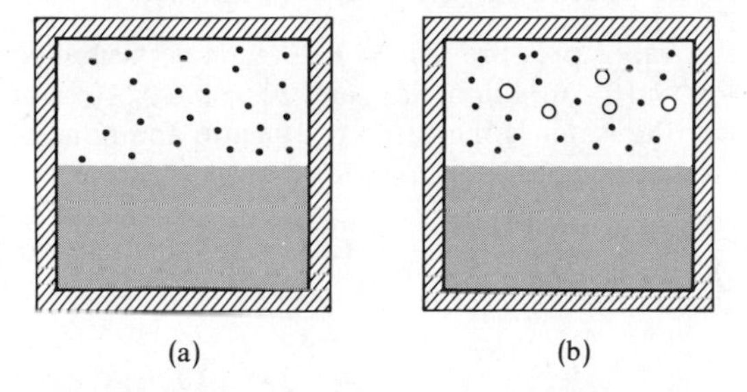

Fig. 8–1 A liquid in equilibrium with its vapor (a) at the vapor pressure, (b) at a higher pressure caused by the presence of an indifferent gas.

Let P represent the *total* pressure, and suppose that a small additional amount of the indifferent gas is pumped in, at constant temperature, increasing the total pressure from P to $P + dP$ and changing the vapor pressure from p to $p + dp$. Since the system is also in equilibrium at the new pressure, the changes $d\mu''$ and $d\mu'''$ must be equal. For the liquid,

$$d\mu'' = dg'' = -s''\,dT + v''\,dP = v''\,dP,$$

since the temperature is constant. Also, since ϕ is a function of temperature only,

$$d\mu''' = \frac{RT}{p}\,dp.$$

Therefore

$$v''\,dP = RT\frac{dp}{p},$$

or

$$\frac{dp}{p} = \frac{v''}{RT}\,dP. \tag{8–32}$$

Let p_0 be the vapor pressure in Fig. 8–1(a), when no indifferent gas is present. In this case, the total pressure P equals p_0. We now integrate Eq. (8–32) from this state to a final state in which the vapor pressure is p and the total pressure is P. The volume v'' can be considered constant, so

$$\int_{p_0}^{p}\frac{dp}{p} = \frac{v''}{RT}\int_{p_0}^{P}dP,$$

and

$$\ln\frac{p}{p_0} = \frac{v''}{RT}(P - p_0). \tag{8–33}$$

It follows that when the total pressure P is increased, the vapor pressure p increases also. That is, as more of the indifferent gas is pumped in, more of the liquid evaporates, contrary to what field might be expected. However, the partial pressure of the vapor phase *by itself* is unaffected by the addition of the indifferent gas, and only the liquid phase feels the additional pressure causing it to evaporate.

The change in vapor pressure, $\Delta p = p - p_0$, is very small since v''/RT is small. For water, $v'' = 18 \times 10^{-3}$ m^3 kilomole^{-1} and $p_0 = 3.6 \times 10^3$ N m^{-2} at 300 K. If the total pressure over the water is increased to 100 atm and none of the indifferent gas dissolves in the water, then

$$\ln \frac{p}{p_0} = \frac{18 \times 10^{-3}}{(8.315 \times 10^3)(300)} (1.01 \times 10^7 - 3.6 \times 10^3)$$

and

$$\ln \frac{p + \Delta p}{p_0} = \frac{\Delta p}{p_0} = 7.29 \times 10^{-2},$$

since $\ln(1 + x) = x$ for $x \ll 1$.

8–4 SURFACE TENSION

The phenomena of *surface tension* and *capillarity* can be explained on the hypothesis that at the outer surface of a liquid there exists a surface layer, a few molecules thick, whose properties differ from those of the bulk liquid within it. The surface film and the bulk liquid can be considered as two phases of the substance in equilibrium, closely analogous to a liquid and its vapor in equilibrium. When the shape of a given mass of liquid is changed in such a way as to increase its surface area, there is a transfer of mass from the bulk liquid to the surface film, just as there is a transfer of mass from liquid to vapor when the volume of a cylinder containing liquid and vapor is increased.

It is found that in order to keep the temperature of the system constant when its surface area is increased, heat must be supplied. Let us define a quantity λ, analogous to the latent heat of vaporization, as the heat supplied per unit increase of area at constant temperature:

$$d'Q_T = \lambda \, dA_T. \tag{8–34}$$

If a film of liquid is formed on a wire frame as in Fig. 3–6, the inward force exerted on the frame as indicated by the short arrows originates in the surface layers as if they were in a state of tension. The force per unit length of boundary is called the *surface tension* σ, and we have shown in Section 3–3 that the work, when the slider is moved down a short distance dx and the area of the film increases by dA, is

$$d'W = -\sigma \, dA.$$

Although the area of the film increases, the surface tension force is found to remain constant if the temperature is constant. That is, the surface tension σ

does not depend on the area but only on the temperature. Thus the film does not act like a rubber membrane, for which the force would increase with increasing area. As the slider is moved down, molecules move from the bulk liquid into the film. The process does not consist of stretching a film of constant mass, but rather of creating an additional area of film whose properties depend only on the temperature.

If the temperature of the system is changed, however, the surface tension changes. Thus surface tension is analogous to vapor pressure, remaining constant for two phases in equilibrium if the temperature is constant, but changing with changing temperature. Unlike the vapor pressure, however, which increases with increasing temperature, the surface tension decreases with increasing temperature, as shown on Fig. 8–2, and becomes zero at the critical temperature, where the properties of liquid and vapor become identical.

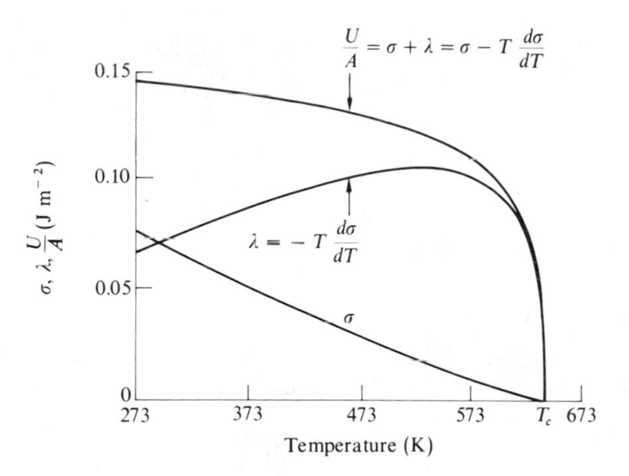

Fig. 8–2 Surface tension σ, "latent heat" λ, and surface energy per unit area U/A, for water, as a function of temperature.

Consider an isothermal process in which the area of a surface film increases by dA_T. The heat flow into the film is $d'Q_T = \lambda\, dA_T$, the work is $d'W_T = -\sigma\, dA_T$, and the increase in internal energy, which in this case is the *surface energy*, is

$$dU_T = d'Q_T - d'W_T = (\lambda + \sigma)\, dA_T.$$

Therefore

$$\frac{dU_T}{dA_T} = \left(\frac{\partial U}{\partial A}\right)_T = \lambda + \sigma. \tag{8–35}$$

Since the work in a process is $-\sigma\, dA$, a surface film is analogous to a PVT

system, for which the work is $P\,dV$. The surface tension σ corresponds to $-P$, and the area A to the volume V. Hence we can write, by analogy with Eq. (6–9),

$$\left(\frac{\partial U}{\partial A}\right)_T = \sigma - T\frac{d\sigma}{dT},$$

where $(\partial\sigma/\partial T)_A$ has been replaced with $d\sigma/dT$, since σ is a function of T only.

From the two preceding equations,

$$\lambda = -T\frac{d\sigma}{dT}, \tag{8–36}$$

which relates the "latent heat" λ to the surface tension σ. Figure 8–2 also shows a graph of λ versus T. (Because σ is a function of temperature only, the same is true of λ.)

Suppose the area of the film is increased isothermally from zero to A, by starting with the slider in Fig. 3–6 at the top of the frame and pulling it down. Since $U = 0$ when $A = 0$, the surface energy, when the area is A, is

$$U = (\lambda + \sigma)A = \left(\sigma - T\frac{d\sigma}{dT}\right)A; \tag{8–37}$$

that is, the surface energy is a function of both T and A. The surface energy per unit area is

$$\frac{U}{A} = \lambda + \sigma = \sigma - T\frac{d\sigma}{dT}.$$

A graph of U/A is also included in Fig. 8–2. Its ordinate at any temperature is the sum of the ordinates of the graphs of λ and σ.

By analogy with the heat capacity at constant volume of a PVT system, the heat capacity at constant area, C_A, is

$$C_A = \left(\frac{\partial U}{\partial T}\right)_A.$$

From Eq. (8–37),

$$\left(\frac{\partial U}{\partial T}\right)_A = A\left[\frac{d\sigma}{dT} - T\frac{d^2\sigma}{dT^2} - \frac{d\sigma}{dT}\right] = -AT\frac{d^2\sigma}{dT^2},$$

and hence,

$$C_A = -AT\frac{d^2\sigma}{dT^2}. \tag{8–38}$$

The specific heat capacity c_A is the heat capacity per unit area:

$$c_A = -T\frac{d^2\sigma}{dT^2}.$$

The internal energy U and Helmholtz function F are related by the equation

$$U = F - T\left(\frac{\partial F}{\partial T}\right)_A.$$

Comparison with Eq. (8–37) shows that the Helmholtz function for a surface film is

$$F = \sigma A;$$

and hence

$$\sigma = \frac{F}{A}; \tag{8–39}$$

that is, the surface tension equals the Helmholtz function per unit area.

The entropy of the film is

$$S = -\left(\frac{\partial F}{\partial T}\right)_A = -A\frac{d\sigma}{dT},$$

and the entropy per unit area is

$$s = -\frac{d\sigma}{dT}. \tag{8–40}$$

8–5 VAPOR PRESSURE OF A LIQUID DROP

The surface tension of a liquid drop causes the pressure inside the drop to exceed that outside. As shown in Section 8–3, this increased pressure results in an increase in vapor pressure, an effect which has an important bearing on the condensation of liquid drops from a supercooled vapor.

Consider a spherical drop of liquid of radius r, in equilibrium with its vapor. Figure 8–3 is an "exploded" view of the drop. The vertical arrows represent the surface tension forces on the lower half of the drop, the total upward force being

$$2\pi r\sigma.$$

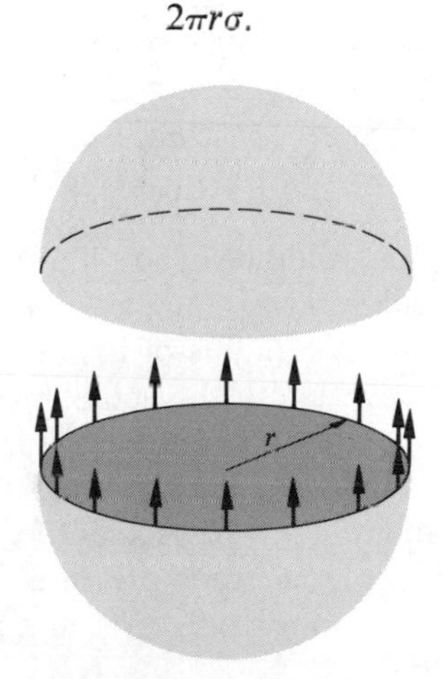

Fig. 8–3 Surface tension forces in a spherical drop.

Let P_i be the internal pressure and P_e the external pressure. The resultant downward force on the lower half of the drop due to these pressures is

$$(P_\text{i} - P_\text{e})\pi r^2;$$

and for mechanical equilibrium,

$$(P_\text{i} - P_\text{e})\pi r^2 = 2\pi r\sigma,$$

or

$$P_\text{i} - P_\text{e} = \frac{2\sigma}{r}.$$

The pressure P_i in the liquid therefore exceeds the external pressure P_e by $2\sigma/r$. The smaller the radius of the drop, the greater the pressure difference.

For complete thermodynamic equilibrium, the pressure P_e must equal the vapor pressure p. We can use Eq. (8–33) to find the vapor pressure p, which will be larger than the vapor pressure p_0 at a plane surface. In Eq. (8–33), the symbol P represented the total pressure of the liquid, which in the present problem is the pressure $P_\text{i} = P_\text{e} + 2\sigma/r = p + 2\sigma/r$, since $P_\text{e} = p$ when the system is in equilibrium. Hence

$$\ln\frac{p}{p_0} = \frac{v''}{RT}\left[(p - p_0) + \frac{2\sigma}{r}\right].$$

In all cases of interest, the difference $(p - p_0)$ between the actual vapor pressure p and the vapor pressure p_0 at a flat surface is small compared with $2\sigma/r$ and can be neglected. Then

$$\ln\frac{p}{p_0} = \frac{2\sigma v''}{rRT},$$

or

$$r = \frac{2\sigma v''}{RT\ln(p/p_0)}, \tag{8–41}$$

and a liquid drop of this radius would be in equilibrium with its vapor at a pressure $P_\text{e} = p$. The equilibrium would *not* be stable, however. Suppose that by the chance evaporation of a few molecules the radius of the drop should decrease. Then the vapor pressure p would increase, and if the actual pressure P_e of the vapor did not change, the vapor pressure would exceed the pressure of the vapor. The system would not be in thermodynamic equilibrium, and the drop would continue to evaporate. On the other hand, if a few molecules of vapor should condense on the drop, its radius would increase, the vapor pressure would decrease, the pressure of the vapor would exceed the vapor pressure, and the drop would continue to grow.

The distinction between "vapor pressure p" and "pressure P_e of the vapor" can be confusing. The term "pressure P_e of the vapor" means the actual pressure exerted by the vapor surrounding the drop. The term "vapor pressure p" is the particular value that the "pressure P_e of the vapor" must have for thermodynamic equilibrium.

For water at 300 K, $\sigma \simeq 70 \times 10^{-3}$ N m^{-1}, $p_0 \simeq 27$ Torr $\simeq 3.6 \times 10^3$ N m^{-2}, and $v'' \simeq 18 \times 10^{-3}$ m^3 kilomole^{-1}. It is found that the pressure P_e of water vapor can be increased to at least 5 times the vapor pressure p_0 over a flat surface before drops of liquid start to form. Setting $p/p_0 = 5$, we find from the values above that

$$r \simeq 6 \times 10^{-10} \text{ m} \simeq 6 \times 10^{-8} \text{ cm}.$$

A drop of this radius contains only about twelve molecules, and there is some question as to whether it is legitimate to speak of it as a sphere with a definite radius and surface tension. However, if a group of this number of molecules should form in the vapor it would continue to grow once it had been formed.

8–6 THE REVERSIBLE VOLTAIC CELL

It was shown in Section 3–3, that when a charge dZ flows through a voltaic cell of emf $\mathscr{E}$, the work is

$$d'W = -\mathscr{E}\, dZ.$$

If there are gaseous products of reaction, $P\,dV$ work must be included also, but we shall neglect any changes in volume and treat the cell as an $\mathscr{E}ZT$ system, corresponding to a PVT system. We also assume, as is nearly true in many cells, that the emf is a function of temperature only, so that

$$\left(\frac{\partial \mathscr{E}}{\partial T}\right)_Z = \frac{d\mathscr{E}}{dT}.$$

Every real cell has an internal resistance R, so that dissipative work at a rate I^2R is done within the cell when there is a current in it. Let the terminals of the cell be connected to a potentiometer. If the voltage across the potentiometer is made just equal to the emf of the cell, the current in the cell is zero. By making the voltage slightly larger or smaller than the emf, the reaction in the cell can be made to go in either direction. Further, since the dissipative work is proportional to the *square* of the current, while the electrical work is proportional to the first power, the former can be made negligible by making the current very small. Hence the cell can be operated as a reversible system in the thermodynamic sense.

It is found, however, that even when the current I is very small so that I^2R heating is negligible, there may still be a heat flow into or out of the cell from its surroundings in an isothermal process. Let us define a quantity ψ as the heat flow per unit charge, so that in an isothermal process,

$$d'Q_T = \psi\, dZ_T.$$

The change in internal energy is then

$$dU_T = d'Q_T - d'W_T = (\psi + \mathscr{E})\, dZ_T,$$

and

$$\frac{dU_T}{dZ_T} = \left(\frac{\partial U}{\partial Z}\right)_T = \psi + \mathscr{E}. \tag{8–42}$$

By analogy with Eq. (6–9),

$$\left(\frac{\partial U}{\partial Z}\right)_T = \mathscr{E} - T\frac{d\mathscr{E}}{dT}, \tag{8–43}$$

and therefore

$$\psi = -T\frac{d\mathscr{E}}{dT}. \tag{8–44}$$

Since $\mathscr{E}$ is a function of T only, the same is true of ψ. The heat flow in an isothermal process is therefore

$$d'Q_T = \psi\, dZ_T = -T\frac{d\mathscr{E}}{dT}\, dZ. \tag{8–45}$$

When the cell "discharges" and does electrical work on the circuit to which it is connected, dZ is a negative quantity. Hence if the emf *increases* with temperature, $d\mathscr{E}/dT$ is positive, $d'Q_T$ is positive, and there is a heat flow *into* the cell from its surroundings. On the other hand, if $d\mathscr{E}/dT$ is negative, then $d'Q_T$ is negative when the cell discharges and there is a heat flow *out* of the cell, even in the absence of any I^2R heating.

The isothermal work is

$$d'W_T = -dU_T + d'Q_T.$$

Thus if $d'Q_T$ is positive, the work is greater than the decrease in internal energy; and if $d'Q_T$ is negative, the work is less than the decrease in internal energy. In the former case, the cell absorbs heat from its surroundings and "converts it into work." Of course, there is no conflict with the second law because this is not the *sole* result of the process. In the latter case, a portion of the decrease in internal energy appears as a flow of heat to the surroundings.

In a finite isothermal process in which a change ΔZ flows through the cell, the heat flow is

$$Q_T = -T\frac{d\mathscr{E}}{dT}\Delta Z_T, \tag{8–46}$$

The work is

$$W_T = -\mathscr{E}\,\Delta Z_T, \tag{8–47}$$

and the change in internal energy is

$$\Delta U_T = \left(\mathscr{E} - T\frac{d\mathscr{E}}{dT}\right)\Delta Z. \tag{8–48}$$

In physical chemistry, Eq. (8–48) is most useful when looked on as a method of measuring heat of reaction. As a specific example, the Daniell* cell consists of a zinc electrode in a solution of zinc sulfate, and a copper electrode in a solution of copper sulfate. When the cell discharges, zinc goes into solution and copper is deposited on the copper electrode. The net chemical effect is the disappearance of Zn and Cu^{++} and the appearance of Zn^{++} and Cu, as represented by

$$\mathrm{Zn} + \mathrm{Cu}^{++} \rightarrow \mathrm{Zn}^{++} + \mathrm{Cu}.$$

By forcing a current through the cell in the opposite direction the process can be reversed, that is, copper goes into solution and zinc is deposited.

The same chemical reaction can be made to take place in a purely chemical manner, quite apart from a Daniell cell. Thus if zinc powder is shaken into a solution of copper sulfate, all the zinc will dissolve (i.e., become ions in solution) and all the copper ions will become metal atoms, provided the original amounts of the two substances are chosen properly. If the process takes place at constant volume, no work is done and the heat liberated equals the change in internal energy, given by Eq. (8–48).

Since emf's can be measured very precisely, then (provided two reacting substances can be combined to form a voltaic cell) the heat of reaction can be computed from measurements of the emf and its rate of change with temperature more precisely than it can be found by direct experiment.

For example, when 1 kilomole of copper and zinc react directly at 273 K, the internal energy change as measured experimentally by calorimetric methods is 232×10^6 J. When the substances are combined to form a voltaic cell at 273 K, the observed emf is 1.0934 V and its rate of change with temperature is -0.453×10^{-3} V K^{-1}. Because the ions are divalent, the charge ΔZ passing through the cell is 2 faradays† per kilomole, or $2 \times 9.649 \times 10^7$ C kilomole^{-1}. Then the internal energy change is found to be

$$\Delta U = 234.8 \times 10^6 \text{ J kilomole}^{-1}.$$

8–7 BLACKBODY RADIATION

The principles of thermodynamics can be applied not only to material substances but also to the radiant energy within an evacuated enclosure. If the walls of the enclosure are at a uniform temperature T, and the enclosure contains at least a speck of a *complete absorber* or *blackbody* (a substance which absorbs 100% of the radiant energy incident on it, at any wavelength), the radiant energy within the enclosure is a mixture of electromagnetic waves of different energies and of all possible frequencies from zero to infinity. Suppose that an opening is made in the walls of the enclosure, small enough so that the radiant energy escaping through

* John F. Daniell, English chemist (1790–1845).

† Michael Faraday, English physical chemist (1791–1867).

the opening does not appreciably affect that within the enclosure. It is found experimentally that the rate at which radiant energy is emitted from the opening, per unit area, is a function only of the temperature T of the walls of the enclosure and does not depend on their nature, or on the volume V or shape of the enclosure. The rate of radiation of energy through the opening is proportional to the radiant energy per unit volume *within* the enclosure, or to the *radiant energy density* $\mathbf{u}$, where

$$\mathbf{u} = \frac{U}{V}.$$

Hence we conclude that the energy density $\mathbf{u}$ is also a function only of the temperature T:

$$\mathbf{u} = \mathbf{u}(T).$$

It follows from electromagnetic theory that if the radiant energy in the enclosure is isotropic (the same in all directions) it exerts on the walls of the enclosure a pressure P equal to one-third of the energy density:

$$P = \frac{1}{3}\mathbf{u}. \tag{8–49}$$

The radiation pressure, like the energy density, is a function of T only and is independent of the volume V.

The energy density, the frequency, and the temperature are found *experimentally* to be related by an equation known as *Planck's law*, according to which the energy density $\Delta\mathbf{u}_\nu$ in an interval of frequencies between ν and $\nu + \Delta\nu$, and at a temperature T, is given by

$$\Delta\mathbf{u}_\nu = \frac{c_1\nu^3}{\exp(c_2\nu/T) - 1}\,\Delta\nu, \tag{8–50}$$

where c_1 and c_2 are constants whose values depend only on the system of units employed. The dependence of the *total* energy density on temperature can be found by integrating Planck's equation over all frequencies from zero to infinity, but the principles of thermodynamics enable us to find the *form* of this dependence without a knowledge of the exact form of Planck's equation. To do this, we again make use of Eq. (6–9), which is derived from the combined first and second laws and which we now write in extensive form:

$$\left(\frac{\partial U}{\partial V}\right)_T = T\left(\frac{\partial P}{\partial T}\right)_V - P. \tag{8–51}$$

Since $U = \mathbf{u}V$, and $\mathbf{u}$ is a function of T only,

$$\left(\frac{\partial U}{\partial V}\right)_T = \mathbf{u}. \tag{8–52}$$

Also, since both P and $\mathbf{u}$ are functions of T only,

$$\left(\frac{\partial P}{\partial T}\right)_V = \frac{1}{3}\left(\frac{\partial \mathbf{u}}{\partial T}\right)_V = \frac{1}{3}\frac{d\mathbf{u}}{dT}. \tag{8–53}$$

Hence Eq. (8–51) becomes

$$\mathbf{u} = \frac{1}{3}T\frac{d\mathbf{u}}{dT} - \frac{1}{3}\mathbf{u},$$

$$\frac{d\mathbf{u}}{\mathbf{u}} = 4\frac{dT}{T},$$

$$\mathbf{u} = \sigma T^4, \tag{8–54}$$

where σ is a constant.

The energy density is therefore proportional to the 4th power of the thermodynamic temperature, a fact which was discovered experimentally by Stefan* before the theory had been developed by Planck and which is called *Stefan's law*, or the *Stefan-Boltzmann*† law. The best experimental value of the Stefan-Boltzmann constant σ is

$$\sigma = 7.561 \times 10^{-16}\ \mathrm{J\ m^{-3}\ K^{-4}}. \tag{8–55}$$

From Eqs. (8–49) and (8–54), the equation of state of the radiant energy in an evacuated enclosure is

$$P = \frac{1}{3}\mathbf{u} = \frac{1}{3}\sigma T^3. \tag{8–56}$$

The total energy U in a volume V is

$$U = \mathbf{u}V = \sigma V T^4. \tag{8–57}$$

The heat capacity at constant volume V is

$$C_V = \left(\frac{\partial U}{\partial T}\right)_V = 4\sigma V T^3. \tag{8–58}$$

To find the entropy, imagine that the temperature of the walls of an enclosure at constant volume is increased from $T = 0$ to $T = T$. Then

$$S = \int_0^T \frac{1}{T} C_V\, dT = \int_0^T 4\sigma V T^2\, dT,$$

and hence

$$S = \frac{4}{3}\sigma V T^3. \tag{8–59}$$

---

* Josef Stefan, Austrian physicist (1835–1893).

† Ludwig Boltzmann, Austrian physicist (1844–1906).

The Helmholtz function is

$$F = U - TS = \sigma V T^4 - \frac{4}{3}\sigma V T^4,$$

and

$$F = -\frac{1}{3}\sigma V T^4. \tag{8–60}$$

The Gibbs function is

$$G = F + PV = -\frac{1}{3}\sigma V T^4 + \frac{1}{3}\sigma V T^4,$$

and hence

$$G = 0. \tag{8–61}$$

We shall return to a discussion of blackbody radiation in Section 13–3 and show how Planck's law, and the value of the Stefan-Boltzmann constant, can be determined by the methods of statistics and the principles of quantum theory.

8–8 THERMODYNAMICS OF MAGNETISM

We showed in Section 3–3 that in a process in which the magnetic moment M of a paramagnetic system is changed by dM, the work is

$$d'W = -\mathscr{H}\, dM,$$

where $\mathscr{H}$ is the external magnetic field intensity.

The magnetic systems of primary interest in thermodynamics are paramagnetic crystals, whose volume change in a process can be neglected and for which the "$P\,dV$" work is negligible compared with $-\mathscr{H}\,dM$. Such crystals have an internal energy U, and also a magnetic potential energy

$$E_p = -\mathscr{H}M. \tag{8–62}$$

As described in Section 3–13, the appropriate energy function is therefore the *total* energy E:

$$E = U + E_p = U - \mathscr{H}M, \tag{8–63}$$

$$dE = dU - \mathscr{H}\, dM - M\, d\mathscr{H}.$$

The combined first and second laws state that

$$T\,dS = dU + d'W = dU - \mathscr{H}\, dM. \tag{8–64}$$

Hence in terms of E,

$$T\,dS = dE + M\, d\mathscr{H}. \tag{8–65}$$

Comparison with Eq. (7–23),

$$T\,dS = dH - V\,dP, \tag{8–66}$$

shows that the total energy E is the magnetic analogue of the enthalpy H of a PVT system, and some authors speak of it as the "magnetic enthalpy" and represent it by H^*. There is an important distinction, however. The enthalpy H of a PVT system is defined as

$$H = U + PV,$$

and the total energy E of a magnetic system as

$$E = U - \mathscr{H}M.$$

In the latter equation, the term $-\mathscr{H}M$ is the potential energy of the system in a conservative external magnetic field and is a joint property of the system and the source of the field, while no such significance attaches to the product PV. Thus the correspondence between Eqs. (8–65) and (8–66) is a mathematical analogy only. But since the equations do have the same *form*, we can take over all of the equations previously derived for the enthalpy H, replacing H with E, V with $-M$, and P with $\mathscr{H}$.

Thus the heat capacity at constant $\mathscr{H}$, corresponding to C_P, is

$$C_{\mathscr{H}} = \left(\frac{\partial E}{\partial T}\right)_{\mathscr{H}}. \tag{8–67}$$

The heat capacity at constant M, corresponding to C_V, is

$$C_M = \left(\frac{\partial U}{\partial T}\right)_M. \tag{8–68}$$

The first and second $T\,dS$ equations become

$$T\,dS = C_M\,dT - T\left(\frac{\partial \mathscr{H}}{\partial T}\right)_M dM, \tag{8–69}$$

$$T\,dS = C_{\mathscr{H}}\,dT + T\left(\frac{\partial M}{\partial T}\right)_{\mathscr{H}} d\mathscr{H}. \tag{8–70}$$

In Section 7–2 we have defined a function F^*, corresponding to the Helmholtz function $F = U - TS$, as

$$F^* = E - TS. \tag{8–71}$$

Then

$$dF^* = dE - T\,dS - S\,dT,$$

and making use of Eq. (8–65), we have

$$dF^* = -S\,dT - M\,d\mathscr{H}. \tag{8–72}$$

Therefore

$$\left(\frac{\partial F^*}{\partial T}\right)_{\mathscr{H}} = -S, \qquad \left(\frac{\partial F^*}{\partial \mathscr{H}}\right)_T = -M. \tag{8–73}$$

The methods of statistics, as we shall show later, lead directly to an expression for F^* as a function of T and $\mathscr{H}$. Then from the second part of Eq. (8–73) we can find M as a function of T and $\mathscr{H}$, which is the *magnetic equation of state* of the system. The first equation gives S as a function of T and $\mathscr{H}$. The energy E is then found from Eq. (8–71),

$$E = F^* + TS,$$

and the internal energy U is

$$U = E + \mathscr{H}M. \tag{8–74}$$

Thus all properties of the system can be found from the expression for F^* as a function of T and $\mathscr{H}$.

The dependence of the entropy on the magnetic intensity can be determined by the method used to derive the Maxwell relations. Applying Eq. (7–39) to Eq. (8–72) we obtain

$$\left(\frac{\partial S}{\partial \mathscr{H}}\right)_T = \left(\frac{\partial M}{\partial T}\right)_{\mathscr{H}}. \tag{8–75}$$

For a paramagnetic salt obeying Curie's law, $(\partial M/\partial T)_{\mathscr{H}} < 0$ and the entropy of a paramagnetic salt decreases as the magnetic intensity increases.

In our discussion of the third law in Section 7–7, it was stated that all processes taking place in a condensed system at $T = 0$ K proceed with no change in entropy. If these processes include the increase in magnetic intensity in a paramagnetic crystal, it follows that at $T = 0$ K,

$$\left(\frac{\partial S}{\partial \mathscr{H}}\right)_T = 0. \tag{8–76}$$

Figure (8–4) is a plot of the entropy of a magnetic system as a function of temperature for values of the applied intensity $\mathscr{H}$ equal to zero and to $\mathscr{H}_1$. The form of these curves will be calculated in Section 13–4.

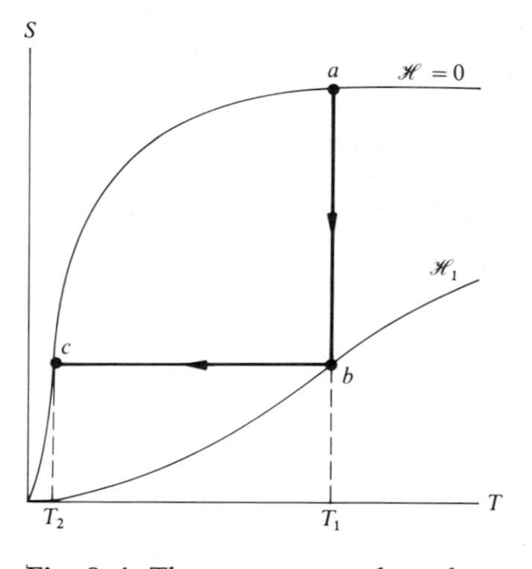

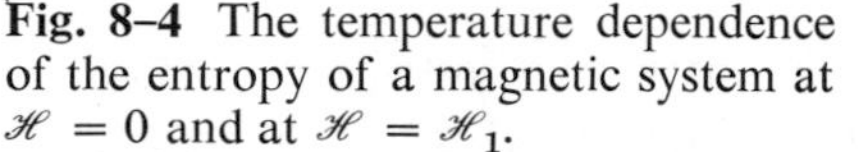
Fig. 8–4 The temperature dependence of the entropy of a magnetic system at $\mathscr{H} = 0$ and at $\mathscr{H} = \mathscr{H}_1$.

Substituting Eq. (8–76) into Eq. (8–75) we obtain that $(\partial M/\partial T)_{\mathscr{H}} = 0$ at $T = 0$. However from Curie's law

$$M = \frac{C_C \mathscr{H}}{T},$$

and $(\partial M/\partial T)_{\mathscr{H}}$ approaches infinity as $T \to 0$. The conclusion is that Curie's law cannot hold at $T = 0$ and that a transition to an ordered magnetic state must take place.

The production of low temperatures by adiabatic demagnetization of a paramagnetic salt can be understood with the help of Fig. 8–4. Suppose that initially the magnetic intensity is zero and that the temperature of the salt has been reduced to a low value T_1 by contact with a bath of liquid helium. The state of the system is then represented by point *a*. The magnetic field is now increased isothermally and reversibly, in the process *a-b*, to a value $\mathscr{H}_1$. In this process there is a heat flow out of the salt into the helium bath. The entropy of the system decreases while its temperature remains constant at T_1. In the isothermal process *a-b* in which $dT = 0$, Eq. (8–70) yields

$$d'Q_T = T\,dS_T = T\left(\frac{\partial M}{\partial T}\right)_{\mathscr{H}} d\mathscr{H}_T.$$

At constant $\mathscr{H}$, $(\partial M/\partial T)_{\mathscr{H}}$ is negative. Then since $\mathscr{H}$ increases, $d'Q_T$ is negative and there is a heat flow *out* of the system to the surroundings.

The next step is to isolate the system thermally from the surroundings and perform the reversible adiabatic process *b-c*, in which the magnetic field is reduced to zero while the entropy remains constant. The final temperature T_2, from Fig. 8–4, is evidently less than the initial temperature T_1. In this process, since $dS = 0$, Eq. (8–70) becomes

$$dT_S = -\frac{T}{C_{\mathscr{H}}}\left(\frac{\partial M}{\partial T}\right)_{\mathscr{H}} d\mathscr{H}_S,$$

and since $(\partial M/\partial T)_{\mathscr{H}}$ and $d\mathscr{H}_S$ are both negative, dT_S is negative also. Temperatures near 10^{-3} K have been attained in this way.

The processes *a-b* and *b-c* in Fig. 8–4 are exactly analogous to those in which a gas is first compressed isothermally and reversibly, and then allowed to expand to its original volume, reversibly and adiabatically. The temperature drop in the adiabatic expansion corresponds to the temperature drop from T_1 to T_2, in process *b-c* in Fig. 8–4.

Process *b-c*, in Fig. 8–4, is commonly described as a "reversible adiabatic demagnetization," or as an "isentropic demagnetization." Suppose, however, that such a process is carried out in a temperature interval in which C_M is negligible, so that

$$C_{\mathscr{H}} = \left(\frac{\partial E_p}{\partial T}\right)_{\mathscr{H}} = -\mathscr{H}\left(\frac{\partial M}{\partial T}\right)_{\mathscr{H}}.$$

Then from Eq. (8–70), in an isentropic process in which $dS = 0$,

$$\mathscr{H}\left(\frac{\partial M}{\partial T}\right)_{\mathscr{H}} dT_S = T\left(\frac{\partial M}{\partial T}\right)_{\mathscr{H}} d\mathscr{H}_S,$$

and

$$\frac{dT_S}{T} = \frac{d\mathscr{H}_S}{\mathscr{H}},$$

$$\left(\frac{\mathscr{H}}{T}\right)_S = \text{constant}. \tag{8–77}$$

The ratio $\mathscr{H}/T$ is therefore constant in the isentropic process in which the magnetic field is reduced from $\mathscr{H}_1$ to zero. Hence since the magnetic moment M is a function of $\mathscr{H}/T$, the magnetic moment is constant also and the term "demagnetization" is inappropriate.

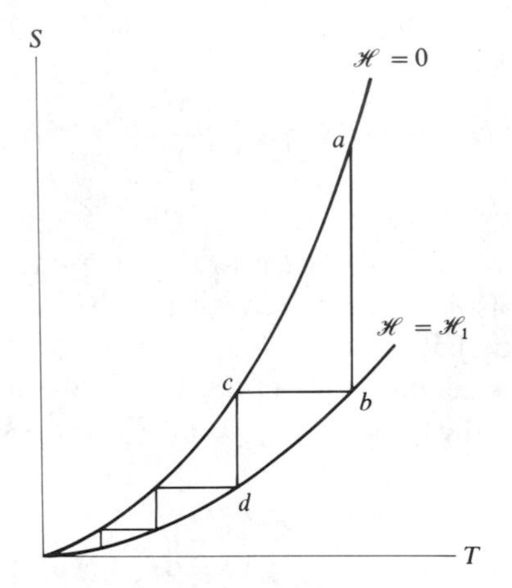

Fig. 8–5 The unattainability of the absolute zero of temperature by a finite series of isothermal magnetizations and adiabatic demagnetizations.

Suppose that a series of isothermal magnetizations from $\mathscr{H} = 0$ to $\mathscr{H} = \mathscr{H}_1$, represented by the vertical lines in Fig. 8–5, are each followed by adiabatic demagnetizations, represented by the horizontal lines. In order to carry out the isothermal magnetizations, in which there is a heat flow out of the crystal, reservoirs at lower and lower temperatures are required, so that the processes become more and more difficult experimentally as the temperature decreases. It will be seen that every adiabatic demagnetization process intersects the curve $\mathscr{H} = 0$ at a temperature above $T = 0$. This is an example of the unattainability statement of the third law. We leave it as a problem to show that if the entropy is not zero at $T = 0$ for $\mathscr{H} = 0$, the absolute zero of temperature could be reached in a finite number of processes in violation of the unattainability statement of the third law.

8–9 ENGINEERING APPLICATIONS

The prospect of continuously "converting heat into work" has intrigued man since ancient times. The credit for some of the most significant contributions to the science of thermodynamics is due to the successful achievement of this conversion, so important to the evolution of our modern civilization. The power cycle, which is the instrument for the continuous conversion of heat into work, presents an illuminating application of the first and second laws that is always exacting and often can be very subtle. This section is devoted to a thermodynamic analysis of a power cycle in which the working substance undergoes a change of phase. Specifically, steam is used as the working substance for the purpose of discussion, but the general principles are applicable to all other similar substances.

Figure 8–6 is a diagram of the s-P-T surface for the liquid and vapor phase of water substance. The surface resembles a P-v-T surface. It can be drawn to scale because the relative *entropy* change between liquid and vapor phases is much smaller than the relative *volume* change. Lines have been constructed on the surface at constant P, T, and s.

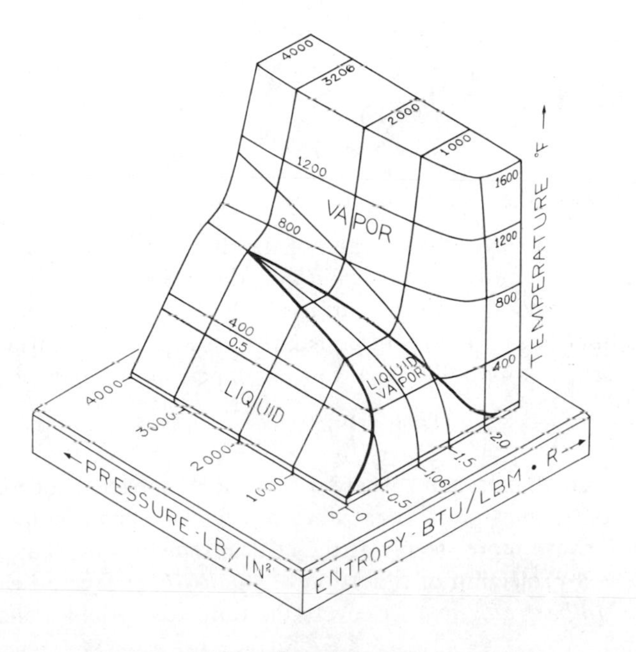

Fig. 8–6 The s-P-T surface for water.

The numerical values of P, T, and s are given, in Fig. 8–6, in the archaic set of units still employed by mechanical engineers in the United States. The unit of pressure is 1 pound-force per square inch, the unit of energy is 1 Btu, and the unit of mass is 1 pound-mass. On the temperature axis, temperatures are expressed in degrees Fahrenheit, but the unit of specific entropy is 1 Btu per pound-mass, per rankine. It is little wonder that engineering students in this country lose sight of the *principles* of thermodynamics because of the welter of conversion factors involved in numerical calculations.

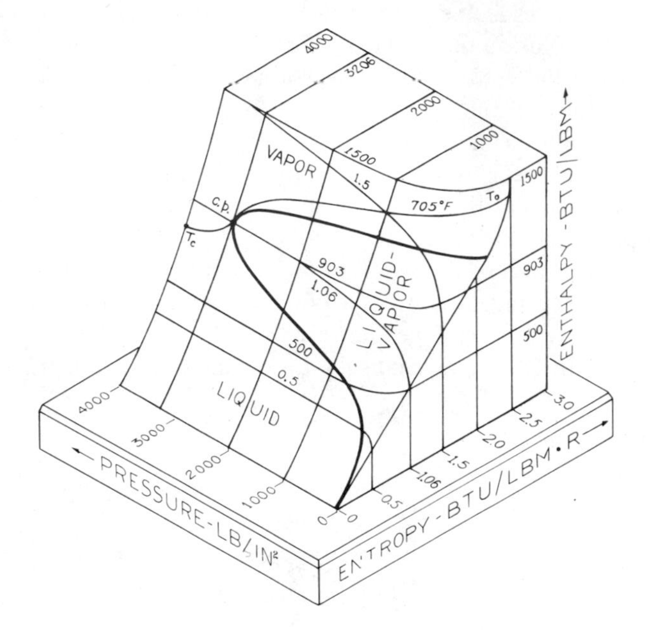

Fig. 8–7 The *h-s-P* surface for water.

Figure 8–7 is a drawing, also to scale, of the thermodynamic surface obtained by plotting the specific enthalpy vertically and the pressure and specific entropy horizontally. The heavy line on the surface is the boundary of the liquid-vapor region and the light lines are lines of constant h, s, and P. Isobaric lines on the surface have a slope at any point equal to the temperature at that point, since

$$\left(\frac{\partial h}{\partial s}\right)_P = T.$$

Hence in the liquid-vapor region, where a reversible isobaric process is also isothermal, the isobaric lines are straight lines having a constant slope equal to T. The lines slope upward more steeply as the critical temperature is approached.

Figure 8–8 is a projection of a portion of the *h-s-P* surface on the *h-s* plane, and is called a *Mollier* diagram*. It covers the range of variables encountered in

* Richard Mollier, German engineer (1863–1935).

most engineering calculations. The practical utility of the diagram lies in the fact that in any process at constant pressure, such as the conversion of liquid water to water vapor in the boiler of a steam engine, the heat flow is equal to the difference in enthalpy h between the endpoints of the process, and this difference can be read directly from a Mollier diagram.

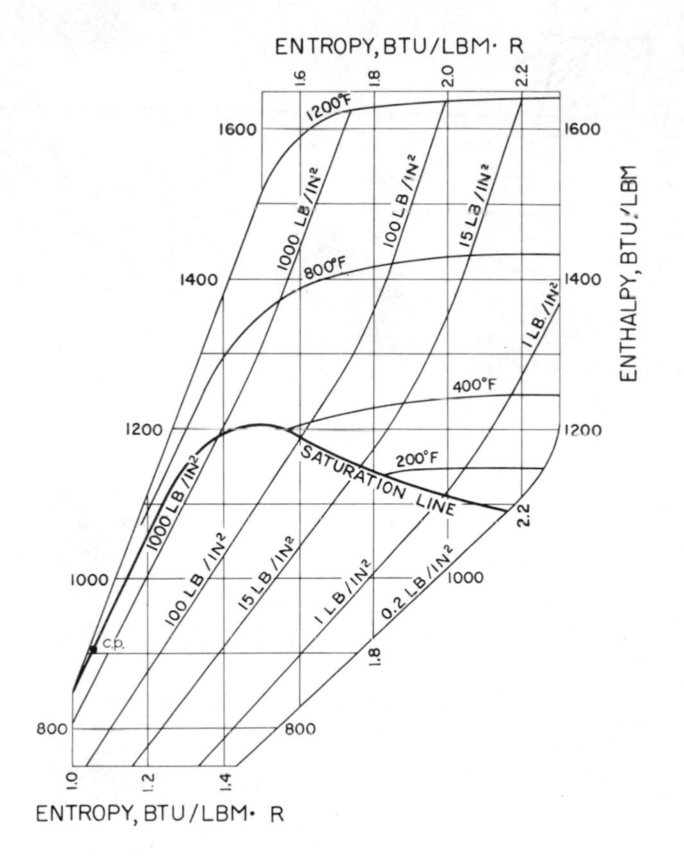

Fig. 8–8 The Mollier diagram for water.

In our earlier discussions of Carnot cycles, it has been tacitly assumed that the substance carried through the cycle underwent no changes in phase. However, a Carnot cycle is any reversible cycle bounded by two isothermals and two adiabatics, and the shaded areas *bcfg* in Fig. 8–9 represent a Carnot cycle operated in the liquid-vapor region. In part (a) of the figure, the cycle is shown on a *P-v-T* surface, and projected on the *P-v* plane. Part (b) shows the same cycle on the *s-P-T* surface and projected on the *T-s* plane, and in part (c) it is shown on the *h-s-P* surface and projected on the *h-s* plane (a Mollier diagram).

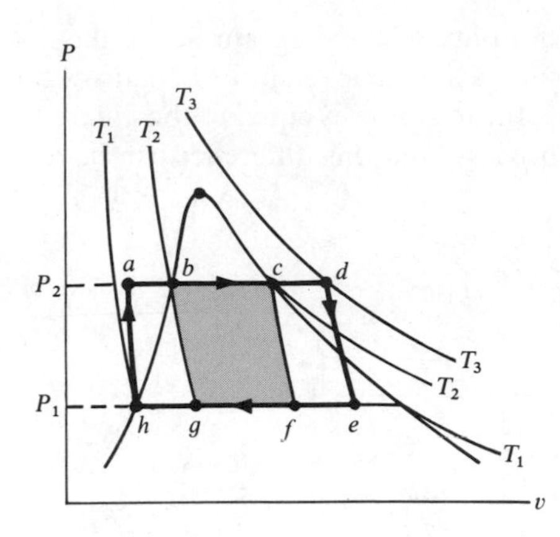

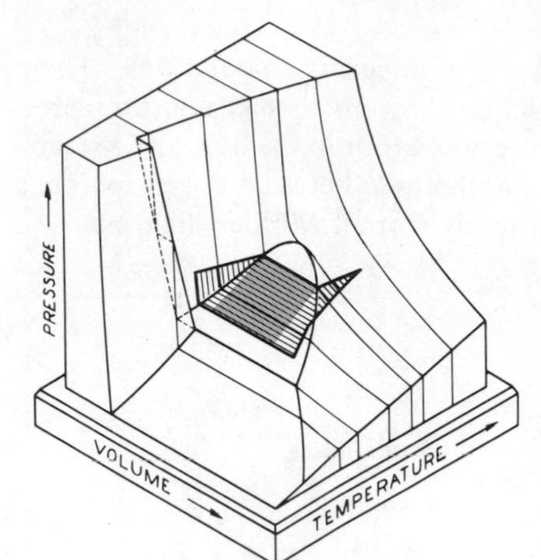

(a)

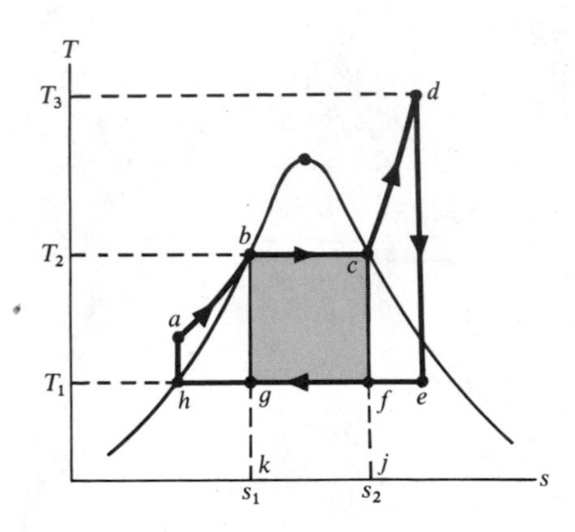

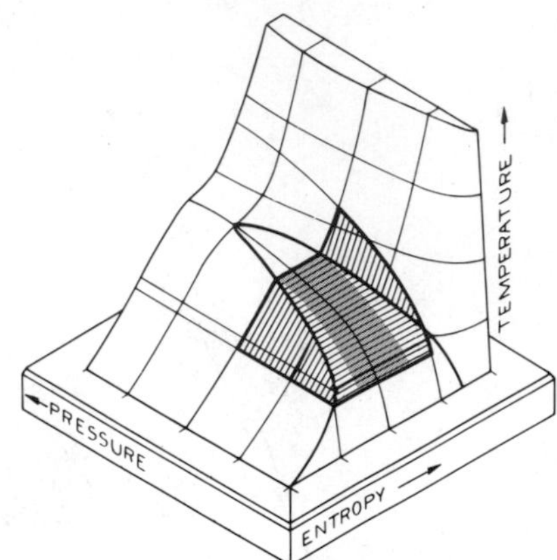

(b)

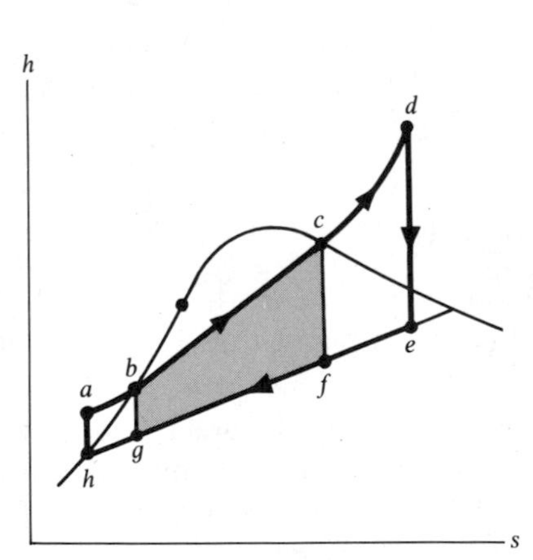

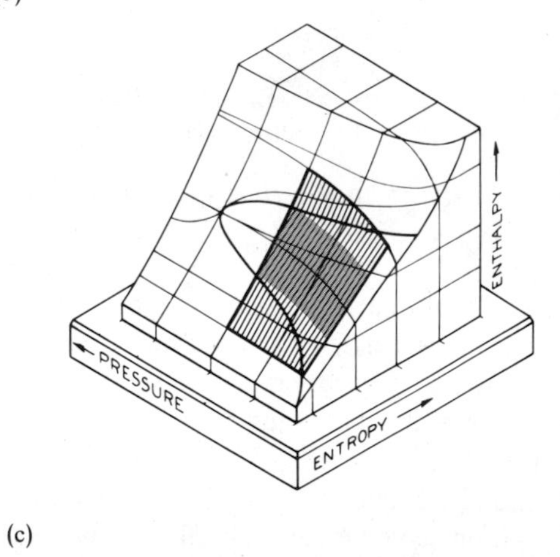

(c)

Fig. 8–9 The Carnot cycle *bcfg* in the liquid vapor region and the Rankine cycle *abcdefgh* with superheat.

Starting with saturated liquid at point b, we carry out a reversible isothermal expansion at the temperature T_2 until the liquid is completely vaporized (point c). During this part of the cycle heat q_2 is withdrawn from a reservoir at temperature T_2. An adiabatic expansion of the vapor lowers the temperature to T_1 (point f). If the material is water-substance, this adiabatic expansion carried us back into the liquid-vapor region. In other words, some of the saturated vapor condenses. (Not all substances behave in this way. For some, the slope of the adiabatic line is less than that of the saturation line and the point corresponding to f lies in the vapor region.) An isothermal compression is now carried out at the temperature T_1 to the state represented by point g, and heat q_1 is rejected to a reservoir. The cycle is completed by an adiabatic compression to point b, during which the remainder of the vapor condenses and the temperature increases to T_2. Note that in the *T-s* diagram of Fig. 8–9(b), the Carnot cycle projects as a rectangle, bounded by two isothermals and two adiabatics.

Since areas in a *T-s* diagram represent heat absorbed or liberated, the area *bcjk* in Fig. 8–9(b) represents the heat q_2 absorbed in the reversible expansion at temperature T_2, the area *gfjk* represents the heat q_1 rejected at temperature T_1, and, from the first law, the area *bcfg* represents the net work w done in the cycle. The thermal efficiency of the cycle is therefore

$$\eta = \frac{w}{q_2} = \frac{\text{area } bcfg}{\text{area } bcjk}$$

$$= \frac{(T_2 - T_1)(s_2 - s_1)}{T_2(s_2 - s_1)} = \frac{T_2 - T_1}{T_2},$$

as must be the case for *any* Carnot cycle operated between temperatures T_2 and T_1.

In the Mollier diagram of Fig. 8–9(c), reversible adiabatics are represented by vertical lines, and isotherms and isobars (which are the same in the liquid-vapor region) by straight lines sloping upward to the right. Since the heat flowing into a system in any reversible isobaric process is equal to the increase in enthalpy of the system, the heat q_2 supplied in the isothermal-isobaric expansion from b to c is equal to $h_c - h_b$. The heat q_1 given up in the isothermal compression from f to g is $h_f - h_g$. The net work w done in the cycle is equal to the difference between the magnitudes of q_2 and q_1. The thermal efficiency is therefore

$$\eta = \frac{w}{q_2} = \frac{h_c - h_b - h_f + h_g}{h_c - h_b}. \tag{8–78}$$

The advantage of the Mollier diagram is that heat, work, and efficiency can all be determined from the *ordinates* of points in the cycle, obviously a simpler procedure than measurements of *area* which must be made on a *T-s* diagram. Of course, the values of h at points b, c, f, and g may be taken from tables instead of being read from a graph.

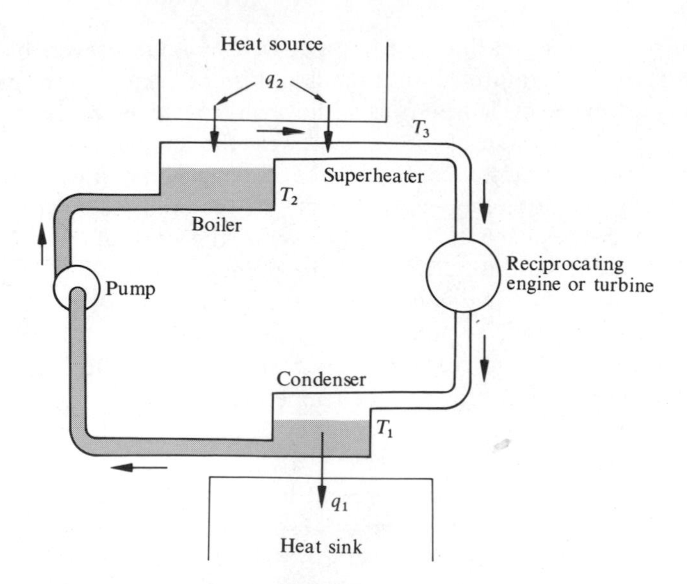

Fig. 8–10 Schematic diagram of processes in a reciprocating steam engine or turbine.

In both the *reciprocating steam engine* and the *turbine*, liquid water and water vapor go through essentially the same sequence of states. The boiler in Fig. 8–10 receives heat from a heat source maintained at a high temperature by the combustion of fossil fuel, or by a nuclear reactor. In the boiler, saturated liquid is converted to saturated vapor at a temperature determined by the pressure in this part of the system. This temperature is very much less than that of the heat source. For example, if the pressure in the boiler is 1000 lb in^{-2} (6.9×10^6 N m^{-2}), the temperature is 544°F (558 K), while the flame temperature in a source in which fuel is burned may be of the order of 3500°F (2200 K). The saturated steam is led from the boiler to the superheater, where it receives more heat from the source and its temperature increases. The superheater is connected directly to the boiler, thus the pressure of the superheated steam does not rise above boiler pressure. In principle, the temperature of the superheated steam could be increased to that of the source, but a limit of about 1000°F (811 K), called the *metallurgical limit*, is set by the fact that above this temperature the materials available for piping are not strong enough to withstand the high pressure.

The superheated steam then flows to the reciprocating engine or turbine, where it delivers mechanical work and at the same time undergoes a drop in temperature and pressure. A portion is usually condensed in this part of the cycle also. The mixture of saturated liquid and vapor then flows to the condenser, where the remaining vapor is liquefied and the heat of condensation is given up to a heat sink, which may be the atmosphere or cooling water from a river or the ocean. The pressure in this part of the system is determined by the temperature of the heat sink. The condensed liquid is then forced into the boiler by the pump. This completes the cycle.

The reciprocating engine and the turbine differ only in the means by which internal energy is abstracted from the flowing steam and converted to mechanical work. In the former, a mass of steam in a cylinder expands against a piston. In the latter, the steam flows through nozzles, as in Fig. 3–14, acquiring kinetic energy in the process. The rapidly moving steam then impinges on the buckets in the turbine rotor and gives up its kinetic energy. The process is approximately adiabatic in both devices but is not completely reversible and hence is not isentropic.

Note that as far as the steam cycle itself is concerned, the sequence of states is the same whether the heat source is a furnace in which fuel is burned, or is a nuclear reactor.

The *Rankine cycle* is a reversible cycle which corresponds more nearly than does the Carnot cycle to the sequence of states assumed by the liquid and vapor in a reciprocating steam engine or turbine. We consider first a cycle in which the steam is not superheated. Starting at point b in Fig. 8–9(c), which corresponds to the boiler in Fig. 8–10, saturated liquid is converted reversibly to saturated vapor at a temperature T_2 and pressure P_2 (point c). The vapor then expands reversibly and adiabatically to the pressure P_1 and temperature T_1 (point f). This stage corresponds to the passage of steam through the engine or turbine. The mixture of vapor and liquid is then completely liquefied at the temperature T_1 (point h) corresponding to the process in the condenser of Fig. 8–10. The liquid is then compressed reversibly and adiabatically to boiler pressure P_2 (point a). This operation is performed by the pump in Fig. 8–10. As we have seen, the temperature of a liquid increases only slightly in an adiabatic compression, so that heat must be supplied to the compressed liquid along the line ab in Fig. 8–9(c) to raise its temperature to T_2. In Fig. 8–10, this heating takes place after the liquid has been pumped into the boiler. If the cycle is to be *reversible*, however, the heat must be supplied by a series of heat reservoirs, ranging in temperature from that at point a, slightly above T_1, to T_2. The average temperature at which heat is supplied is therefore less than T_2, so the Rankine cycle, although reversible, has a lower thermal efficiency than the Carnot cycle in which heat is taken in only at the temperature T_2.

The thermal efficiency of the Rankine cycle can be determined directly from the Mollier diagram, Fig. 8–9(c), by the same method used for the Carnot cycle. Heat q_2 is supplied along the path a-b-c and heat q_1 is rejected along the path f-h. Although process a-b-c is not isothermal, it is isobaric (see Fig. 8–9a), and the heat q_2 supplied is equal to the enthalpy difference $h_c - h_a$. The heat q_1 rejected is $h_f - h_h$ and the net work w equals the difference between q_2 and q_1. The efficiency is therefore

$$\eta = \frac{w}{q_2} = \frac{h_c - h_a - h_f + h_h}{h_c - h_a}. \tag{8–79}$$

Note that although the expression for the efficiency in terms of enthalpy differences has the same form as that for the Carnot cycle, Eq. (8-79) does *not* reduce to $(T_2 - T_1)/T_2$, as is obvious from a comparison of the graphs of the Carnot and Rankine cycles. As stated above, the efficiency of the Rankine cycle is less than that of a Carnot cycle operating between temperatures T_2 and T_1.

It was mentioned in Section 5-8, in connection with the general subject of entropy and irreversibility, that irreversible processes in a heat engine result in a decrease in efficiency. We can now see how irreversibility affects the efficiency of a Rankine cycle. If the expansion of the steam in a reciprocating engine or turbine is *reversible* as well as adiabatic, it is also isentropic, and process *c-f* in Fig. 8-9(b) is a vertical line of constant entropy. If the expansion is *irreversible*, the entropy increases and at the end of the expansion the state of the system is represented by a point to the right of point *f*. The decrease in enthalpy in the process, from Fig. 8-9(c), is therefore less in the irreversible than in the reversible expansion. Now apply the energy equation of steady flow to a turbine. The elevations of intake and exhaust can be assumed the same, the velocities at intake and exhaust can be considered equal, and the process is very nearly adiabatic, even if it is not isentropic. The shaft work is therefore equal to the enthalpy difference between intake and exhaust, and the efficiency of the irreversible cycle is less than that of the reversible since the turbine delivers less mechanical work for the same heat input.

In practically all steam cycles the vapor is superheated to a temperature T_3 higher than that of the saturated vapor T_2 before it is expanded adiabatically (see Fig. 8-10). The corresponding Rankine cycle is then represented by the process *b-c-d-e-h-a-b* in Fig. 8-9(c). The superheating stage is represented by the segment *cd* in this figure. There are two reasons for superheating. One is that the average temperature at which heat is supplied is thereby increased above the temperature of vaporization, with a resulting increase in efficiency. The other, which is actually of greater importance, can be seen from an examination of Fig. 8-9(c). If the adiabatic expansion starts from the state of saturated vapor, point *c*, the state of the steam at the end of the expansion is represented by point *f*. If the expansion starts at point *d*, the state of the steam at the end of the expansion is represented by point *e*. The "moisture content" of the steam, that is, the fractional amount in the *liquid* phase, is greater at point *f* than at point *e*. If the moisture content is too great, mechanical wear on the turbine buckets becomes excessive. Hence superheating must be carried to a sufficiently high temperature to keep the moisture content down to a safe value.

In Fig. 8-9(c), heat q_2 is absorbed along the path *a-b-c-d*, and because this is isobaric, we have $q_2 = h_d - h_a$. Since $q_1 = h_e - h_h$, the efficiency is

$$\eta = \frac{w}{q_2} = \frac{h_d - h_a - h_e + h_h}{h_d - h_a}. \qquad (8\text{-}80)$$

PROBLEMS

8–1 A volume V is divided into two parts by a frictionless diathermal partition. There are n_A moles of an ideal gas A on one side of the partition and n_B moles of an ideal gas B on the other side. (a) Calculate the change in entropy of the system which occurs when the partition is removed. (b) As the properties of gas A approach those of gas B, the entropy of mixing appears to remain unchanged. Yet we know that if gas A and gas B are identical, there can be no change in entropy as the partition is removed. This is Gibb's paradox. Can you explain it?

8–2 A container of volume V is divided by partitions into three parts containing one kilomole of helium gas, two kilomoles of neon gas, and three kilomoles of argon gas, respectively. The temperature of each gas is initially 300 K and the pressure is 2 atm. The partitions are removed and the gases diffuse into each other. Calculate (a) the mole fraction and (b) the partial pressure of each gas in the mixture. Calculate the change (c) of the Gibbs function and (d) of the entropy of the system in the mixing process.

8–3 For a two-component open system $dU = T\,dS - P\,dV + \mu_1\,dn_1 + \mu_2\,dn_2$. (a) Derive a similar expression for dG, and (b) derive Maxwell relations for this system from it.

8–4 (a) Show that

$$-S\,dT + V\,dP - \sum_i n_i\,d\mu_i = 0. \tag{8–81}$$

This is known as the Gibbs-Duhem* equation. (b) For a two-component system use the Gibbs-Duhem equation to show that

$$x\left(\frac{\partial \mu_a}{\partial x}\right)_{T,P} + (1 - x)\left(\frac{\partial \mu_b}{\partial x}\right)_{T,P} = 0, \tag{8–82}$$

where $x = n_a/(n_a + n_b)$. This equation expresses the variation of the chemical potential with composition. [*Hint:* Express μ in terms of P, T and x and note that $(\partial \mu_a/\partial P)_{T,x} = v_a$, etc.]

8–5 Consider a mixture of alcohol and water in equilibrium with their vapors. (a) Determine the number of degrees of freedom for the system and state what they are. (b) Show that for each constituent

$$-s_i''\,dT + v_i''\,dP + \left(\frac{\partial \mu_i''}{\partial x''}\right)_{T,P} dx'' = -s_i'''\,dT + v_i'''\,dP + \left(\frac{\partial \mu_i'''}{\partial x'''}\right)_{T,P} dx''',$$

where x'' is the mole fraction of one of the constituents in the liquid and x''' is the mole fraction of the same constituent in the vapor phase. (c) Using the equation of part (b) and Eq. (8–82), show that

$$\left(\frac{\partial P}{\partial T}\right)_{x''} = \frac{x'''(s_a''' - s_a'') + (1 - x''')(s_b''' - s_b'')}{x'''(v_a''' - v_a'') + (1 - x''')(v_b''' - v_b'')},$$

where x'' is held constant artificially.

---

* Pierre M. M. Duhem, French physicist (1861–1916).

8–6 The direction in which a chemical reaction occurs depends upon the value of the thermodynamic equilibrium constant K, which can be defined as

$$\Delta G_T(\text{reaction}) = \Delta G_T^0(\text{reaction}) + RT \ln K_P,$$

where ΔG_T is the change in Gibbs function for the reaction and must be equal to zero at equilibrium; and ΔG_T^0 is the change in Gibbs function for the reaction taking place at one atmosphere and at constant temperature. (a) For the reaction of ideal gases

$$n_A A + n_B B \rightleftarrows n_C C + n_D D,$$

where $n_A A$ is n_A moles of A, etc., show that

$$K_P = \frac{(p_C^{n_C} \times p_D^{n_D})}{(p_A^{n_A} \times p_B^{n_B})},$$

where p_A is the partial pressure of A in the mixture, etc. (b) For the reaction $\frac{1}{2}N_2 + \frac{3}{2}H_2 \leftrightarrows NH_3$ show that K is 0.0128 if the total pressure is 50 atm and the mole fraction of NH_3 is 0.151 of the equilibrium mixture. (c) How does K_P change with pressure and temperature?

8–7 To make baking soda ($NaHCO_3$), a concentrated aqueous solution of Na_2CO_3 is saturated with CO_2. The reaction is given as

$$2Na^+ + CO_3^= + H_2O + CO_2 \rightleftarrows 2NaHCO_3.$$

Thus Na^+ ions, $CO_3^=$ ions, H_2O, CO_2, and $NaHCO_3$ are present in arbitrary amounts, except that all the Na^+ and $CO_3^=$ are from Na_2CO_3. Find the number of degrees of freedom of this system.

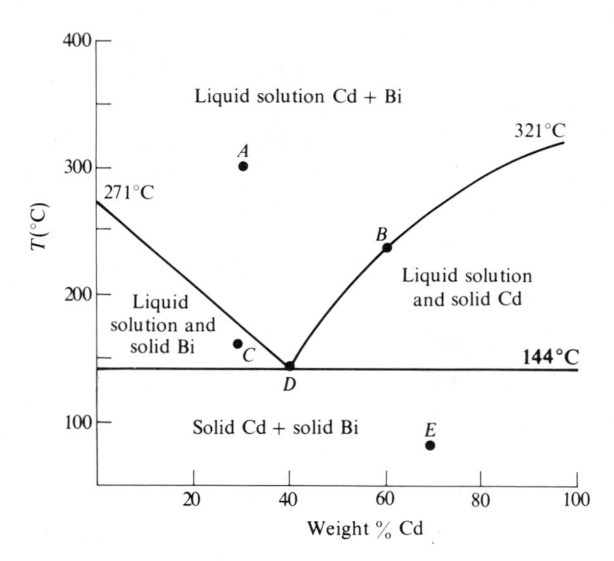

Figure 8–11

8–8 A phase diagram is a temperature-composition diagram for a system of two constituents in various phases. An idealized phase diagram for the cadmium-bismuth system is shown in Fig. 8–11 for $P = 1$ atm. (a) Determine the number of degrees of freedom for the system at each lettered point and state what they are. (b) Draw a sketch of a temperature versus time curve for cooling the system at 80 weight per cent Cd from 350°C to room temperature. (c) The freezing point of a solvent is lowered by the addition of solute, according to the relation $\Delta T_f = km$ where k is the freezing point constant, and m is the number of kilomoles of the solute per kilogram of the solvent. Calculate the freezing point constant of bismuth.

8–9 (a) Show that for a liquid containing a nonvolatile solute in equilibrium with its vapor at a given temperature T and pressure P

$$g''' = \mu'' = g'' + RT \ln (1 - x)$$

where x is the mole fraction of the solute. This assumes that the solute and solvent mix as ideal gases. (b) For a pure substance show that at constant pressure

$$d\left(\frac{g}{T}\right) = h\, d\left(\frac{1}{T}\right).$$

(c) Use part (b) to show that for a small change in x at constant pressure, part (a) reduces to

$$(h''' - h'')\, d\left(\frac{1}{T}\right) = R\, d \ln (1 - x).$$

(d) In the limit of small x

$$dT = \frac{RT^2}{l_{23}}\, dx,$$

where l_{23} is the latent heat of vaporization. This shows that the boiling temperature is elevated if a solute is added to a liquid. (e) Show how the result in part (d) can be used to determine molecular weights of solutes.

8–10 (a) The vapor pressure of water at 20°C, when the total pressure equals the vapor pressure, is 17.5 Torr. Find the change in vapor pressure when the water is open to the atmosphere. Neglect any effect of the dissolved air. (b) Find the pressure required to increase the vapor pressure of water by 1 Torr.

8–11 If the total pressure on a solid in equilibrium with its vapor is increased, show that the vapor pressure of the solid increases.

8–12 The equation of state for a surface film can be written as $\sigma = \sigma_0(1 - T/T_c)^n$ where $n = 1.22$ and σ_0 is a constant. (a) Assume that this equation holds for water and use the data on Fig. 8–2 to determine σ_0. (b) Determine values for λ, c_A and s at $T = 373$ K. (c) Calculate the temperature change as the area of the film is increased from 0 to 2×10^{-3} m^2 adiabatically.

8–13 Let a soap film be carried through a Carnot cycle consisting of an isothermal increase in area at a temperature T, an infinitesimal adiabatic increase in area in which the temperature decreases to $T - dT$, and returning to the initial state by an isothermal and an infinitesimal adiabatic decrease in area as shown in Fig. 8–12. (a) Calculate the work done by the film during the cycle. (b) Calculate the heat absorbed by the film in the cycle. (c) Derive Eq. (8–36) by considering the efficiency of the cycle. (d) Plot the cycle on a T-S diagram.

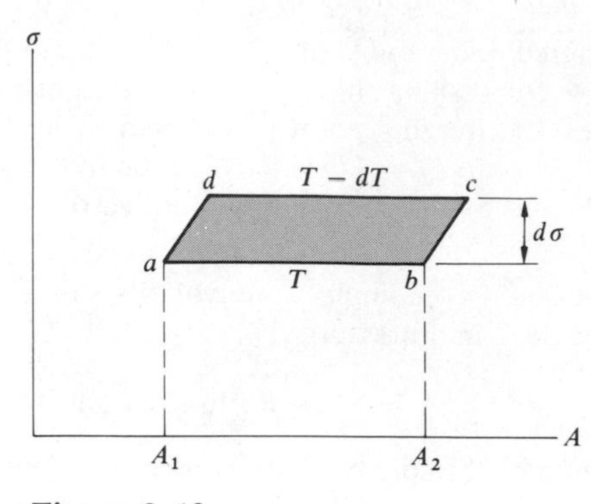

Figure 8–12

8–14 Suppose that below a critical temperature T_c, the Helmholtz function of a film is to be expressed as

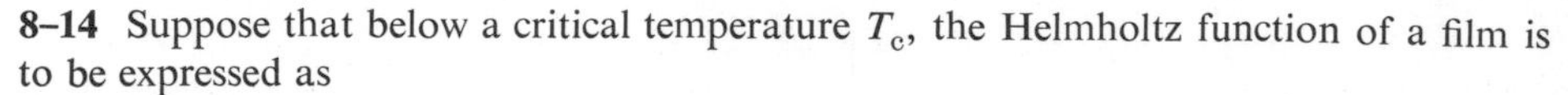

$$F = AB\left(1 - \frac{T}{T_c}\right)^n$$

where B, T_c, and n are constants depending upon the film and A is the area of the film. (a) What experimental information will determine the values B, T_c, and n? (b) Is there enough information to specify all the properties of the film? (c) Is the specification, as far as it goes, sensible?

8–15 Consider a rubber band as a one-dimensional system. (a) Derive an expression for the difference between the specific heat capacity at constant tension $c_{\mathscr{F}}$ and that at constant length c_l. (b) Find the ratio $c_{\mathscr{F}}/c_l$. (c) A rubber band heated at constant tension becomes *shorter*. Use this fact to show that if the tension in a rubber band is released adiabatically its temperature drops. (This can be checked experimentally by sensing the temperature of a rubber band with your lip while it is under tension and just after the tension is released.)

8–16 Show that the pressure P_i inside a bubble of radius r in a liquid which is under an external pressure P_e is given by $P_i - P_e = 2\sigma/r$.

8–17 The temperature dependence of the emf $\mathscr{E}$ of a reversible cell is given by $\mathscr{E} = 3.2 + 0.007t$ where t is the Celsius temperature of the cell. This cell discharges 200 mA for 30 s when $t = 27°C$. Calculate (a) the entropy change, (b) the heat absorbed, (c) the work done, (d) and the internal energy change of the cell during the process.

8–18 Show that when a charge ΔZ flows reversibly through a voltaic cell of emf $\mathscr{E}$ at constant temperature and pressure, (a) $\Delta G = \mathscr{E}\,\Delta Z$, and (b) $\Delta H = \Delta Z\, d(\mathscr{E}/T)/d(1/T)$. (c) Calculate ΔG and ΔH for the cell undergoing the process described in the previous problem and compare with the answers for parts (b) and (d) of that problem.

8–19 Calculate the total work done to electrolyze acidic water to produce 1 kilomole of H_2 and $\frac{1}{2}$ kilomole of O_2 at 1 atm and at 300 K. The emf used is 1.2 V. Assume that the gases are ideal.

8–20 Let the radiant energy in a cylinder be carried through a Carnot cycle, similar to that shown on Fig. 8–12, consisting of an isothermal expansion at the temperature T, an infinitesimal adiabatic expansion in which the temperature drops to $T - dT$, and returning to the original state by an isothermal compression and an infinitesimal adiabatic compression. Assume $P = \mathbf{u}/3$ and that $\mathbf{u}$ is a function of T alone. (a) Plot the cycle in the P-V plane. (b) Calculate the work done by the system during the cycle. (c) Calculate the heat flowing into the system during the cycle. (d) Show that $\mathbf{u}$ is proportional to T^4 by considering the efficiency of the cycle.

8–21 Show that the heat added during an isothermal expansion of blackbody radiation is four times larger than that expected for the heat added during the expansion of an ideal gas of photons obeying the same equation of state. The factor of four arises because the number of photons is not conserved but increases proportionally to the volume during an isothermal expansion.

8–22 The walls of an evacuated insulated enclosure are in equilibrium with the radiant energy enclosed. The volume of the enclosure is changed suddenly from 100 to 50 cm^3. If the initial temperature of the walls is 300 K, compute (a) the final temperature of the walls, (b) the initial and final pressure exerted on the walls by the radiant energy, and (c) the change of entropy of the radiant energy.

8–23 Show that the internal energy U of an ideal paramagnet is a function of temperature only.

8–24 In a certain range of temperature T and magnetic intensity $\mathscr{H}$ the function F^* of a magnetic substance is given by

$$F^* = -aT - \frac{b\mathscr{H}^2}{2T},$$

where a and b are constants. (a) Obtain the equation of state and sketch the magnetization as a function of temperature at constant magnetic intensity. (b) If the magnetic intensity is increased adiabatically, will the temperature of the substance rise or fall?

8–25 The refrigerator for an adiabatic demagnetization experiment is to be made from 40 g of chromium potassium alum [$CrK(SO_4)_2 \cdot 12H_2O$] which has the following properties: the molecular weight is 499.4 g mole^{-1}; the density is 1.83 g cm^{-3}; the Curie constant per gram is 3.73×10^{-3} K g^{-1}; and the lattice specific heat capacity is $4.95 \times 10^{-4}\, RT^3$. (a) Assuming that the salt obeys Curie's law, calculate the heat flow during an isothermal magnetization at 0.5 K and 10^4 Oe using a He^3 refrigerator and a superconducting magnet. (b) Calculate the change in E_p, E, U, and F^* during the process of part (a). (c) An adiabatic demagnetization to zero-applied magnetic intensity does not reach 0 K because of local effective magnetic fields in the material. Calculate the magnitude of these fields if the salt can be demagnetized adiabatically to 0.005 K. (d) Calculate the ratio of $C_{\mathscr{H}}$ of the magnetic system to the lattice heat capacity of the salt at 0.5 K.

8–26 Show that if the graph for $\mathscr{H} = 0$ on Fig. (8–5) intersects the vertical axis at a point above that for $\mathscr{H} = \mathscr{H}_1$, the unattainability statement of the third law would be violated.

8–27 Since the magnetic induction B inside a superconductor is zero, for a long cylindrical sample, the magnetization $M/\mu_0 V$ is equal to the negative of the applied magnetic intensity $\mathscr{H}$ for $\mathscr{H}$ less than some critical intensity $\mathscr{H}_c$. For $\mathscr{H}$ greater than $\mathscr{H}_c$ the superconductor becomes a normal metal and $M = 0$. (a) Sketch a graph of the magnetization as a function of the applied intensity. Show that in the transition from the superconducting to the normal state (b) the heat of transformation l is given by $-T\mu_0\mathscr{H}_c(d\mathscr{H}_c/dT)$

and (c) the difference in the specific heat capacities of the superconductor and the normal metal is given by

$$c_s - c_n = \frac{\mu_0 T}{2} \frac{d^2(\mathscr{H}_c^2)}{dT^2}.$$

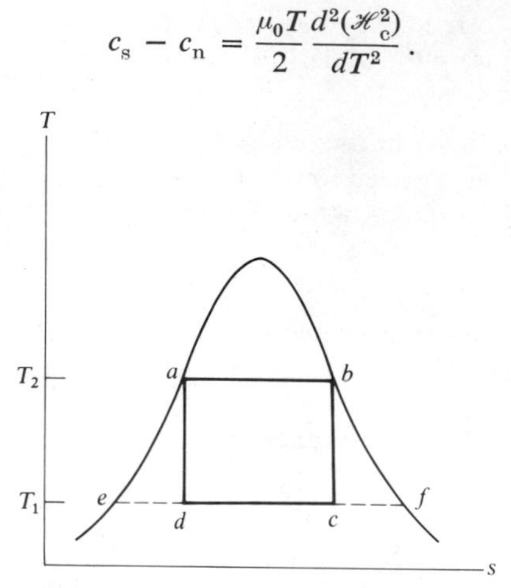

Figure 8–13

8–28 Figure 8–13, which is similar to Fig. 8–9(b), shows a Carnot cycle in the liquid-vapor region. The working substance is 1 kg of water, and $T_2 = 453$ K, $T_1 = 313$ K. Steam tables list values of T, P, u, s, and h at points on the saturation lines and these are tabulated below, in MKS units, for points a, b, e, and f. We wish to make a complete analysis of the cycle.

Point	t(°C)	T(K)	P(N m^{-2})	u(J kg^{-1})	s(J kg^{-1} K^{-1})	h(J kg^{-1})
a	180	453	10×10^5	7.60×10^5	2140	7.82×10^5
b	180	453	10×10^5	25.8×10^5	6590	27.7×10^5
e	40	313	$.074 \times 10^5$	1.67×10^5	572	1.67×10^5
f	40	313	$.074 \times 10^5$	24.3×10^5	8220	25.6×10^5

(a) Show that in the process a-b,

$$q_{ab} = h_b - h_a, \qquad w_{ab} = h_b - h_a - u_b + u_a.$$

(b) Show that in the process b-c,

$$q_{bc} = 0, \qquad w_{bc} = u_b - u_c.$$

(c) Show that in the process c-d,

$$q_{cd} = h_d - h_c, \qquad w_{cd} = h_d - h_c - u_d + u_c.$$

(d) Show that in the process d-a,

$$q_{da} = 0, \qquad w_{da} = u_d - u_a.$$

(e) Let x_2 and x_1 represent the fraction of the mass of the system in the vapor phase at points c and d, respectively. Show that

$$x_2 = \frac{s_b - s_e}{s_f - s_e}, \qquad x_1 = \frac{s_a - s_e}{s_f - s_e}.$$

(f) Show that

$$u_c = u_e + x_2(u_f - u_e), \qquad h_c = h_e + x_2(h_f - h_e),$$
$$u_d = u_e + x_1(u_f - u_e), \qquad h_d = h_e + x_1(h_f - h_e).$$

(g) Compute in joules the "expansion work" in the cycle, along the path a-b-c.
(h) Compute in joules the "compression work," along the path c-d-a, and find the ratio of expansion work to compression work.
(i) Compute from (g) and (h) the net work done in the cycle.
(j) Compute from (i) and (a) the efficiency of the cycle, and show that it is equal to $(T_2 - T_1)/T_2$.
(k) In any real engine there are unavoidable friction losses. To estimate the effect of these, assume that in the expansion stroke 5% of the work done by the system is lost, and that in the compression stroke 5% more work must be done than computed in part (h). Compute the net work delivered per cycle, and the efficiency.

8–29 A steam turbine operates in a reversible Rankine cycle. Superheated steam enters the turbine at a pressure of 100 lb in^{-2} and a temperature of 800°F. The pressure of the exhaust steam is 15 lb in^{-2}. (a) Find from Fig. 8–8 the work done per pound of steam. (b) If as a result of irreversible processes the specific entropy of the exhaust steam is 2 Btu lb^{-1} deg F^{-1} at the exhaust pressure of 15 lb in^{-2}, how much work is done per pound of steam?

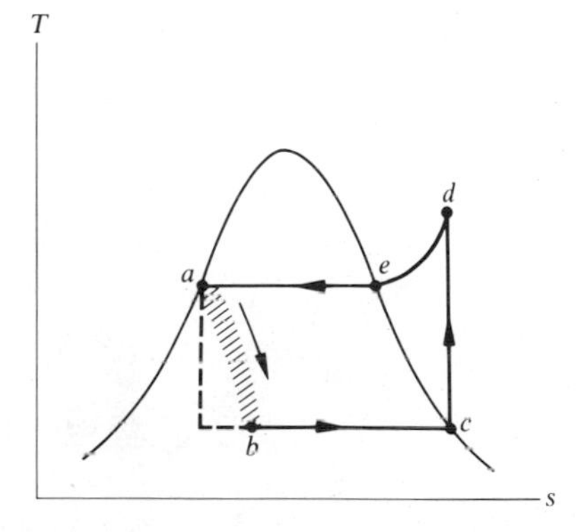

Figure 8–14

8–30 Figure 8–14 represents a refrigeration cycle in which the adiabatic compression stage, cd, takes place in the vapor region. The expansion stage from d to a is at constant pressure and the irreversible expansion from a to b takes place through a throttling valve.

(a) Sketch the cycle in an *h-s* diagram. (b) Show that the coefficient of performance of the cycle is given by

$$c = \frac{h_c - h_a}{h_d - h_c}.$$

(c) In a typical cycle using Freon-12 as a working substance, the specific enthalpies at points *d*, *c*, and *a* are, respectively, 90.6, 85.0, and 36.2 Btu lb^{-1}. The measured coefficient of performance of the cycle was 2.4. Compare with the value computed from the equation above, which assumes that all processes except *a-b* are reversible.

9

Kinetic theory

9–1 INTRODUCTION

The subject of thermodynamics deals with the conclusions that can be drawn from certain experimental laws, and with the applications of these conclusions to relations between properties of materials such as heat capacities, coefficients of expansion, compressibilities, and so on. It makes no hypotheses about the nature of matter and is purely an empirical science.

Although thermodynamic principles can predict many relations between the properties of matter, such as the difference between the specific heat capacities c_P and c_v, or the variation of these quantities with pressure, it is not possible to derive from thermodynamic considerations alone the absolute magnitudes of the heat capacities, or the equation of state of a substance.

We can go beyond the limitation of pure thermodynamics only by making hypotheses regarding the nature of matter, and by far the most fruitful of such hypotheses, as well as one of the oldest, is that matter is not continuous in structure but is composed of particles called molecules. In particular, the molecular theory of gases has been very completely developed, because the problems to be solved are much simpler than those encountered in dealing with liquids and solids.

The properties of matter in bulk are predicted, starting with a molecular theory by means of two different lines of attack. The first, called the *kinetic* or *dynamic* theory, applies the laws of mechanics to the individual molecules of a system, and from these laws derives, for example, expressions for the pressure of a gas, its internal energy and its specific heat capacity. The method of *statistical thermodynamics* ignores detailed considerations of molecules as individuals, and applies considerations of *probability* to the very large number of molecules that make up any piece of matter. We shall see that the methods of statistical thermodynamics provide a further insight into the concept of entropy and the principle of increase of entropy.

Both kinetic theory and statistical thermodynamics were first developed on the assumption that the laws of mechanics, deduced from the behavior of matter in bulk, could be applied without change to particles like molecules and electrons. As the sciences progressed, it became evident that in some respects this assumption was not correct; that is, conclusions drawn from it by logical methods were not in accord with experimental facts. The failure of small-scale systems to obey the same laws as large-scale systems led to the development of quantum theory and quantum mechanics, and statistical thermodynamics is best treated today from the viewpoint of quantum mechanics.

This chapter and the next will be devoted to the kinetic aspects of molecular theory, and the following chapters to statistical thermodynamics. As we go along, we shall make many references to concepts and equations that have already been discussed in the preceding chapters on thermodynamics, and we shall see how a much deeper insight into many questions can be attained with the help of molecular theory as a background.

9–2 BASIC ASSUMPTIONS

In thermodynamics, the equation of state of a system expresses the relation between its measurable macroscopic properties. The simplest equation of state is that of an ideal gas; and although kinetic theory is limited neither in concept nor in application to ideal gases, we shall begin by showing how the equation of state of an ideal gas can be derived on the basis of a molecular model with the following assumptions:

1. Any macroscopic volume of a gas contains a very large number of molecules. This assumption is justified by all experimental evidence. The number of molecules in a kilomole (Avogadro's* number N_A) is 6.03×10^{26}. Experimental methods for arriving at this number are discussed in later chapters. At standard conditions, 1 kilomole of an ideal gas occupies 22.4 m³. Hence at standard conditions there are approximately 3×10^{25} molecules in a cubic meter, 3×10^{19} in a cubic centimeter, and 3×10^{16} in a cubic millimeter.

2. The molecules are separated by distances that are large compared with their own dimensions and are in a state of continuous motion. The diameter of a molecule, considered as a sphere, is about 2 or 3×10^{-10} m. If we imagine one molal volume at standard conditions to be divided into cubical cells with one molecule per cell, the volume of each cell is 30×10^{-27} m³. The length of one side of such a cell is about 3×10^{-9} m, which means that the distance between molecules is of this order of magnitude, about 10 times the molecular diameter.

3. To a first approximation, we assume that molecules exert no forces on one another except when they collide. Therefore between collisions with other molecules or with the walls of the container, and in the absence of external forces, they move in straight lines.

4. Collisions of molecules with one another and with the walls are perfectly elastic. The walls of a container can be considered perfectly smooth, so that there is no change in tangential velocity in a collision with the walls.

5. In the absence of external forces, the molecules are distributed uniformly, throughout the container. If N represents the total number of molecules in a container of volume V, the average number of molecules per unit volume, $\mathbf{n}$, is

$$\mathbf{n} = N/V.$$

The assumption of uniform distribution then implies that in any element of volume ΔV, wherever located, the number of molecules ΔN is

$$\Delta N = \mathbf{n}\Delta V.$$

Obviously, the equation above its not correct if ΔV is too small, since the number of molecules N, although large, is finite, and one can certainly imagine a volume

* Count Amedeo Avogadro, Italian physicist (1776–1856).

element so small that it contains no molecules, in contradiction to the equation above. However, it is possible to divide a container into volume elements large enough so that the number of molecules per unit volume within them does not differ appreciably from the average, and at the same time small enough compared with the dimensions of physical apparatus that they can be treated as infinitesimal in the mathematical sense and the methods of differential and integral calculus can be applied to them. For example, a cube 1/1000 mm on a side is certainly small in comparison with the volume of most laboratory apparatus, yet at standard conditions it contains approximately 30 million molecules.

6. The *directions* of molecular velocities are assumed to be distributed uniformly. To put this assumption in analytic form, imagine that there is attached to each molecule a vector representing the magnitude and direction of its velocity. Let us transfer all these vectors to a common origin and construct a sphere of arbitrary radius r with center at the origin. The velocity vectors, prolonged if necessary, intersect the surface of the sphere in as many points as there are molecules and the assumption of uniform distribution in direction means that these points are distributed uniformly over the surface of the sphere. The average number of these points per unit area is

$$\frac{N}{4\pi r^2},$$

and the number in any element of area ΔA is

$$\Delta N = \frac{N}{4\pi r^2}\,\Delta A,$$

wherever the element is located. As in the preceding paragraph, the area must be large enough (that is, it must include a large enough range of directions) so that the surface density of points within it does not differ appreciably from the average. Because of the large number of molecules, the range of directions can be made very small and still include a large number of points.

Let us carry this description of velocity directions one step further. Any arbitrary direction in space can be specified with reference to a polar coordinate system by the angles θ and ϕ, as in Fig. 9–1. The area ΔA of a small element on the surface of a sphere of radius r is, very nearly,

$$\Delta A = (r \sin\theta\,\Delta\theta)(r\,\Delta\phi) = r^2 \sin\theta\,\Delta\theta\,\Delta\phi.$$

The number of points in this area, or the number of molecules $\Delta N_{\theta\phi}$ having velocities in a direction between θ and $\theta + \Delta\theta$, ϕ and $\phi + \Delta\phi$, is

$$\Delta N_{\theta\phi} = \frac{N}{4\pi r^2}\,r^2 \sin\theta\,\Delta\theta\,\Delta\phi = \frac{N}{4\pi}\sin\theta\,\Delta\theta\,\Delta\phi.$$

When both sides of this equation are divided by the volume V occupied by the gas, we get

$$\Delta \mathbf{n}_{\theta\phi} = \frac{\mathbf{n}}{4\pi} \sin\theta \, \Delta\theta \, \Delta\phi, \tag{9–1}$$

where $\Delta \mathbf{n}_{\theta\phi}$ is the number density of molecules with velocities having directions between θ and $\theta + \Delta\theta$, and ϕ and $\phi + \Delta\phi$.

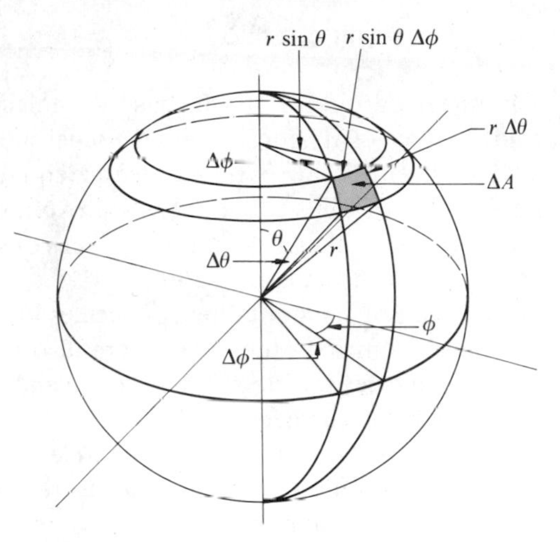

Fig. 9–1 Polar coordinates.

Consider, finally, the magnitudes of the molecular velocities, or the *speeds* of the molecules. It is clear that not all molecules have the same speed, although this simplifying assumption is often made. Even if we could start them off in this way, intermolecular collisions would very quickly bring about differences in speed. We shall show in Section 12–2 how to calculate the number that have speeds in any specified range, but for the present we shall assume that the speed can have any magnitude from zero to infinity*, and we represent by ΔN_v the number of molecules with speeds between v and $v + \Delta v$. Geometrically this number equals the number of velocity vectors *terminating* within a thin spherical shell in Fig. 9–1, between spheres of radii $r_1 = v$ and $r_2 = v + \Delta v$. As a result of collisions, the speed of any one molecule is continually changing, but we assume that in the equilibrium state the number of molecules with speeds in any specified range remains constant.

* It would be better to say, from zero to the speed of light. However, as we shall show, the number of molecules with speeds exceeding even a small fraction of the speed of light is so small for ordinary gases that for mathematical simplicity we may as well make the assumption above.

9–3 MOLECULAR FLUX

Because of the continuous random motion of the molecules of a gas, molecules are continually arriving at every portion of the walls of the container, and also at each side of any imagined surface *within* the gas. Let ΔN represent the total number of molecules arriving from all directions and with all speeds at one side of an element of surface of area ΔA during a time interval Δt. The *molecular flux* Φ at the surface is defined as the total number of molecules arriving at the surface, per unit area and per unit time. Thus,

$$\Phi = \frac{\Delta N}{\Delta A\, \Delta t}. \tag{9–2}$$

If the surface is an imagined one *within* the gas, all molecules arriving at the surface, from either side, will cross it, and if there is no net motion of the gas as a whole, the molecular fluxes on either side of the surface are equal and are in opposite directions. Thus at either side of the surface there are two molecular fluxes, one consisting of molecules arriving at that side and the other consisting of molecules that have crossed the surface from the other side.

If the surface is at the wall of the container, molecules arriving at the surface do not cross it but rebound from it. Hence there are also two fluxes at such a surface, one consisting of molecules arriving at the surface and the other consisting of molecules rebounding from the surface.

In Fig. 9–2, the shaded area ΔA represents a small element of surface, either within the gas or at a wall. Construct the normal to the area, and some reference plane containing the normal. We first ask, how many molecules arrive at the surface during a time interval Δt, travelling in the particular direction θ, ϕ, and with a specified speed v. (To avoid continued repetition, let it be understood that this means the number of molecules with directions between θ and $\theta + \Delta\theta$, ϕ and $\phi + \Delta\phi$, and with speeds between v and $v + \Delta v$.)

Construct the slant cylinder shown in Fig. 9–2, with axis in the direction θ, ϕ, and of length $v\,\Delta t$, equal to the distance covered by a molecule with speed v in time Δt. Then the number of $\theta\phi v$-molecules that arrive at the surface during the time Δt is equal to the number of $\theta\phi v$-molecules in the cylinder, where a $\theta\phi v$-molecule means one with speed v, traveling in the θ, ϕ direction.

To show that this is correct, we can see first that all $\theta\phi v$-molecules in the cylinder will reach the surface during the time Δt. (We are ignoring any collisions with other molecules that may be made on the way to the surface, so that the molecules are considered as geometrical points. In Section 10–3 we shall see how to take such collisions into account.) There are, of course, many other types of molecules in the cylinder. Some of these will reach the surface element during the time Δt and others will not. Those that do *not* are either not traveling toward the element (that is, they are not $\theta\phi$-molecules) or are not traveling fast enough to reach the element during the time Δt (that is, their speed is less than v). Those within the cylinder that *do* reach the surface during the time Δt are necessarily $\theta\phi$-molecules, but unless they have a speed v they are not $\theta\phi v$-molecules.

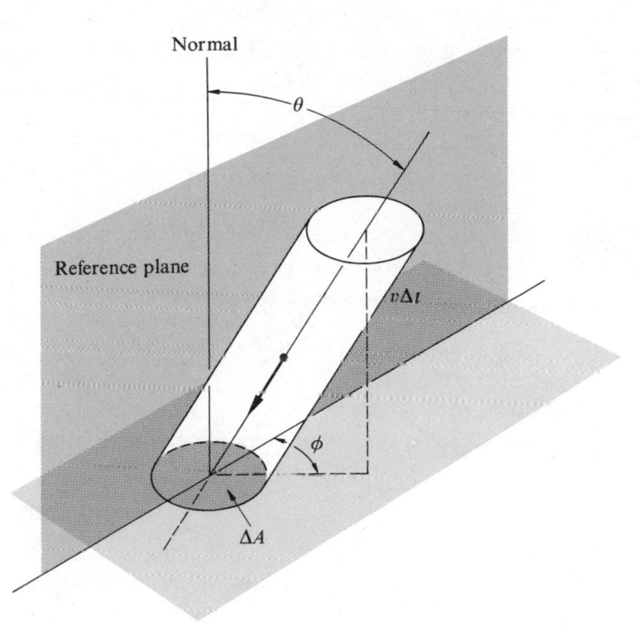

Fig. 9–2 Only the $\theta\phi v$-molecules in the cylinder will arrive at the area ΔA during a time Δt.

Many other molecules, not in the cylinder, will arrive at the element during the time Δt. Some of these will have a speed v, but they are not $\theta\phi$-molecules since they come in from other directions. Therefore *all* $\theta\phi v$-molecules in the cylinder, and *only* those molecules, will reach the surface during the time Δt, traveling in the $\theta\phi$-direction with speed v.

Let $\Delta \mathbf{n}_v$ represent the number density of molecules with speeds between v and $v + \Delta v$. Then from Eq. (9–1) the number density of $\theta\phi v$-molecules is

$$\Delta \mathbf{n}_{\theta\phi v} = \frac{1}{4\pi} \Delta \mathbf{n}_v \sin\theta \, \Delta\theta \, \Delta\phi. \tag{9–3}$$

The volume of the slant cylinder in Fig. 9–2 is

$$\Delta V = (\Delta A \cos\theta)(v \, \Delta t).$$

The number of $\theta\phi v$-molecules in the cylinder is therefore

$$\Delta N_{\theta\phi v} = \frac{1}{4\pi} v \, \Delta \mathbf{n}_v \sin\theta \cos\theta \, \Delta\theta \, \Delta\phi \, \Delta A \, \Delta t,$$

and the flux $\Delta\Phi_{\theta\phi v}$ of $\theta\phi v$-molecules is

$$\Delta\Phi_{\theta\phi v} = \frac{\Delta N_{\theta\phi v}}{\Delta A \, \Delta t} = \frac{1}{4\pi} v \, \Delta \mathbf{n}_v \sin\theta \cos\theta \, \Delta\theta \, \Delta\phi. \tag{9–4}$$

The flux $\Delta\Phi_{\theta v}$ due to molecules arriving at an angle θ with speed v, but including all angles ϕ, is found by replacing $\Delta\phi$ with $d\phi$ and integrating over all values of ϕ from 0 to 2π. The result is

$$\Delta\Phi_{\theta v} = \frac{1}{2} v \, \Delta\mathbf{n}_v \sin\theta \cos\theta \, \Delta\theta. \tag{9-5}$$

The flux $\Delta\Phi_\theta$ due to molecules arriving at the angle θ, including all angles ϕ and all speeds v, is found by summing the expression for $\Delta\Phi_{\theta v}$ over all values of v. Thus

$$\Delta\Phi_\theta = \frac{1}{2} \sin\theta \cos\theta \, \Delta\theta \sum v \, \Delta\mathbf{n}_v. \tag{9-6}$$

The flux $\Delta\Phi_v$ of molecules with speed v, including all angles θ and ϕ, is found by replacing $\Delta\theta$ with $d\theta$ in Eq. (9-5) and integrating over all values of θ from zero to $\pi/2$. This gives

$$\Delta\Phi_v = \frac{1}{4} v \, \Delta\mathbf{n}_v. \tag{9-7}$$

Finally, the total flux Φ, including all speeds and all angles, is obtained either by summing $\Delta\Phi_v$ over all values of v, or by replacing $\Delta\theta$ with $d\theta$ in Eq. (9-6) and integrating over θ from zero to $\pi/2$. The result is

$$\Phi = \frac{1}{4} \sum v \, \Delta\mathbf{n}_v. \tag{9-8}$$

Let us express this result in terms of the *average* or *arithmetic mean* speed $\bar{v}$. This quantity is found by adding together the speeds of all the molecules, and dividing by the total number of molecules:

$$\bar{v} = \frac{\sum v}{N},$$

where the sum is over all *molecules*. But if there are ΔN_1 molecules with speeds v_1. ΔN_2 molecules with speeds v_2, etc., the sum of the speeds can also be found by multiplying the speed v_1 by the number of molecules ΔN_1 having that speed, multiplying v_2 by the number ΔN_2 having speed v_2, and so on, and adding these products. The average speed is then the sum of all such products, divided by the total number of molecules. That is,

$$\bar{v} = \frac{v_1 \, \Delta N_1 + v_2 \, \Delta N_2 + \cdots}{N} = \frac{1}{N} \sum v \, \Delta N_v, \tag{9-9}$$

where the sum is now over all *speeds*. When numerator and denominator are divided by the volume V, we have

$$\bar{v} = \frac{1}{\mathbf{n}} \sum v \, \Delta\mathbf{n}_v.$$

It follows that

$$\sum v \, \Delta\mathbf{n}_v = \bar{v}\mathbf{n}, \tag{9-10}$$

and hence from Eq. (9–8) the molecular flux Φ, including all molecules arriving at one side of the element and coming in from all directions and with all speeds, is

$$\Phi = \frac{1}{4}\bar{v}\mathbf{n}. \tag{9–11}$$

As a numerical example, the number of molecules per cubic meter, $\mathbf{n}$, is approximately 3×10^{25} molecules m^{-3} at standard conditions. We shall show later that the average speed of an oxygen molecule at 300 K is approximately 450 m s^{-1}. The molecular flux in oxygen at standard conditions is therefore

$$\Phi = \frac{1}{4}\mathbf{n}\bar{v} \approx \frac{1}{4} \times 3 \times 10^{25} \times 450 \approx 3.3 \times 10^{27} \text{ molecules m}^{-2}\text{ s}^{-1}.$$

It is sometimes useful to put Eq. (9–4) in the following form. Consider the area ΔA in Fig. 9–2 to be located at the origin in Fig. 9–1 and to lie in the x-y plane. The molecules arriving at the area in the $\theta\phi$-direction are those coming in within the small cone in Fig. 9–1, whose base is the shaded area ΔA on the spherical surface in that diagram. This area is

$$\Delta A = r^2 \sin\theta\,\Delta\theta\,\Delta\phi,$$

and the solid angle of the cone, $\Delta\omega$ is

$$\Delta\omega = \frac{\Delta A}{r^2} = \sin\theta\,\Delta\theta\,\Delta\phi.$$

Hence from Eq. (9–4) the flux $\Delta\Phi_{\theta\phi v}$ can be written

$$\Delta\Phi_{\theta\phi v} = \frac{1}{4\pi}v\,\Delta\mathbf{n}_v \cos\theta\,\Delta\omega = \Delta\Phi_{\omega v}; \tag{9–12}$$

and the flux per unit solid angle, of molecules with speed v, is

$$\frac{\Delta\Phi_{\omega v}}{\Delta\omega} = \frac{1}{4\pi}v\,\Delta\mathbf{n}_v \cos\theta. \tag{9–13}$$

The total flux per unit solid angle, including all speeds, is

$$\frac{\Delta\Phi_{\omega}}{\Delta\omega} = \frac{1}{4\pi}\bar{v}\mathbf{n}\cos\theta. \tag{9–14}$$

If we consider a number of small cones with apexes at ΔA in Fig. 9–1, the greatest number of molecules arrives with direction in the cone centered about the normal, since $\cos\theta$ has its maximum value for this cone, and the number decreases to zero for cones tangent to ΔA, where $\theta = 90°$.

If the area ΔA is a hole in the wall of a thin-walled container, small enough so that leakage through the hole does not appreciably affect the equilibrium of the gas, then every molecule coming up to the hole will escape through it. The distribution of directions of the molecules emerging from the hole is also given by Eq. (9–14). The number emerging per unit solid angle is a maximum in the direction normal to the plane of the hole and decreases to zero in the tangential direction.

9–4 EQUATION OF STATE OF AN IDEAL GAS

Figure 9–3 shows a $\theta\phi v$-molecule before and after a collision with the wall of a vessel containing a gas. From our assumption of perfect elasticity, the magnitude of the velocity v is the same before and after the collision, and from the assumption that the wall is perfectly smooth, the tangential component of velocity is also unaltered by the collision. It follows that the angle of reflection is equal to the angle of incidence and the normal component of velocity is reversed in the collision, from $v \cos \theta$ to $-v \cos \theta$.

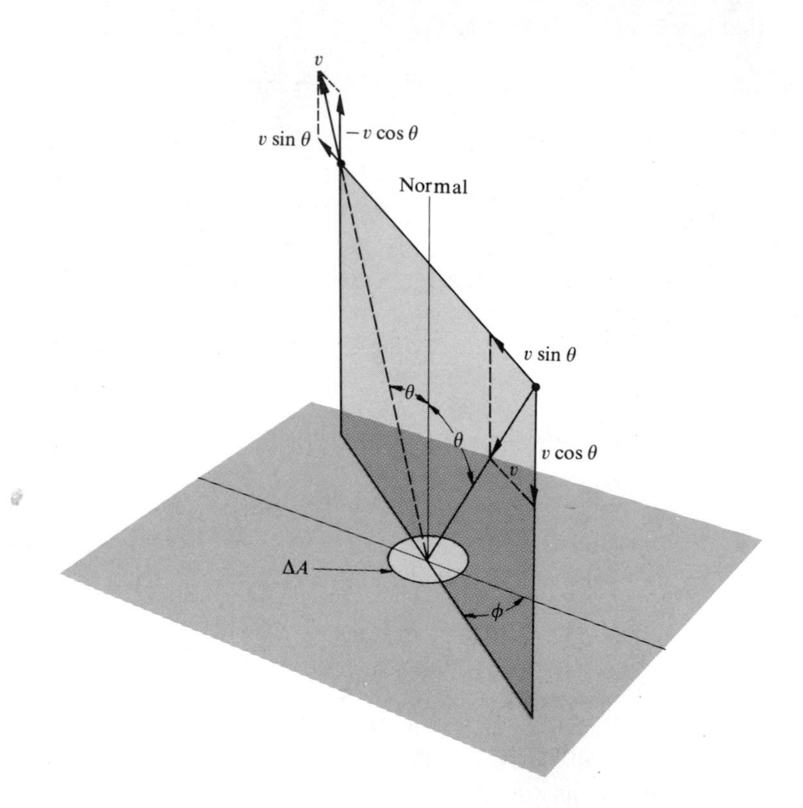

Fig. 9–3 Change in velocity in an elastic collision.

The force exerted on the wall by any one molecule in a collision is an impulsive force of short duration. The details of its variation with time are unknown; but it is not necessary to know them because from Newton's second law we can set the average force per unit area exerted on the surface, or the *average pressure*, equal to the average rate of change of momentum per unit area.

If m is the mass of a colliding molecule, the change in the normal component of momentum in a $\theta\phi v$-collision is

$$mv \cos \theta - (-mv \cos \theta) = 2mv \cos \theta. \tag{9–15}$$

The change in momentum depends on θ and v, but not on the angle ϕ. Hence we need the number of θv-molecules arriving at the surface per unit area, and per unit time, or the flux $\Delta\Phi_{\theta v}$ given by Eq. (9-5).

The rate of change of momentum per unit area due to all molecules arriving at an angle θ with speed v, or the pressure $\Delta P_{\theta v}$, equals the product of $\Delta\Phi_{\theta v}$ and the change in momentum of a θv-molecule:

$$\Delta P_{\theta v} = (\tfrac{1}{2}v\, \Delta \mathbf{n}_v \sin \theta \cos \theta\, \Delta\theta)(2mv \cos \theta) = mv^2\, \Delta \mathbf{n}_v \sin \theta \cos^2 \theta\, \Delta\theta.$$

To find the pressure ΔP_v due to molecules of speed v coming in at all values of θ, we integrate over θ from 0 to $\pi/2$. This gives

$$\Delta P_v = \frac{1}{3} mv^2\, \Delta \mathbf{n}_v.$$

Finally, summing over all values of v, we have for the total pressure P,

$$P = \frac{1}{3} m \sum v^2\, \Delta \mathbf{n}_v. \tag{9–16}$$

The same reasoning as that above can be applied to any imagined surface in the interior of the gas. The molecular flux $\Delta\Phi_{\theta\phi v}$ is the same for all surfaces, wherever located. Molecules approaching an *internal* surface from one side pass through it and do not rebound, but the flux across the surface from the opposite side carries the same momentum away from the surface as do the molecules rebounding from a wall of the container. That is, for every $\theta\phi v$-molecule crossing the surface from one side, there will be another $\theta\phi v$-molecule crossing from the other side, and Fig. 9–3 will apply to any surface within the gas, except that the black circles in Fig. 9–3 do not represent the *same* molecule.

Hence the net flux of momentum, at right angles to any surface, is the same as at the boundary wall; and if we consider the pressure as the flux of momentum, the pressure has the same value at *all* points, both within the gas and at its surface.

Equation (9–16) is more conveniently expressed as follows. The average value of the *square* of the speed of all molecules, or the *mean square speed*, is found by squaring all the speeds, adding these quantities, and dividing by the total number of molecules:

$$\overline{v^2} = \frac{\sum v^2}{N}.$$

Just as in calculating the average speed, we can obtain Σv^2 more conveniently by multiplying v_1^2 by ΔN_1, v_2^2 by ΔN_2, etc., and adding these products. That is

$$\overline{v^2} = \frac{\sum v^2 \,\Delta N_v}{N}, \qquad \text{or} \qquad \overline{v^2} = \frac{\sum v^2 \,\Delta n_v}{n}.$$

Then

$$\sum v^2 \,\Delta \mathbf{n}_v = \mathbf{n}\overline{v^2}$$

and

$$P = \frac{1}{3}\,\mathbf{n}m\overline{v^2}. \tag{9-17}$$

Since the mean kinetic energy of a single molecule is $\frac{1}{2}m\overline{v^2}$, the right side of Eq. (9-17) equals two-thirds of the total kinetic energy per unit volume or two-thirds of the *kinetic energy density;* and Eq. (9-17) thus expresses the pressure in terms of the kinetic energy density.

It will be shown in Section 12-2 that the average value of the square of the speed, v^2, is always greater than the square of the average speed, $(\bar{v})^2$.

Since $\mathbf{n}$ represents the number of molecules per unit volume, N/V, we can write the preceding equation as

$$PV = \frac{1}{3}\,Nm\overline{v^2}.$$

This begins to look like the equation of state of an ideal gas,

$$PV = nRT,$$

where n represents the number of kilomoles, equal to the total number of molecules divided by the number of molecules per kilomole, or Avogadro's number N_A. We can therefore write the equation of state of an ideal gas as

$$PV = N\frac{R}{N_A}\,T.$$

The quotient R/N_A occurs frequently in kinetic theory. It is called the universal gas constant *per molecule*, or *Boltzmann's constant*, and is represented by k:

$$k \equiv \frac{R}{N_A}. \tag{9-18}$$

Since R and N_A are universal constants, k is a universal constant also. That is, its magnitude depends only on the system of units employed. In the MKS system,

$$k = \frac{R}{N_A} = \frac{8.314 \times 10^3}{6.022 \times 10^{26}} = 1.381 \times 10^{-23}\, J \text{ molecule}^{-1}\, \text{K}^{-1}.$$

In terms of the Boltzmann constant, the equation of state of an ideal gas becomes

$$PV = NkT.$$

This will agree with the equation derived from kinetic theory, Eq. (9–17), if we set

$$NkT = \frac{1}{3} N m\overline{v^2},$$

or

$$\overline{v^2} = \frac{3kT}{m}. \qquad (9\text{–}19)$$

The theory has thus led us to a goal we did not deliberately set out to seek; namely, it has given us a molecular interpretation of the concept of absolute temperature T, as a quantity proportional to the mean square speed of the molecules of an ideal gas. It is even more significant to write Eq. (9–19) as

$$\frac{1}{2} m\overline{v^2} = \frac{3}{2} kT. \qquad (9\text{–}20)$$

The product of one-half the mass of a molecule and the mean square speed is the same as the mean translational kinetic energy, and we see from the preceding equation that the mean translational kinetic energy of a gas molecule is proportional to the absolute temperature. Furthermore, since the factor $3k/2$ is the same for all molecules, the mean kinetic energy depends only on the temperature and not on the pressure or volume or species of molecule. That is, the mean kinetic energies of the molecules of H_2, He, O_2, Hg, etc., are all the same at the same temperature, despite the disparities in their masses.

We can compute from Eq. (9–20) what this energy is at any temperature. Let $T = 300$ K. Then

$$\frac{3}{2} kT = \frac{3}{2} \times 1.38 \times 10^{-23} \times 300 = 6.21 \times 10^{-21} \text{ J}.$$

If the molecules are oxygen, the mass m is 5.31×10^{-26} kg, and the mean square speed is

$$\overline{v^2} = \frac{2 \times 6.21 \times 10^{-21}}{5.31 \times 10^{-26}} = 23.4 \times 10^4 \text{ m}^2\, \text{s}^{-2}.$$

The square root of this quantity, or the *root-mean-square* speed is

$$v_{\text{rms}} = \sqrt{\overline{v^2}} = 482 \text{ m s}^{-1} = 1607 \text{ ft s}^{-1} = 1100 \text{ mi hr}^{-1}.$$

By way of comparison, the speed of sound in air at standard conditions is about 350 m s^{-1} or 1100 ft s^{-1} and the speed of a .30 cal rifle bullet is about 2700 ft s^{-1}.

The speed of a compressional wave in a fluid is given by

$$v = \sqrt{1/\kappa_s \rho}$$

which, for an ideal gas, is equivalent to

$$v = \sqrt{\gamma kT/m},$$

where $\gamma = c_P/c_v$. Since the root-mean-square speed of a molecule is

$$v_{\rm rms} = \sqrt{3kT/m}, \tag{9–21}$$

we see that the two are nearly equal but that the speed of a sound wave is somewhat smaller than the rms molecular speed, as would be expected.

When electrons and ions are accelerated by an electric field, it is convenient to express their energies in *electron-volts* (abbreviated eV), where by definition

$$1 \text{ electron-volt} = 1.602 \times 10^{-19} \text{ J}.$$

An electron-volt is equal to the energy acquired by a particle of charge $e = 1.602 \times 10^{-19}$ C accelerated through a potential difference of 1 V.

At a temperature of 300 K,

$$\frac{3}{2} kT = 6.21 \times 10^{-21} \text{ J} \simeq 0.04 \text{ eV}.$$

or

$$kT = 0.026 \text{ eV} \simeq \tfrac{1}{40} \text{ eV}.$$

Hence at a temperature of 300 K, the mean kinetic energy of a gas molecule is only a few hundredths of an electron-volt.

9–5 COLLISIONS WITH A MOVING WALL

We now examine the nature of the mechanism by which an expanding gas does work against a moving piston, and show that if the process is adiabatic, the work is done at the expense of the kinetic energy of the molecules (that is, the internal energy of the gas) and the temperature of the gas decreases. Figure 9–4 represents a gas in a cylinder provided with a piston. Let the piston move up with speed u, small compared with molecular speeds and small enough so that the gas remains practically in an equilibrium state. From the thermodynamic viewpoint, then, the process is reversible.

When a molecule collides elastically with a *stationary* wall, the magnitude of the normal component of velocity is unchanged. If the wall is moving, the magnitude of the *relative* velocity is unchanged.

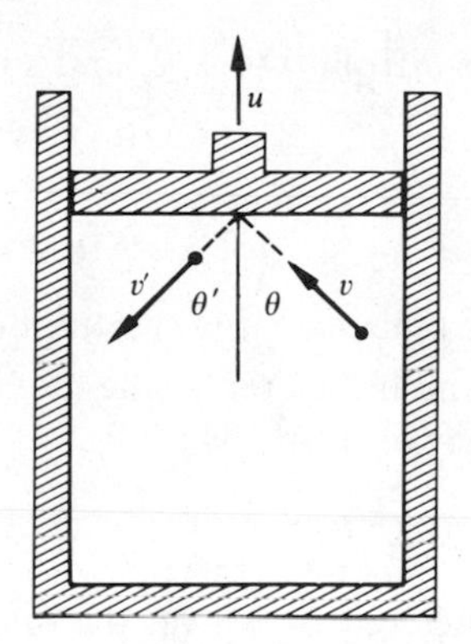

Fig. 9–4 Collisions with a moving wall.

To illustrate by a simple numerical example, if a particle approaches a stationary wall normally with a speed of 15 m s^{-1}, referred to a coordinate system fixed in the laboratory, it rebounds with a speed of 15 m s^{-1}. If the wall is moving away from the particle with a speed of 5 m s^{-1}, and if the particle has a speed of 20 m s^{-1}, both relative to the laboratory coordinate system, the molecule is again approaching the wall with a relative velocity of 15 m s^{-1}. After the collision the magnitude of the velocity of the particle relative to the wall will again be 15 m s^{-1}, but since the particle is now moving in a direction opposite to that of the wall, its speed in the laboratory coordinate system is only 10 m s^{-1}.

In general, if the normal component of the velocity before collision is $v \cos \theta$, where θ is the angle between v and the normal to the wall, the velocity component after collision, $v' \cos \theta'$, is equal to $v \cos \theta - 2u$. The loss of kinetic energy in the collision is

$$\frac{1}{2} m(v \cos \theta)^2 - \frac{1}{2} m(v \cos \theta - 2u)^2 \simeq 2mvu \cos \theta,$$

since by hypothesis $u \ll v$. The kinetic energy of the molecule can decrease even if the collision is perfectly elastic, because in the collision the molecule exerts a force against a *moving* wall and hence does work on the wall.

The loss of kinetic energy depends on θ and v but not on ϕ. By Eq. (9–5) the number of θv-collisions with a wall, per unit area and per unit time, is

$$\Delta\Phi_{\theta v} = \frac{1}{2} v \, \Delta \mathbf{n}_v \sin \theta \cos \theta \, \Delta\theta.$$

Multiplying this by the loss in kinetic energy in such a collision, we obtain for the loss in kinetic energy per unit area and per unit time, by molecules making θv-collisions,

$$muv^2 \, \Delta \mathbf{n}_v \sin \theta \cos^2 \theta \, \Delta\theta.$$

Finally, after integrating over θ from 0 to $\pi/2$, and summing over all values of v, we get

$$\frac{1}{3}\,\mathbf{n}m\overline{v^2}u$$

for the total loss of molecular kinetic energy, per unit area and per unit time. But $\frac{1}{3}\mathbf{n}m\overline{v^2}$ equals the pressure P, and if the area of the moving piston is A, the decrease of molecular kinetic energy per unit time is

$$PAu = Fu. \tag{9–22}$$

The product Fu (force times velocity) gives the rate at which mechanical work is done on the piston or the power developed by the expanding gas, and we see that this is just equal to the rate of decrease of molecular kinetic energy. If the molecules do not receive energy from any other source, their kinetic energy, and hence the temperature of the gas, decreases. Note that it is not correct to say that the temperature *of a molecule* decreases. From the molecular point of view, temperature is an attribute of the assembly of molecules as a whole, namely, a quantity proportional to the mean kinetic energy. An individual molecule can have more or less kinetic energy but it does not have a higher or lower temperature.

The derivation above was based on the assumption that the piston velocity, u, was very much smaller than the molecular velocities, and it does not hold if the piston is pulled up rapidly. In particular, if the piston velocity is very much greater than the molecular velocities, no molecules (or at least very few) will be able to overtake the piston and collide with it. Then there is no loss of kinetic energy and no decrease in temperature, intermolecular forces being neglected. Such a process is equivalent to an expansion into a vacuum, as in the Joule experiment, where we showed on thermodynamic grounds that the work and the change in internal energy were both zero.

9–6 THE PRINCIPLE OF EQUIPARTITION OF ENERGY

Suppose we have a mixture of gases that do not react chemically with one another, and that the temperature and density are such that their behavior approximates that of an ideal gas. It is found experimentally that the total pressure of the mixture is the sum of the pressures that each gas alone would exert if a mass of each, equal to the mass of that gas in the mixture, occupied the entire volume of the mixture. The pressure that would be exerted by each gas if present alone is called its *partial pressure* and the experimental law above is *Dalton's law of partial pressures*. If the gases are distinguished by subscripts, we can then write

$$p_1V = N_1kT, \qquad p_2V = N_2kT, \quad \text{etc.,}$$

where p_1, p_2, etc. are the partial pressures of the constituent gases, N_1, N_2, etc.

are the numbers of molecules of each constituent, and V and T are the volume and temperature, common to all of the gases.

Let m_1, m_2, etc. represent the masses of the molecules of the constituents and $\overline{v_1^2}$, $\overline{v_2^2}$, etc., the respective mean square speeds. By the methods of Section 9–4, considering the collisions of each type of molecule with the walls and computing the pressure produced by each, we would find

$$p_1V = \frac{1}{3} N_1 m_1 \overline{v_1^2}, \qquad p_2V = \frac{1}{3} N_2 m_2 \overline{v_2^2}, \quad \text{etc.}$$

Equating corresponding expressions for p_1V, p_2V, etc., gives

$$\frac{1}{2} m_1 \overline{v_1^2} = \frac{3}{2} kT, \qquad \frac{1}{2} m_2 \overline{v_2^2} = \frac{3}{2} kT, \quad \text{etc.}$$

The terms on the left side of the preceding equation are the mean translational kinetic energies of the molecules of the various gases, and we conclude that in a mixture of gases the mean kinetic energies of the molecules of each gas are the same. That is, in a mixture of hydrogen and mercury vapor, although the masses of the molecules are in the ratio of 2 to 200, the mean translational kinetic energy of the hydrogen molecules equals that of the mercury molecules.

The example above is one illustration of the *principle of equipartition of energy*. We know now that this principle is not a universal law of nature but, rather, a limiting case under certain special conditions. However, it has been a very fruitful principle in the development of molecular theories.

Let us give another example. The translational kinetic energy associated with the x-component of the velocity of a molecule of mass m is $\frac{1}{2}mv_x^2$, with corresponding expressions for the y- and z-components. The mean square value of the velocities of a group of molecules is

$$\overline{v^2} = \overline{v_x^2} + \overline{v_y^2} + \overline{v_z^2}.$$

Since the x-, y-, and z-directions are all equivalent, the mean square values of the components of velocity must be equal, so that

$$\overline{v_x^2} = \overline{v_y^2} = \overline{v_z^2},$$

and

$$\overline{v^2} = 3\overline{v_x^2} = 3\overline{v_y^2} = 3\overline{v_z^2}.$$

The mean kinetic energy per molecule, associated with any one component of velocity, say v_x, is therefore

$$\frac{1}{2} m\overline{v_x^2} = \frac{1}{6} m\overline{v^2} = \frac{1}{2} kT.$$

Since the mean *total* translational kinetic energy per molecule is $3kT/2$, it follows that the translational kinetic energy associated with each component of velocity is just one-third of the total.

Each independent quantity that must be specified to determine the energy of a molecule is called a *degree of freedom.* Since the translational kinetic energy of a molecule is determined by the three velocity components of its center of mass, it has three translational degrees of freedom. We see that the average translational kinetic energy per molecule is divided equally among them. In other words, we have equipartition of the energy among the three translational degrees of freedom.

Molecules, however, are not geometrical points but are of finite size. They have moments of inertia, as well as mass, and can therefore have kinetic energy of rotation as well as of translation. Furthermore, we would expect them to rotate because of the random collisions with other molecules and with the walls. Since the angular velocity vector of a rotating molecule can have a component along all three coordinate axes, a molecule would be expected to have three rotational degrees of freedom or, if it is a rigid body, six degrees of freedom in all. However, molecules are not perfectly rigid structures and can also be expected to oscillate or vibrate as the result of collisions with other molecules, giving rise to still more degrees of freedom. (It may be mentioned at this point that rotations and vibrations of molecules are facts that are as well established as most of our other information about molecular properties. The best experimental method of studying rotations and vibrations consists of a spectroscopic analysis of the light emitted or absorbed by molecules in the infrared.) Without committing ourselves to any specific number, let us say that in general a molecule has f degrees of freedom, of which 3 only are translational, however complex the molecule.

We shall show in Section 12–5, on the basis of the Boltzmann statistics, that if the energy associated with any degree of freedom is a quadratic function of the variable specifying the degree of freedom, the mean value of the corresponding energy equals $kT/2$. For example, the kinetic energy associated with the velocity component v_x is a quadratic function of v_x, and, as shown above, its mean value equals $kT/2$. Similarly for rotation, where the kinetic energy is $I\omega^2/2$, the mean rotational kinetic energy is $kT/2$; and for a harmonic oscillator, where the potential energy is $Kx^2/2$ (K being the force constant), the mean potential energy is $kT/2$. Hence all of the degrees of freedom for which the energy is a quadratic function have associated with them, on the average, equal amounts of energy; and if all degrees of freedom are of this nature, the total energy is shared equally among them. This is the general statement of the principle of equipartition of energy. The mean *total* energy of a molecule with f degrees of freedom, assuming the equipartition principle holds, is therefore

$$\bar{\epsilon} = \frac{f}{2} kT, \tag{9–23}$$

and the total energy of N molecules is

$$N\bar{\epsilon} = \frac{f}{2} NkT = \frac{f}{2} nRT, \tag{9–24}$$

where n is the number of moles and R the universal gas constant.

9–7 CLASSICAL THEORY OF SPECIFIC HEAT CAPACITY

In thermodynamics, the change in internal energy U of a system, between two equilibrium states, is defined by the equation

$$U_a - U_b = W_{\text{ad}},$$

where W_{ad} is the work in any adiabatic process between the states. Only *changes* in internal energy are defined.

Starting with a molecular model of a system, we can identify the internal energy with the sum of the energies of the individual molecules. In the preceding section we have derived a theoretical expression for the total energy associated with the f degrees of freedom of each of the N molecules of a gas. We therefore set this equal to the internal energy U:

$$U = \frac{f}{2} NkT = \frac{f}{2} nRT. \tag{9–25}$$

The specific internal energy per mole is

$$u = \frac{U}{n} = \frac{f}{2} RT. \tag{9–26}$$

How can we test the validity of the assumptions made in the foregoing derivation? The most direct way is from measurements of specific heat capacities. The molal specific heat capacity at constant volume is

$$c_v = \left(\frac{\partial u}{\partial T}\right)_v.$$

Hence, if the hypothesis above is correct, we should have

$$c_v = \frac{d}{dT}\left(\frac{f}{2} RT\right) = \frac{f}{2} R. \tag{9–27}$$

We also know from thermodynamic reasoning that for an ideal gas,

$$c_P = c_v + R.$$

Hence

$$c_P = \frac{f}{2} R + R = \frac{f+2}{2} R, \tag{9–28}$$

and

$$\gamma = \frac{c_P}{c_v} = \frac{\dfrac{f+2}{2}}{\dfrac{f}{2}} = \frac{f+2}{f}. \tag{9–29}$$

Table 9–1 Molal specific heat capacities of a number of gases, at temperatures near room temperature. The quantities measured experimentally are c_P and γ. The former is determined by use of a continuous flow calorimeter and the latter is obtained from measurements of the velocity of sound in the gas.

Gas	γ	c_P/R	c_v/R	$\dfrac{c_P - c_v}{R}$
He	1.66	2.50	1.506	.991
Ne	1.64	2.50	1.52	.975
A	1.67	2.51	1.507	1.005
Kr	1.69	2.49	1.48	1.01
Xe	1.67	2.50	1.50	1.00
H_2	1.40	3.47	2.47	1.00
O_2	1.40	3.53	2.52	1.01
N_2	1.40	3.50	2.51	1.00
CO	1.42	3.50	2.50	1.00
NO	1.43	3.59	2.52	1.07
Cl_2	1.36	4.07	3.00	1.07
CO_2	1.29	4.47	3.47	1.00
NH_3	1.33	4.41	3.32	1.10
CH_4	1.30	4.30	3.30	1.00
Air	1.40	3.50	2.50	1.00

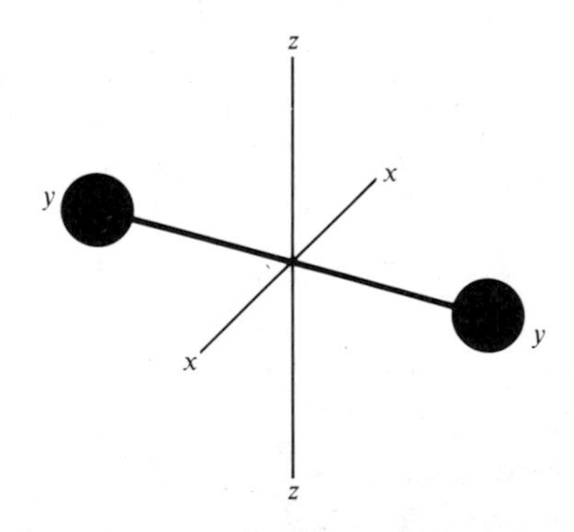

Fig. 9–5 A dumbbell molecule.

Thus while the principles of thermodynamics could give us only an expression for the *difference* between the specific heat capacities at constant pressure and constant volume, molecular theory, together with the equipartition principle, predicts the actual magnitudes of the specific heat capacities and their ratio γ, in terms of the number of degrees of freedom f and the experimentally determined universal constant R. Note that, according to the theory, c_v, c_P, and γ are all constants independent of the temperature.

Consider first a gas whose atoms are monatomic and for which the energy is wholly kinetic energy of translation. Since there are three translational degrees of freedom, $f = 3$, and we would expect

$$c_v = \frac{f}{2}R = \frac{3}{2}R = 1.5R,$$

$$c_P = \frac{f+2}{2}R = \frac{5}{2}R = 2.5R,$$

and

$$\gamma = \frac{c_P}{c_v} = \frac{5}{3} = 1.67.$$

This is in good agreement with the values of c_v and γ for the monatomic gases listed in Table 9–1. Furthermore, the specific heat capacities of these gases are found to be practically independent of temperature, in agreement with the theory.

Consider next a diatomic molecule having the dumbbell structure shown in Fig. 9–5. Its moment of inertia about the x- and z-axes is very much greater than that about the y-axis, and if the latter can be neglected, the molecule has two rotational degrees of freedom, the two quantities specifying the rotational kinetic energy being the components of angular velocity about the x- and z-axes. Also, since the atomic bond is not perfectly rigid, the atoms can vibrate along the line joining them. This introduces two vibrational degrees of freedom, since the vibrational energy is part kinetic and part potential and is specified by the velocity and the separation of the atoms. We might therefore expect seven degrees of freedom for a diatomic molecule (3 for translation, 2 for rotation, and 2 for vibration). For $f = 7$, the theory predicts

$$c_v = \frac{7}{2}R = 3.5R, \qquad \gamma = \frac{9}{7} = 1.29.$$

These values are not in good agreement with those observed for the diatomic gases listed in Table 9–1. However, letting $f = 5$, we find

$$c_v = \frac{5}{2}R = 2.5R, \qquad \gamma = \frac{7}{5} = 1.40.$$

These are almost exactly equal to the average values of c_v and γ for the diatomic molecules in the second part of the table (Cl_2 is an interesting exception). Thus, near room temperature, these molecules behave as if either the rotational or vibrational degrees of freedom, but not both, shared equally with the translational degrees of freedom in the total molecular energy.

As the number of atoms in a molecule increases, the number of degrees of freedom can be expected to increase; and the theory predicts a decreasing ratio of specific heat capacities, in general agreement with experiment.

The main features of the theory are fairly well borne out. It predicts that γ is never greater than 1.67 or less than 1 and this is in fact true. However, if we insert in Eq. (9–29) the measured values of γ and solve for f, the result is in general not exactly an integer. Now a molecule either has a degree of freedom or it has not. Degrees of freedom are counted, not weighed. It is meaningless to speak of a fraction of a degree of freedom, and the simple concept of equipartition is certainly not the whole story.

When we examine the temperature variation of specific heat capacities, the divergences between experiment and the simple theory above become even more apparent. Except for gases whose atoms are monatomic, the specific heat capacities of all gases increase with increasing temperature and decrease as the temperature is lowered. In fact, at a temperature of 20 K, the specific heat capacities of hydrogen (the only diatomic gas that remains a gas at very low temperatures) decreases to $\frac{3}{2}R$, the value predicated by theory for a monatomic gas. Thus at this low temperature neither the rotational nor the vibrational degrees of freedom of the hydrogen molecule appear to share at all in the change of internal energy associated with a change in temperature. All of the difficulties mentioned above are removed, however, when the principles of quantum mechanics and of statistics are taken into consideration. These are discussed in Section 12–7.

The pressure of a gas depends on its *translational* kinetic energy, and regardless of its molecular complexity a molecule has only three translational degrees of freedom, and its translational kinetic energy equals $3kT/2$. Then if U_{tr} represents this portion of the internal energy,

$$U_{tr} = \frac{3}{2} NkT.$$

The pressure P equals NkT/V, so

$$P = \frac{2}{3}\frac{U_{tr}}{V} = \frac{2}{3}\mathbf{u}_{tr}, \tag{9–30}$$

where $\mathbf{u}_{tr}$ is the translational energy per unit volume, or the energy density; and, as pointed out earlier, the pressure equals two-thirds of the translational energy density.

9–8 SPECIFIC HEAT CAPACITY OF A SOLID

The molecules of a solid, unlike those of a gas, are constrained to oscillate about fixed points by the relatively large forces exerted on them by other molecules. Let us imagine that each executes harmonic motion. Each has three degrees of freedom, considered as a mass point, but the *potential* energy associated with its motion, which could be neglected for the widely separated molecules of a gas, is on the average just equal to the kinetic energy, if the motion is simple harmonic. Hence, if the equipartition principle is valid for solids, we must assign an energy kT to each degree of freedom ($kT/2$ for kinetic energy, $kT/2$ for potential energy) rather than just $kT/2$ as for the molecules of a gas. The total energy of N molecules is then

$$U = 3NkT, \tag{9–31}$$

and the molal specific heat capacity at constant volume, from the theory, is

$$c_v = 3R = 3 \times 8.31 \times 10^3 = 24.9 \times 10^3 \text{ J kilomole}^{-1} \text{ K}^{-1}. \tag{9–32}$$

This is in agreement with the empirical law of Dulong and Petit which states that at temperatures which are not too low, the molal specific heat capacities at constant volume of all pure substances in the solid state are very nearly equal to $3R$. Again we have reasonably good agreement with experiment at high temperatures. At low temperatures the agreement is definitely bad, since, as we have seen, the specific heat capacities of all substances must approach zero as the temperature approaches absolute zero. This is another problem to which the classical theory does not provide the right answer and in which the methods of quantum mechanics must be used.

One other discrepancy between simple theory and experiment should be pointed out here. There is good reason to believe that in metals, which are electrical conductors, each atom parts with one or more of its outer electrons and that these electrons form a sort of electron cloud or electron gas, occupying the volume of the metal and constrained by electrical forces at the metal surfaces in much the same way that ordinary gases occupy a containing vessel. This electron gas has translational degrees of freedom which are quite independent of the metallic ions forming the crystal lattice, and it should have a molal specific heat capacity equal to that of any other monatomic gas, namely, $3R/2$. That is, as the temperature of the metal is increased, energy must be supplied to make the electrons move faster as well as to increase the amplitude of vibrations of the metallic ions. The latter should have a specific heat capacity of $3R$, so the total heat capacity of a metal should be at least $3R + 3R/2 = 9R/2$. Actually, metals obey the Dulong-Petit law as well as do nonconductors, so apparently the electrons do not share in the thermal energy. This was a very puzzling thing for many years, but again it has a very satisfactory explanation when quantum methods are used.

PROBLEMS

9–1 (a) Compute the number of molecules per unit volume in a gas at 300 K when the pressure is 10^{-3} Torr. (b) How many molecules are there in a cube of 1 mm on a side under these conditions?

9–2 The model used in this chapter assumes that the molecules are uniformly distributed throughout the container. What must be the size of a cubical element of volume in the container so that the number of particles in each volume element may vary by 0.1% when the gas is at standard conditions? (From a study of statistics it can be shown that the probable deviation of the number of particles in each volume element from the average number of particles, N, is given by $N^{1/2}$).

9–3 (a) In Fig. 9–1, let $\phi = 45°$, $\Delta\phi = 0.01$ radian, $\theta = 60°$, and $\Delta\theta = 0.01$ radian. What fraction of the molecules of a gas have velocity vectors within the narrow cone which intercepts the shaded area ΔA? (b) Consider a second cone intercepting the *same area* on the spherical surface, but for which $\phi = 90°$, $\theta = 0$. Sketch this cone and compare the number of velocity vectors included within it with those in the cone of part (a).

9–4 (a) Approximately what fraction of the molecules of a gas have velocities for which the angle ϕ in Fig. 9–1 lies between 29.5° and 30.5°, while θ lies between 44.5° and 45.5°? (b) What fraction have speeds for which ϕ lies between 29.5° and 30.5°, regardless of the value θ? [*Note:* Angles must be expressed in radians.]

9–5 Suppose that the number of molecules in a gas having speeds between v and $v + \Delta v$ is given by $\Delta N_v = N\,\Delta v/v_0$ for $v_0 > v > 0$ and $\Delta N_v = 0$ for $v > v_0$. (a) Find the fraction of molecules having speeds between $0.50\,v_0$ and $0.51\,v_0$. (b) Find the fraction having the speeds in part (a) in the direction described in part (a) and part (b) of the previous problem. (c) Find the flux of molecules described in part (b) of this problem arriving at a surface, if the gas is at standard conditions.

9–6 Calculate $\bar{v}$ and $v_{\rm rms}$ for the following distributions of six particles: (a) all have speeds of 20 m s^{-1}; (b) three have speeds of 5 m s^{-1} and three have speeds of 20 m s^{-1}; (c) four have speeds of 5 m s^{-1} and two have speeds of 20 m s^{-1}; (d) three are at rest and three have speeds of 20 m s^{-1}; (e) one has a speed of 5 m s^{-1}, two have speeds of 7 m s^{-1}, two have speeds of 15 m s^{-1} and one has a speed of 20 m s^{-1}.

9–7 The speed distribution function of a group of N particles is given by $\Delta N_v = kv\,\Delta v$ for $v_0 > v > 0$ and $\Delta N_v = 0$ for $v > v_0$. (a) Draw a graph of the distribution function. (b) Show that the constant $k = 2N/v_0^2$. (c) Compute the average speed of the particles. (d) Compute the root-mean-square speed of the particles.

9–8 (a) Derive Eq. (9–7) beginning with Eq. (9–4). (b) For a gas at standard conditions, find $\Delta\Phi_v$ for molecules obeying the speed distribution law of the previous problem and having speeds between $0.50\,v_0$ and $0.51\,v_0$. (c) Determine Φ for molecules having the same speed distribution.

9–9 What form would Eq. (9–17) take if several kinds of molecules were present in a gas? Does the answer agree with Dalton's law?

9–10 Derive an expression equivalent to Eq. (9–17) for a two-dimensional gas, i.e., one whose molecules can move only in a plane. (The concept corresponding to pressure, or force per unit area, becomes force per unit length.)

9–11 (a) Compute the rms speed of a gas of helium atoms at 300 K. (b) At what temperature will oxygen molecules have the same rms speed? (c) Through what potential difference must a singly ionized oxygen molecule be accelerated to have the same speed?

9–12 (a) How many molecular impacts are made per second on each square centimeter of a surface exposed to air at a pressure of 1 atm and at 300 K? The mean molecular weight of air is 29. (b) What would be the length of a cylinder 1 cm^2 in cross section containing the number of air molecules at 1 atm and 300 K which collide with a surface 1 cm^2 in one second?

9–13 A cubical box of 0.1 m on a side contains 3×10^{22} molecules of O_2 at 300 K. (a) On the average, how many collisions does each molecule make with the walls of the box in one second? (b) What pressure does the oxygen exert on the walls of the box?

9–14 A closed vessel contains liquid water in equilibrium with its vapor at 100°C and 1 atm. One gram of water vapor at this temperature and pressure occupies a volume of 1670 cm^3. The heat of vaporization at this temperature is 2250 J g^{-1}. (a) How many molecules are there per cm^3 of vapor? (b) How many vapor molecules strike each cm^2 of liquid surface per second? (c) If each molecule which strikes the surface condenses, how many evaporate from each cm^2 per second? (d) Compare the mean kinetic energy of a vapor molecule with the energy required to transfer one molecule from the liquid to the vapor phase.

9–15 When a liquid and its vapor are in equilibrium, the rates of evaporation of the liquid and condensation of the vapor are equal. Assume that every molecule of the vapor striking the liquid surface condenses, and assume that the rate of evaporation is the same when the vapor is rapidly pumped away from the surface, as when liquid and vapor are in equilibrium. The vapor pressure of mercury at 0°C is 185×10^{-6} Torr and the latent heat of vaporization is about 340 J g^{-1}. Compute the rate of evaporation of mercury into a vacuum, in g cm^{-2} s^{-1}, at a temperature of (a) 0°C, (b) 20°C.

9–16 A thin-walled vessel of volume V contains N particles which slowly leak out of a small hole of area A. No particles enter the volume through the hole. Find the time required for the number of particles to decrease to $N/2$. Express your answer in terms of A, V, and $\bar{v}$.

9–17 The pressure in a vacuum system is 10^{-3} Torr. The external pressure is 1 atm and $T = 300$ K. There is a pinhole in the walls of the system, of area 10^{-10} cm^2. Assume that every molecule "striking" the hole passes through it. (a) How many molecules leak into the system in 1 hour? (b) If the volume of the system is 2 liters, what rise in pressure would result in the system? (c) Show that the number of molecules that leak out is negligible.

9–18 A vessel of volume $2V$ is divided into compartments of equal volume by a thin partition. The left side contains initially an ideal gas at a pressure P_0, and the right side is initially evacuated. A small hole of area A is punched in the partition. Derive an expression for the pressure P_l on the left side as a function of time. Assume the temperature to remain constant and to be the same on both sides of the partition.

9–19 An insulated chamber containing liquid helium in equilibrium with its vapor is maintained at 1.2 K. It is separated from a second insulated chamber maintained at 300 K, by a thin insulating partition with a small hole in it. The helium vapor is allowed to fill both chambers. If the vapor pressure of the helium at 1.2 K is P_0, show that the pressure P in the other chamber is $P_0\sqrt{300/1.2}$. (The ratio of P/P_0 is called the *thermomolecular pressure ratio* and is important in vapor pressure thermometry when the pressure is so low that the particles do not collide in a distance long compared to linear dimensions of the apparatus.)

9–20 An ideal monatomic gas is confined to an insulated cylinder fitted with an insulated piston. (a) By considering collisions of the molecules of the gas with the quasistatically moving piston, show that $PV^{5/3}$ = constant. (b) Determine the pressure dependence of the rms speed of the molecules in an adiabatic compression or expansion.

9–21 A molecule consists of four atoms at the corners of a tetrahedron. (a) What is the number of translational, rotational, and vibrational degrees of freedom for this molecule? (b) On the basis of the equipartition principle, what are the values of c_v and γ for a gas composed of these molecules?

9–22 Under the action of suitable radiation a diatomic molecule splits into two atoms. The ratio of the number of dissociated molecules to the total number of molecules is α. Plot $\gamma(= c_P/c_v)$ as a function of α at a temperature where the vibrational modes of the diatomic molecule are excited.

9–23 Find the total translational kinetic energy and the rms speed of the molecules of 10 liters of helium gas at an equilibrium pressure of 10^5 N m^{-2}.

9–24 (a) Find the specific heat capacity at constant volume for a gas of H_2 molecules and H_2O molecules. (b) How do the specific heat capacities change if the gas is liquefied or solidified?

10

Intermolecular forces. Transport phenomena

10–1 INTERMOLECULAR FORCES

In the preceding chapter, the molecules of a gas were treated as geometrical points that exerted no forces on each other. We now wish to take such forces into account.

The force between any pair of molecules is of electrical origin; and because of the complicated structure of an atom or molecule, it is not expressible by any simple law. In general, at relatively large separations, the force is one of attraction, referred to as a van der Waals force, which decreases rapidly with increasing separation. When two molecules approach so closely that their electron clouds overlap, the force becomes one of repulsion that rises very rapidly as the separation becomes smaller. Thus the intermolecular force must have the general form of the solid curve in Fig. 10–1.

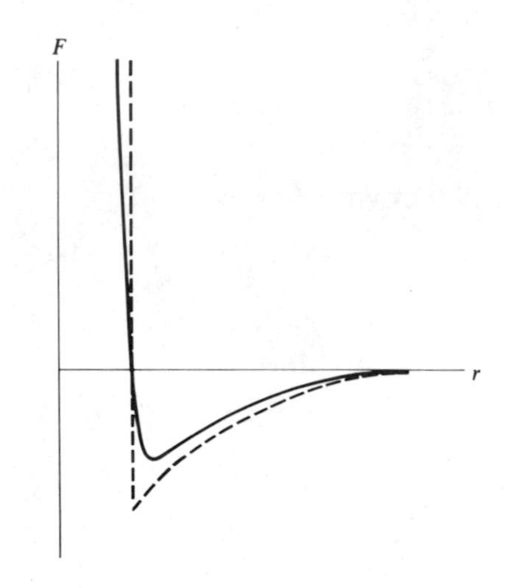

Fig. 10–1 Intermolecular forces.

The simplest approximation to this law is to treat the molecules as elastic *hard spheres*, for which the force of repulsion becomes infinite when the surfaces of the spheres come into contact. If we include a force of attraction when the molecules are not in contact, the force law has the form of the dotted curve in Fig. 10–1.

10–2 THE VAN DER WAALS EQUATION OF STATE

We have made extensive use of the van der Waals equation of state in earlier chapters, not so much because of any great accuracy of this equation in describing the properties of real gases but because it shows in a general way, through the factor a, how these properties depend on intermolecular forces of attraction, and through the factor b how they depend on molecular sizes.

The latter correction to the equation of state was actually first suggested by Clausius. He reasoned that in the derivation in Section 9–4 one should use not the actual volume V of the container, but the volume available to a single molecule, which will be somewhat less than V because of the volume occupied by the other molecules. If we represent the "unavailable" volume per mole by b, then in a gas consisting of n moles the unavailable volume is nb and we should write

$$P(V - nb) = nRT,$$

or, dividing through by n,

$$P(v - b) = RT. \tag{10–1}$$

This equation was first written down by Hirn.* (Here, the letter v represents the molal specific volume, not the molecular speed.)

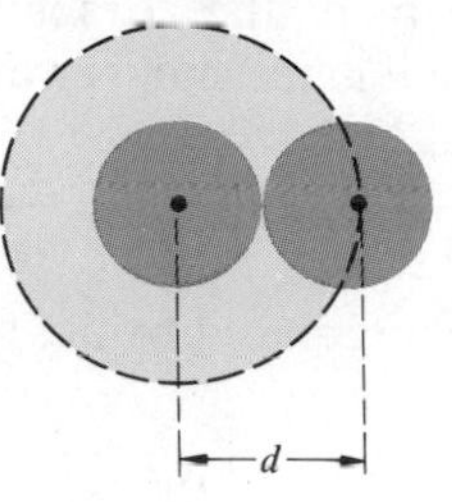

Fig. 10–2 The radius of the sphere of exclusion equals the molecular diameter d.

If the molecules are considered as hard spheres of diameter d, the minimum distance between the centers of two molecules, as shown in Fig. 10–2, is equal to d. In effect, the center of each molecule is excluded by the other from a sphere of radius d, known as the "sphere of exclusion." The volume of this sphere is $4\pi d^3/3$, and to avoid counting each pair twice, we take as the total unavailable volume, for a system of N molecules,

$$\frac{1}{2} N \times \frac{4}{3} \pi d^3.$$

The number of molecules N is the product of the number of moles n and Avogadro's number N_A, so the unavailable volume per mole, or the constant b, is

$$b = \frac{2}{3} N_A \pi d^3. \tag{10–2}$$

This is four times as great as the actual molecular volume per mole, which is

$$\frac{1}{6} N_A \pi d^3. \tag{10–3}$$

Van der Waals, in 1873, included a second correction term in the equation of state to take into account the force of attraction between molecules. Let us assume

* Gustav A. Hirn, French engineer, (1815–1890).

that these forces decrease so rapidly with distance (for example, as $1/r^6$) that they are appreciable only between a molecule and its nearest neighbors. Molecules within the body of the gas are on the average attracted equally in all directions, but those in the outermost layers experience a net inward force. A molecule approaching the wall of the container is therefore slowed down and the average force exerted on the wall, and hence the observed pressure, is somewhat smaller than it would be in the absence of attractive forces.

The reduction in pressure will be proportional both to the number of molecules per unit volume in the outer layer, $\mathbf{n} = N/V$, and to the number per unit volume in the next layer beneath them, which is doing the attracting. Hence the pressure will be reduced by an amount proportional to $\mathbf{n}^2$, or equal to $\alpha\mathbf{n}^2$, where α is a factor dependent on the strength of the attractive force. Since the number of molecules N equals nN_A, where n is the number of moles, then

$$\alpha\mathbf{n}^2 = \alpha\left(\frac{N}{V}\right)^2 = \alpha N_A^2\frac{n^2}{V^2} = \frac{\alpha N_A^2}{v^2} = \frac{a}{v^2}, \tag{10–4}$$

where the product αN_A^2 has been replaced by a. Thus the pressure P given by the Hirn equation,

$$P = \frac{RT}{v - b},$$

should be reduced by a/v^2; and

$$P = \frac{RT}{v - b} - \frac{a}{v^2},$$

or

$$\left(P + \frac{a}{v^2}\right)(v - b) = RT, \tag{10–5}$$

which is the van der Waals equation of state.

Since the molal specific critical volume of a van der Waals gas, v_c, is equal to $3b$, it follows from Eq. (10–2) that

$$v_c = 3b = 2N_A\pi d^3, \tag{10–6}$$

which is 12 times the total molecular volume. The value of b for a van der Waals gas therefore provides a means of estimating molecular diameters, since

$$d = \left(\frac{3b}{2\pi N_A}\right)^{1/3}. \tag{10–7}$$

Thus for helium, for which $b = 23.4 \times 10^{-3}$ m^3 kilomole^{-1}, we have

$$d = \left(\frac{3 \times 23.4 \times 10^{-3}}{2 \times 3.14 \times 6.02 \times 10^{26}}\right)^{1/3} \simeq 2.6 \times 10^{-10}\ \text{m} = 2.6 \times 10^{-8}\ \text{cm}.$$

Other methods of estimating molecular diameters will be described in Section 10–4. Values of a and b for several gases are given in Table 2–1.

10–3 COLLISION CROSS SECTION. MEAN FREE PATH

In deriving the expression for the pressure exerted by a gas, the molecules were treated as geometrical points which could fly freely from one wall of a container to the other without colliding with other molecules. One of the objections raised in the early development of kinetic theory was that if molecules acted in this way, a small amount of gas released in a large room would spread throughout the room practically instantaneously, whereas we know that when the stopper is removed from a bottle of perfume, a considerable time elapses before the odor can be detected even at a point only a few feet away, in the absence of air currents. It was soon realized that this relatively slow *diffusion* of one gas in another resulted from molecular collisions such as that shown in Fig. 10–3, which cause a molecule to move in an irregular, zigzag path.

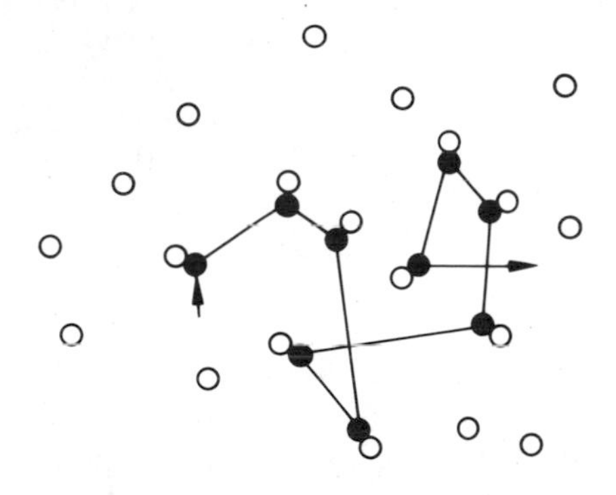

Fig. 10–3 Molecular free paths.

We again assume that a molecule is a hard sphere. Let us refer to one of the colliding molecules as the "target" molecule and to the other as the "bullet" molecule. Then a collision occurs whenever the distance between the centers of the molecules becomes equal to the molecular diameter d, as in Fig. 10–2.

Since it is only the center-to-center distance that determines a collision, it does not matter whether the target is large and the bullet small, or vice versa. We may therefore consider the bullet molecule to shrink to a point at its center, and the target molecule to occupy the entire sphere of exclusion, of radius d.

Now consider a thin layer of gas of dimensions L, L, and Δx, as in Fig. 10–4. The layer contains (equivalent) target molecules, represented by the shaded circles. We then imagine that a very large number N of bullet molecules, represented by the black dots, is projected toward the face of the layer—like pellets from a shotgun—in such a way that they are distributed at random over the face of the layer. If the thickness of the layer is so small that no target molecule can hide behind another, the layer presents to the bullet molecules the appearance of Fig. 10–4.

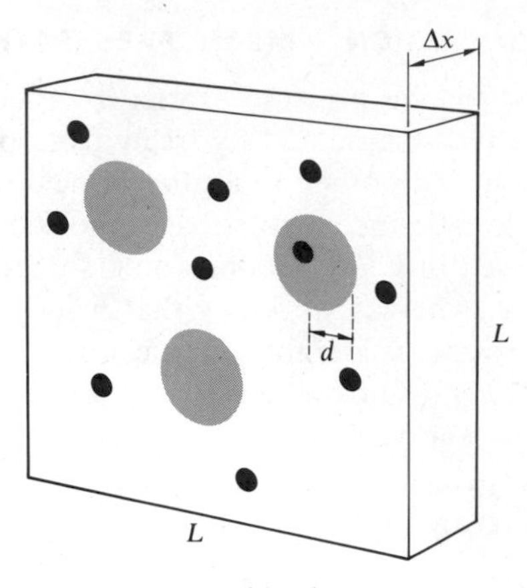

Fig. 10–4 A thin layer of gas of "target" molecules being bombarded by "bullet" molecules.

Most of the bullet molecules will pass through the layer, but some will collide with target molecules. The ratio of the number of collisions, ΔN, to the total number of bullet molecules, N, is equal to the ratio of the area presented by the target molecules to the total area presented by the layer:

$$\frac{\Delta N}{N} = \frac{\text{target area}}{\text{total area}}.$$

The target area σ of a single (equivalent) molecule is the area of a *circle* of radius d, the exclusion radius:

$$\sigma = \pi d^2. \tag{10–8}$$

This area is called the *microscopic collision cross section* of one (equivalent) molecule. The total *target* area is the product of this and the number of target molecules in the layer. If there are $\mathbf{n}$ target molecules per unit volume, this number is $\mathbf{n}L^2\,\Delta x$, so the total target area is

$$\mathbf{n}\sigma L^2\,\Delta x.$$

The *total* area of the layer is L^2, so

$$\frac{\Delta N}{N} = \frac{\mathbf{n}\sigma L^2\,\Delta x}{L^2} = \mathbf{n}\sigma\,\Delta x. \tag{10–9}$$

The quantity $\mathbf{n}\sigma$ is called the *macroscopic collision cross section* of the (equivalent) molecules. Since the number density $\mathbf{n}$, in the MKS system, is the number of molecules per cubic meter and the collision cross section σ is the number of square meters per molecule, the unit of the product $\mathbf{n}\sigma$ is 1 square meter per cubic meter ($1\ \text{m}^2\,\text{m}^{-3} = 1\ \text{m}^{-1}$). More generally, in any system, the unit of macroscopic collision cross section is a *reciprocal length*, not an area.

Each of the ΔN collisions diverts a molecule from its original path or *scatters* it out of the beam, and decreases the number remaining in the beam. Let us, therefore, interpret ΔN not as a "number of collisions," but as the *decrease* in the number N, and write

$$\Delta N = -N\mathbf{n}\sigma\,\Delta x,$$

or

$$\frac{\Delta N}{N} = -\mathbf{n}\sigma\,\Delta x.$$

In reality, N decreases in stepwise fashion as individual molecules make collisions, but if N is very large we can consider it a continuous function of x and write

$$\frac{dN}{N} = -\mathbf{n}\sigma\,dx.$$

Then

$$\ln N = -\mathbf{n}\sigma x + \text{constant};$$

and if $N = N_0$, when $x = 0$,

$$N = N_0 \exp(-\mathbf{n}\sigma x). \tag{10–10}$$

This is known as the *survival equation.* It represents the number of molecules N, out of an initial number N_0, that has not yet made a collision after traveling a distance x.

Inserting the expression for N in Eq. (10–9), we obtain

$$\Delta N = N_0\mathbf{n}\sigma \exp(-\mathbf{n}\sigma x)\,\Delta x. \tag{10–11}$$

In this equation, N is the number of molecules making their first collision after having traveled a distance between x and $x + \Delta x$.

Let us calculate the average distance traveled by a group of N_0 molecules before they make their first collision. This average distance is known as the *mean free path, l.* To calculate it, we multiply x by the number of particles ΔN that travels the distance x before colliding, sum over all values of x, and divide by the total number N_0. Replacing the sum by an integral, we have

$$l = \frac{\sum x\,\Delta N}{N_0} = \mathbf{n}\sigma \int_0^\infty x \exp(-\mathbf{n}\sigma x)\,dx.$$

The definite integral equals $1/\mathbf{n}^2\sigma^2$, so

$$l = \frac{1}{\mathbf{n}\sigma}, \tag{10–12}$$

and the mean free path is inversely proportional to the macroscopic collision cross section. Since the unit of macroscopic collision cross section is the reciprocal of the unit of length, the unit of mean free path is the unit of length. Note that the mean free path does not depend on the molecular speed.

The concept of mean free path may be visualized by thinking of a man shooting bullets aimlessly into a thick forest. All of the bullets will eventually hit trees, but some will travel farther than others. It is easy to see that the average distance traveled will depend inversely on both the denseness of the woods (**n**) and on the size of the trees (σ).

A common experimental technique is to project into a gas a beam of particles (either neutral or having an electric charge) and to measure the quantity N_0 and the number N remaining in the beam after a distance x. The exponential decrease predicted by Eq. (10–10) is found to be well obeyed, and we may now reverse the reasoning by which this equation was derived. That is, since N_0, N, and x are all measurable experimentally, Eq. (10–10) can be solved for $\mathbf{n}\sigma$ or l, and we can consider these quantities to be *defined* by Eq. (10–10), quite independently of any theory of molecular collisions.

Although we derived the equations above by considering a beam of molecules projected into a gas, the mean free path is the same if the group is considered to consist of the molecules of a gas moving at random among the other molecules and making collisions with them. The motion of a single molecule is then a zigzag path as suggested in Fig. 10–3, and we can understand why it is that although the average molecular speed is very large, a molecule wanders away from a given position only relatively slowly.

As an example, suppose the molecular diameter d equals 2×10^{-10} m. At standard conditions, there are about 3×10^{25} molecules m^{-3} in a gas. The macroscopic collision cross section is then

$$\mathbf{n}\sigma = \mathbf{n}\pi d^2 \approx 3 \times 10^{25} \times 3.14 \times 4 \times 10^{-20} \approx 40 \times 10^5 \text{ m}^{-1},$$

and the mean free path is

$$l = \frac{1}{\mathbf{n}\sigma} \approx 2.5 \times 10^{-7} \text{ m},$$

which is smaller than the wavelength of visible light. The average intermolecular separation at standard conditions is about 3×10^{-9} m, so the mean free path is much larger than the average intermolecular separation, and Fig. 10–3 is therefore misleading.

Since the number of molecules per unit volume, **n**, is inversely proportional to the pressure, the mean free path increases as the pressure is decreased. A moderately good "vacuum" system will reduce the pressure to 10^{-3} Torr, which is about 10^{-6} atm. The mean free path is then a million times that at atmospheric pressure, or of the order of 25 cm.

More complete theories of the mean free path take into account the relative motion of all the molecules of a gas, that is, they consider the "target" molecules, as well as the "bullet" molecules, to be in motion. The only change in the end result is to introduce a small correction factor in Eq. (10–12). The inverse dependence on the number of molecules per unit volume and on the collision cross section remains unaltered. On the assumption that all molecules have the same speed, Clausius obtained the result

$$l = \frac{3}{4}\frac{1}{\mathbf{n}\sigma} = \frac{0.75}{\mathbf{n}\sigma}.$$

If the molecules have a Maxwellian velocity distribution (see Section 12–2),

$$l = \frac{1}{\sqrt{2}}\frac{1}{\mathbf{n}\sigma} = \frac{0.707}{\mathbf{n}\sigma}.$$

However, we shall continue to use the simpler result of Eq. (10–12).

In the preceding discussion, the target molecules and bullet molecules were considered identical hard spheres, each of diameter d. One often wishes to know the mean free path of an electron, moving among the neutral or ionized molecules of a gas in a plasma, or among the fixed metallic ions in a metallic conductor. The "diameter" of an electron is so much smaller than that of a molecule that the electron can be considered a geometrical point, and the center-to-center distance in a collision (see Fig. 10–2) becomes $d/2$ rather than d, where d is the molecular diameter. Furthermore, the velocities of the electrons are so much greater than those of the molecules that the latter can be assumed at rest, and the correction for relative velocities need not be made. From the considerations above, the *electronic mean free path* l_e is

$$l_e = 4\frac{1}{\mathbf{n}\sigma}, \tag{10–13}$$

where $\mathbf{n}$ is the number density of *molecules* and $\mathbf{n}\sigma$ is the macroscopic collision cross section of electrons with molecules or ions.

In terms of the mean free path, the survival equation can be written

$$N = N_0 \exp(-\mathbf{n}\sigma x) = N_0 \exp(-x/l). \tag{10–14}$$

Figure 10–5 is a graph of this equation, in which the dimensionless ratio N/N_0 is plotted as a function of x/l. The ordinate of the curve is the fractional number of molecules with free paths longer than any fraction of the mean free path. Note that the fraction with free paths longer than the mean is $\exp(-1)$ or 37%, while the number with free paths shorter than the mean is 63%.

An interesting aspect of the theory of distribution of free paths is that the N_0 molecules considered originally are not necessarily just starting out on their free paths after having made a collision. We merely make a random selection of a large number of molecules at any instant and inquire into their *future* without asking questions about their *past*. Sometimes, however, it is the past rather than the future that is of interest. That is, we may fix our attention on a group of molecules at some instant and instead of asking, as we did above, how far each will travel on the average before it makes its *next* collision, ask how far each has traveled on the average since making its *last previous* collision. The same reasoning as that above shows that this average distance is also the free path l, and that the distribution of "past" free paths is the same as the distribution of "future" free paths. Hence when we consider a large number of molecules in a gas at any instant, the average distance they have yet to travel before their next collision is equal to the average distance they have already traveled since their last collisions, and both distances are equal to the mean free path l. We shall make use of this fact in the next section, in calculating the average distance above or below a plane at which molecules make their last collision before crossing the plane.

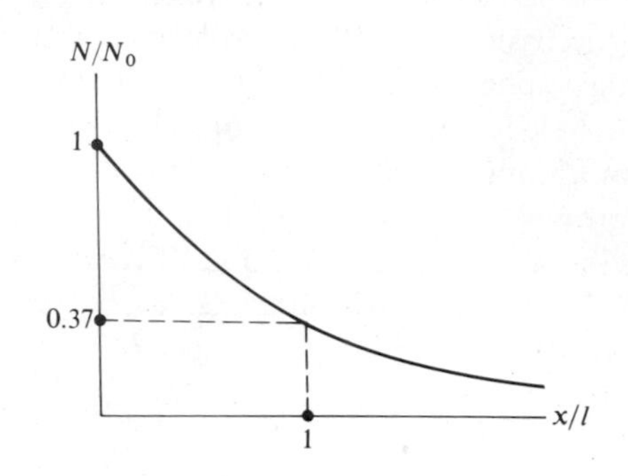

Fig. 10–5 Graph of the survival equation.

This result raised the following interesting question. If the average distance traveled by the group *before* we consider it is l, and the average distance *after* we consider it is also l, why is the mean free path not equal to $2l$ rather than l?

Another important concept is that of *collision frequency* z, the average number of collisions per unit time made by a molecule with other molecules. In a time interval Δt, a molecule travels an average distance $\bar{v}\,\Delta t$ along its zigzag path. The average number of collisions it makes in this time is $\bar{v}\,\Delta t/l$, and hence the collision frequency is

$$z = \frac{\bar{v}}{l} = \bar{v}\mathbf{n}\sigma. \tag{10–15}$$

From the values of $\bar{v}$, $\mathbf{n}$, and σ for oxygen molecules at room temperature, we find

$$z \approx 5.5 \times 10^9 \text{ collisions s}^{-1}.$$

The *mean free time* τ, or the average time between collisions, is the reciprocal of the collision frequency z and hence

$$\tau = \frac{1}{z} = \frac{l}{\bar{v}} = \frac{1}{\bar{v}\mathbf{n}\sigma}. \tag{10–16}$$

For oxygen molecules at room temperature,

$$\tau \approx \frac{1}{5.5 \times 10^9} \approx 1.8 \times 10^{-10}\ \text{s}.$$

The preceding results form the basis of the theory of metallic conduction developed by Drude* in 1900. We assume that the free electrons in a metallic conductor can be considered an ideal gas and that their average random speed $\bar{v}$ is the same as that of gas molecules of the same mass, at the same temperature. (We shall show in Chapter 13 that this is not a very good assumption). If the electric field intensity in the conductor is E, the force F on each electron, of (negative) charge e, is $F = eE$. As a result of this force, the electrons have an acceleration a opposite to the direction of the field and of magnitude

$$a = \frac{F}{m} = \frac{eE}{m}.$$

The electrons do not accelerate indefinitely, however, because of collisions with the fixed metallic ions. We assume that at each such collision an electron is brought to rest and makes a fresh start losing all memory of its previous velocity. In the mean free time τ between collisions, an electron acquires a velocity opposite to the field equal to $a\tau$, and its average velocity between collisions, or the drift velocity u, is

$$u = \frac{1}{2}\,a\tau = \frac{1}{2}\left(\frac{eE}{m}\right)\frac{l_e}{\bar{v}}.$$

This drift velocity is superposed on the random "thermal" velocity , $\bar{v}$ but in an actual conductor it is very small compared with the random velocity. Note that in the expression for the mean free path l_e, we should use Eq. (10–13).

The current density J in the metal (the current per unit of cross-sectional area) is the product of the number density $\mathbf{n}_e$ of electrons, their charge e, and the drift velocity u:

$$J = \mathbf{n}_e e u = \left(\frac{n_e e^2 l_e}{2mv}\right)E.$$

The resistivity ρ of the metal is defined as the ratio of the electric intensity E to the current density J: $\rho = E/J$. Hence

$$\rho = \frac{2m\bar{v}}{\mathbf{n}_e e^2 l_e}. \tag{10–17}$$

In a given metal at a given temperature, all quantities on the right side of the preceding equation are constants so that the Drude theory predicts that under these conditions the resistivity of a metallic conductor is a constant independent of E.

* Paul K. L. Drude, German physicist (1863–1906).

In other words, the current density J is directly proportional to the electric intensity E, and the metal, in agreement with experiment, obeys Ohm's law.

A more familiar statement of Ohm's law is that at a given temperature, the potential difference V between two points of a conducting wire is directly proportional to the current I in the wire, or that $V = IR$, where R is a constant independent of I. The total current I in a conductor of constant cross-sectional area A is $I = JA$. If the length of the conductor is L, the potential difference V between its ends is $V = EL$, so the equation $\rho j = E$ can be written

$$\rho \frac{I}{A} = \frac{V}{L},$$

or,

$$V = \frac{\rho L}{A} I = IR,$$

where the *resistance* $R = \rho L/A$.

We shall show in Chapter 12 that the average random velocity $\bar{v}$ in a gas is proportional to $T^{1/2}$, so the theory predicts that the resistivity ρ should increase with the square root of the temperature. However, experimentally the resistance of metallic conductors increases linearly with increasing temperature, so the Drude theory is far from complete.

10–4 COEFFICIENT OF VISCOSITY

In the next three sections, we give an elementary treatment of three properties of a gas described by the general term of *transport phenomena*. These are its viscosity, thermal conductivity, and coefficient of diffusion, and they can be explained in terms of the transport across some imagined surface within the gas of momentum, energy, and mass, respectively. Consider first the coefficient of viscosity.

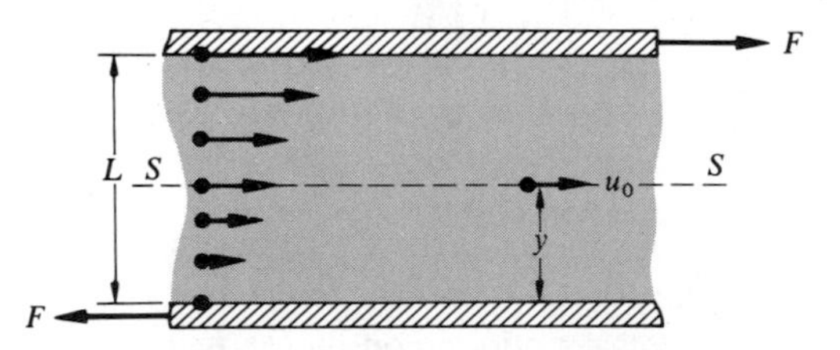

Fig. 10–6 Viscous flow between a stationary lower plate and a moving upper plate.

It appears contradictory at first sight that a gas consisting of widely separated molecules making perfectly elastic collisions with one another should exhibit any viscosity or internal friction. Every real gas, however, is viscous; and we now show that this property is another necessary consequence of our simple model and does not require the assignment of any new properties to the molecules.

Figure 10–6 represents a portion of two large plates separated by a layer of gas of thickness L. Because of the viscosity of the gas, a force F must be exerted on the upper plate to drag it to the right at constant velocity relative to the lower, stationary plate. (An equal and opposite force must be exerted on the lower plate to keep it at rest.) The molecules in the layer of gas have a forward velocity component u which increases uniformly with the distance y above the lower plate. The *coefficient of viscosity* of the gas, η, is defined by the equation

$$\frac{F}{A} = \eta \frac{du}{dy}, \tag{10–18}$$

where A is the area of either plate and du/dy is the *velocity gradient* at right angles to the plates.

In the MKS system, the unit of F/A is 1 newton per square meter and the unit of the velocity gradient du/dy is 1 meter per second, per meter. The unit of the coefficient of viscosity η is therefore 1 newton per square meter, per meter per second per meter, which reduces to 1 N s m^{-2}. The corresponding cgs unit is 1 dyne s cm^{-2} and is called 1 *poise* in honor of Poiseuille.* (1 poise = 10 N s m^{-2}.)

The forward velocity u of the molecules is superposed on their large random velocities, so that the gas is not in thermodynamic equilibrium. However, in most practical problems the random velocities are so much larger than any forward velocity that we can use the results previously derived for an equilibrium state.

The dotted line S-S in Fig. 10–6 represents an imagined surface within the gas at an arbitrary height y above the lower plate. Because of their random motions, there is a molecular flux Φ across the dotted surface, both from above and from below. We shall assume that at its last collision before crossing the surface, each molecule acquires a flow velocity toward the right, corresponding to the particular height at which the collision was made. Since the flow velocity above the dotted surface is greater than that below the surface, molecules crossing from above transport a greater momentum (toward the right) across the surface than do the molecules crossing from below. There results a net rate of transport of momentum across the surface, and from Newton's second law we can equate the net rate of transport of momentum, per unit area, to the viscous force per unit area.

Thus the viscosity of a gas arises not from any "frictional" forces between its molecules, but from the fact that they carry momentum across a surface as a result

* Jean-Louis M. Poiseuille, French physician (1799–1869).

of their random motion. The process is analogous to that of two freight trains of open-top coal cars moving in the same direction on parallel tracks at slightly different speeds, with a gang of laborers in each car, each laborer shoveling coal from his car into the opposite car on the other track. The cars in the slower train are continually being struck by pieces of coal traveling slightly faster than the cars, with the result that there is a net forward force on that train. Conversely, there is a net backward force on the faster train, and the effect is the same as if the sides of the cars were rubbing together and exerting forces on one another through the mechanism of sliding friction.

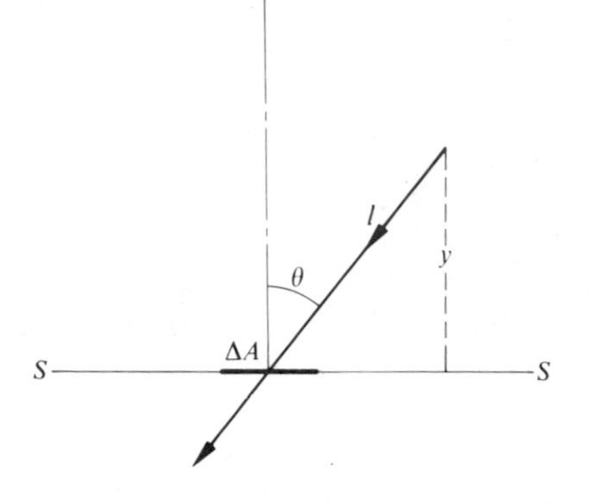

Fig. 10–7 The last mean free path before the molecule crosses the surface started a distance $y = l \cos \theta$ from the surface.

Let us compute the average height $\bar{y}$ above (or below) the surface at which a molecule made its last collision before crossing. In Section 9–3, we assumed that the molecules were geometrical points and that all $\theta\phi v$-molecules in the slant cylinder of Fig. 9–2 would arrive at the area ΔA without having made a collision. This cannot be correct, because on the average each molecule travels only a distance l without colliding with another molecule. These molecular collisions will not affect the *total* flux of $\theta\phi v$-molecules arriving at the surface, because for every collision that scatters a $\theta\phi v$-molecule out of the number originally in the cylinder, there will be another collision that results in an identical $\theta\phi v$-molecule at essentially the same point. However, as explained in the preceding section, molecules arriving at the surface will on the average have started their last free paths before reaching the surface at a distance l away from the surface. The *perpendicular* distance y from the surface, for any θ-molecule (see Fig. 10–7) is $y = l \cos \theta$. The average value of y, or $\bar{y}$, is found by multiplying $l \cos \theta$ by the flux $\Delta\Phi_\theta$, summing over all values of θ, and dividing by the total flux Φ. From Eq. (9–6), replacing $\Sigma v\mathbf{n}_v$ with $\bar{v}\mathbf{n}$,

$$\Delta\Phi_\theta = \frac{1}{2}\,\bar{v}\mathbf{n}\sin\theta\cos\theta\,\Delta\theta;$$

and from Eq. (9–11),

$$\Phi = \frac{1}{4}\,v\mathbf{n}.$$

Therefore, replacing $\Delta\theta$ with $d\theta$ and integrating over θ from zero to $\pi/2$,

$$\bar{y} = \frac{\frac{1}{2}\,\bar{v}\mathbf{n}l\int_0^{\pi/2}\sin\theta\cos^2\theta\,d\theta}{\frac{1}{4}\,\bar{v}\mathbf{n}} = \frac{2}{3}\,l. \tag{10–19}$$

Hence on the average, a molecule crossing the surface makes its last collision before crossing at a distance equal to two-thirds of a mean free path above (or below) the surface.

Let u_0 represent the forward velocity of the gas at the plane *S-S*. At a distance $2l/3$ above the surface, the forward velocity is

$$u = u_0 + \frac{2}{3}\,l\,\frac{du}{dy},$$

since the forward velocity gradient du/dy can be considered constant over a distance of the order of a free path. The forward momentum of a molecule with this velocity is

$$mu = m\left(u_0 + \frac{2}{3}\,l\,\frac{du}{dy}\right).$$

Hence the net momentum $\vec{G}\!\downarrow$ in the direction of flow, carried across the surface per unit time and per unit area by the molecules crossing from above, is the product of the momentum mu and the *total* flux Φ:

$$\vec{G}\!\downarrow\; = \frac{1}{4}\,\mathbf{n}m\bar{v}\left(u_0 + \frac{2}{3}\,l\,\frac{du}{dy}\right).$$

Similarly, the momentum carried across the surface by the molecules crossing from below is

$$\vec{G}\!\uparrow\; = \frac{1}{4}\,\mathbf{n}\bar{v}m\left(u_0 - \frac{2}{3}\,l\,\frac{du}{dy}\right).$$

The net rate of transport of momentum per unit area is the difference between these quantities, or

$$\vec{G} = \frac{1}{3}\,\mathbf{n}m\bar{v}l\,\frac{du}{dy}, \tag{10–20}$$

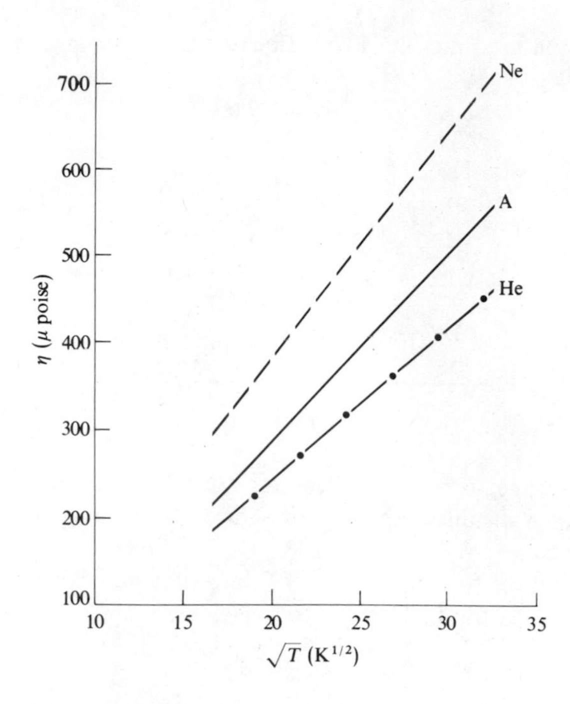

Fig. 10–8 The viscosity of helium, argon, and neon is almost a linear function of $\sqrt{T}$.

Table 10–1 Values of the mean free path and molecular diameter of some gases determined from viscosity measurements. The values of l and d in this table were calculated using Eq. (10–13) for l.

Gas	η (15°C) (N s m^{-2})	l (15°C, 1 atm) (m)	d (m)
He	19.4×10^{-6}	18.6×10^{-8}	2.18×10^{-10}
Ne	31.0	13.2	2.60
A	22.0	6.66	3.64
H_2	8.71	11.8	2.74
N_2	17.3	6.28	3.76
O_2	20.0	6.79	3.60
CO_2	14.5	4.19	4.60
NH_3	9.7	4.51	4.44
CH_4	10.8	5.16	4.14

and from Newton's second law this equals the viscous force per unit area. Hence, by comparison with the definition of the coefficient of viscosity in Eq. (10–18), we have

$$\eta = \frac{1}{3}\,\mathbf{n}m\bar{v}l = \frac{1}{3}\frac{m\bar{v}}{\sigma}. \tag{10–21}$$

An unexpected conclusion from this equation is that the viscosity of a gas is independent of the pressure or density, and is a function of temperature alone through the dependence of $\bar{v}$ on T. Experiment bears this out, however, except at very low pressures where the mean free path becomes of the order of the dimensions of the apparatus. The theory above would not be expected to hold under these conditions, where a molecule could go bouncing from one wall to the other without making a large number of collisions on the way.

We shall show in Section 12–2 that the mean speed $\bar{v}$ is given by

$$\bar{v} = \sqrt{\frac{8}{\pi}\frac{kT}{m}},$$

so that

$$\eta = \frac{1}{3}\frac{m\bar{v}}{\sigma} = \frac{1}{3}\sqrt{\frac{8k}{\pi}}\frac{\sqrt{mT}}{\sigma}. \tag{10–22}$$

Thus for molecules of a given species, the theory predicts that η is proportional to $\sqrt{T}$, and that for different species at a given temperature it is proportional to $\sqrt{m}/\sigma$.

Figure 10–8 shows some experimental values of the viscosities of helium, neon, and argon, plotted as functions of $\sqrt{T}$. The graphs are very nearly straight lines, but they curve upward slightly, indicating that the viscosity increases with temperature at a somewhat greater rate than predicted by the "hard-sphere" theory. This can be explained by realizing that the molecules are not truly rigid spheres and that a "collision" is more like that between two soft tennis balls than between two billiard balls. The higher the temperature, the greater the average molecular kinetic energy and the more the molecules become "squashed" in a collision. Thus the center-to-center distance in a collision, and the corresponding collision cross section σ, will be slightly smaller, the higher the temperature, with a corresponding increase in η.

As for the dependence of viscosity on the cross section σ, Eq. (10–22) is as a matter of fact one of the relations used to "measure" collision cross sections and the corresponding hard-sphere diameters d. Some values of d computed from viscosity measurements, are given in Table 10–1.

10–5 THERMAL CONDUCTIVITY

The thermal conductivity of a gas is treated in the same way as its viscosity. Let the upper and lower plates in Fig. 10–6 be at rest but at different temperatures, so that there is a *temperature gradient* rather than a velocity gradient in the gas. (It is difficult to prevent conductive heat flow in a gas from being masked by convection currents. The gas layer must be thin, and the upper plate must be at a higher temperature than the lower.) If dT/dy is the temperature gradient normal to a surface within the gas, the thermal conductivity λ is defined by the equation

$$H = -\lambda \frac{dT}{dy}, \tag{10–23}$$

where H is the heat flow or heat current per unit area and per unit time across the surface. The negative sign is included because if dT/dy is positive the heat current is downward and is negative.

In the MKS system, the unit of H is 1 joule per square meter per second and the unit of the temperature gradient dT/dy is 1 kelvin per meter. The unit of thermal conductivity λ is therefore 1 joule per square meter per second, per kelvin per meter, which reduces to $1 \text{ J m}^{-1} \text{ s}^{-1} \text{ K}^{-1}$.

From the molecular viewpoint, we consider the thermal conductivity of a gas to result from the net flux of *molecular kinetic energy* across a surface. The *total* kinetic energy per mole of the molecules of an ideal gas is simply its internal energy u, which in turn equals $c_v T$. The average kinetic energy of a single molecule is therefore $c_v T$ divided by Avogadro's number, N_A, and if we define a "molecular heat capacity" c_v^* as $c_v^* = c_v/N_A$, the average molecular kinetic energy is $c_v^* T$.

We assume as before that each molecule crossing the surface made its last collision at a distance $2l/3$ above or below the surface, and that its kinetic energy corresponds to the temperature at that distance. If T_0 is the temperature at the surface S-S, the kinetic energy of a molecule at a distance $2l/3$ below the surface is

$$c_v^* T = c_v^* \left(T_0 - \frac{2}{3} l \frac{dT}{dy} \right).$$

The energy transported in an upward direction, per unit area and per unit time, is the product of this quantity and the molecular flux Φ:

$$H\uparrow = \frac{1}{4} \mathbf{n} \bar{v} c_v^* \left(T_0 - \frac{2}{3} l \frac{dT}{dy} \right).$$

In the same way, the energy transported by molecules crossing from above is

$$H\downarrow = \frac{1}{4} \mathbf{n} \bar{v} c_v^* \left(T_0 + \frac{2}{3} l \frac{dT}{dy} \right).$$

The net rate of transport per unit area, which we identify with the heat current H, is

$$H = -\frac{1}{3}\,\mathbf{n}\bar{v}c_v^*\,l\,\frac{dT}{dy}, \tag{10–24}$$

and by comparison with Eq. (10–23) we see that the thermal conductivity λ is

$$\lambda = \frac{1}{3}\,\mathbf{n}\bar{v}c_v^*\,l = \frac{1}{3}\frac{\bar{v}c_v^*}{\sigma}. \tag{10–25}$$

Thus the thermal conductivity, like the viscosity, should be independent of density. This is also in good agreement with experiments down to pressures so low that the mean free path becomes of the same order of magnitude as the dimensions of the container.

The ratio of thermal conductivity to viscosity is

$$\frac{\lambda}{\eta} = \frac{c_v^*}{m} = \frac{c_v}{mN_{\mathrm{A}}} = \frac{c_v}{M},$$

and

$$\frac{\lambda M}{\eta c_v} = 1, \tag{10–26}$$

where M is the molecular weight of the gas. Therefore the theory predicts that for all gases this combination of experimental properties should equal unity. Some figures are given in Table 10–2 for comparison. The ratio does have the right order of magnitude, but we see again that the hard-sphere model for molecules is inadequate.

Table 10–2 Values of the thermal conductivity λ, molecular weight M, viscosity η, and specific heat capacity c_v of a number of gases

Gas	λ(0°C) (J m^{-1} s^{-1} K^{-1})	M (kg kilomole^{-1})	η(0°C) (N s m^{-2})	c_v (J kilomole^{-1} K^{-1})	$\frac{\lambda M}{\eta c_v}$
He	0.141	4.003	18.6×10^{-6}	12.5×10^3	2.43
Ne	.0464	20.18	29.7	12.7	2.48
A	.163	39.95	21.3	12.5	2.45
H_2	.168	2.016	8.41	20.1	2.06
N_2	.241	28.02	16.6	20.9	1.95
O_2	.245	32.00	19.2	21.0	1.94
CO_2	.145	44.01	13.7	28.8	1.62
NH_3	.218	17.03	9.2	27.6	1.46
CH_4	.305	16.03	10.3	27.4	1.73
Air	.241	29.	17.2	20.9	1.94

10–6 DIFFUSION

The vessel in Fig. 10–9 is initially divided by a partition, on opposite sides of which are two different gases A and B at the same temperature and pressure, so that the number of molecules per unit volume is the same on both sides. If the partition is removed, there is no large scale motion of the gas in either direction, but after a sufficiently long time has elapsed, one finds that both gases are uniformly distributed throughout the entire volume. This phenomenon, as a result of which each gas gradually permeates the other, is called *diffusion.* It is not restricted to gases but occurs in liquid and solids as well. Diffusion is a consequence of random molecular motion and occurs whenever there is a *concentration gradient* of any molecular species, that is, when the number of particles of one kind per unit volume on one side of a surface differs from that on the other side. The phenomenon can be described as a transport of *matter*, (that is, of molecules) across a surface.

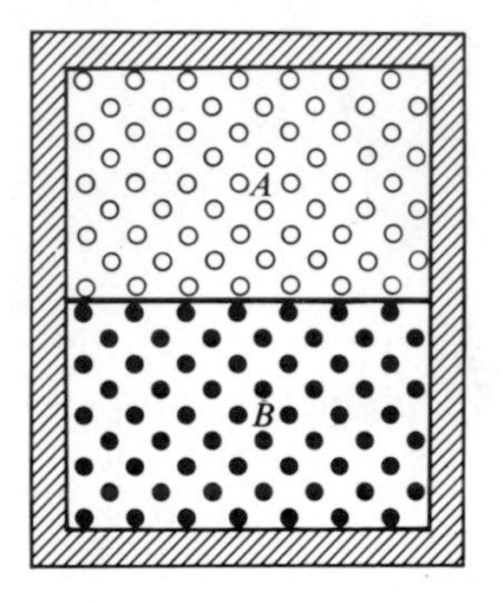

Fig. 10–9 A vessel containing two different gases separated by a partition.

The phenomenon of diffusion may be complicated by the fact that when more than one type of molecule is present the rates of diffusion of one into the other are not the same. We can simplify the problem and still bring out the essential ideas by considering the diffusion of molecules of a single species into others of the same species, known as *self-diffusion.*

If all of the molecules of a system were *exactly* alike, any calculation of self-diffusion among them would be of academic interest only, since there would be no experimental method by which the diffusing molecules could be distinguished from the others. However, molecules that are isotopes of the same element, or molecules whose nuclei have been made radioactive, differ only in their nuclear structure and are essentially identical as far as collision cross sections are concerned.

(Their mean kinetic energies will differ slightly because of differences in mass.) It is thus possible to "tag" certain molecules so that they can be distinguished from others, and yet treat the problem as if the molecules were all alike.

Consider an imagined horizontal surface S-S within the vessel of Fig. 10–9, at some stage of the diffusion process. The vessel contains a mixture of tagged and untagged molecules, the total number of molecules per unit volume being the same at all points so that the pressure is uniform. We assume the temperature to be uniform also. Let $\mathbf{n}^*$ represent the number of tagged molecules per unit volume at any point. We shall assume that $\mathbf{n}^*$ is a function of y only, where the y-axis is normal to the surface S-S. If $d\mathbf{n}^*/dy$ is positive, the downward flux of tagged molecules across the surface is then greater than the upward flux. If Γ represents the net flux of tagged molecules across the surface, per unit time and per unit area, the *coefficient of self-diffusion* D is defined by the equation

$$\Gamma = -D\frac{d\mathbf{n}^*}{dy}. \tag{10–27}$$

The negative sign is included since if $d\mathbf{n}^*/dy$ is positive, the net flux Γ is downward and is negative.

In the MKS system, the unit of Γ is 1 molecule per square meter per second and the unit of the concentration gradient $d\mathbf{n}^*/dy$ is 1 molecule per cubic meter, per meter. The unit of the diffusion coefficient D is therefore 1 molecule per square meter per second, per molecule per cubic meter, per meter, which reduces to 1 $m^2\ s^{-1}$.

We assume as before that each molecule makes its last collision before crossing at a perpendicular distance $2l/3$ away from the surface. If $\mathbf{n}_0^*$ is the number of tagged molecules per unit volume at the surface S-S, the number per unit volume at a distance $2l/3$ below the surface is

$$\mathbf{n}^* = \mathbf{n}_0^* - \frac{2}{3}l\frac{d\mathbf{n}^*}{dy}.$$

In the expression previously derived for the flux Φ, we must replace $\mathbf{n}$ by $\mathbf{n}^*$, and the upward flux $\Gamma\uparrow$ is then

$$\Gamma\uparrow = \frac{1}{4}\bar{v}\left(\mathbf{n}_0^* - \frac{2}{3}l\frac{d\mathbf{n}^*}{dy}\right).$$

In the same way, the downward flux is

$$\Gamma\downarrow = \frac{1}{4}\bar{v}\left(\mathbf{n}_0^* + \frac{2}{3}l\frac{d\mathbf{n}^*}{dy}\right).$$

The net flux Γ is the difference between these, so

$$\Gamma = -\frac{1}{3}\bar{v}l\frac{d\mathbf{n}^*}{dy}. \tag{10–28}$$

Comparison with Eq. (10–24) shows that

$$D = \frac{1}{3}\bar{v}l = \frac{1}{3}\frac{\bar{v}}{\mathbf{n}\sigma}, \tag{10–29}$$

where $\mathbf{n}$ is the *total* number of molecules per unit volume.

The phenomenon of diffusion through fine capillary pores in a ceramic material is one of the methods used to separate the isotopes U^{235} and U^{238}. Naturally occurring uranium is converted to the hexafluoride UF_6, a gas, and the mixture of isotopes flows by diffusion through a porous barrier. The phenomenon is more complicated than the simple case described above because the free path is no longer small compared with the dimensions of the capillaries, and collisions with the walls become an important factor. However, we can see qualitatively that because of the slightly smaller mass of U^{235} compared with U^{238}, the mean speed $\bar{v}$ of the hexafluoride molecules containing U^{235} will be slightly greater than for the others. The diffusion coefficient is slightly greater also, so that this component is slightly enriched in the gas that has diffused through the pores.

The operation of a nuclear reactor is also dependent on the phenomenon of diffusion. The neutrons in a reactor behave like a gas that is continuously being generated throughout the reactor by fission processes and which diffuses through the reactor and eventually escapes from the surface. In order that the reactor may operate successfully, conditions must be such that the rate of generation of neutrons is at least as great as the loss by diffusion, plus the losses due to collisions in which the neutrons are absorbed.

10–7 SUMMARY

Let us compare the three results obtained in the preceding sections. We can write Eqs. (10–20), (10–24), and (10–28) as

$$\vec{G} = \left(\frac{1}{3}\,\mathbf{n}\bar{v}l\right)\frac{d(mu)}{dy},$$

$$H = -\left(\frac{1}{3}\,\mathbf{n}\bar{v}l\right)\frac{d(c_v^*T)}{dy},$$

$$\Gamma = -\left(\frac{1}{3}\,\mathbf{n}\bar{v}l\right)\frac{d(\mathbf{n}^*/\mathbf{n})}{dy}.$$

The last equation is obtained by multiplying numerator and denominator of Eq. (10–26) by $\mathbf{n}$.

The product (mu) in the first equation is the *flow momentum* of a gas molecule, the product (c_v^*T) in the second is the *kinetic energy* of a molecule, and the ratio $(\mathbf{n}^*/\mathbf{n})$ in the third is the *concentration* of tagged molecules.

The corresponding expressions for the coefficients of viscosity, of thermal conductivity, and of self-diffusion, are

$$\eta = \frac{1}{3}\,\mathbf{n}m\bar{v}l = \frac{1}{3}\frac{\bar{v}m}{\sigma},$$

$$\lambda = \frac{1}{3}\,\mathbf{n}c_v^*\bar{v}l = \frac{1}{3}\frac{\bar{v}c_v^*}{\sigma},$$

$$D = \frac{1}{3}\,\bar{v}l = \frac{1}{3}\frac{\bar{v}}{\mathbf{n}\sigma}.$$

PROBLEMS

10–1 How are the assumptions of kinetic theory given in Section 9–2 changed in the development of the Hirn and the van der Waals equations of state?

10–2 The critical temperature of CO_2 is 31.1°C and the critical pressure is 73 atm. Assume that CO_2 obeys the van der Waals equation. (a) Show that the critical density of CO_2 is 0.34 g cm^{-3}. (b) Show that the diameter of a CO_2 molecule is 3.2×10^{-10} m.

10–3 Using the data of the previous problem, (a) find the microscopic collision cross section for a CO_2 molecule. (b) If one kilomole of CO_2 occupies 10 m^3, find the mean free path of the CO_2 molecules. (c) If the mean speed of the CO_2 molecules is 500 m s^{-1}, compute the average number of collisions made per molecule in one second.

10–4 Find the pressure dependence at constant temperature of the mean free path and the collision frequency.

10–5 A beam of molecules of radius 2×10^{-10} m strikes a gas composed of molecules whose radii are 3×10^{-10} m. There are 10^{24} gas molecules per m^3. Determine (a) the exclusion radius, (b) the microscopic collision cross section, (c) the macroscopic collision cross section, (d) the fraction of the beam scattered per unit distance it travels in the gas, (e) the fraction of molecules left in the beam after it travels 10^{-6} m in the gas, (f) the distance the beam travels in the gas before half of the molecules are scattered out, (g) the mean free path of the beam in the gas.

10–6 A group of oxygen molecules start their free paths at the same instant. The pressure is such that the mean free path is 3 cm. After how long a time will half of the group still remain unscattered. Assume all particles have a speed equal to the rms speed. The temperature is 300 K.

10–7 Bowling pins with an effective diameter of 10 cm are placed randomly on a bowling green with an average density of 10 pins per square meter. A large number of 10-cm diameter bowling balls are bowled at the pins. (a) What is the ratio of the mean free path of the bowling ball to the average distance between pins? (b) What fraction of the bowling balls will travel at least 3 meters without striking a pin?

10–8 The mean free path in a certain gas is 5 cm. Consider 10,000 mean free paths. How many are longer than (a) 5 cm? (b) 10 cm? (c) 20 cm? (d) How many are longer than 3 cm but shorter than 5 cm? (e) How many are between 4.5 and 5.5 cm long? (f) How many are between 4.9 and 5.1 cm long? (g) How many are exactly 5 cm long?

10–9 A large number of throws are made with a single die. (a) What is the average number of throws between the appearances of a six? At any stage of the process, what is the average number of throws (b) before the next appearance of a six, (c) since the last appearance of a six? (d) How do you answer the question raised in Section 10–3; that is, why is the mean free path l and not $2l$?

10–10 The mean free path of a helium atom in helium gas at standard conditions is 20×10^{-8} m. What is the radius of a helium atom?

10–11 A beam of electrons is projected from an electron gun into a gas at a pressure P, and the number remaining in the beam at a distance x from the gun is determined by allowing the beam to strike a collecting plate and measuring the current to the plate. The electron current emitted by the gun is 100 μA, and the current to the collector when $x =$ 10 cm and $P = 100$ N m^{-2} (about 1 Torr) is 37 μA. (a) What is the electron mean free path? (b) What current would be collected if the pressure were reduced to 50 N m^{-2}?

10–12 A singly charged oxygen molecule starts a free path in a direction at right angles to an electric field of intensity 10^4 V m^{-1}. The pressure is one atmosphere and the temperature 300 K. (a) Compute the distance moved in the direction of the field in a time equal to that required to traverse one mean free path. (b) What is the ratio of the mean free path to this distance? (c) What is the average speed in the direction of the field? (d) What is the ratio of the rms speed to this speed? (e) What is the ratio of the energy of thermal agitation to the energy gained from the field in one mean free path?

10–13 The resistance of 2 m of 0.01 cm diameter copper wire is measured to be 3 Ω. The density of copper is 8.9×10^3 kg m^{-3} and its atomic weight is 64. (a) Determine the mean free time τ between collisions of the electrons with the copper ion cores. (b) Determine the mean free path of the electrons assuming that $\bar{v}$ for an electron is given by $(8kT/\pi m)^{1/2}$. How many atomic distances is this, assuming copper is cubic? (c) Determine the ratio of the diameter of the copper ion cores to the atomic distance. [Parts (b) and (c) do not give correct answers because electron speeds are approximately 10^3 times as large as those given by $(8kT/\pi m)^{1/2}$, Section 13–6.] (d) Determine the average length of time it takes an electron to move the length of the wire when the current through the wire is 0.333 A.

10–14 Satellites travel in a region where the mean free path of the particles in the atmosphere is much greater than the characteristic size of the body. Show that the force per unit area on the satellite due to this rarefied gas is $4\mathbf{n}mv^2/3$, where $\mathbf{n}$ is the number density of particles in the atmosphere, m is their mass, and v is the speed of the satellite. [*Hint:* Since the satellite speed is much greater than the speed of sound, assume that the satellite is moving through a stationary cloud of particles.]

10–15 Calculate the coefficient of friction of a disc gliding on an air table with a speed of 1 m s^{-1}. The diameter of the disc is 0.1 m and its mass is 0.3 kg. Assume that it glides 10^{-4} m above the table. The diameter of a nitrogen molecule is about 4×10^{-10} m.

10–16 The viscosity of carbon dioxide over a range of temperatures is given in the table below. (a) Compute the ratio $\eta/\sqrt{T}$ at each temperature and (b) determine the diameter of the CO_2 molecule. (c) Compare that diameter with the diameter of A and Ne taken from Fig. 10–8.

t°C	−21	0	100	182	302
$\eta(10^6$ N s m$^{-2})$	12.9	14.0	18.6	22.2	26.8

10–17 (a) Derive an expression for the temperature dependence of the thermal conductivity of an ideal gas. (b) Calculate the thermal conductivity of helium (considered as an ideal gas) at 300 K.

10–18 (a) From the data in Table 10–2 determine the self-diffusion coefficient of helium at standard conditions in two ways. (b) How does the self-diffusion coefficient depend upon pressure at constant temperature, upon temperature at constant pressure, and upon the mass of the diffusing particle.

10–19 A tube 2 m long and 10^{-4} m^2 in cross section contains CO_2 at atmospheric pressure and at a temperature of 300 K. The carbon atoms in one-half of the CO_2 molecules are the radioactive isotope C^{14}. At time $t = 0$, all of the molecules at the extreme left end of the tube contain radioactive carbon, and the number of such molecules per unit volume decreases uniformly to zero at the other end of the tube. (a) What is the initial concentration gradient of radioactive molecules? (b) Initially, how many radioactive molecules per second cross a cross section at the midpoint of the tube from left to right? (c) How many cross from right to left? (d) What is the initial net rate of diffusion of radioactive molecules across the cross section, in molecules per second and micrograms per second?

10–20 Given that the density of air is 1.29 kg m^{-3}, $\bar{v} = 460$ m s^{-1}, and $l = 6.4 \times 10^{-8}$ m at standard conditions, determine the coefficients of viscosity, (b) diffusion, and (c) thermal conductivity. Assume that air is a diatomic ideal gas.

10–21 There is a small uniform pressure gradient in an ideal gas at constant temperature so that there is a mass flow in the direction of the gradient. Using the mean free path approach show that the rate of flow of mass in the direction of the pressure gradient per unit area and per unit pressure gradient is $m\bar{v}l/3kT$.

11

Statistical thermodynamics

11–1 INTRODUCTION

The methods of statistical thermodynamics were first developed during the latter part of the last century, largely by Boltzmann in Germany and Gibbs in the United States. With the advent of the quantum theory in the early years of the present century, Bose* and Einstein†, and Fermi‡ and Dirac§ introduced certain modifications of Boltzmann's original ideas and succeeded in clearing up some of the unsatisfactory features of the Boltzmann statistics.

The statistical approach has a close connection with both thermodynamics and kinetic theory. For those systems of particles in which the energy of the particles can be determined, one can derive by statistical means the equation of state of a substance and its energy equation. Statistical thermodynamics provides an additional interpretation of the concept of entropy.

Statistical thermodynamics (also called statistical mechanics), unlike kinetic theory, does not concern itself with detailed considerations of such things as collisions of molecules with one another or with a surface. Instead, it takes advantage of the fact that molecules are very numerous and *average* properties of a large number of molecules can be calculated even in the absence of any information about specific molecules. Thus an actuary for an insurance company can predict with high precision the average life expectancy of all persons born in the United States in a given year, without knowing the state of health of any one of them.

Statistical methods can be applied not only to molecules but to photons, to elastic waves in a solid, and to the more abstract entities of quantum mechanics called wave functions. We shall use the neutral term "particle" to designate any of these.

11–2 ENERGY STATES AND ENERGY LEVELS

The principles of classical mechanics, or Newtonian mechanics, describe correctly the behavior of matter in bulk, or of *macroscopic* systems. On a molecular or *microscopic* scale, classical mechanics does not apply and must be replaced by *quantum mechanics*. The principles of quantum mechanics lead to the result that the energy of a particle, not acted on by some conservative force field such as a gravitational, electric, or magnetic field, cannot take on any arbitrary value, or cannot change in a continuous manner. Rather, the particle can exist only in some one of a number of states having specified energies. The energy is said to be *quantized*.

A knowledge of quantum mechanics will not be assumed in this book. We shall try to make plausible some of its predictions; others will simply be stated and

---

* Satyendranath Bose, Indian physicist (1894–1974).

† Albert Einstein, German physicist, (1879–1955).

‡ Enrico Fermi, Italian physicist (1901–1954).

§ Paul A. M. Dirac, English physicist (1902–).

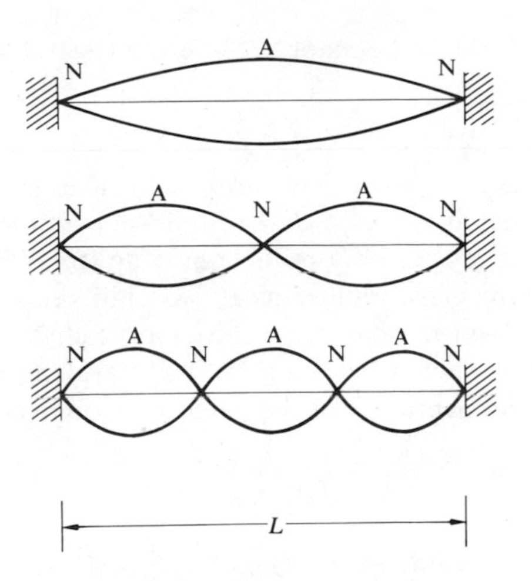

Fig. 11-1 Three of the possible stationary waves in a stretched string fixed at both ends.

the reader will have to take them on faith or refer to texts devoted to this subject. In any event, as far as the methods of *statistics* are concerned, it is enough to know that quantized energy states *exist*.

In quantum mechanics, also known as wave mechanics, the general method of attacking a problem is to set up and (hopefully) solve an equation known as Schrödinger's* equation. In many problems, this equation is exactly analogous to the wave equation describing the propagation of transverse waves in a stretched string, fixed at both ends. As is well known, the string can vibrate in a steady state in any one of a number of stationary waves, three of which are shown in Fig. 11-1. That is, there may be a node N at each end and an antinode A at the center, or there may be a node at the center as well as at the ends, with antinodes midway between the nodes, and so on. The important result is that there is always an *integral number* of antinodes in the steady-state modes of vibration; one antinode in the upper diagram, two in the next, and so on. The distance between nodes (or antinodes) is one-half a wavelength, so if L is the length of the string, the wavelengths λ of the possible stationary waves are

$$\lambda_1 = 2L, \qquad \lambda_2 = \frac{1}{2}\,2L, \qquad \lambda_3 = \frac{1}{3}\,2L, \quad \text{etc.};$$

or in general

$$\lambda_j = \frac{1}{n_j}\,2L,$$

* Erwin Schrödinger, Austrian physicist (1887-1961).

where n_j is an integer equal to the number of antinodes and can have some one of the values

$$n_j = 1, 2, 3, \ldots .$$

A stationary wave is equivalent to two traveling waves propagating in opposite directions, the waves being reflected and re-reflected at the ends of the string. This is analogous to the motion of a particle moving freely back and forth along a straight line and making elastic collisions at two points separated by the distance L. According to quantum mechanics, a stationary Schrödinger wave is in fact completely equivalent to such a particle, and the wavelength λ of the stationary wave is related to the momentum p of the particle through the relation

$$p = \frac{h}{\lambda}, \tag{11–1}$$

where h is a universal constant called *Planck's constant*. In the MKS system,

$$h = 6.6262 \times 10^{-34} \text{ J s}.$$

The momentum of the particle is therefore permitted to have only some one of the set of values

$$p_j = n_j \frac{h}{2L}. \tag{11–2}$$

If a particle is free to move in any direction within a cubical box of side length L whose sides are parallel to the x, y, z axes of a rectangular coordinate system, the x, y, and z components of its momentum are permitted to have only the values

$$p_x = n_x \frac{h}{2L}, \qquad p_y = n_y \frac{h}{2L}, \qquad p_z = n_z \frac{h}{2L},$$

where n_x, n_y, and n_z are integers called *quantum numbers*, each of which can have some one of the values 1, 2, 3, etc. Each set of quantum numbers therefore corresponds to a certain *direction* of the momentum. Then if p_j is the resultant momentum corresponding to some set of quantum numbers n_x, n_y, n_z,

$$p_j^2 = p_x^2 + p_y^2 + p_z^2 = (n_x^2 + n_y^2 + n_z^2)\frac{h^2}{4L^2};$$

or, if we let $(n_x^2 + n_y^2 + n_z^2) = n_j^2$,

$$p_j^2 = n_j^2 \frac{h^2}{4L^2}.$$

The kinetic energy ϵ of a particle of mass m, speed v, and momentum $p = mv$ is

$$\epsilon = \frac{1}{2} mv^2 = \frac{p^2}{2m}.$$

The energy ϵ_j corresponding to the momentum p_j is therefore

$$\epsilon_j = \frac{p_j^2}{2m} = n_j^2 \frac{h^2}{8mL^2}. \tag{11-3}$$

The values of n_x, n_y, and n_z are said to define the *state* of a particle, and the energies corresponding to the different possible values of n_j^2 are the possible energy *levels*. The energy levels depend only on the values of n_j^2 and not on the individual values of n_x, n_y, and n_z. In other words, the energy depends only on the *magnitude* of the momentum p_j and not on its direction, just as in classical mechanics. In general, a number of different states (corresponding to different directions of the momentum) will have the same energy. The energy level is then said to be *degenerate*, and we shall use the symbol g_j to designate the degeneracy of level j, that is, the number of states having the same energy ϵ_j.

The volume V of a cubical box of side length L equals L^3, so $L^2 = V^{2/3}$; and Eq. (11–3) can be written, for a free particle in a cubical box,

$$\epsilon_j = n_j^2 \frac{h^2}{8m} V^{-2/3}. \tag{11-4}$$

The same result applies to a container of any shape whose dimensions are large compared with the wavelength of the Schrödinger waves. The energy of the jth level therefore depends on the quantum number n_j and on the volume V. If the volume is *decreased*, the value of a given ϵ_j *increases*.

As an example, consider a 1-liter volume of helium gas. When numerical values of h, m, and V are inserted in Eq. (11–4), we find that

$$\frac{h^2}{8m} V^{-2/3} \simeq 8 \times 10^{-40}\ \text{J} \simeq 5 \times 10^{-21}\ \text{eV}.$$

We have shown that at room temperature the mean kinetic energy of a gas molecule is about 1/40 eV or 2.5×10^{-2} eV. Hence for a molecule with this kinetic energy,

$$n_j^2 \simeq \frac{2.5 \times 10^{-2}}{5 \times 10^{-21}} \simeq 5 \times 10^{18},$$

$$n_j \simeq 2.2 \times 10^9.$$

Thus for the vast majority of the molecules of a gas at ordinary temperatures, the quantum numbers n_j are very large indeed.

The lowest energy level ($j = 1$) is that for which $n_x = n_y = n_z = 1$. Then $n_1^2 = 3$ and

$$\epsilon_1 = \frac{3h^2}{8m} V^{-2/3}.$$

There is only one *state* (one set of quantum numbers n_x, n_y, n_z) having this energy. The lowest level is therefore *nondegenerate* and $g_1 = 1$. The x, y, and z components of the corresponding momentum p_1 are all equal, and each equals $h/2L$.

In the next level ($j = 2$) we might have any one of the following states:

n_x	n_y	n_z
2	1	1
1	2	1
1	1	2

Thus in the first of these states, for example, the components of momentum are

$$p_{2x} = 2\frac{h}{2L}, \qquad p_{2y} = \frac{h}{2L}, \qquad p_{2z} = \frac{h}{2L}.$$

In each state, $n_2^2 = (n_x^2 + n_y^2 + n_z^2) = 6$, and in this level,

$$\epsilon_2 = \frac{6h^2}{8m} V^{-2/3}.$$

Since three different states have the same energy, the degeneracy $g_2 = 3$.

The preceding discussion of the energy levels and degeneracies of a free particle in a box is only one example of energy quantization. Other constraints also leading to energy quantization will be discussed later.

Figure 11–2 represents in a schematic way the concepts of energy states, energy levels, and the degeneracy of a level. The energy *levels* can be thought of as a set of shelves at different elevations, while the energy *states* correspond to a set of boxes on each shelf. The *degeneracy* g_j of level j is the number of boxes on the corresponding shelf. If a number of marbles are distributed among the various boxes, the number in any one box is the number in a particular *state*. Those marbles

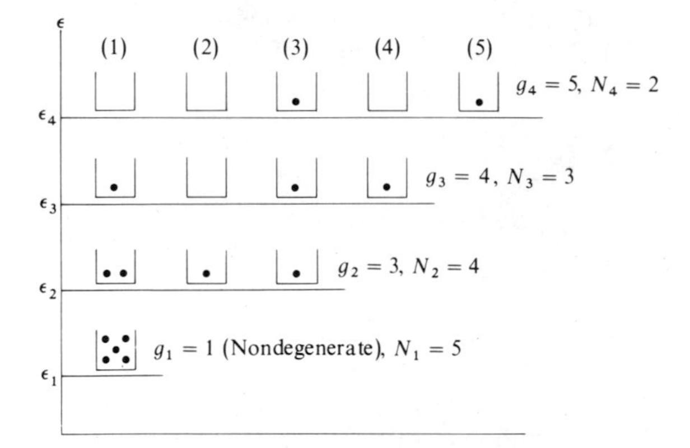

Fig. 11–2 A schematic representation of a set of energy levels ϵ_j, their degeneracies g_j and their occupation numbers N_j.

in the boxes on any one shelf are in different states, but all have the same energy. The total number of marbles in the boxes at any level j is called the *occupation number* N_j of that level.

Evidently, the sum of the occupation numbers N_j over all levels equals the total number of particles N:

$$\sum_j N_j = N. \tag{11–5}$$

Also, since the particles in those states included in any level j all have the same energy ϵ_j, the total energy of the particles in level j is $\epsilon_j N_j$, and the total energy E of the system is

$$\sum_j \epsilon_j N_j = E. \tag{11–6}$$

If the system is in an external conservative force field such as a gravitational, electric, or magnetic field, the total energy E will consist in part of the potential energy E_p of the system. If the potential energy is zero, the total energy E is then the internal energy U and

$$\sum_j \epsilon_j N_j = U. \tag{11–7}$$

11–3 MACROSTATES AND MICROSTATES

A number N of identical entities is called an assembly. The entities may be single particles, or they may themselves be identical assemblies of particles, in which case one has an assembly of assemblies, or an *ensemble*. We shall for the most part consider only assemblies of single particles, and shall refer to them as an *assembly* or simply as a *system*.

If the distribution of the particles of the system among its energy states is known, the macroscopic properties of the system can be determined. Thus a central problem of statistical mechanics is to determine the possible distributions of particles among energy *levels* and energy *states*.

The description of a single-particle assembly depends on whether the particles of which it consists are *distinguishable* or *indistinguishable*. Suppose the assembly is a sample of gas and the individual molecules are the particles. Since there is no way in which the molecules can be labeled, the particles are indistinguishable. On the other hand, if the assembly is a crystal, the molecules can be labeled in accord with the positions they occupy in the crystal lattice and can be considered distinguishable.

Whether the particles are distinguishable or not, a specification of the number of particles N_j in each energy *level* is said to define a *macrostate* of the assembly. For example, the macrostate of the assembly in Fig. 11–2 is specified by the set of occupation numbers $N_1 = 5$, $N_2 = 4$, $N_3 = 3$, $N_4 = 2$.

If the particles are *indistinguishable*, a specification of the total number of particles in each energy *state* is said to define a *microstate* of the assembly. Thus if the energy states in each level in Fig. 11–2 are numbered (1), (2), (3), etc., up to the number of states g_j in the level; and if the particles are indistinguishable, the microstate of the assembly is specified by saying that in level 4 there is one particle in each of the states (3) and (5), and there are no particles in states (1), (2), and (4); in level 3 there is one particle in states (1), (3), and (4), and no particle in state (2); in level 2 there are two particles in state (1) and one particle in each of states (2) and (3); and in level 1 there are five particles in the only state in this level.

If one or both of the particles in level 4 were in states other than (3) and (5), the microstate would be different, but the macrostate would be unchanged since we would still have $N_4 = 2$. Evidently, many different microstates will correspond to the same macrostate.

If the particles are *distinguishable*, a specification of the energy state of *each particle* is said to define a microstate of the assembly. That is, we must specify not only *how many* particles are in each state, but *which* particles they are. Thus suppose that the particles in Fig. 11–2 are distinguishable and are lettered a, b, c, etc., and that in level 4 particle a is in state (3) and particle b is in state (5); in level 3, particle c is in state (1) and particles d and e are in states (3) and (4) respectively, and so on. The preceding specification, including all levels, describes the microstate of the assembly. In contrast to an assembly of indistinguishable particles, in which the microstate would be the same no matter *which* particles occupied states (3) and (5) in level 4, the microstate is now considered different if particles a and b are interchanged between these states. Also, the microstate would be different if, say, particles c and d in level 3 were interchanged with a and b in level 4. In each such interchange we have a different specification of the energy states of the particles and hence a different microstate; although the macrostate does not change because the occupation numbers of the levels are the same.

If there is more than one particle in a given energy state, an interchange of the *order* in which the letters designating the particles is written is not considered to change the microstate. Thus suppose the two particles in state (1) of level 2 are lettered p and q. The microstate is considered the same if the letters are written in the order pq or qp.

The number of microstates that are considered different, for a given set of occupation numbers, is evidently much greater if the particles are distinguishable than if they are indistinguishable.

The possible macrostates and microstates of an assembly of particles is analogous to a table of ages of groups of individuals. As an example let us take the number of children in each grade of an elementary school having a total enrollment of 368 children.

Grade	K	1	2	3	4	5
Children	60	70	62	61	62	53

The grades correspond to the energy levels of the system and the specification of the number of children in each grade defines the macrostate of the system. A different macrostate with the same total number of children would be

Grade	K	1	2	3	4	5
Children	52	57	60	73	62	64

.

The change in distribution may have several macroscopic consequences: needs for different numbers of teachers, different equipment, different numbers of textbooks, etc.

The grades could be further subdivided into classes, that is, in the first macrostate described there may be 3 first grade classes and 2 second grade classes. These classes would correspond to the degenerate energy states of each level. There would be 3 degenerate states in level 1, etc.

If the children were considered as indistinguishable particles (a bad pedagogic practice), then a microstate of the system would be

Class	K	1(a)	1(b)	1(c)	2(a)	2(b)
Children	60	22	25	23	30	32

, etc.

A different microstate of the same macrostate of the system would be

Class	K	1(a)	1(b)	1(c)	2(a)	2(b)
Children	60	20	25	25	30	32

, etc.

Although the number of children in each *class* was changed, the number of children in each *grade* remained constant.

However, the distribution

Class	K	1(a)	1(b)	1(c)	2(a)	2(b)
Children	60	22	27	23	30	30

, etc.

would correspond to a different macrostate since the number of children in each grade was changed, even though the total number of children in the school remained constant.

When the children are considered distinguishable particles, the microstate is different, if Evelyn is in 1(a) and Mildred is in 1(b), or vice versa, or if both are in 1(b). However, in the last case the microstate is the same if Mildred's name appears on the class list before Evelyn's or after it.

11–4 THERMODYNAMIC PROBABILITY

In the preceding section, no restriction was imposed on the possible ways in which the particles of an assembly might be distributed among the energy states. In an isolated, closed system, however, the energy E and the total number of particles N are both constant. Hence the only possible microstates of such a system are those that satisfy these conditions.

As time goes on, interactions between the particles of an isolated, closed system will result in changes in the numbers of particles occupying the energy states, and, if the particles are distinguishable, will result in changes in the energy state of each particle. These interactions might be collisions of the molecules of a gas between themselves or with the walls of the container, or an energy interchange between the oscillating molecules of a crystal. Every such interchange results in a change in the microstate of the assembly, but every possible microstate must satisfy the conditions of constant N and E.

The fundamental postulate of statistical thermodynamics is that *all possible microstates of an isolated assembly are equally probable.* The postulate can be interpreted in two different ways. Consider a time interval t that is long enough so that each possible microstate of an isolated, closed system occurs a large number of times. Let Δt be the total time during which the system is in some one of its possible microstates. The postulate then asserts that *the time interval* Δt *is the same for all microstates.*

Alternatively; one can consider a very large number $\mathcal{N}$ of replicas of a given assembly (an ensemble). At any instant, let $\Delta\mathcal{N}$ be the number of replicas which are in some one of the possible microstates. The postulate then asserts that *the number* $\Delta\mathcal{N}$ *is the same for all microstates.* The postulate does not seem to be derivable from any more fundamental principle, and of course it cannot be verified by experiment. Its justification lies in the correctness of the conclusions drawn from it.

In terms of the example of the previous section, if all microstates were equally probable and the population of the school were limited to exactly 368 children, over many, many years each distribution of children among classes would occur as often as any other. Alternatively, if in a given year one looked at many elementary schools having a population of 368 children, each distribution of children among classes would occur with the same frequency. In each case, the examples given in the previous section would occur the same number of times.

The number of equally probable microstates that correspond to a given macrostate k is called the *thermodynamic probability* $\mathcal{W}_k$ of the macrostate. (The symbol $\mathcal{W}$ comes from the German word for probability, *Wahrscheinlichkeit.* Other symbols are often used, and the quantity is also known as the *statistical count.*) For most macrostates of an assembly of a large number of particles, the thermodynamic probability is a very large number. The total number Ω of possible *microstates* of an assembly, or the thermodynamic probability of the *assembly,* equals the sum over all macrostates of the thermodynamic probability of each macrostate:

$$\Omega = \sum_k \mathscr{W}_k.$$

The principles of quantum mechanics lead to expressions for the possible different ways in which the particles may be distributed among the energy states of a single assembly at one instant of time. In other words, quantum mechanics determines the microstate at each instant for a single assembly or of each of the large number of replicas of an assembly at one instant. The calculation of $\mathscr{W}_k$ for three different cases is carried out in Sections 11–5, 11–6, and 11–7.

The observable properties of a macroscopic system depend on the *time average* values of its microscopic properties. Thus the pressure of a gas depends on the time average value of the rate of transport of momentum across an area. By the fundamental postulate, the observable properties of a macroscopic system will also depend upon the average value of the microscopic properties of a large number of replicas of an assembly taken at one instant.

Thus the primary goal of a statistical theory is to derive an expression for the average number of particles $\bar{N}_j$ in each of the permitted energy levels j of the assembly. The expression to be derived is called the average *occupation number* of the level j.

Let N_{jk} be the occupation number of level j in macrostate k. The *group average* value of the occupation number of level j, $\bar{N}_j^g$, is found by multiplying N_{jk} by the number of replicas in macrostate k, summing over all macrostates and dividing by the total number of replicas, $\mathscr{N}$. The total number of replicas of a given assembly that are in macrostate k equals the product of the number of replicas $\Delta\mathscr{N}$ that are in some microstate and the number of microstates $\mathscr{W}_k$ included in the macrostate. Therefore

$$\bar{N}_j^g = \frac{1}{\mathscr{N}} \sum_k N_{jk} \mathscr{W}_k \Delta\mathscr{N}.$$

However,

$$\mathscr{N} = \sum_k \mathscr{W}_k \Delta\mathscr{N},$$

and since $\Delta\mathscr{N}$ is the same for all macrostates, we can cancel it from the numerator and denominator. The group average is

$$\bar{N}_j^g = \frac{\sum_k N_{jk} \mathscr{W}_k}{\sum_k \mathscr{W}_k} = \frac{1}{\Omega} \sum_k N_{jk} \mathscr{W}_k. \tag{11–8}$$

Similarly, we can calculate the time average of the occupation number of level j, $\bar{N}_j^t$. As explained above, the postulate that all microstates are equally probable means that over a sufficiently long period of time t, each microstate exists for the same time interval Δt. The total time the assembly is found in macrostate k is then the product of the time interval Δt and the number $\mathscr{W}_k$ of microstates in macrostate k. The sum of these products over all macrostates equals the total time t:

$$t = \sum_k \mathscr{W}_k \Delta t.$$

The *time average* value of the occupation number of level j, $\overline{N}_j^t$, is found by multiplying the occupation number N_{jk} of level j in macrostate k by the time $\mathscr{W}_k \Delta t$ that the assembly spends in macrostate k, summing these products over all macrostates, and dividing by the total time t. The time average is therefore

$$\overline{N}_j^t = \frac{1}{t}\sum_k N_{jk}\mathscr{W}_k \Delta t = \frac{\sum_k N_{jk}\mathscr{W}_k \Delta t}{\sum_k \mathscr{W}_k \Delta t}.$$

Since Δt is the same for all microstates, we can cancel it from numerator and denominator, giving

$$\overline{N}_j^t = \frac{\sum_k N_{jk}\mathscr{W}_k}{\sum_k \mathscr{W}_k} = \frac{1}{\Omega}\sum_k N_{jk}\mathscr{W}_k. \tag{11–9}$$

Comparison of Eqs. (11–8) and (11–9) shows that if all microstates are equally probable, the *time average* value of an occupation number is equal to the *group average*, and we can represent either by $\overline{N}_j$.

The values of the average occupation numbers of the energy levels are calculated for different cases in the next three sections. The general expressions for the $\overline{N}_j$, the distribution functions for these cases, are derived in Sections 11–9 to 11–12.

11–5 THE BOSE-EINSTEIN STATISTICS

The thermodynamic probability $\mathscr{W}_k$ of a macrostate of an assembly depends on the particular statistics obeyed by the assembly. We consider first the statistics developed by Bose and Einstein, which for brevity we shall refer to as the B-E statistics. In the B-E statistics, the particles are considered indistinguishable, and there is no restriction on the number of particles that can occupy any energy state. The energy states, however, are distinguishable. Let the particles be lettered a, b, c, etc. (Although the particles are indistinguishable, we assign letters to them temporarily as an aid in explaining how the thermodynamic probability is computed.) In some one arrangement of the particles in an arbitrary level j, we might have particles a and b in state (1) of that level, particle c in state (2), no particles in state (3), particles d, e, f, in state (4), and so on. This distribution of particles among states can be represented by the following mixed sequence of numbers and letters:

$$[(1)ab]\ [(2)c]\ [(3)]\ [(4)def] \cdots \tag{11–10}$$

where in each bracketed group the letters following a number designate the particles in the state corresponding to the number.

If the numbers and letters are arranged in all possible sequences, each sequence will represent a possible distribution of particles among states, provided the sequence begins with a number. There are therefore g_j ways in which the sequences can begin, one for each of the g_j states, and in each of these sequences the remaining $(g_j + N_j - 1)$ numbers and letters can be arranged in any order.

The number of different sequences in which N distinguishable objects can be arranged is $N!$ (N factorial). There are N choices for the first term in a sequence. For each of these there are $(N - 1)$ choices for the second, $(N - 2)$ choices for the third, and so on down to the last term, for which only one choice remains. The total number of possible sequences is therefore

$$N(N - 1)(N - 2) \cdots 1 = N!$$

As an example, the three letters a, b, and c can be arranged in the following sequences:

abc, *acb*, *bca*, *bac*, *cba*, *cab*.

We see that there are six possible sequences, equal to 3!.

Using the example of the previous section, the number $\mathscr{W}$ of different sequences in which the 70 children of the first grade can be lined up is 70!. It is shown in Appendix C that Stirling's* approximation for the natural logarithm of the factorial of a large number x is

$$\ln x! = x \ln x - x.$$

Hence

$$\ln 70! = 70 \ln 70 - 70 = 245$$

$$\log_{10} 70! = 245/2.303 = 106$$

$$70! = 10^{106}$$

The number of different possible sequences of the $(g_j + N_j - 1)$ numbers and letters is therefore $(g_j + N_j - 1)!$ and the total number of possible sequences of g_j numbers and N_j letters is

$$g_j[(g_j + N_j - 1)!]. \tag{11-11}$$

Although each of these sequences represents a possible distribution of particles among the energy states, many of them represent the *same* distribution. For example, one of the possible sequences will be the following:

$$[(3)]\ [(1)ab]\ [(4)def]\ [(2)c] \cdots .$$

This is the same distribution as (11-10), since the same states contain the same particles, and it differs from (11-10) only in that the bracketed *groups* appear in a different sequence. There are g_j groups in the sequence, one for each state, so the number of different sequences of groups is $g_j!$ and we must divide (11-11) by $g_j!$ to avoid counting the same distribution more than once.

---

* James Stirling, Scottish mathematician (1696-1770).

Also, since the particles are actually indistinguishable, a different sequence of *letters* such as

$$[(1)ca]\ [(2)e]\ [(3)]\ [(4)bdf] \cdots$$

also represents the same distribution as (11–10) because any given state contains the same *number* of particles. The N_j letters can be arranged in sequence in $N_j!$ different ways, so (11–11) must also be divided by $N_j!$. Hence the number of different distributions for the jth level is

$$w_j = \frac{g_j[(g_j + N_j - 1)!]}{g_j!\, N_j!},$$

which may be more conveniently written as

$$w_j = \frac{(g_j + N_j - 1)!}{(g_j - 1)!\, N_j!}, \tag{11–12}$$

since

$$g_j! = g_j(g_j - 1)!.$$

As a simple example, suppose that an energy level j includes 3 states ($g_j = 3$) and 2 particles ($N_j = 2$). The possible distributions of the particles among the states are shown in Fig. 11–3 in which, since the particles are indistinguishable, they are represented by dots instead of letters. The number of possible distributions, from Eq. (11–12), is

$$w_j = \frac{(3 + 2 - 1)!}{(3 - 1)!2!} = \frac{4!}{2!2!} = 6,$$

in agreement with Fig. 11–3.

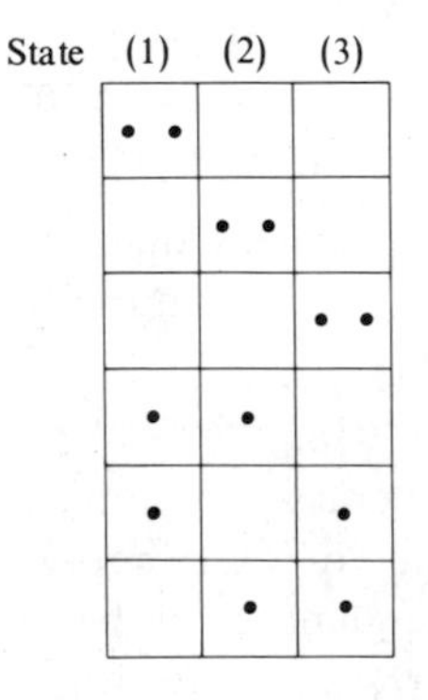

Fig. 11–3 The possible distributions of two indistinguishable particles among three energy states, with no restriction on the number of particles in each state.

If a level is nondegenerate, that is, if there is only one state in the level and $g_j = 1$, then there is only one possible way in which the particles in the level can be arranged, and hence $w_j = 1$. But if $g_j = 1$, Eq. (11–12) becomes

$$w_j = \frac{N_j!}{0!\,N_j!} = 1.$$

It follows that we must set $0! = 1$, which may be considered as a convention that is necessary in order to get the right answer. A further discussion can be found in Appendix C.

Also, if a level j is unoccupied and $N_j = 0$,

$$w_j = \frac{(g_j - 1)!}{(g_j - 1)!\,(0)!} = 1$$

and $w_j = 1$ for that level.

For each of the possible distributions in any level, we may have any one of the possible distributions in each of the other levels, so the total number of possible distributions, or the thermodynamic probability $\mathscr{W}_{\text{B-E}}$ of a macrostate in the B-E statistics is the product over all levels of the values of w_j for each level, or,

$$\mathscr{W}_{\text{B-E}} = \mathscr{W}_k = \prod_j w_j = \prod_j \frac{(g_j + N_j - 1)!}{(g_j - 1)!\,N_j!}, \tag{11–13}$$

where the symbol $\prod_j$ means that one is to form the *product* of all terms following it, for all values of the subscript j. It corresponds to the symbol $\sum_j$ for the *sum* of a series of terms.

If an assembly includes two levels p and q, with $g_p = 3$ and $N_p = 2$, as in the preceding example, and with $g_q = 2$, $N_q = 1$, the thermodynamic probability of the macrostate $N_p = 2$, $N_q = 1$, is

$$\mathscr{W}_{\text{B-E}} = \frac{4!}{2!2!} \cdot \frac{2!}{1!1!} = 6 \times 2 = 12,$$

and there are 12 different ways in which three indistinguishable particles can be distributed among the energy states of the assembly.

We next calculate the thermodynamic probabilities of those macrostates that are accessible to a given system and the average occupation numbers of the permitted energy levels. Although all microstates of an isolated, closed system are equally *probable*, the only *possible* microstates are those in which the total number of particles equals the number N of particles in the system, and in which the total energy of the particles equals the energy U of the system. As an example, suppose that we have a system of just six particles, that the permitted energy levels are equally spaced, and that there are three energy states in each level so that $g_j = 3$. We shall take the reference level of energy as that of the lowest level, so that $\epsilon_0 = 0$, $\epsilon_1 = \epsilon$, $\epsilon_2 = 2\epsilon$, etc. We also assume that the total energy U of the system equals 6ϵ.

If the particles are indistinguishable and the system obeys the B-E statistics, the only possible *macrostates* consistent with the conditions $N = 6$, $U = 6\epsilon$, are

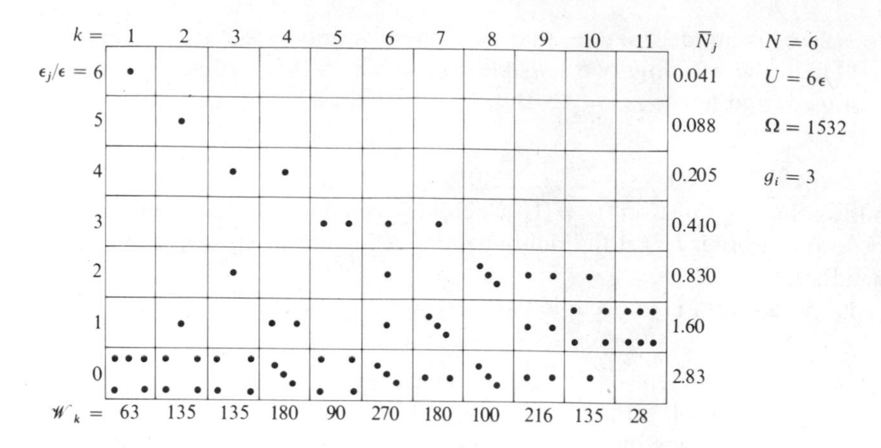

Fig. 11–4 The eleven possible macrostates of an assembly of 6 particles obeying Bose-Einstein statistics. The energy levels are equally spaced and have a degeneracy $g_j = 3$ in each level. The total energy of the system is $U = 6\epsilon$. The thermodynamic probability of each macrostate is given at the bottom and the average occupation number of each level is printed on the right of the diagram.

shown in the columns of Fig. 11–4. Each horizontal row corresponds to an energy *level* (the three states in each level are not shown in the figure). The dots represent the number of particles in each level. The columns could represent either the macrostates of a single system at different times, or the macrostates of a number of replicas of the system at a given instant. If we consider the figure to represent these replicas, then out of a large number $\mathscr{N}$ of replicas there would be a number $\Delta\mathscr{N}$ in each macrostate, but since all of these numbers $\Delta\mathscr{N}$ would be equal, we can consider that each macrostate occurs just once.

The diagram can be constructed as follows. The macrostate represented by the first column is obtained by first placing one particle in level 6, with energy 6ϵ. The remaining five particles must then be placed in the lowest level with energy zero, so that the total energy of the system is 6ϵ. Evidently, there can be no particles in levels higher than the sixth. In the second column, we place one particle in level 5, one particle in level 1, and the remaining four particles in the lowest level, and so on.

The thermodynamic probability $\mathscr{W}_k$ of each macrostate, calculated from Eq. (11–13), is given under the corresponding column. Thus for macrostate $k = 1$, since $g_j = 3$ in all levels and all occupation numbers are zero except in level 6, where $N_6 = 1$, and in level 0, where $N_0 = 5$,

$$\mathscr{W}_1 = \frac{(3 + 1 - 1)!}{2!\,1!} \cdot \frac{(3 + 5 - 1)!}{2!\,5!} = 3 \times 21 = 63.$$

That is, the single particle in level 6 could be in any one of three states, and in the lowest level the remaining five particles could be distributed in 21 different ways among the three states, making a total of 63 different possible arrangements.

The total number of possible microstates of the system, or the thermodynamic probability of the system, is

$$\Omega = \sum_k \mathscr{W}_k = 1532.$$

The average occupation numbers of each level, calculated from Eq. (11–8), are given at the right of the corresponding level. In level 2, for example, we see that macrostate 3 includes 135 microstates, in each of which there is one particle in level 2. Macrostate 6 includes 270 microstates in each of which there is also one particle in level 2, and so on. The average occupation number of level 2 is therefore

$$\bar{N}_2 = \frac{1}{\Omega}\sum_k N_{2k}\mathscr{W}_k = \frac{1272}{1532} = 0.83.$$

In any macrostate k in which level 2 is unoccupied, the corresponding value of N_k is zero and the product $N_{2k}\mathscr{W}_k$ for that level is zero. Note that although the *actual* occupation number of any level in any macrostate must be an integer or zero, the *average* occupation number is not necessarily an integer.

The *most probable macrostate* in Fig. 11–4, that is, the one with the largest number of microstates (270), is the sixth. The occupation number of each level for this macrostate is roughly the same as the average occupation number for the assembly. It can be shown (Appendix D) that when the number of particles in an assembly is very large, the occupation numbers in the most probable state are very nearly the same as the average occupation numbers.

11–6 THE FERMI-DIRAC STATISTICS

The statistics developed by Fermi and Dirac, which for brevity we call the F-D statistics, applies to indistinguishable particles that obey the *Pauli* exclusion principle*, according to which there can be no more than one particle in each permitted energy state. (It is as if every particle were aware of the occupancy of all states, and could only take a state unoccupied by any other particle.) Thus the arrangements in the upper three rows of Fig. 11–3, in which there are two particles in each state, would not be permitted in the F-D statistics. Evidently, the number of particles N_j in any level cannot exceed the number of states g_j in that level.

To calculate the thermodynamic probability of a macrostate, we again temporarily assign numbers to energy states of a level and letters to the particles, and we represent a possible arrangement of the particles in a level by a mixed sequence of numbers and letters. A possible arrangement might be the following:

$$[(1)a]\ [(2)b]\ [(3)]\ [(4)c]\ [(5)]\cdots \tag{11–14}$$

meaning that states (1), (2), (4), ... are occupied with their quota of one particle each while states (3), (5), ... are empty. For a given sequence of numbers, we

* Wolfgang Pauli, Austrian physicist (1900–1958).

first select some arbitrary sequence of *letters*. There are g_j possible locations for the first letter, following any one of the g_j numbers. This leaves only $(g_j - 1)$ possible locations for the second letter, $(g_j - 2)$ locations for the third, down to $[g_j - (N_j - 1)]$ or $(g_j - N_j + 1)$ locations for the last letter. Since for any one location of any one letter we may have any one of the possible locations of each of the others, the total number of ways in which a given sequence of N_j letters can be assigned to the g_j states is

$$g_j(g_j - 1)(g_j - 2)\cdots(g_j - N_j + 1) = \frac{g_j!}{(g_j - N_j)!}, \tag{11-15}$$

since

$$g_j! = g_j(g_j - 1)(g_j - 2)\cdots(g_j - N_j + 1)(g_j - N_j)!.$$

Because the particles are indistinguishable, a state is occupied regardless of the particular letter that follows the number representing the state, and since there are $N_j!$ different sequences in which the N_j letters can be written, we must divide Eq. (11-15) by $N_j!$. Again, although the states are distinguishable, a different sequence of states does not change the distribution. Therefore we do not need to consider other sequences of letters and for level j,

$$w_j = \frac{g_j!}{(g_j - N_j)!\,N_j!}. \tag{11-16}$$

If a level j includes 3 states ($g_j = 3$) and two particles ($N_j = 2$), then

$$w_j = \frac{3!}{(3-2)!2!} = \frac{3!}{1!2!} = 3.$$

The possible arrangements are shown in Fig. 11-5, which corresponds to the lower three rows of Fig. 11-3, the upper three being excluded.

Finally, since for every arrangement in any one level we may have any one of the possible arrangements in the other levels, the thermodynamic probability

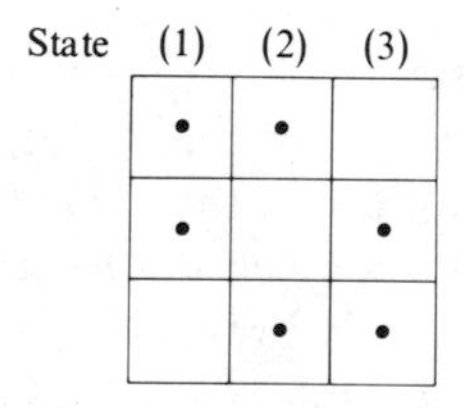

Fig. 11-5 The possible distributions of two indistinguishable particles among three energy states, with no more than one particle in each state.

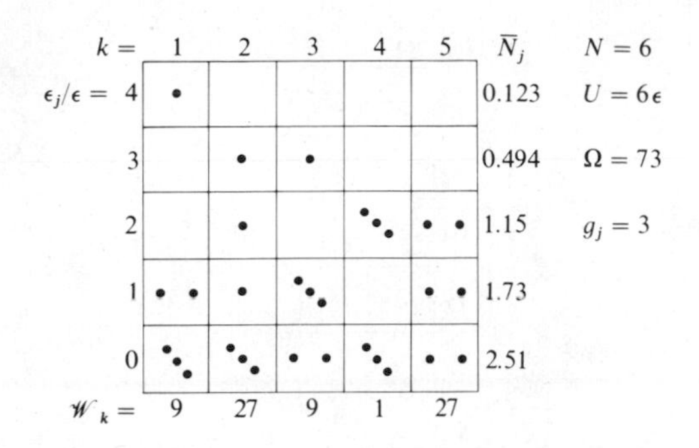

Fig. 11–6 The five possible macrostates of an assembly of 6 particles obeying Fermi-Dirac statistics. The energy levels are equally spaced and have a degeneracy of $g_j = 3$ each. The total energy of the system is $U = 6\epsilon$. The thermodynamic probability of each macrostate is given at the bottom, and the average occupation number of each level is printed on the right of the diagram.

$\mathscr{W}_{\text{F-D}}$ of a macrostate in the F-D statistics is

$$\mathscr{W}_{\text{F-D}} = \mathscr{W}_k = \prod_j w_j = \prod_j \frac{g_j!}{(g_j - N_j)!\, N_j!}. \tag{11–17}$$

Figure 11–6 shows the possible macrostates of a system of six particles obeying the F-D statistics in which, as in Fig. 11–4, the energy levels are equally spaced and the degeneracy of each level is $g_j = 3$. In comparison with Fig. 11–4, macrostates 1, 2, 3, 5, 10, and 11 of that figure are excluded because there can be no more than three particles in each level. Thus there are only five possible macrostates, each with energy 6ϵ. The thermodynamic probability of each macrostate, calculated from Eq. (11–17), is written under the corresponding column. Thus in macrostate 1,

$$\mathscr{W}_1 = \frac{3!}{(3-1)!\,1!} \cdot \frac{3!}{(3-2)!\,2!} \cdot \frac{3!}{(3-3)!\,3!} = 3 \times 3 \times 1 = 9.$$

That is, there are three possible locations of the single particle in level 4 (in any one of the three states), three ways in which the two particles in level 1 can be distributed among the three states (as in Fig. 11–5) and only one way in which the three particles in level zero can be distributed among the three states (one in each state).

The total number of possible macrostates is

$$\Omega = \sum_k \mathscr{W}_k = 73.$$

The average occupation numbers of each level, calculated from Eq. (11–8) are given at the right of the corresponding level. These may be compared with the occupation numbers in Fig. 11–4.

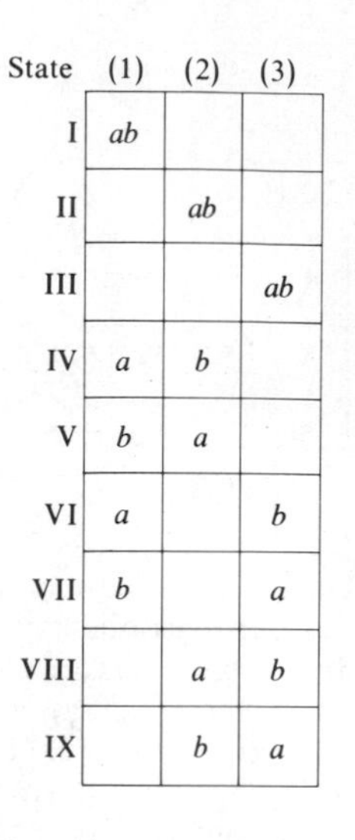

State	(1)	(2)	(3)
I	*ab*		
II		*ab*	
III			*ab*
IV	*a*	*b*	
V	*b*	*a*	
VI	*a*		*b*
VII	*b*		*a*
VIII		*a*	*b*
IX		*b*	*a*

Fig. 11–7 The possible arrangements of two distinguishable particles *a* and *b* among three energy states, with no restriction on the number of particles per state.

11–7 THE MAXWELL-BOLTZMANN STATISTICS

In the Maxwell-Boltzmann statistics, which for brevity we call M-B statistics, the particles of an assembly are considered distinguishable, but as in the B-E statistics there is no restriction on the number of particles that can occupy the same energy state. We consider an assembly of N particles and a macrostate specified by the occupation numbers $N_1, N_2, \ldots, N_j, \ldots$. The degeneracies of the levels are respectively $g_1, g_2, \ldots, g_j, \ldots$. Since the particles are distinguishable, two arrangements are considered different if a level contains different particles, even though the occupation number of the level may be the same. That is, an arrangement in which the particles in a level are a, b, and c is different from one in which the particles are a, b, and d or p, q, and r. Consider first any level j, including g_j states and some specified set of N_j particles. The first particle may be placed in any one of the g_j states. But since there is no restriction on the number of particles per state, the second particle can also be placed in any one of the g_j states, making a total of g_j^2 possible locations for the first two particles. Since there are N_j particles in the level, the total number of possible distributions in this level is

$$w_j = g_j^{N_j}. \tag{11–18}$$

For example, if level j includes three states ($g_j = 3$) and the two particles a and b ($N_j = 2$), the possible arrangements of the particles are shown in Fig. 11–7, and we see that there are nine. An interchange of the letters a and b between different *states*, as in arrangements IV and V, VI and VII, VIII and IX, is considered to give rise to a different microstate since the particles a and b are in different states. On the other hand, a change in the order of the letters *within* a given state does not change the microstate since it leaves the same particles in the same state. That is, in arrangements I, II, and III we could equally well have designated the particles as ba instead of ab. Note that if the particles are indistinguishable and are represented by dots instead of letters, arrangements IV and V correspond to the same microstates, as do arrangements VI and VII, and VIII and IX, leaving only six different arrangements as in Fig. 11–3. From Eq. (11–18), the number of different arrangements is

$$g_j^{N_j} = 3^2 = 9,$$

in agreement with Fig 11–7.

Since for any distribution of particles in one level we may have any one of the possible distributions in each of the other levels, the total number of distributions including all levels, with a specified set of particles in each level, is

$$\prod_j w_j = \prod_j g_j^{N_j}. \tag{11–19}$$

But $\prod_j w_j$ is not equal to $\mathscr{W}_k$ as in the other statistics since an interchange of particles between *levels* (as well as an interchange between states in the same level) will also give rise to a different microstate. (If the particles are indistinguishable, an interchange between levels does not result in a different microstate.) Thus for example, if particle b in Fig. 11–7 were interchanged with particle c from some other level so that the two particles in level j were a and c instead of a and b, we would have another nine different arrangements of particles in this level. The question then is, out of a total of N particles, in how many different ways can the particles be distributed among the energy levels, with given numbers of particles N_1, N_2, N_3, etc., in the various levels?

Imagine that the N letters representing the particles are written down in all possible sequences. We have shown that there are $N!$ such sequences. Let the first N_1 letters in each sequence represent the particles in level 1, the next N_2 letters those in level 2, and so on. Out of the $N!$ possible sequences, there will be a number in which the *same* letters appear in the first N_1 places, but in a different order. Whatever the order in which the letters appear, the same particles are assigned to level 1, so we must divide $N!$ by the number of different sequences in which the same letters appear in the first N_1 places, which is $N_1!$. In the same way, we must also divide by $N_2!$, $N_3!$, etc., so that the total number of ways in which N particles can be distributed among the levels, with N_1 particles in level 1, N_2 particles in level 2, and so on, is

$$\frac{N!}{N_1!\,N_2!\cdots} = \frac{N!}{\prod_j N_j!}. \tag{11–20}$$

The total number of different distributions, or the thermodynamic probability $\mathscr{W}_{\text{M-B}}$ of a macrostate in the M-B statistics, is therefore the product of (11–19)

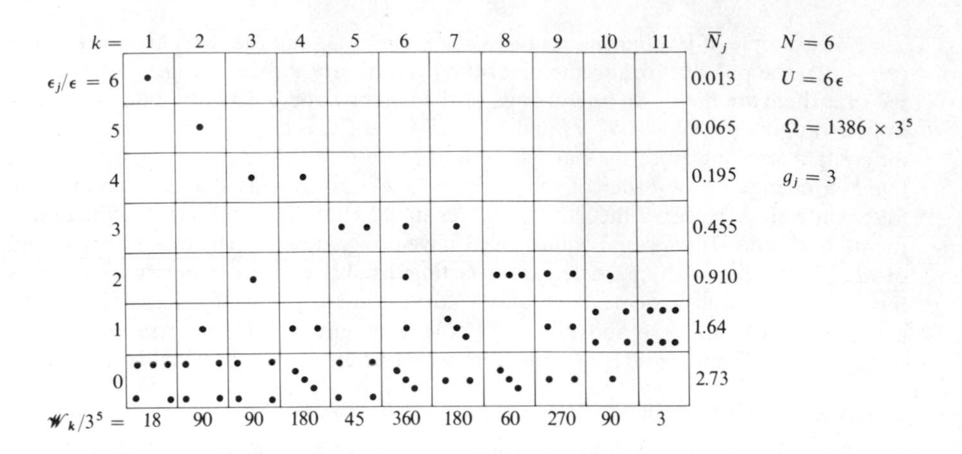

Fig. 11–8 The eleven possible macrostates of an assembly of 6 particles obeying Maxwell-Boltzmann statistics. The energy levels are equally spaced and have a degeneracy of $g_j = 3$ each. The total energy of the system is $U = 6\epsilon$. The thermodynamic probability of each macrostate is given at the bottom, and the average occupation number of each level is printed on the right of the diagram.

and (11–20):

$$\mathscr{W}_{\text{M-B}} = \frac{N!}{\prod_j N_j!} \prod_j g_j^{N_j} = N! \prod_j \frac{g_j^{N_j}}{N_j!}. \tag{11–21}$$

Figure 11–8 shows the possible macrostates of an assembly of 6 particles obeying the M-B statistics. As in Figs. 11–4 and 11–6, the energy levels are presumed to be equally spaced and the degeneracy of each level is $g_j = 3$. Although each particle could be designated by a letter, the dots represent only the occupation numbers N_j of the respective levels. The figure is identical with Fig. 11–4 for the B-E statistics, but it represents a much greater number of microstates because of the possible interchanges of particles between the states in any level, and between various levels. The thermodynamic probability of each macrostate, calculated from Eq. (11–21), is given under the corresponding column. The values of $\mathscr{W}_k$ have been divided by 3^5. Thus for macrostate $k = 1$, in which only levels zero and six are occupied,

$$\mathscr{W}_1 = 6!\frac{3^5}{5!}\frac{3^1}{1!} = 18 \times 3^5,$$

$$\mathscr{W}_1/3^5 = 18.$$

The total number of possible microstates is

$$\Omega = \sum_k \mathscr{W}_k = 1386 \times 3^5 = 3.37 \times 10^5.$$

The average occupation number of each level is given at the right of the corresponding row.

11–8 THE STATISTICAL INTERPRETATION OF ENTROPY

In the three preceding sections, the average occupation numbers of the energy levels of a system were calculated for particles obeying the Bose-Einstein, Fermi-Dirac, and Maxwell-Boltzmann statistics. It was stated in Section 11–4 that the thermodynamic variables of a system were related to the average occupation numbers of its energy levels. In this section we derive the connection and begin by asking what property of a statistical model of a system can be associated with its entropy.

For two equilibrium states of an *open PVT* system in which the temperature, pressure, and chemical potential are the same but in which the energy, volume, and number of particles are different, the principles of thermodynamics lead to the result that the entropy difference between the states is given by

$$T\,\Delta S = \Delta U + P\,\Delta V - \mu\,\Delta N. \tag{11–22}$$

From the statistical point of view, changes in the energy of an assembly, in its volume, and in the number of particles result in changes in the total number of possible microstates in which the system can exist. For example, if the energy U of the system in Fig. 11–4 is increased from 6ϵ to 7ϵ, the number of possible microstates increases from 1532 to 2340 and the average occupation numbers of each level change. (See Problem 11–9.)

However, entropy is an extensive property and the total entropy S of two independent systems is the *sum* of the entropies S_1 and S_2 of the individual systems:

$$S = S_1 + S_2.$$

On the other hand, if Ω_1 and Ω_2 are the thermodynamic probabilities of the systems, and since for every microstate of either system the other may be in any one of its possible microstates, the number Ω of possible microstates of the two systems is the product of Ω_1 and Ω_2:

$$\Omega = \Omega_1\Omega_2. \tag{11–23}$$

It follows that the entropy cannot be simply proportional to the thermodynamic probability; and to find the form of the functional relationship between S and Ω such that the conditions above are satisfied, we assume that S is some unknown function of Ω, say $S = J(\Omega)$. Then since $S = S_1 + S_2$, and $\Omega = \Omega_1\Omega_2$,

$$J(\Omega_1) + J(\Omega_2) = J(\Omega_1\Omega_2).$$

Now take the partial derivatives of both sides of this equation, first with respect to Ω_1 with Ω_2 constant, and then with respect to Ω_2 with Ω_1 constant. Since $J(\Omega_1)$ is a function of Ω_1 only, its partial derivative with respect to Ω_1 is equal to its total derivative:

$$\frac{\partial J(\Omega_1)}{\partial \Omega_1} = \frac{dJ(\Omega_1)}{d\Omega_1}.$$

The partial derivative of $J(\Omega_2)$ with respect to Ω_1 is zero, since Ω_2 is constant.

On the right side, the partial derivative of $J(\Omega_1\Omega_2)$, with respect to Ω_1, equals the total derivative of $J(\Omega_1\Omega_2)$ with respect to its argument $(\Omega_1\Omega_2)$, multiplied by the partial derivative of its argument with respect to Ω_1, which is simply the constant Ω_2. Then if we represent by $J'(\Omega_1\Omega_2)$ the derivative of $J(\Omega_1\Omega_2)$ with respect to its argument, we have

$$\frac{dJ(\Omega_1)}{d\Omega_1} = \Omega_2 J'(\Omega_1\Omega_2).$$

In the same way,

$$\frac{dJ(\Omega_2)}{d\Omega_2} = \Omega_1 J'(\Omega_1\Omega_2).$$

It follows from these equations that

$$\Omega_1 \frac{dJ(\Omega_1)}{d\Omega_1} = \Omega_2 \frac{dJ(\Omega_2)}{d\Omega_2};$$

and since Ω_1 and Ω_2 are independent, the equation can be satisfied only if each side equals the same constant $k_{\rm B}$. Then for any arbitrary system,

$$\Omega \frac{dJ(\Omega)}{d\Omega} = k_{\rm B},$$

$$dJ(\Omega) = k_{\rm B} \frac{d\Omega}{\Omega},$$

$$J(\Omega) = k_{\rm B} \ln \Omega;$$

and hence

$$S = k_{\rm B} \ln \Omega. \tag{11-24}$$

Thus the only function of Ω which satisfies the condition that entropies are *additive* while thermodynamic probabilities are *multiplicative* is the logarithm.

This equation provides the connecting link between statistical and classical thermodynamics. The numerical value of the proportionality constant $k_{\rm B}$ must be chosen so that the classical and statistical values of the entropy will agree. We shall show in Section 11-15 that $k_{\rm B}$ turns out to be none other than the Boltzmann constant $k = R/N_{\rm A}$.

From a statistical point of view the entropy of a system consisting of a very large number of particles is proportional to the natural logarithm of the total number of microstates available to the system. If we could prepare an assembly so that energetically only one microstate is available to it, $\Omega = 1$, $\ln \Omega = 0$, and the entropy would be zero. This system is perfectly ordered since the state of each particle can be uniquely specified. If more energy states become available to the system, Ω becomes greater than 1 and the entropy is larger than zero. In this case it is not possible to specify uniquely the state of each particle since the state of a particle may be different when the system is in different microstates. Thus the system becomes more disordered as more microstates become available to it.

The entropy of the system may be thought of as a measure of the disorder of the system.

This statistical interpretation of entropy allows additional insight into the meaning of the absolute zero of temperature. According to the Planck statement of the third law (Section 7–7) the entropy of a system in internal equilibrium approaches zero as the temperature approaches zero. Therefore systems in internal equilibrium must be perfectly ordered at absolute zero.

Does the quantity $k_B \ln \Omega$ have the other properties of entropy? We give some *qualitative* answers.

1. If there is a reversible flow of heat $d'Q_r$ into a system at a temperature T, the entropy of the system increases by $dS = d'Q_r/T$. If the system is at constant volume so that the work in the process is zero, the increase dU in internal energy of the system equals $d'Q_r$. But for an assembly of noninteracting particles, the values of the energy levels depend upon the volume; and if the volume is constant, these values do not change. If the energy of an assembly increases, more of the higher energy levels become available to the particles of the assembly, with a corresponding increase in the number of available microstates or the thermodynamic probability Ω. Hence both S and $\ln \Omega$ increase when the energy of the system is increased.

2. The entropy of an ideal gas increases in an *irreversible* free expansion from a volume V_1 to a volume V_2. There is no change in internal energy in the process, and no work is done, but the permitted energy levels become more closely spaced because of the increase in volume. For a constant total energy, more microstates become available as the spacing of the energy levels decreases, and again both S and $\ln \Omega$ increase in the irreversible free expansion.

3. In a reversible *adiabatic* expansion of an ideal gas, the entropy S remains constant. There is no heat flow into the gas, and the work in the expansion is done at the expense of the internal energy, which decreases in the process. If the spacing of the energy levels did not change, a reduction in internal energy would result in a smaller number of available microstates with a corresponding decrease in $\ln \Omega$, but because of the increase in volume the energy levels become more closely spaced, and the resulting increase in $\ln \Omega$ just compensates for the decrease arising from a decrease in internal energy. The result is that $\ln \Omega$, like S, remains constant.

Many other examples could be cited, and it turns out in fact that complete agreement between thermodynamics and statistics results from the assumption that the entropy S, whose change dS is defined in thermodynamics by the relation

$$dS = \frac{d'Q_r}{T},$$

has its statistical counterpart in the logarithm of the thermodynamic probability Ω of an assembly of a very large number of particles, or in the logarithm of the total number of microstates available to the assembly. Thus if $S = k_B \ln \Omega$, the entropy difference between two neighboring states of an assembly is $dS = k_B\, d(\ln \Omega)$.

Additional insight into the connection between statistical and classical thermodynamics can be gained by considering two neighboring states of a closed system, in which the values of the internal energy U, the energy levels ϵ_j, and the average occupation numbers $\overline{N}_j$ are slightly different. Since the energy U is given by $\sum_j \epsilon_j \overline{N}_j$, the energy difference between the states is then

$$dU = \sum_j \epsilon_j\, d\overline{N}_j + \sum_j \overline{N}_j\, d\epsilon_j; \tag{11–25}$$

that is, the difference in energy results in part from the differences $d\overline{N}_j$ in the average occupation numbers, and in part from the differences $d\epsilon_j$ in the energy levels.

If the values of the energy levels are functions of some extensive parameter X, such as the volume V, then

$$d\epsilon_j = \frac{d\epsilon_j}{dX}\, dX, \tag{11–26}$$

and

$$\sum_j \overline{N}_j\, d\epsilon_j = \left[\sum_j \overline{N}_j \frac{d\epsilon_j}{dX}\right] dX.$$

Let us define a quantity Y as

$$Y \equiv -\sum_j \overline{N}_j \frac{d\epsilon_j}{dX}. \tag{11–27}$$

Then

$$\sum_j \overline{N}_j\, d\epsilon_j = -Y\, dX. \tag{11–28}$$

If, for example, the parameter X is the volume V, the quantity Y is the pressure P and

$$Y\, dX = P\, dV.$$

The energy difference dU is then

$$dU = \sum_j \epsilon_j\, d\overline{N}_j - Y\, dX.$$

For two states in which the value of the parameter X is the same, $dX = 0$, and

$$dU_X = \sum_j \epsilon_j\, d\overline{N}_j.$$

The principles of thermodynamics lead to the result that when X is constant,

$$dU_X = T\, dS,$$

and hence

$$\sum_j \epsilon_j\, d\overline{N}_j = T\, dS. \tag{11–29}$$

Thus the equation

$$dU = \sum_j \epsilon_j\, d\overline{N}_j + \sum_j \overline{N}_j\, d\epsilon_j$$

is the statistical form of the combined first and second law of thermodynamics for a closed system:

$$dU = T\,dS - Y\,dX.$$

If the system is taken from one state to the other by a *reversible* process, then

$$T\,dS = d'Q_{\text{r}}, \qquad \text{and} \qquad Y\,dX = d'W_{\text{r}}.$$

Hence in such a process,

$$dU = d'Q_{\text{r}} - d'W_{\text{r}}$$

and

$$\sum_j \epsilon_j\,d\overline{N}_j = d'Q_{\text{r}}, \qquad \sum_j \overline{N}_j\,d\epsilon_j = -d'W_{\text{r}}. \tag{11–30}$$

It is sometimes assumed that the sum $\sum_j \epsilon_j\,d\overline{N}_j$ is always equal to the heat flow $d'Q$ into the system and sum $\sum_j \overline{N}_j\,d\epsilon_j$ is always equal to the negative of the work $-d'W$. We see that this is the case for a *reversible* process only, and only for such a process can we identify the sums in Eq. (11–25) with the heat flow and the work.

11–9 THE BOSE-EINSTEIN DISTRIBUTION FUNCTION

If a system consists of only a relatively small number of particles, as in Fig. 11–4, the average values of the occupation numbers of the energy levels can be calculated without much difficulty, when the total number of particles and the total energy are fixed. When the number is very large, as in the statistical model of a macroscopic system, direct calculations are impossible. We now show how to derive a general expression for the average occupation numbers when the total number of particles is very large. Such an expression is called a *distribution function*. The procedure is first to derive a general relation for the relative values of $\ln \Omega$ for two systems having the same set of energy levels, but in the second system the number of particles is less than that in the first by some small number n, where $n \ll N$, and in which the energy is less than in the first by $n\epsilon_r$, where ϵ_r is the energy of some arbitrary level r. Thus if unprimed symbols refer to the first system and primed symbols to the second system,

$$N' = N - n, \qquad U' = U - n\epsilon_r. \tag{11–31}$$

These conditions can always be met, since we can control independently the number of particles in the system, and its energy. The difference in the values of $k_{\text{B}} \ln \Omega$ is then equated to the entropy difference between the systems, using Eq. (11–24).

The only way in which Eqs. (11–31) can be satisfied is that in every macrostate of the primed system the occupation numbers of all levels, with the exception of level r, are the same in both systems, while the occupation number of level r in the primed system is less than that in the unprimed system by the number n. That is, to satisfy Eqs. (11–31) we must have in every macrostate k,

$$N'_j = N_j\,(j \neq r), \qquad N'_r = N_r - n. \tag{11–32}$$

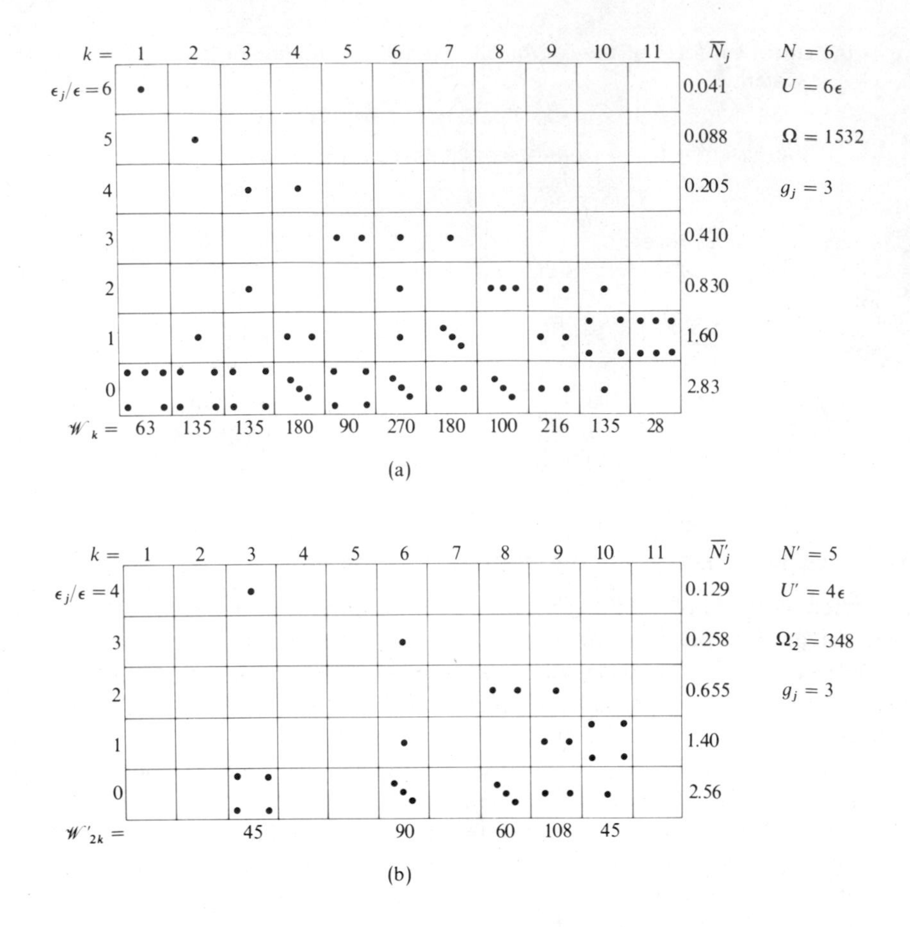

Fig. 11–9 (a) The possible macrostates of an assembly of 6 particles obeying B-E statistics when $U = 6\epsilon$. (b) The possible macrostates when one particle is removed from level 2 of the assembly of part (a). The thermodynamic probability of each macrostate is given at the bottom and the average occupation number of each level is printed on the right of the diagram.

The result is equivalent to the removal of n particles from level r in the unprimed system, without changing the occupation numbers of the other levels.

We consider first a system obeying the Bose-Einstein statistics, and illustrate the relation between corresponding macrostates by taking as an example of the unprimed system that of Fig. 11–4, shown again in part (a) of Fig. 11–9. The number of particles is $N = 6$, the energy $U = 6\epsilon$, and we let n have its smallest possible value, $n = 1$. The number of particles in the primed system is $N' = N - 1 = 5$, and level 2 has been selected as the arbitrary level r so that the energy of the primed system is $U' = U - 2\epsilon = 4\epsilon$.

The only possible macrostates of the primed system are shown in part (b) of Fig. 11–9. There can evidently be no macrostates of the primed system corresponding to a macrostate of the unprimed system in which level 2 is unoccupied. Thus there are only five possible macrostates, and it will be seen that in each of these the occupation number of level 2 is one less than in the corresponding macrostate of the unprimed system, the occupation numbers of all other levels being the same in both systems.

The thermodynamic probability $\mathscr{W}_k$ of macrostate k in the unprimed system is

$$\mathscr{W}_k = \prod_j \frac{(g_j + N_{jk} - 1)!}{(g_j - 1)!\, N_{jk}!}. \tag{11–33}$$

In the primed system,

$$\mathscr{W}'_{rk} = \prod_j \frac{(g_j + N'_{jk} - 1)!}{(g_j - 1)!\, N'_{jk}!}. \tag{11–34}$$

The double subscript rk means that $\mathscr{W}'_{rk}$ is the thermodynamic probability of macrostate k in the primed system, and that level r has been selected as the arbitrary level from which one particle has been removed. The double subscript jk means that N_{jk} and N'_{jk} are, respectively, the occupation numbers of level j in macrostate k, in the unprimed and primed systems.

The fact that there are no macrostates in the primed system corresponding to states in the unprimed system in which level r is unoccupied is equivalent to stating that for such macrostates the thermodynamic probability $\mathscr{W}'_{rk}$ is zero. But if $N_{rk} = 0$, then $N'_{rk} = 0 - 1 = -1$, and the rth term in the product in Eq. (11–34) becomes

$$\frac{(g_r - 2)!}{(g_r - 1)!\,(-1)!} = \frac{1}{(g_r - 1)(-1)!}.$$

Hence in order that $\mathscr{W}'_{rk}$ shall be zero, and provided that $g_r > 1$, we must adopt the convention that $(-1)! = \infty$. For a more general discussion, see Appendix C.

The ratio of thermodynamic probabilities is

$$\frac{\mathscr{W}'_{rk}}{\mathscr{W}_k} = \prod_j \frac{(g_j + N'_{jk} - 1)!}{(g_j + N_{jk} - 1)!} \frac{N_{jk}!}{N'_{jk}!}.$$

In all levels except level r, $N'_{jk} = N_{jk}$ so that all terms in the product above will cancel between numerator and denominator, with the exception of level r in which $N'_{rk} = N_{rk} - 1$. Therefore, since

$$N_{rk}! = N_{rk}(N_{rk} - 1)! = N_{rk}N'_{rk}!$$

and

$$(g_r + N_{rk} - 1)! = (g_r + N'_{rk})! = (g_r + N'_{rk})(g_r + N'_{rk} - 1)!,$$

it follows that

$$\frac{\mathscr{W}'_{rk}}{\mathscr{W}_k} = \frac{N_{rk}}{g_r + N'_{rk}},$$

or

$$N_{rk}\mathscr{W}_k = (g_r + N'_{rk})\mathscr{W}'_{rk};$$

and summing over all macrostates,

$$\sum_k N_{rk}\mathscr{W}_k = g_r \sum_k \mathscr{W}'_{rk} + \sum_k N'_{rk}\mathscr{W}'_{rk}.$$

The term on the left, from Eq. (11–8), equals $\bar{N}_r\Omega$. On the right, the term $g_r \sum_k \mathscr{W}'_{rk}$ equals $g_r\Omega'_r$ and the last term equals $\bar{N}'_r\Omega'_r$. Therefore

$$\bar{N}_r\Omega = (g_r + \bar{N}'_r)\Omega'_r$$

and

$$\frac{\bar{N}_r}{g_r + \bar{N}'_r} = \frac{\Omega'_r}{\Omega}. \tag{11–35}$$

In a macroscopic system in which the occupation numbers are very large, the removal of one particle from a level will make only a relatively small change in the average occupation number of the level, and to a good approximation we can set $\bar{N}'_r = \bar{N}_r$ so that

$$\frac{\bar{N}_r}{g_r + \bar{N}_r} = \frac{\Omega'_r}{\Omega_r}. \tag{11–36}$$

Taking the natural logarithms of both sides, we have

$$\ln \frac{\bar{N}_r}{g_r + \bar{N}_r} = \ln \frac{\Omega'_r}{\Omega}.$$

But

$$\ln \frac{\Omega'_r}{\Omega} = \ln \Omega'_r - \ln \Omega;$$

and since by Eq. (11–24), $S = k_B \ln \Omega$,

$$\ln \frac{\bar{N}_r}{g_r + \bar{N}_r} = \frac{S' - S}{k_B} = \frac{\Delta S}{k_B}. \tag{11–37}$$

From the principles of thermodynamics, the entropy difference ΔS between two states of a nonisolated open system in which the volume (or the appropriate extensive variable) is constant is related to the energy difference ΔU, the difference ΔN in the number of particles, and the temperature T, by Eq. (8–11):

$$T\,\Delta S = \Delta U - \mu\,\Delta N,$$

where μ is now the chemical potential *per particle*. For the two states we are considering,

$$\Delta U = -\epsilon_r, \qquad \Delta N = -1,$$

and hence

$$\Delta S = \frac{\mu - \epsilon_r}{T}.$$

Then from Eq. (11–37), since level r was arbitrarily chosen and could be any level j,

$$\ln \frac{\bar{N}_j}{g_j + \bar{N}_j} = \frac{\mu - \epsilon_j}{k_B T};$$

and

$$\frac{g_j + \bar{N}_j}{\bar{N}_j} = \frac{g_j}{\bar{N}_j} + 1 = \exp \frac{\epsilon_j - \mu}{k_B T},$$

which can be written as

$$\frac{\bar{N}_j}{g_j} = \frac{1}{\exp\left(\dfrac{\epsilon_j - \mu}{k_B T}\right) - 1}. \tag{11–38}$$

This equation is the *Bose-Einstein distribution function.* It expresses the average occupation number per state in any level j, $\bar{N}_j/g_j$, in terms of the energy ϵ_j of the state, the chemical potential μ, the universal constant k_B, and the temperature T. Of course, to apply the equation to a particular system we must know the expression for the energies ϵ_j of the permitted energy levels, and for the chemical potential μ. Another derivation of Eq. (11–38) is found in Appendix D.

11–10 THE FERMI-DIRAC DISTRIBUTION FUNCTION

To derive the distribution function in the F-D statistics, we again consider two assemblies in which the numbers of particles are respectively N and $N' = N - 1$. In any pair of corresponding macrostates, $N'_{jk} = N_{jk}$ in all levels except an arbitrary level r; and in level r, $N'_{rk} = N_{rk} - 1$. The corresponding energies are U and $U' = U - \epsilon_r$.

Part (a) of Fig. 11–10 is the same as Fig. 11–6 and shows the possible macrostates of an assembly of $N = 6$ particles and $U = 6\epsilon$, for an assembly obeying the F-D statistics and in which the energy levels are equally spaced and $g_j = 3$ in each level. Part (b) is the corresponding diagram for an assembly of $N' = 5$ particles and one in which level 2 has been chosen as the arbitrary level r so that $U' = U - 2\epsilon = 4\epsilon$. Again it will be seen that in every pair of corresponding macrostates the occupation numbers are the same in all levels except level 2, and that in this level $N'_{2k} = N_{2k} - 1$.

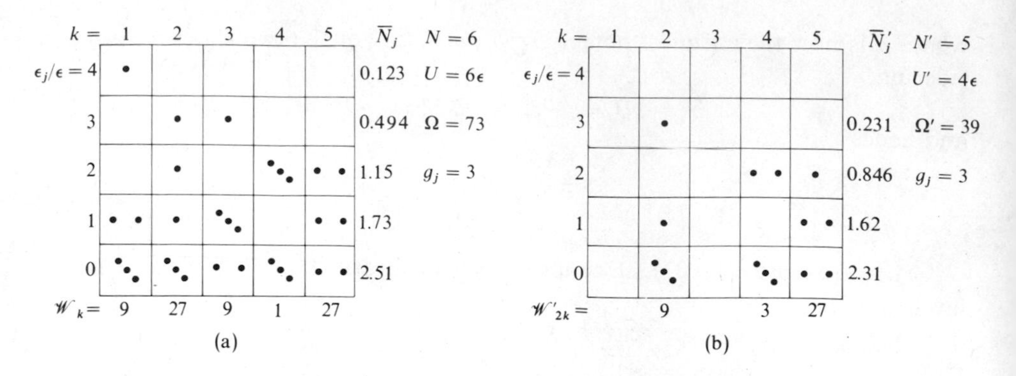

Fig. 11–10 (a) The possible macrostates of an assembly of 6 particles obeying F-D statistics when $U = 6\epsilon$. (b) The possible macrostates when one particle is removed from level 2 of the assembly of part (a). The thermodynamic probability of each macrostate is given at the bottom and the average occupation number of each level is printed on the right of the diagram.

The thermodynamic probabilities of corresponding macrostates in the unprimed and primed assemblies are

$$\mathscr{W}_k = \prod_j \frac{g_j!}{(g_j - N_{jk})!\, N_{jk}!},$$

$$\mathscr{W}'_{rk} = \prod_j \frac{g_j!}{(g_j - N'_{jk})!\, N'_{jk}!}.$$

Then

$$\frac{\mathscr{W}'_{rk}}{\mathscr{W}_{rk}} = \prod_j \frac{(g_j - N_{jk})!\, N_{jk}!}{(g_j - N'_{jk})!\, N'_{jk}!},$$

which after cancellation reduces to

$$\frac{\mathscr{W}'_{rk}}{\mathscr{W}_k} = \frac{N_{rk}}{g_r - N'_{rk}}$$

or

$$N_{rk}\mathscr{W}_k = (g_r - N'_{rk})\mathscr{W}'_{rk}.$$

Summing over all values of k, we have

$$\sum_k N_{rk}\mathscr{W}_k = g_r \sum_k \mathscr{W}'_{rk} - \sum_k N'_{rk}\mathscr{W}'_{rk},$$

and

$$\frac{\bar{N}_r}{g_r - \bar{N}'_r} = \frac{\Omega'_r}{\Omega}. \tag{11–39}$$

Here we can let $\bar{N}'_r = \bar{N}_r$ since if the states are degenerate enough, N_r and N'_r can be much larger than one. By the same reasoning as in the B-E statistics

$$\frac{\bar{N}_j}{g_j} = \frac{1}{\exp\left(\frac{\epsilon_j - \mu}{k_B T}\right) + 1}, \tag{11–40}$$

which is the *Fermi-Dirac distribution function.* It differs from the B-E distribution in that we have + 1 in the denominator instead of −1.

11–11 THE CLASSICAL DISTRIBUTION FUNCTION

In many systems of indistinguishable particles, the average number of particles $\bar{N}_j$ in a level is very much less than the number of states g_j in the level, so that the average number of particles per state, $\bar{N}_j/g_j$, is very small. The denominator in Eqs. (11–38) and (11–40) must then be very large; we can neglect the 1; and both the B-E and F-D distribution functions reduce to

$$\frac{\bar{N}_j}{g_j} = \exp\frac{\mu - \epsilon_j}{k_B T}, \tag{11–41}$$

which is the *classical distribution function.*

11–12 COMPARISON OF DISTRIBUTION FUNCTIONS FOR INDISTINGUISHABLE PARTICLES

The distribution functions for indistinguishable particles can all be represented by the single equation,

$$\frac{\bar{N}_j}{g_j} = \frac{1}{\exp\left(\frac{\epsilon_j - \mu}{k_B T}\right) + a}, \tag{11–42}$$

where $a = -1$ in the B-E statistics, $a = +1$ in the F-D statistics, and $a = 0$ in the classical statistics.

The curves in Fig. 11–11 are graphs of the average number of particles per state, $\bar{N}_j/g_j$, at a given temperature, for the B-E and F-D statistics, plotted as functions of the dimensionless quantity $(\epsilon_j - \mu)/k_B T$. (The energy therefore increases toward the right.) The ordinates of the curves have a meaning, of course, only at those abscissas at which the energy ϵ_j has some one of its permitted values. When $\bar{N}_j/g_j$ is very small, the B-E and F-D distributions very nearly coincide, and both reduce to the classical distribution.

Note that when $\epsilon_j = \mu$, the value of $\bar{N}_j/g_j$ in the B-E statistics becomes infinite, and for levels in which ϵ_j is less than μ, it is negative and hence meaningless. That is, in this statistics, the chemical potential must be less than the energy of the lowest permitted energy level. The particles like to concentrate in levels for which ϵ_j is only slightly greater than μ.

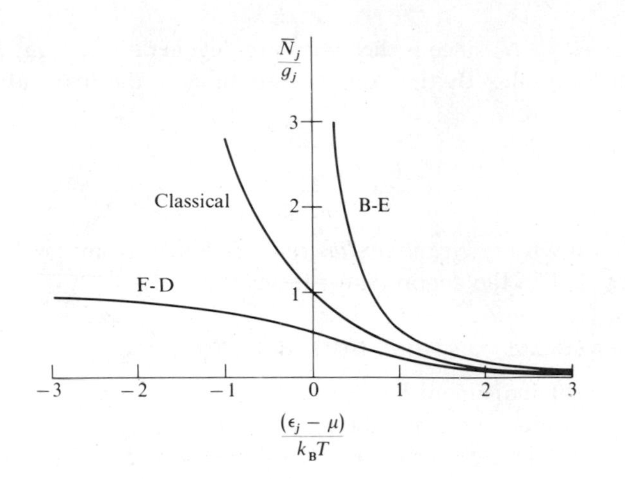

Fig. 11–11 Graphs of the Bose-Einstein, Fermi-Dirac, and classical distribution functions.

In the F-D statistics, on the other hand, all levels are populated down to the lowest and as ϵ_j decreases, $\bar{N}_j/g_j$ approaches 1. That is, the low-energy levels are very nearly uniformly populated with one particle per state.

The curve for classical statistics has no meaning except when $(\epsilon_j - \mu)/kT$ is large. It is drawn on Fig. 11–11 for comparison only. If the ordinate of Fig. 11–11 is taken as $\bar{N}_j/Ng_j$ instead of $\bar{N}_j/g_j$, this curve is the distribution function for M-B statistics which is developed in the next section.

11–13 THE MAXWELL-BOLTZMANN DISTRIBUTION FUNCTION

The distribution function in the M-B statistics is derived in the same way as in the B-E and F-D statistics. Part (a) of Fig. 11–12 is the same as Fig. 11–8, in which the dots represent the occupation numbers of an assembly of $N = 6$ particles and of energy $U = 6\epsilon$. Part (b) shows the possible macrostates of an assembly of $N' = N - 1 = 5$ particles, and in this assembly level 2 has been chosen for the arbitrary level r so that $U' = U - 2\epsilon = 4\epsilon$. The only possible macrostates of the primed assembly are those in which level 2 is occupied in the unprimed assembly. In any pair of corresponding macrostates, the occupation numbers are the same in all levels except level 2; and in level 2, $N'_{2k} = N_{2k} - 1$.

The thermodynamic probabilities of corresponding macrostates in the unprimed and primed assemblies are

$$\mathscr{W}_k = N! \prod_j \frac{g_j^{N_j}}{N_j!},$$

$$\mathscr{W}'_k = N'! \prod_j \frac{g_j^{N'_j}}{N'_j!}.$$

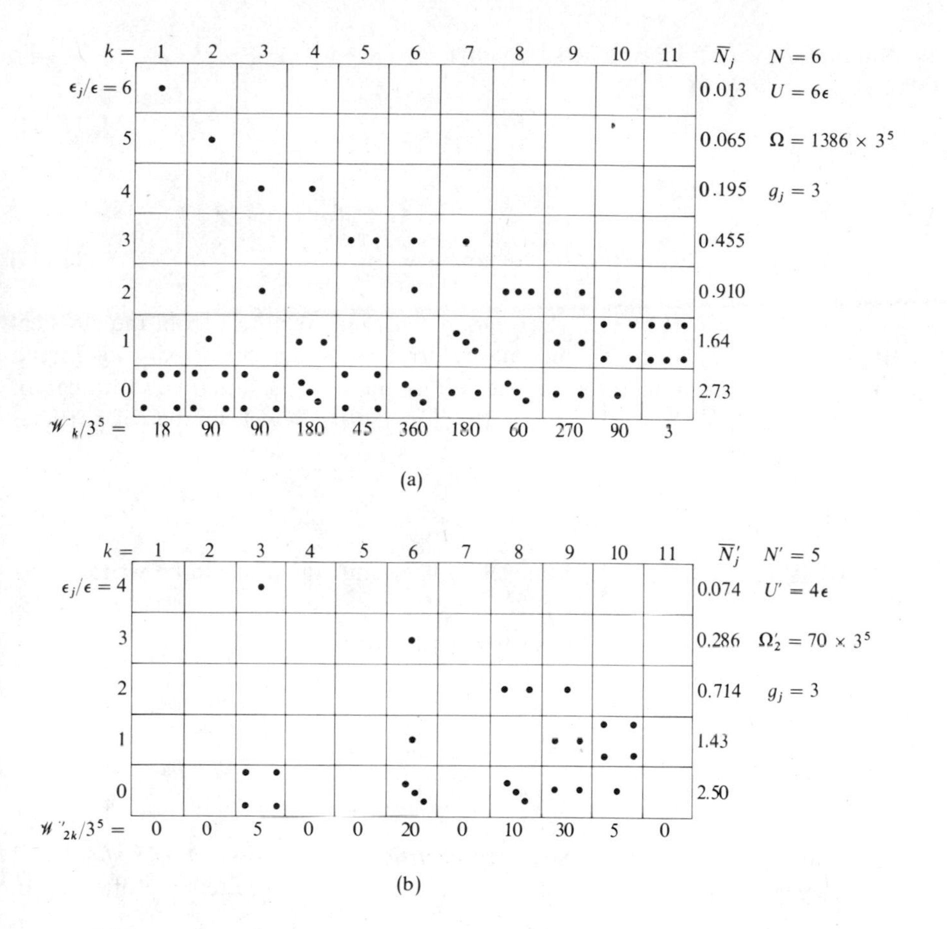

Fig. 11–12 (a) The possible macrostates of an assembly of 6 particles obeying M-B statistics when $U = 6\epsilon$. (b) The possible macrostates when one particle is removed from level 2 of the assembly of part (a). The thermodynamic probability of each macrostate is given at the bottom and the average occupation number of each level is printed on the right of the diagram.

Then

$$\frac{\mathcal{W}'_{rk}}{\mathcal{W}_k} = \frac{N'!}{N!} \prod_j \frac{g_j^{N'_{jk}} N_{jk}!}{g_j^{N_{jk}} N'_{jk}!},$$

which simplifies to

$$\frac{\mathcal{W}'_{rk}}{\mathcal{W}_k} = \frac{N_{rk}}{N g_r}$$

or

$$N_{rk} \mathcal{W}_k = N g_r \mathcal{W}'_{rk}.$$

Summing over all macrostates, we have

$$\frac{\bar{N}_r}{Ng_r} = \frac{\Omega'_r}{\Omega}, \tag{11-43}$$

and by the same procedure as before,

$$\frac{\bar{N}_j/N}{g_j} = \exp\frac{\mu - \epsilon_j}{k_B T}, \tag{11-44}$$

which is the *Maxwell-Boltzmann distribution function*. It differs from the classical distribution function, which is sometimes referred to as the "corrected" Boltzmann function, in that the numerator on the left is the average *fractional* number of particles in level j, $\bar{N}_j/N$, so that the left side is the *fractional* number of particles per state in any level.

11–14 THE PARTITION FUNCTION

The distribution function in the Maxwell-Boltzmann statistics can be written

$$\bar{N}_j = N\left(\exp\frac{\mu}{k_B T}\right) g_j \exp\frac{-\epsilon_j}{k_B T}.$$

Since $\sum_j \bar{N}_j = N$, and the chemical potential μ does not depend on j, it follows that

$$\sum_j \bar{N}_j = N = N\left(\exp\frac{\mu}{k_B T}\right) \sum_j g_j \exp\frac{-\epsilon_j}{k_B T}.$$

The sum in the last term is called the *partition function* or *sum over states* and will be represented by Z. (German *Zustandssumme*) Other letters are often used.

$$Z \equiv \sum_j g_j \exp\frac{-\epsilon_j}{k_B T}. \tag{11-45}$$

The partition function depends only on the temperature T and on the parameters that determine the energy levels. It follows from the two preceding equations that in the M-B statistics,

$$\exp\frac{\mu}{k_B T} = \frac{1}{Z}, \tag{11-46}$$

and hence the M-B distribution function can be written

$$\frac{\bar{N}_j}{g_j} = \frac{N}{Z}\exp\frac{-\epsilon_j}{k_B T}. \tag{11-47}$$

Thus in a given system, the average number of particles per state in any level decreases exponentially with the energy ϵ_j of the level; and the lower the temperature T, the more rapid is the rate of decrease.

The classical distribution function can be written

$$\bar{N}_j = \left(\exp \frac{\mu}{k_B T}\right) g_j \exp \frac{-\epsilon_j}{k_B T},$$

and summing over all values of j, we have

$$\sum_j \bar{N}_j = N = \left(\exp \frac{\mu}{k_B T}\right) \sum_j g_j \exp \frac{-\epsilon_j}{k_B T}.$$

Then if the partition function Z is defined in the same way as in the M-B statistics, we have

$$\exp \frac{\mu}{k_B T} = \frac{N}{Z}, \tag{11–48}$$

and the classical distribution function can be written

$$\frac{\bar{N}_j}{g_j} = \frac{N}{Z} \exp \frac{-\epsilon_j}{k_B T}, \tag{11–49}$$

which has the same form as the M-B distribution.

Because of the form of the B-E and F-D distribution functions, these cannot be expressed in terms of a single-particle partition function, and we shall discuss them later.

11–15 THERMODYNAMIC PROPERTIES OF A SYSTEM

The importance of the partition function Z is that in Maxwell-Boltzmann and classical statistics, all the thermodynamic properties of a system can be expressed in terms of $\ln Z$ and its partial derivatives. Thus the first step in applying the methods of statistics to such a system is to evaluate the partition function of the system.

It will be recalled that all thermodynamic properties of a system are also completely determined by its *characteristic equation;* that is, the Helmholtz function expressed in terms of X and T or the Gibbs function expressed in terms of Y and T. Here X and Y stand for some related pair of variables such as the volume V and the pressure P.

Thus we begin by deriving expressions for the Helmholtz and Gibbs functions in terms of $\ln Z$. As shown in Section 8–1, these functions are related to the chemical potential μ by the equation

$$\mu = \left(\frac{\partial G}{\partial N}\right)_{T,Y} = \left(\frac{\partial F}{\partial N}\right)_{T,X}. \tag{11–50}$$

For a system obeying M-B statistics, the chemical potential of the system is related to the partition function by Eq. (11–46):

$$\mu = -k_B T \ln Z. \tag{11–51}$$

In classical statistics, the chemical potential is given by Eq. (11–48):

$$\mu = -k_B T(\ln Z - \ln N). \tag{11–52}$$

The partition function, $Z = \Sigma g_j \exp(-\epsilon_j/k_B T)$, is a function of the temperature of the system and of the parameters that determine the energy levels of the system (such as the volume V or the magnetic intensity $\mathscr{H}$). Thus Eqs. (11–51) and (11–52) express μ in terms of X or Y.

Consider first a system of indistinguishable particles obeying the classical statistics and one in which the energy levels are functions of an extensive parameter X. Then the partition function is a function of X and T, and as these are the "natural" variables of the Helmholtz function F, we have from Eqs. (11–50) and (11–52),

$$\left(\frac{\partial F}{\partial N}\right)_{T,X} = -k_B T(\ln Z - \ln N). \tag{11–53}$$

The right side of this equation is constant when X and T are constant. Integrating at constant X and T yields

$$F = -Nk_B T(\ln Z - \ln N + 1), \tag{11–54}$$

since $\int N \ln N\, dN = N \ln N - N$. Equation (11–53) would be satisfied if any function $f(T, X)$ were added to the right side of Eq. (11–54), but since F must be zero when $N = 0$, it follows that $f(T, X) = 0$. Equation (11–54) is an expression for F in terms of N, T, and X; therefore all the thermodynamic properties of the system can be determined by the methods of Section 7–2.

The entropy S is given by $S = -(\partial F/\partial T)_{N,X}$ so that

$$S = Nk_B T\left(\frac{\partial \ln Z}{\partial T}\right)_X + Nk_B(\ln Z - \ln N + 1). \tag{11–55}$$

Since $U = F + TS$, the internal energy is

$$U = Nk_B T^2\left(\frac{\partial \ln Z}{\partial T}\right)_X. \tag{11–56}$$

The expression for the entropy can now be rewritten as

$$S = \frac{U}{T} + Nk_B(\ln Z - \ln N + 1). \tag{11–57}$$

The intensive variable Y associated with the extensive variable X is given by $Y = -(\partial F/\partial X)_{N,T}$, so that

$$Y = Nk_B T\left(\frac{\partial \ln Z}{\partial X}\right)_T, \tag{11–58}$$

which is the equation of state of the system, expressing Y as a function of N, T, and X. Thus all thermodynamic properties of this system can be determined if Z is known as a function of X and T.

For a one-component system, the Gibbs function $G = \mu N$, so that from Eq. (11–52)

$$G = -Nk_B T(\ln Z - \ln N). \tag{11–59}$$

But in general for the variables X and Y,

$$G = U - TS - YX = F + YX,$$

and

$$G - F = YX.$$

From Eqs. (11–54) and (11–59),

$$G - F = Nk_B T,$$

so that for any system obeying the classical statistics and in which the energy levels are functions of a single extensive parameter X,

$$YX = Nk_B T. \tag{11–60}$$

In the special case in which the parameter X is the volume V and Y is the pressure P,

$$PV = Nk_B T.$$

This is the equation of state of an ideal gas as derived from kinetic theory, provided that the universal constant k_B, which was introduced earlier only as the proportionality constant in the equation $S = k_B \ln \Omega$, is equal to the Boltzmann constant $k = R/N_A$. Since k_B is a *universal* constant, which in this special case is equal to R/N_A, it must equal R/N_A regardless of the nature of an assembly. In the future we shall, for simplicity, drop the subscript B and write simply $S = k \ln \Omega$.

It is at first surprising that we obtain only the ideal gas equation of state. However, the partition function can only be given by the sum over single particle states when the particles do not interact. This is the same condition needed to derive the ideal gas law from kinetic theory.

In terms of this notation, the expressions for the thermodynamic properties of a system obeying classical statistics and a system in which the energy levels are determined by the extensive parameter X are given by

$$F = -NkT(\ln Z - \ln N + 1), \tag{11–61}$$

$$U = NkT^2\left(\frac{\partial \ln Z}{\partial T}\right)_X, \tag{11–62}$$

$$S = \frac{U}{T} + Nk(\ln Z - \ln N + 1), \tag{11–63}$$

and

$$Y = NkT\left(\frac{\partial \ln Z}{\partial X}\right)_T. \tag{11–64}$$

It is left as an exercise (Problem 11–30) to show that for a system of distinguishable particles obeying M-B statistics and in which the energy levels are determined by an extensive parameter X, the expressions for U and Y are unchanged, but the expressions for F and S are

$$F = -NkT \ln Z \tag{11–65}$$

and

$$S = \frac{U}{T} + Nk \ln Z. \tag{11–66}$$

These expressions differ from those for indistinguishable particles by a term proportional to $N \ln N - N$. (See Problem 11–31).

As a second example, consider a system of distinguishable particles obeying the M-B statistics and for which the energy levels are functions of an intensive parameter Y. Then Z is a function of Y and T; and since these are the "natural" variables of the Gibbs function, we have, from Eqs. (11–50) and (11–51),

$$\left(\frac{\partial G}{\partial N}\right)_{T,Y} = -kT \ln Z. \tag{11–67}$$

The right side of this equation is constant when T and Y are constant. Integrating at constant T and Y yields

$$G = -NkT \ln Z. \tag{11–68}$$

The arbitrary function $g(T, Y)$ which should be added to the right side of Eq. (11–68) is again zero since $G = 0$ when $N = 0$. This equation appears at first to contradict Eq. (11–65) since $F \neq G$. However, Eq. (11–65) is derived for a system in which the energy levels are functions of an extensive parameter X, whereas Eq. (11–68) applies to a system in which the energy levels depend upon an intensive parameter Y.

The entropy is now given by $S = -(\partial G/\partial T)_{N,Y}$, and hence

$$S = NkT\left(\frac{\partial \ln Z}{\partial T}\right)_Y + Nk \ln Z. \tag{11–69}$$

The enthalpy H equals $G + TS$, so

$$H = NkT^2\left(\frac{\partial \ln Z}{\partial T}\right)_Y, \tag{11–70}$$

and Eq. (11–69) can be written

$$S = \frac{H}{T} + Nk \ln Z. \tag{11–71}$$

The equation of state is given by

$$X = \left(\frac{\partial G}{\partial T}\right)_{N,T} = -NkT\left(\frac{\partial \ln Z}{\partial Y}\right)_T. \tag{11–72}$$

If the parameter Y is the intensity of a conservative field of force, the only energy of the particle is its *potential* energy (gravitational, magnetic, or electric). The internal energy of the system is then zero, and its total energy E is its potential energy E_p only. If X represents the extensive variable associated with the intensive variable Y, the potential energy $E_p = YX$. Then since the enthalpy H is defined as $H = U + YX$, and $U = 0$, it follows that

$$E = E_p = H,$$

and Eqs. (11–70) and (11–71) can be written

$$E = NkT^2\left(\frac{\partial \ln Z}{\partial T}\right)_Y, \tag{11–73}$$

and

$$S = \frac{E}{T} + Nk \ln Z. \tag{11–74}$$

It has been assumed thus far in this section that the energy levels were functions either of a single extensive variable X or a single intensive variable Y. We now consider the more general case of a *multivariable* system in which the energy levels are functions of more than one independent variable. We restrict the discussion to systems whose energy levels are functions of *two* variables only, one of which is an extensive variable X_1 while the other is an intensive variable Y_2, which we consider to be the intensity of a conservative field of force.

If the system is described by either the Maxwell-Boltzmann or classical statistics, we can still define the partition function as

$$Z = \sum_j g_j \exp\left(\frac{-\epsilon_j}{kT}\right).$$

The only difference is that the ϵ_j's are now functions of *both* X_1 and Y_2, and the partition function is a function of T, X_1, and Y_2. Since the system has both an internal energy U and a potential energy $E_p = Y_2X_2$, its total energy E is

$$E = U + E_p = U + Y_2X_2,$$

and we therefore make use of the generalized Helmholtz function F^*, defined by Eq. (7–34) as

$$F^* = E - TS = U - TS + Y_2X_2.$$

The chemical potential is now

$$\mu = \left(\frac{\partial F^*}{\partial N}\right)_{T,X_1,Y_2}.$$

If the system obeys the classical statistics,

$$\mu = -kT(\ln Z - \ln N),$$

and, integrating at constant T, X_1, Y_2,

$$F^* = -NkT(\ln Z - \ln N + 1), \tag{11–75}$$

setting the arbitrary function of X_1, Y_2, and T equal to zero as before.

The variables Y_1 and X_2, associated with the variables X_1 and Y_2, are given by

$$Y_1 = -\left(\frac{\partial F^*}{\partial X_1}\right)_{N,T,Y_2} = NkT\left(\frac{\partial \ln Z}{\partial X_1}\right)_{T,Y_2}, \tag{11–76}$$

$$X_2 = \left(\frac{\partial F^*}{\partial Y_2}\right)_{N,T,X_1} = -NkT\left(\frac{\partial \ln Z}{\partial Y_2}\right)_{T,X_1}. \tag{11–77}$$

The system thus has *two* equations of state, expressing Y_1 and X_2 in terms of N, T, X_1, and Y_2.

The entropy S is

$$S = -\left(\frac{\partial F^*}{\partial T}\right)_{N,X_1,Y_2} = NkT\left(\frac{\partial \ln Z}{\partial T}\right)_{X_1,Y_2} + Nk(\ln Z - \ln N + 1). \tag{11–78}$$

The total energy E equals $F^* + TS$, so

$$E = NkT^2\left(\frac{\partial \ln Z}{\partial T}\right)_{X_1,Y_2}, \tag{11–79}$$

and hence

$$S = \frac{E}{T} + Nk(\ln Z - \ln N + 1). \tag{11–80}$$

If the system obeys the Maxwell-Boltzmann statistics,

$$\mu = -kT \ln Z;$$

and by similar reasoning,

$$F^* = -NkT \ln Z \tag{11–81}$$

The variables Y_1 and X_2 are again given by Eqs. (11–75) and (11–76). The entropy is

$$S = NkT\left(\frac{\partial \ln Z}{\partial T}\right)_{X_1,Y_2} + Nk \ln Z. \tag{11–82}$$

The total energy is

$$E = NkT^2\left(\frac{\partial \ln Z}{\partial T}\right)_{X_1,Y_2}, \tag{11–83}$$

so one can also write

$$S = \frac{E}{T} + Nk \ln Z. \tag{11–84}$$

In either statistics, the potential energy $E_p = Y_2X_2$ and the internal energy U is

$$U = E - E_p = E - Y_2X_2. \tag{11–85}$$

Specific examples of the general relations derived in this section will be discussed in the next two chapters.

PROBLEMS

11–1 Using quantum mechanics, show that the energy levels of a one-dimensional infinite square well of width L are also given by Eq. (11–3).

11–2 (a) Tabulate the values of the quantum numbers n_x, n_y, n_z for the twelve lowest energy levels of a free particle in a container of volume V. (b) What is the degeneracy g of each level? (c) Find the energy of each level in units of $h^2/8mV^{2/3}$. (d) Are the energy levels equally spaced?

11–3 Calculate the value of n_j in which an oxygen atom confined to a cubical box 1 cm on a side will have the same energy as the lowest energy available to a helium atom confined to a cubical box 2×10^{-10} m on a side.

11–4 Five indistinguishable particles are to be distributed among the four equally spaced energy levels shown in Fig. 11–2 with no restriction on the number of particles in each energy state. If the total energy is to be $12\epsilon_1$, (a) specify the occupation number of each level for each macrostate, and (b) find the number of microstates for each macrostate, given the energy states represented in Fig. 11–2.

11–5 (a) Find the number of macrostates for an assembly of four particles distributed among two energy levels one of which is two-fold degenerate. (b) Find the thermodynamic probability of each macrostate if there is no restriction on the number of particles in each energy state and the particles are indistinguishable, (c) distinguishable. (d) Calculate the thermodynamic probability of the assembly for parts (b) and (c).

11–6 In the poker game *seven-card stud*, seven cards are dealt to each player. He makes the best hand out of five of those cards. The cards are well shuffled between each deal. (a) How many different seven-card hands can be made in a deck of 52 cards? (b) If there are four players, how many different ways can the cards be dealt if the players are distinguishable? (c) How many different five-card hands can be made from a seven-card hand?

11–7 For the example illustrated in Fig. 11–4, find (a) the thermodynamic probability $\mathscr{W}_k$ of each macrostate, (b) the total number of microstates of the assembly Ω, (c) the average occupation number of each level, and (d) the sum of the average occupation numbers.

11–8 Do Problem 11–7 for a system of seven indistinguishable particles obeying B-E statistics and having a total energy $U = 6\epsilon$.

11–9 (a) Construct a diagram similar to Fig. 11–6, but having eight energy levels. Show the possible macrostates of the system if the energy $U = 7\epsilon$ for six indistinguishable particles, obeying B-E statistics. (b) Calculate the thermodynamic probability of each macrostate, and (c) show that the total number of possible microstates Ω is 2340. (d) Find the average occupation number of each level.

11–10 (a) Suppose that in the F-D statistics, level j includes three states (1), (2), (3), and two particles a and b. If the particular sequence of numbers (1), (2), and (3), is selected, write down the possible different sequences of letters and numbers, and show that this agrees with Eq. (11–15). (b) How many different sequences of numbers are possible? (c) What is the total number of different possible sequences of letters and numbers?

11–11 Show that in the Fermi-Dirac statistics, if level j is fully occupied with one particle per state, $\mathscr{W}_k = 1$ and there is only one way of distributing the particles among the energy states of that level.

11–12 Do Problem 11–9 for six indistinguishable particles obeying F-D statistics. In this case $\Omega = 162$.

11–13 Do Problem 11–9 for six distinguishable particles obeying M-B statistics. In this case $\Omega = 5.77 \times 10^5$.

11–14 There are 30 distinguishable particles distributed among three nondegenerate energy levels labeled 1, 2, 3, such that $N_1 = N_2 = N_3 = 10$. The energies of the levels are $\epsilon_1 = 2$ eV, $\epsilon_2 = 4$ eV, $\epsilon_3 = 6$ eV. (a) If the change in the occupation number of level 2, $\delta N_2 = -2$, find δN_1 and δN_3 such that $\delta E = 0$. (b) Find the thermodynamic probability of the macrostate before and after the change.

11–15 Six distinguishable particles are distributed over three nondegenerate energy levels. Level 1 is at zero energy; level 2 has an energy ϵ; and level 3 has an energy 2ϵ. (a) Calculate the total number of microstates for the system. (b) Calculate the number of microstates such that there are three particles in level 1, two in level 2, and one in level 3. (c) Find the energy of the distribution for which $\mathscr{W}_k$ is largest. (d) Calculate the total number of microstates if the total energy of the six particles is 5ϵ.

11–16 Five particles are distributed among the states of the four equally spaced energy levels shown on Fig. 11–2 such that the total energy is $12\epsilon_1$. Calculate the thermodynamic probability of each macrostate and the average occupation number of each level if the particles obey (a) B-E, (b) F-D, (c) M-B statistics.

11–17 Calculate the change in the entropy of each of the systems illustrated in Figs. 11–4, 11–6, and 11–8 when an additional energy level is available to the particles and the total energy is increased to 7ϵ. [See Problems 11–9, 11–12, and 11–13.]

11–18 The internal energy of the six indistinguishable particles of Fig. 11–4 is increased reversibly from 6ϵ to 7ϵ without work being done, but only the levels up through level 6 can be occupied. (a) Show explicitly that $d'Q_r = \sum_j \epsilon_j \, d\bar{N}_j$ and (b) find the increase in the entropy of the system.

11–19 (a) Construct a diagram similar to part (b) of Fig. 11–9, but in which level 3 is selected as the arbitrary level r so that $U' = 6\epsilon - 3\epsilon = 3\epsilon$. Note that every possible macrostate of the primed system corresponds to a macrostate of the unprimed system and that with the exception of level 3 the occupation numbers of all levels are the same in each pair of corresponding macrostates. (b) How many possible macrostates are there for the primed system? (c) How many microstates? (d) Calculate the average occupation number of the levels of the primed system. (e) Use Eq. (11–35) to calculate the average occupation number of level 3 of the primed system. (f) Calculate the change in the entropy of the unprimed system upon removing one particle from level 3.

11–20 Fill in the steps of the derivation of (a) Eq. (11–39) and (b) Eq. (11–40).

11–21 (a) Construct a diagram similar to part (b) of Fig. 11–10 but in which level 3 is selected as the arbitrary level r so that $U' = 3\epsilon$. (b) Calculate the number of microstates available to the primed system. (c) Calculate the average occupation number of the levels of the primed system. (d) Use Eq. (11–39) to calculate the average occupation number of level 3 of the primed system. (e) Calculate the change in the entropy of the unprimed system upon removing one particle from level 3.

11–22 Show that Eq. (11–13) for $\mathscr{W}_{\text{B-E}}$ and Eq. (11–17) for $\mathscr{W}_{\text{F-D}}$ both reduce to

$$\mathscr{W}_{\text{c}} = \prod_j \frac{g_j^{N_j}}{N_j!} \tag{11–86}$$

in the limit that $g_j \gg N_j$. This is the thermodynamic probability of a system obeying classical statistics.

11–23 By a method similar to that in Section 11–9, show that Eq. (11–86) of the previous problem leads to the distribution function of Eq. (11–41).

11–24 Show that Eq. (11–13) for $\mathscr{W}_{\text{B-E}}$, Eq. (11–17) for $\mathscr{W}_{\text{F-D}}$ and Eq. (11–86) (Problem 11–22) for classical statistics can all be represented by

$$\mathscr{W} = \prod_j \frac{g_j(g_j - a)(g_j - 2a)\cdots[g_j - (N_j - 1)a]}{N_j!},$$

where a has the values given in Section 11–12.

11–25 Fill in the steps of the derivation of the Maxwell-Boltzmann distribution function done in Section 11–13.

11–26 Derive the Maxwell-Boltzmann distribution function by the method of Section 11–13 but assume that n particles are removed from the level r of the unprimed system, where $n \ll N$.

11–27 (a) Construct a diagram similar to part (b) of Fig. 11–12 but one in which level 3 is selected as the arbitrary level r so that $U' = 3\epsilon$. (b) Calculate the number of microstates available to the primed system. (c) Calculate the average occupation number of the levels of the primed system. (d) Calculate the change in the entropy of the unprimed system upon removing one particle from level 3.

11–28 Substitute the Maxwell-Boltzmann distribution function into Eq. (11–29), the expression for the entropy change of a system in a reversible process, to obtain

$$S = -k \sum_j \bar{N}_j \ln \frac{\bar{N}_j}{g_j}.$$

11–29 Seven distinguishable particles are distributed over two energy levels. The upper level is nondegenerate and has an energy 10^{-3} eV higher than the lower level which is two-fold degenerate. (a) Calculate the internal energy and entropy of the system if it is prepared to have two particles in the upper level. (b) If there is no change in the system when it is brought into contact with a reservoir at a temperature T, calculate the temperature of the reservoir. (c) Write the partition function for this system. (d) Repeat parts (a), (b), and (c) for the case that the degenerate level has an energy 10^{-3} eV higher than the nondegenerate level.

11–30 (a) Derive Eqs. (11–65) and (11–66) for a system obeying M-B statistics and in which the energy levels are determined by an extensive parameter X. (b) Show that the expressions for the internal energy U and the intensive parameter Y for this system are still given by Eqs. (11–62) and (11–64).

11–31 (a) Using Eqs. (11–21) and (11–86) (Problem 11–22) for the thermodynamic probability of a macrostate of a system of N particles obeying M-B and classical statistics respectively, show that $\Omega_{\text{M-B}} = N!\Omega_{\text{c}}$. (b) Use the result of part (a) to show that the entropies of the two systems are related by $S_{\text{M-B}} = S_c + Nk_B(\ln N - 1)$ and that the Helmholtz functions are related by $F_{\text{M-B}} = F_c + Nk_BT(\ln N - 1)$.

11–32 Show that for a system of N particles obeying M-B or classical statistics the average number of particles in the level j is given by

$$\bar{N}_j = -Nk_BT\left(\frac{\partial \ln Z}{\partial \epsilon_j}\right)_T. \tag{11–87}$$

11–33 (a) Derive an expression for the enthalpy of a system if the partition function depends on X and T. (b) Derive an expression for the internal energy of a system if the partition function depends on Y and T.

11–34 Consider a system of N distinguishable particles distributed in two nondegenerate levels separated by an energy ϵ and in equilibrium with a reservoir at a temperature T. Calculate (a) the partition function, (b) the fraction N_1/N and N_2/N of particles in each state, (c) the internal energy U of the system, (d) the entropy S of the system, (e) the specific heat capacity c_v of the system. (f) Make sketches of N_1/N, N_2/N, U, S, and c_v as a function of T.

11–35 Consider a system of N distinguishable particles each having a magnetic moment μ, distributed over two nondegenerate levels having energies $\mu\mathscr{H}_0/2$ and $-\mu\mathscr{H}_0/2$, when the magnetic intensity is $\mathscr{H}_0$. The particles in the upper level have their magnetic moments antiparallel to the field and those in the lower level are aligned parallel to the field. The system is prepared to have one-third of all the particles in the upper level and is isolated. (a) Find the energy and the net magnetic moment of the system. (b) Calculate the change of the energy and the change of the net magnetic moment of the isolated system when the magnetic intensity is reversibly reduced to $\mathscr{H}_0/2$. (c) Calculate the change in the net magnetic moment of the system when the magnetic intensity is reversibly reduced to $\mathscr{H}_0/2$ but the energy of the system remains constant.

11–36 The system of the previous problem is in thermal equilibrium with a reservoir at a temperature T. (a) Show that the partition function is given by

$$Z = 2\cosh\frac{\mu\mathscr{H}_0}{2k_BT}.$$

(b) Derive expressions for U, E, S, F^*, and M for this system and sketch curves of these properties as a function of T for a fixed value of $\mathscr{H}_0$. (c) Use Eq. (11–87) (Problem 11–32) to find how the number of particles in each level varies with $\mathscr{H}_0$ and T.

11–37 The M-B statistics and the F-D statistics can be developed by calculating collision probabilities for elastic collisions between two particles. If two particles obeying M-B statistics initially have energies ϵ_1 and ϵ_2 and after the collision ϵ_3 and ϵ_4, then

$$\epsilon_3 + \epsilon_4 = (\epsilon_1 - \delta) + (\epsilon_2 + \delta).$$

The number of collisions per unit time F is proportional to the probability $f(\epsilon_i)$ that each initial state is occupied:

$$F_{1,2} = cf(\epsilon_1)f(\epsilon_2).$$

Similarly, $F_{3,4} = cf(\epsilon_3)f(\epsilon_4)$. In equilibrium, $F_{1,2} = F_{3,4}$. (a) Show that $f(\epsilon_i) = e^{-\epsilon_i/kT}$ solves this equation. (b) Use similar reasoning to derive the F-D statistics. Here, however, the initial states must be filled and the final states must be empty. Therefore the number of collisions per unit time is

$$F_{1,2} = cf(\epsilon_1)f(\epsilon_2)[1 - f(\epsilon_3)][1 - f(\epsilon_4)].$$

Show that the equation $F_{1,2} = F_{3,4}$ can be solved by

$$\frac{1 - f(\epsilon_i)}{f(\epsilon_i)} = ce^{\epsilon_i/kT}$$

which yields an equation of the form of Eq. (11–40).

11–38 Another way to derive the distribution functions is to define a *grand partition function* $\mathcal{Y}$

$$\mathcal{Y} = \sum_{n=0}^{H} \exp\left[\frac{n(\mu - \epsilon)}{kT}\right],$$

and calculate values of $\bar{n}$:

$$\bar{n} = \frac{1}{\mathcal{Y}} \sum_{n=0}^{H} n \exp \frac{n(\mu - \epsilon)}{kT}.$$

(a) Show that

$$\bar{n} = -\frac{d}{d\dfrac{(\epsilon - \mu)}{kT}} \ln \mathcal{Y}.$$

(b) Show that $H = 1$ gives the Fermi-Dirac distribution function. (c) Show that $H = \infty$ gives the Bose-Einstein distribution function.

12

Applications of statistics to gases

12–1 THE MONATOMIC IDEAL GAS

We next apply the general relations derived in the preceding chapter to the special case of a monatomic ideal gas consisting of N identical molecules each of mass m. The molecules are *indistinguishable*, and as we shall show later, the average number of molecules in each of the possible energy states, except at extremely low temperatures where all real gases have liquefied, is extremely small. The proper statistics is therefore the *classical* statistics (Section 11–11).

The first step is to calculate the partition function,

$$Z = \sum_j g_j \exp \frac{-\epsilon_j}{kT}.$$

This requires a knowledge of the energy ϵ_j and the degeneracy g_j of each level. We assume that the molecules do not interact except at the instant of a collision, so that each is essentially an independent particle and has the same set of energy levels as does a single particle in a box. It was shown earlier that the principles of quantum mechanics lead to the result that the energy levels of such a particle are given by Eq. (11–4).

$$\epsilon_j = \frac{n_j^2 h^2 V^{-2/3}}{8m}, \tag{12–1}$$

where $n_j^2 = n_x^2 + n_y^2 + n_z^2$, and n_x, n_y, n_z are integers each of which can equal 1, 2, 3, . . . , etc.

The degeneracy g_j of a level, or the number of energy states in the level, can readily be calculated when the quantum numbers are small, as in the example in Section 11–2. In many instances, however, the energy levels of an assembly are very closely spaced relative to the value of the energy itself. We can then subdivide the energy levels into groups of width $\Delta\epsilon_j$, including those levels with energies between ϵ_j and $\epsilon_j + \Delta\epsilon_j$. We refer to each of these groups as a *macrolevel*. Let $\mathscr{G}_j$ represent the total number of possible *states* in all energy levels up to and including the energy ϵ_j. The number of possible states $\Delta\mathscr{G}_j$ *within* the macrolevel is equal to the number of states in all levels included in the macrolevel. That is, $\Delta\mathscr{G}_j$ is the degeneracy of the macrolevel, but it arises in part from the grouping together of a large number of levels, while the numbers g_j are fixed by the nature of the assembly.

Imagine that the quantum numbers n_x, n_y, n_z are marked off on three mutually perpendicular axes, as suggested in the two-dimensional diagram of Fig. 12–1. Every triad of integral values of n_x, n_y, n_z determines a point in what can be called "n-space," and each such point corresponds to a possible state, provided the quantum numbers are positive. We can think of each point as located at the center of a cubical cell, each of whose sides is of unit length and whose volume is therefore unity.

The quantum number n_j corresponds to a vector in n-space from the origin to any point, since $n_j^2 = n_x^2 + n_y^2 + n_z^2$. In a system of given volume, the energy

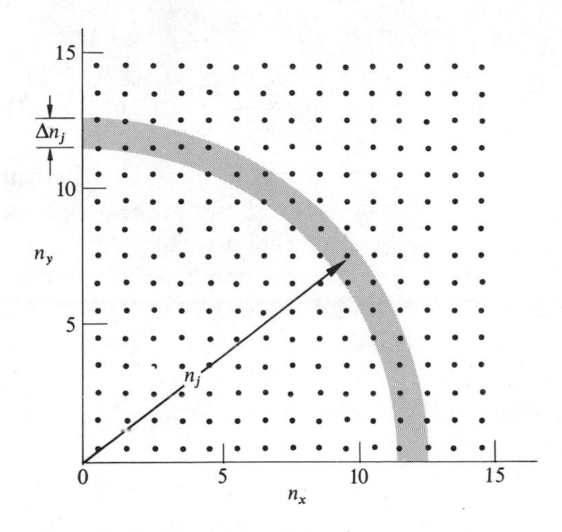

Fig. 12–1 Quantum states in n-space.

depends only on n_j, so that all states of equal energy lie on a spherical surface of radius n_j with center at the origin. Since n_x, n_y, and n_z are all positive, and since there is one point per unit volume of n-space, the total number $\mathscr{G}_j$ of possible states, in all levels up to and including the energy ϵ_j, is equal to the volume of one octant of a sphere of radius n_j. That is,

$$\mathscr{G}_j = \frac{1}{8} \times \frac{4}{3}\pi n_j^3 = \frac{\pi}{6} n_j^3. \tag{12–2}$$

The spherical surface will of course cut through some of the unit cells and it is not certain whether a point representing an energy state lies inside or outside the surface. However, when n_j is a large number, as is the case for the vast majority of molecules of a gas at ordinary temperatures, the uncertainty becomes negligibly small.

The number of states in the macrolevel between ϵ_j and $\epsilon_j + \Delta\epsilon_j$, or the degeneracy $\Delta\mathscr{G}_j$ of the macrolevel, is

$$\Delta\mathscr{G}_j = \frac{\pi}{6} \times 3n_j^2 \Delta n_j = \frac{\pi}{2} n_j^2 \Delta n_j. \tag{12–3}$$

Geometrically, this corresponds to the number of points in a thin spherical shell of radius n_j and thickness Δn_j. The degeneracy therefore increases with the *square* of the quantum number n_j, for equal values of Δn_j.

The partition function Z for this system is written

$$Z = \sum_j \Delta\mathscr{G}_j \exp\frac{-\epsilon_j}{kT},$$

and on inserting the expressions for $\Delta\mathscr{G}_j$ and ϵ_j, we have

$$Z = \frac{\pi}{2}\sum_j n_j^2 \exp\left(-\frac{h^2V^{-2/3}}{8mkT}n_j^2\right)\Delta n_j. \tag{12–4}$$

This sum can be interpreted graphically as follows. Let the values of n_j be marked off on a horizontal axis, and for brevity represent the coefficient of Δn_j in Eq. (12–4) by $f(n_j)$. At each value of n_j, we construct a vertical line of length $f(n_j)$, as in Fig. 12–2. Each product $f(n_j)\,\Delta n_j$ then corresponds to the area of a rectangle such as that shown shaded in Fig. 12–2, and the value of Z corresponds to the sum of all such areas over values of n_j from $j = 1$ to $j = \infty$, since there is no upper limit to the permissible values of n_j. To a sufficiently good approximation, this sum is equal to the area under a continuous curve through the tops of the vertical lines, between the limits of 0 and ∞, so

$$Z = \frac{\pi}{2}\int_0^\infty n_j^2 \exp\left(-\frac{h^2V^{-2/3}}{8mkT}n_j^2\right)dn_j. \tag{12–5}$$

The value of the definite integral can be found from Table 12–1, and finally,

$$Z = V\left(\frac{2\pi mkT}{h^2}\right)^{3/2}. \tag{12–6}$$

The partition function therefore depends both on the temperature T and the volume V, which corresponds to the general extensive variable X in Section 11–15.

The Helmholtz function F is given by Eq. (11–63) as

$$F = -NkT(\ln Z - \ln N + 1),$$

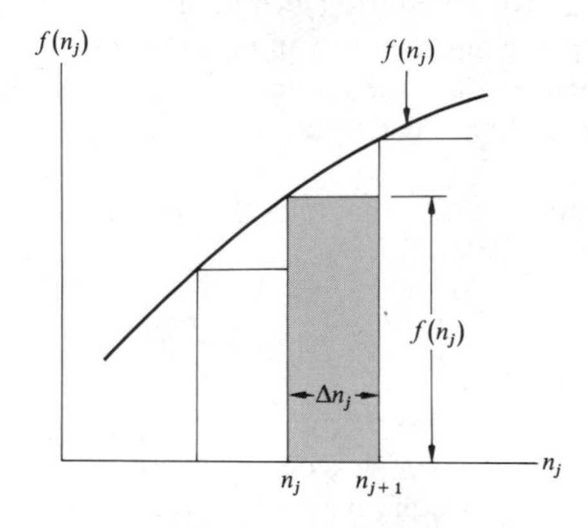

Fig. 12–2 The partition function Z is equal to the total area under the step function, and is very nearly equal to the area under the continuous curve.

Table 12–1 $f(n) = \int_0^\infty x^n e^{-ax^2}\, dx.$

n	$f(n)$	n	$f(n)$
0	$\frac{1}{2}\sqrt{\frac{\pi}{a}}$	1	$\frac{1}{2a}$
2	$\frac{1}{4}\sqrt{\frac{\pi}{a^3}}$	3	$\frac{1}{2a^2}$
4	$\frac{3}{8}\sqrt{\frac{\pi}{a^5}}$	5	$\frac{1}{a^3}$
6	$\frac{15}{16}\sqrt{\frac{\pi}{a^7}}$	7	$\frac{3}{a^4}$

If n is even, $\int_{-\infty}^{+\infty} x^n e^{-ax^2}\, dx = 2f(n)$.

If n is odd, $\int_{-\infty}^{+\infty} x^n e^{-ax^2}\, dx = 0$.

and the pressure P, which corresponds to the intensive variable Y, is

$$P = NkT\left(\frac{\partial \ln Z}{\partial V}\right)_T. \tag{12–7}$$

Since by Eq. (12–6),

$$\ln Z = \ln V + \frac{3}{2}\ln\left(\frac{2\pi mkT}{h^2}\right), \tag{12–8}$$

it follows that

$$\left(\frac{\partial \ln Z}{\partial V}\right)_T = \frac{1}{V}. \tag{12–9}$$

Consequently

$$P = \frac{NkT}{V} = \frac{nRT}{V}, \tag{12–10}$$

which is just the equation of state of an ideal gas as derived from kinetic theory.

The internal energy U is

$$U = NkT^2\left(\frac{\partial \ln Z}{\partial T}\right)_V = \frac{3}{2}NkT = \frac{3}{2}nRT, \tag{12–11}$$

which also agrees with the results of kinetic theory for a monatomic gas having three degrees of freedom.

The heat capacity at constant volume is

$$C_V = \left(\frac{\partial U}{\partial T}\right)_V = \frac{3}{2}Nk = \frac{3}{2}nR, \tag{12–12}$$

and the molal specific heat capacity is

$$c_v = \frac{C_V}{n} = \frac{3}{2}R. \tag{12–13}$$

The entropy is

$$S = \frac{U}{T} + Nk(\ln Z - \ln N + 1),$$

and after inserting the expressions for $\ln Z$ and U, we have

$$S = Nk\left[\frac{5}{2} + \ln\frac{V(2\pi mkT)^{3/2}}{Nh^3}\right]. \tag{12–14}$$

The principles of thermodynamics define only *differences* in entropy; the expression for the entropy itself contains an undetermined constant. There are no undetermined constants in Eq. (12–14) and the methods of statistics therefore lead to an expression for the entropy itself.

Using Eq. (12–13), the molal specific entropy can be written

$$s = c_v \ln T + R\ln V + R\left[\ln\frac{(2\pi mk)^{3/2}}{Nh^3} + \frac{5}{2}\right]. \tag{12–15}$$

This agrees with the thermodynamic expression for s in its dependence on V and T, and contains no undetermined constants. Equation (12–15) is known as the *Sackur*-Tetrode† equation* for the absolute entropy of a monatomic ideal gas.

12–2 THE DISTRIBUTION OF MOLECULAR VELOCITIES

In the chapters describing the kinetic theory of gases, a number of results were obtained which involved the average or root-mean-square speed of the molecules, but at that time we could say nothing as to how the molecular speeds were distributed around these average values. (We use the term "speed" to mean the *magnitude* of the velocity.) The methods of statistics, however, lead directly to the expression for the occupation numbers of the energy levels and hence to the speed distribution. An expression for the distribution was first worked out by Maxwell, before the development of statistical methods, and later by Boltzmann and is referred to as the *Maxwell-Boltzmann* distribution.

As in the previous section, we express the distribution in terms of the average occupation number of a macrolevel, including an energy interval between ϵ_j and

---

* Otto Sackur, German chemist (1880–1914).
† Hugo M. Tetrode, Dutch physicist (1895–1931).

$\epsilon_j + \Delta\epsilon_j$. Let $\mathcal{N}$ represent the total number of molecules with energies up to and including the energy ϵ_j. The average number of molecules included in the macrolevel, or the average occupation number of the macrolevel, is then $\Delta\mathcal{N}_j$. The quantities $\Delta\mathcal{N}_j$ and $\Delta\mathcal{G}_j$ then correspond to the occupation number $\bar{N}_j$ and degeneracy g_j of a single energy level and both the M-B and classical distribution functions can be written

$$\Delta\mathcal{N}_j = \frac{N}{Z}\Delta\mathcal{G}_j \exp\left(\frac{-\epsilon_j}{kT}\right). \tag{12–16}$$

Because we are interested in the distribution in *speed* rather than in *energy*, we express the degeneracy $\Delta\mathcal{G}_j$ in terms of the speed v_j instead of the quantum number n_j. We have from Eqs. (12–1) and (12–3),

$$\varepsilon_j = \frac{n_j^2 h^2 V^{-2/3}}{8m} = \frac{1}{2}mv_j^2.$$

$$\Delta\mathcal{G}_j = \frac{\pi}{2}n_j^2\,\Delta n_j.$$

It follows from these equations that

$$\Delta\mathcal{G}_v = \frac{4\pi m^3 V}{h^3}v^2\,\Delta v. \tag{12–17}$$

For simplicity, we have dropped the subscript j from v, and written $\Delta\mathcal{G}_v$ to indicate that the degeneracy is expressed in terms of v. Finally, taking the expression for Z from Eq. (12–6), we have

$$\Delta\mathcal{N}_v = \frac{4N}{\sqrt{\pi}}\left(\frac{m}{2kT}\right)^{3/2} v^2 \exp\left(-\frac{mv^2}{2kT}\right)\Delta v. \tag{12–18}$$

The quantity $\mathcal{N}_v$ represents the average total number of molecules with all speeds up to and including v, and $\Delta\mathcal{N}_v$ is the average number with speeds between v and $v + \Delta v$.

It is helpful to visualize the distribution in terms of "velocity space." Imagine that at some instant a vector v is attached to each molecule representing its velocity in magnitude and direction, and that these vectors are then transferred to a common origin, resulting in a sort of spiny sea urchin. The velocity of each molecule is represented by the point at the tip of the corresponding velocity vector. Figure 12–3 shows one octant of this velocity space. Geometrically speaking, the quantity $\mathcal{N}_v$ represents the average total number of representative points within a sphere of radius v, and $\Delta\mathcal{N}_v$ the number within a spherical shell of radius v and thickness Δv.

The coefficient of Δv in Eq. (12–18), equal to the ratio $\Delta\mathcal{N}_v/\Delta v$, depends only on the *magnitude* of v, or on the *speed*. It is called the *Maxwell-Boltzmann speed distribution function* and is plotted as a function of v on Fig. 12–4. The number of

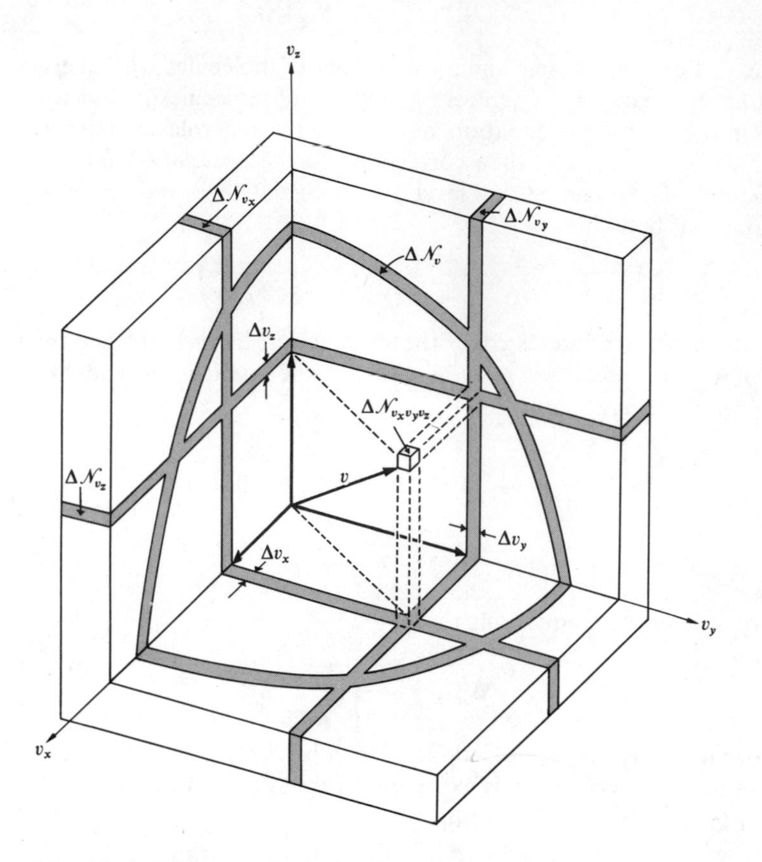

Fig. 12–3 Diagram of velocity space.

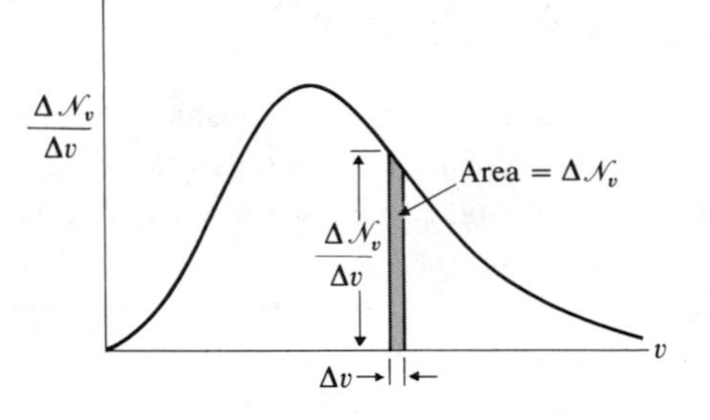

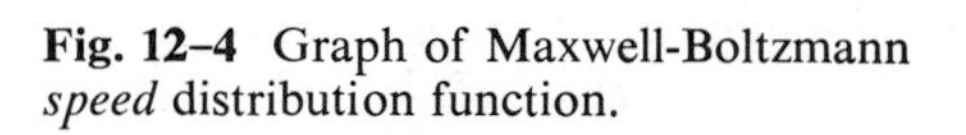

Fig. 12–4 Graph of Maxwell-Boltzmann *speed* distribution function.

velocity vectors $\Delta\mathcal{N}_v$ terminating between v and $v + \Delta v$ is represented in this graph by the *area* of a narrow vertical strip such as the shaded one shown, since the height of the strip is $\Delta\mathcal{N}_v/\Delta v$ and its width is Δv. (Note carefully that the ordinate of the speed distribution function does not represent $\Delta\mathcal{N}_v$.) The distribution function is zero when $v = 0$, since then $v^2 = 0$ and the exponential term equals 1. This means that no molecules (or very few molecules) are at rest. The function rises to a maximum and then decreases because the exponential term decreases more rapidly than v^2 increases.

If velocity space is subdivided into spherical shells of equal thickness, the speed v_m at which the distribution function is a maximum is the radius of that spherical shell which includes the largest number of representative points. The speed v_m is called the *most probable speed.* To find its value, we take the first derivative of the distribution function with respect to v and set it equal to zero. Neglecting the constant terms in Eq. (12–18), this procedure yields:

$$\frac{d}{dv}\left[v^2 \exp\left(\frac{-mv^2}{2kT}\right)\right] = 0.$$

It is left as a problem to show that

$$v_m = \sqrt{2kT/m}. \tag{12–19}$$

The distribution function can now be expressed more compactly in terms of v_m:

$$\frac{\Delta\mathcal{N}_v}{\Delta v} = \frac{4N}{\sqrt{\pi}v_m^3}\, v^2 \exp\left(\frac{-v^2}{v_m^2}\right). \tag{12–20}$$

The distribution function depends on the temperature of the gas through the quantity v_m, which appears both in the exponential function and its coefficient. Figure 12–5 is a graph of the distribution function at three different temperatures.

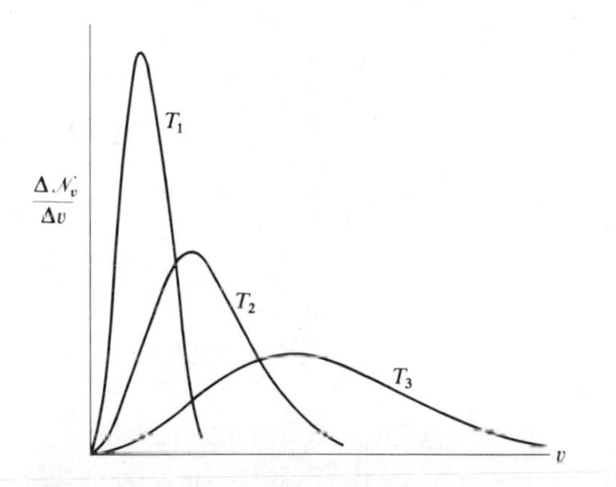

Fig. 12–5 Graph of M-B *speed* distribution function at three different temperatures, $T_3 > T_2 > T_1$.

The most probable speed decreases as the temperature decreases and the "spread" of the speeds becomes smaller. The areas under all three curves are equal, since the area corresponds to the total number of molecules.

As explained in Section 9–3, the average or arithmetic mean speed is

$$\bar{v} = \frac{1}{N}\sum v\,\Delta\mathcal{N}_v.$$

Using Eq. (12–20) and approximating the sum by an integral, we have

$$\bar{v} = \frac{4}{\sqrt{\pi}v_m^3}\int_0^\infty v^3 \exp\left(\frac{-v^2}{v_m^2}\right) dv.$$

The definite integral, from Table 12–1, is $v_m/2$, so

$$\bar{v} = \frac{2}{\sqrt{\pi}}v_m = \sqrt{\frac{8}{\pi}\frac{kT}{m}}. \tag{12–21}$$

The root-mean-square speed is

$$v_{\text{rms}} = \sqrt{\overline{v^2}} = \left(\frac{1}{N}\sum v^2\,\Delta\mathcal{N}_v\right)^{1/2} = \left[\frac{4}{\sqrt{\pi}v_m^3}\int_0^\infty v^4 \exp\left(\frac{-v^2}{v_m^2}\right) dv\right]^{1/2}.$$

The definite integral equals $\frac{3\sqrt{\pi}}{8}v^5$, so

$$v_{\text{rms}} = \frac{3}{2}v_m = \sqrt{3\frac{kT}{m}}, \tag{12–22}$$

which agrees with Eq. (9–19) obtained from kinetic theory. The method used here is far more general than that used to derive Eq. (9–19). The method is applicable to systems more complicated than an ideal gas by changing the dependence of ϵ_j and g_j on the velocity of the particles.

In summary, we have

$$v_m = \sqrt{2\frac{kT}{m}},$$

$$v = \sqrt{\frac{8}{\pi}\frac{kT}{m}} = \sqrt{2.55\frac{kT}{m}},$$

$$v_{\text{rms}} = \sqrt{3\frac{kT}{m}}.$$

The three speeds are shown in Fig. 12–6. The relative magnitudes of the three, at a given temperature, are

$$v_m : \bar{v} : v_{\text{rms}} = 1:1.128:1.224.$$

The quantity $\Delta\mathcal{N}_v$ represents the number of velocity vectors terminating in a spherical shell in velocity space, of "volume" $4\pi v^2\,\Delta v$, between v and $v + \Delta v$. The number of representative points per unit "volume" within the shell, or the

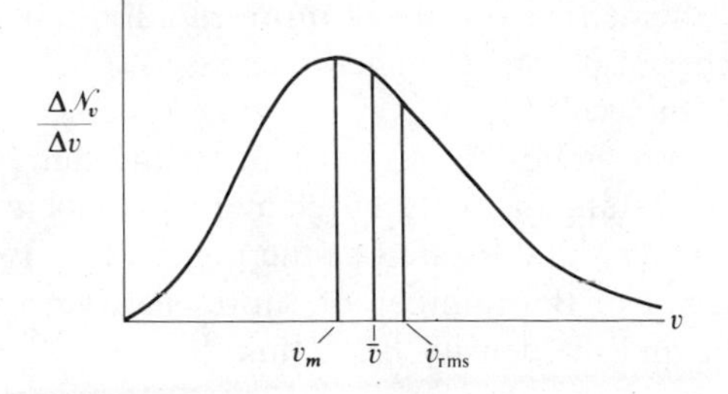

Fig. 12–6 Most probable (v_m), arithmetic mean ($\bar{v}$), and root-mean-square (v_{rms}) speeds.

"density" ρ_v in velocity space, is

$$\rho_v = \frac{\Delta \mathcal{N}_v}{4\pi v^2 \, \Delta v} = N\left(\frac{1}{\sqrt{\pi} v_m}\right)^3 \exp\left(\frac{-v^2}{v_m^2}\right). \tag{12–23}$$

The quantity ρ_v is called the Maxwell-Boltzmann *velocity distribution function.* It is a maximum at the origin, where $v = 0$, and decreases exponentially with v^2 as shown in Fig. 12–7.

Note that although the density is a maximum at the origin, the spherical shell containing the largest number of representative points is that of radius v_m. The reason for this apparent discrepancy is that as we proceed outward from the origin, the volumes of successive spherical shells of equal thickness Δv continually increase, while the number of representative points per unit volume continually decreases. The volume of the innermost shell (which is actually a small sphere of radius Δv) is essentially zero, so that although the density is a maximum for this shell, the number of points within it is practically zero because its volume is so small. In other words, practically none of the molecules is at rest. Beyond the

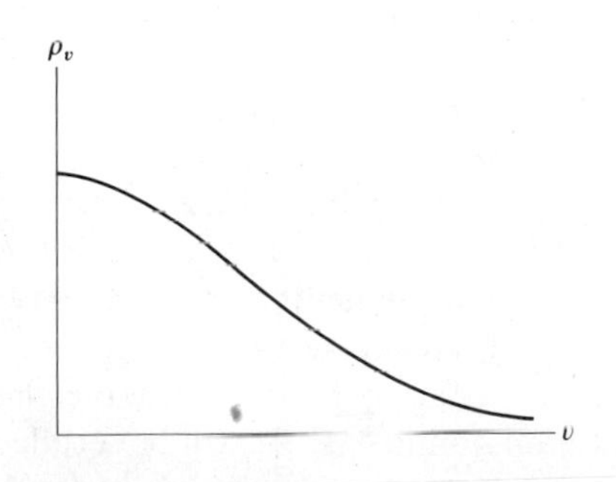

Fig. 12–7 Graph of Maxwell-Boltzmann *velocity* distribution function.

sphere of radius v_m, the density decreases more rapidly than the shell volume increases and the number of points in a shell decreases.

The number of molecules $\Delta\mathcal{N}_{v_xv_yv_z}$ having specified values of all three velocity *components* corresponds, in Fig. 12–3, to the number of representative points within a small rectangular volume element in velocity space having sides of length Δv_x, Δv_y, and Δv_z, and located at the point v_x, v_y, v_z. The volume of the element is $\Delta v_x\,\Delta v_y\,\Delta v_z$ and the number of representative points within it is the product of its volume and the density ρ_v. Thus

$$\begin{aligned}\Delta\mathcal{N}_{v_xv_yv_z} &= \rho_v\,\Delta v_x\,\Delta v_y\,\Delta v_z\\ &= N\left(\frac{1}{\sqrt{\pi}\,v_m}\right)^3\exp\left[\frac{-(v_x^2+v_y^2+v_z^2)}{v_m^2}\right]\Delta v_x\,\Delta v_y\,\Delta v_z,\end{aligned}$$

since $v^2 = v_x^2 + v_y^2 + v_z^2$.

The number of molecules having an x-, y-, or z-component of velocity in some specified interval, regardless of the values of the other components, is represented in Fig. 12–3 by the number of representative points in the thin slices perpendicular to the velocity axes. (The diagram shows only the intersections of these slices with planes perpendicular to the axes.) Thus to find the number of molecules $\Delta\mathcal{N}_{v_x}$ with velocity components between v_x and $v_x + \Delta v_x$, we sum $\Delta\mathcal{N}_{v_xv_yv_z}$ over all values of v_y and v_z. When the sum is replaced with an integral, we have

$$\Delta\mathcal{N}_{v_x} = N\left(\frac{1}{\sqrt{\pi}\,v_m}\right)^3\left[\int_{-\infty}^{\infty}\exp\left(\frac{-v_y^2}{v_m^2}\right)dv_y\int_{-\infty}^{\infty}\exp\left(\frac{-v_z^2}{v_m^2}\right)dv_y\right]\exp\left(\frac{-v_x^2}{v_m^2}\right)\Delta v_x.$$

Each of the integrals, from Table 12–1, equals $\sqrt{\pi}\,v_m$, and therefore

$$\frac{\Delta\mathcal{N}_{v_x}}{\Delta v_x} = N\frac{1}{\sqrt{\pi}\,v_m}\exp\left(\frac{-v_x^2}{v_m^2}\right), \tag{12–24}$$

with similar expressions for v_y and v_z. These are the Maxwell-Boltzmann distribution functions for *one component* of velocity, and that for the x-component is plotted in Fig. 12–8. The slice in Fig. 12–8 containing the largest number of representative points is therefore the one at $v_x = 0$, and the most probable velocity component along any axis is zero.

The distribution represented by Eq. (12–24) and Fig. 12–8 is known as a *gaussian* distribution* and is typical of many sorts of random distributions, not just that of molecular velocity components. This is to be expected, since the treatment that led to Eq. (12–24) is so very general.

We can now show that it is appropriate to use the classical distribution function to describe an ideal monatomic gas. It will be recalled that the Bose-Einstein and Fermi-Dirac distribution function both reduce to the classical distribution

---

* J. Carl F. Gauss, German mathematician (1777–1855).

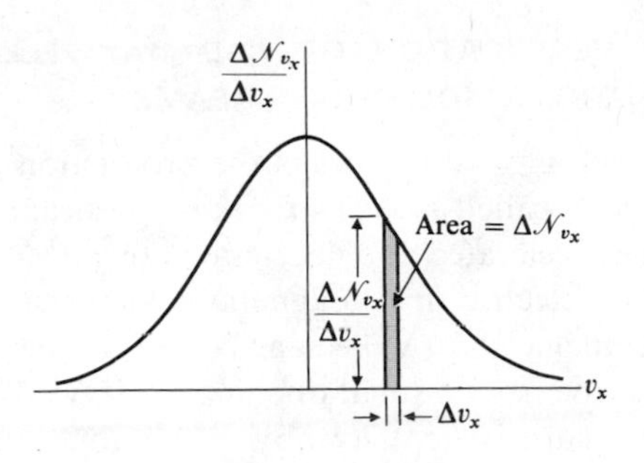

Fig. 12–8 Maxwell-Boltzmann velocity distribution function for a single component of velocity.

function, provided the occupation numbers $\Delta\mathcal{N}_j$ are much *smaller* than the number of states $\Delta\mathcal{G}_j$ in the macrolevel j. In other words, the classical distribution function is applicable provided $\Delta\mathcal{N}_j/\Delta\mathcal{G}_j \ll 1$. According to Eq. (12–16), the general expression for $\Delta\mathcal{N}_j/\Delta\mathcal{G}_j$ in this case is

$$\frac{\Delta\mathcal{N}_j}{\Delta\mathcal{G}_j} = \frac{N}{Z}\exp\left(\frac{-\epsilon_j}{kT}\right),$$

and for an ideal gas,

$$Z = V\left(\frac{2\pi mkT}{h^2}\right)^{3/2}.$$

Therefore

$$\frac{\Delta\mathcal{N}_j}{\Delta\mathcal{G}_j} = \frac{N}{V}\left(\frac{2\pi mkT}{h^2}\right)^{-3/2}\exp\left(\frac{-\epsilon_j}{kT}\right).$$

Let us take as an example helium gas at standard conditions. In a Maxwell-Boltzmann velocity distribution, the energies ϵ_j are grouped around the mean value $3kT/2$. Then ϵ_j/kT is of the order of unity and so is $\exp(-\epsilon_j/kT)$. The number of molecules per unit volume, N/V, is about 3×10^{25} molecules m^{-3} and for helium, $m = 6.7 \times 10^{-27}$ kg. Inserting the values of h, k, m, and T in the preceding equation, we get

$$\frac{\Delta\mathcal{N}_j}{\Delta\mathcal{G}_j} \simeq 4 \times 10^{-6},$$

which is certainly much less than unity. (Only about four states in a million are occupied!) However, as the temperature is lowered, the value of $\Delta\mathcal{N}_j/\Delta\mathcal{G}_j$ increases, and provided the gas can be cooled to very low temperatures without condensing, the classical statistics may cease to be applicable. Conversely, condensation may well be adjusted just when the classical statistics cease to be applicable, and this reflects the essentially quantum-mechanical nature of liquid helium.

12–3 EXPERIMENTAL VERIFICATION OF THE MAXWELL-BOLTZMANN SPEED DISTRIBUTION. MOLECULAR BEAMS

An important technique in atomic physics is the production of a collimated beam of neutral particles in a so-called *molecular beam*. A beam of *charged* particles, electrons or ions, can be accelerated and decelerated by an electric field, and guided and focused by either an electric or a magnetic field. These methods cannot be used if the particles are uncharged. Molecular beams can be produced by allowing molecules of a gas to escape from a small opening in the walls of a container into a region in which the pressure is kept low by continuous pumping. A series of baffles, as in Fig. 12–9, limits the beam to a small cross section. Since one often wishes to work with molecules of a material such as silver, which is a solid at room temperature, the temperature in the container must be great enough to produce a sufficiently high vapor pressure. Hence the container is often a small electric furnace or oven.

We have shown in Section 9–3 that the number of molecules with speed v, striking the surface of a container per unit area and per unit time, is

$$\frac{1}{4} v \, \Delta \mathbf{n}_v \tag{12–25}$$

where $\Delta \mathbf{n}_v$ is the number of molecules per unit volume with speed v.

If the molecules have a Maxwell-Boltzmann speed distribution, the number per unit volume with a speed v is given by Eq. (12–18)

$$\Delta \mathbf{n}_v = \frac{4\mathbf{n}}{\sqrt{\pi}} \left(\frac{m}{2kT} \right)^{3/2} v^2 \exp \left(\frac{-mv^2}{2kT} \right) \Delta v.$$

If there is a hole in a wall of the oven, small enough so that leakage through the hole does not appreciably affect the equilibrium state of the gas in the oven, Eq. (12–25) gives the number with speed v escaping through the hole, per unit area and per unit time. We wish to compute the rms speed of those that escape. Following the standard method, the mean-square speed of the escaping molecules is found by multiplying by v^2 the number that escapes with speed v, integrating over all values of v, and dividing by the total number. The rms speed is the square

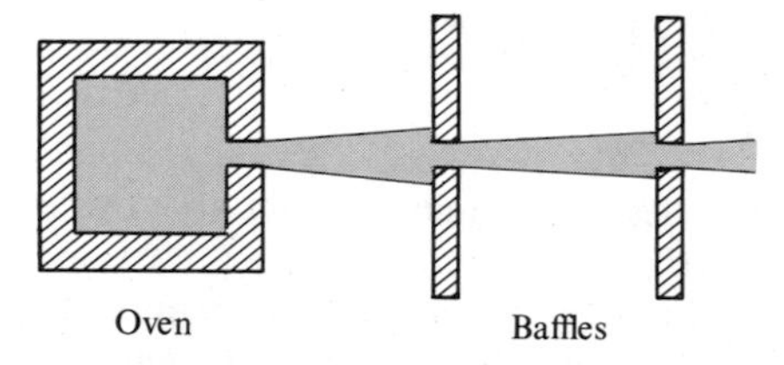

Fig. 12–9 Production of a beam of neutral particles.

root of the result. It is left as a problem to show that

$$v_{\text{rms}} = \sqrt{\frac{4kT}{m}}. \tag{12–26}$$

The rms speed of the molecules *in* the oven is

$$v_{\text{rms}} = \sqrt{\frac{3kT}{m}},$$

so that those escaping have a somewhat higher speed than those in the oven.

The distribution in *direction* of the molecules escaping through the hole is given by Eq. (9–14):

$$\frac{\Delta\Phi_\omega}{\Delta\omega} = \frac{1}{4\pi}\bar{v}\mathbf{n}\cos\theta.$$

That is, the number per unit solid angle in the emerging beam is a maximum in the direction of the normal to the plane of the opening and decreases to zero in the tangential direction.

Direct measurements of the distribution of velocities in a molecular beam have been made by a number of methods. Figure 12–10 is a diagram of the apparatus used by Zartman and Ko in 1930–1934, a modification of a technique developed by Stern in 1920. In Fig. 12–10, O is an oven and S_1 and S_2 are slits defining a molecular beam; C is a cylinder that can be rotated at approximately 6000 rpm about the axis A. If the cylinder is at rest, the molecular beam enters the cylinder through a slit S_3 and strikes a curved glass plate G. The molecules

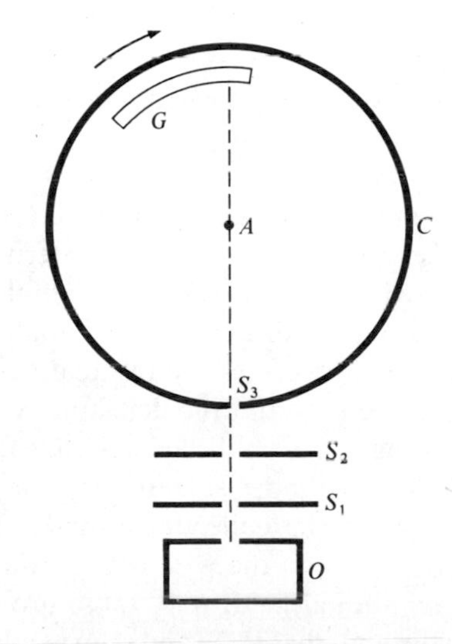

Fig. 12–10 Apparatus used by Zartman and Ko in studying distribution of velocities.

stick to the glass plate, and the number arriving at any portion can be determined by removing the plate and measuring with a recording microphotometer the darkening that has resulted.

Now suppose the cylinder is rotated. Molecules can enter it only during the short time intervals during which the slit S crosses the molecular beam. If the rotation is clockwise, as indicated, the glass plate moves toward the right while the molecules cross the diameter of the cylinder. They therefore strike the plate at the left of the point of impact when the cylinder is at rest, and the more slowly they travel, the farther to the left is this point of impact. The blackening of the plate is therefore a measure of the "velocity spectrum" of the molecular beam.

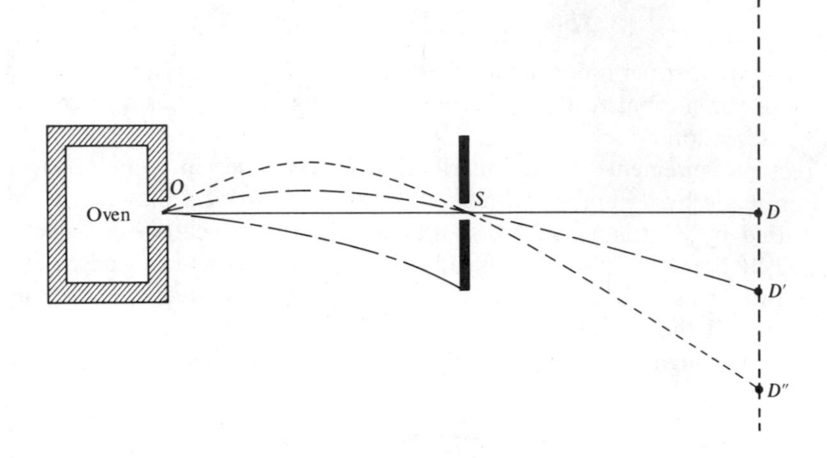

Fig. 12–11 Schematic diagram of apparatus of Estermann, Simpson, and Stern.

A more precise experiment, making use of the free fall of the molecules in a beam, was performed by Estermann, Simpson, and Stern in 1947. A simplified diagram of the apparatus is given in Fig. 12–11. A molecular beam of cesium emerges from the oven slot O, passes through the collimating slit S, and impinges on a hot tungsten wire D. The pressure of the residual gas in the apparatus is of the order of 10^{-8} Torr. Both the slits and the detecting wire are horizontal. The cesium atoms striking the tungsten wire become ionized, reevaporate, and are collected by a negatively charged cylinder surrounding the wire but not shown in the diagram. The ion current to the collecting cylinder then gives directly the number of cesium atoms impinging on the wire per second.

In the absence of a gravitational field, only those atoms emerging in a horizontal direction would pass through the slit S, and they would all strike the collector in the position D regardless of their velocities. Actually, the path of each atom is a parabola, and an atom emerging from the slit O in a horizontal direction, as indicated by the dot and dash line, (with the vertical scale greatly exaggerated) would not pass through the slit S. The dashed line and the dotted line represent the trajectories of two atoms that can pass through the slit S, the velocity along the dashed trajectory being greater than that along the dotted one. Hence as the detector is moved down from the position D, those atoms with velocities corresponding to the dashed trajectories will be collected at D', those with the slower velocity corresponding to the dotted trajectory will be collected at D'', etc. Measurement of the ion current as a function of the vertical height of the collector then gives the velocity distribution.

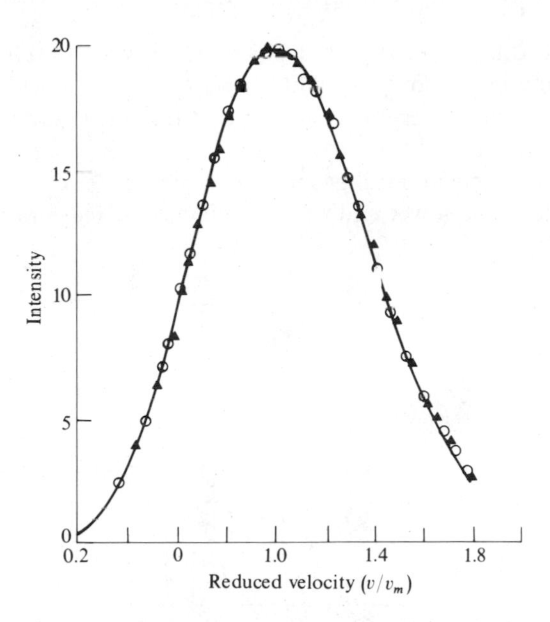

Fig. 12–12 Experimental verification of the Maxwell-Boltzmann speed distribution function. This is Fig. 7 from R. C. Miller and P. Kusch, "Velocity Distribution in Potassium and Thallium Atomic Beams," *Physical Review* 99 (1955): 1314. Reprinted by permission.

In 1955 Miller and Kusch reported a still more precise measurement of the distribution of velocities in a beam of thallium atoms. Their data are shown in Fig. 12–12. The oven, which was controlled to 0.25°C, was made from copper to insure a uniform temperature distribution. The thallium atoms passed through a slit whose dimension parallel to the beam was 0.003 cm to avoid scattering in the neighborhood of the slit. The detector was similar to the previous experiment. As the atoms came out of the slit they had to pass through one of 702 helical slits milled along the surface of cylinder 20 cm in diameter, 25.4 cm in length. Each slit was 0.04 cm wide and 0.318 cm deep. As the cylinder was rotated, only those atoms having an appropriate velocity would pass through the slit without being scattered. With these precautions Miller and Kusch were able to show that the velocity distribution of the thallium atoms agreed with the Maxwell-Boltzmann velocity distribution to within 1% for $0.2 < x < 1.8$, where $x = v/v_m$. This agreement is seen on Fig. 12–12 where the points are the data for two different experiments and the solid line is the theoretical curve computed from the Maxwell-Boltzmann speed distribution.

12–4 IDEAL GAS IN A GRAVITATIONAL FIELD

In the preceding sections, the energy of a gas molecule was considered to be wholly kinetic; that is, any gravitational potential energy of the molecule was ignored. We now take this potential energy into account, so that the gas serves as an example of a multivariable system.

Let us take as a system an ideal gas in a vertical cylinder of cross-sectional area A, as in Fig. 12–13. The lower end of the cylinder is fixed and the upper end is

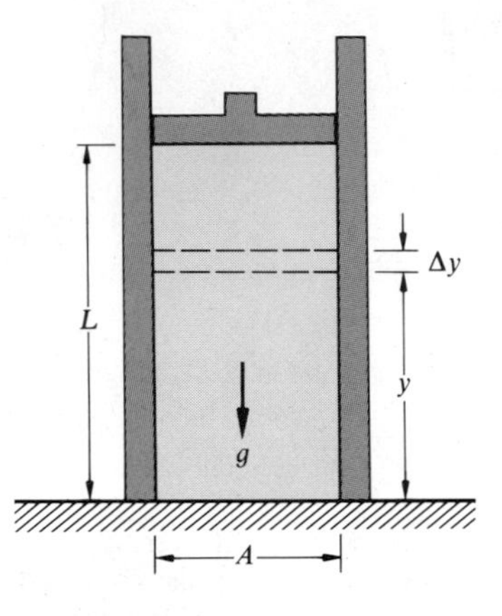

Fig. 12–13 An ideal gas in a cylinder in a gravitational field.

provided with a movable piston. If the piston is at a height L above the bottom of the cylinder, the volume V occupied by the gas is $V = AL$. The origin of space coordinates is at the bottom of the cylinder, with the y-axis vertically upward. The system is in a uniform gravitational field of intensity g, directed vertically downward; but the value of g can be changed by, say, moving the system to another location where g has a different value. The temperature T is assumed to be uniform.

The gas is therefore a multivariable system, described by three independent variables T, L, and g, and it has a gravitational potential energy E_p as well as an internal energy U. The appropriate energy function is therefore the total energy

E, given by

$$E = U + E_p,$$

and from Eq. (7–31),

$$T\,dS = dE + Y_1\,dX_1 - X_2\,dY_2.$$

The extensive variable X_1 is the length L, and the intensive variable Y_2 is the gravitational field intensity g. Let us represent the variable Y_1 by Π, and the variable X_2 by Γ. Then

$$T\,dS = dE + \Pi\,dL - \Gamma\,dg. \tag{12–27}$$

We now use the methods of statistics to find the quantities Π and Γ. The first step is to determine the partition function Z.

A molecule whose vertical coordinate is y has a gravitational potential energy mgy in addition to its kinetic energy $mv^2/2$, and its total energy ϵ is

$$\epsilon = mv^2/2 + mgy.$$

An energy interval between ϵ and $\epsilon + \Delta\epsilon$ includes a kinetic energy interval corresponding to speeds between v and $v + \Delta v$, and a potential energy interval corresponding to elevations between y and $y + \Delta y$. The degeneracy $\Delta\mathscr{G}_v$ of the speed interval, since $V = AL$, is given by Eq. (12–17),

$$\Delta\mathscr{G}_v = \frac{4\pi m^3 AL}{h^3} v^2\,\Delta v. \tag{12–28}$$

The potential energy is not quantized; a molecule may have any arbitrary elevation y and any potential energy mgy. The distribution in potential energy is given by the same expression as that for quantized levels, however, if we set the degeneracy $\Delta\mathscr{G}_y$ of the potential energy interval equal to $\Delta y/L$:

$$\Delta\mathscr{G}_y = \frac{\Delta y}{L}. \tag{12–29}$$

For any one of the possible kinetic energy states, a molecule can have any one of the possible potential energy states. The total number of possible states $\Delta\mathscr{G}$ in the energy interval is therefore the product of $\Delta\mathscr{G}_v$ and $\Delta\mathscr{G}_y$:

$$\Delta\mathscr{G} = \Delta\mathscr{G}_v\,\Delta\mathscr{G}_y.$$

The partition function Z is

$$\begin{aligned} Z &= \sum \Delta\mathscr{G} \exp\left(\frac{-\epsilon}{kT}\right) \\ &= \left[\sum \Delta\mathscr{G}_v \exp\left(\frac{-mv^2}{2kT}\right)\right]\left[\sum \Delta\mathscr{G}_y \exp\left(\frac{-mgy}{kT}\right)\right]. \end{aligned} \tag{12–30}$$

If we designate the sums by Z_v and Z_y, respectively, then

$$Z = Z_v Z_y, \qquad \ln Z = \ln Z_v + \ln Z_y.$$

The first sum in Eq. (12–30) is to be evaluated over all values of v from 0 to ∞, and the second over all values of y from 0 to L. When the expressions for $\Delta\mathscr{G}_v$ and $\Delta\mathscr{G}_y$ are inserted, and the sums replaced with integrals, we find

$$Z_v = AL\left(\frac{2\pi mkT}{h^2}\right)^{3/2}, \tag{12–31}$$

$$Z_y = \frac{kT}{mgL}\left[1 - \exp\left(\frac{-mgL}{kT}\right)\right]. \tag{12–32}$$

Therefore

$$\ln Z = \frac{5}{2}\ln T - \ln g + \ln\left[1 - \exp\left(\frac{-mgL}{kT}\right)\right] + \text{constant}. \tag{12–33}$$

The function F^* is given by Eq. (11–75),

$$F^* = -NkT(\ln Z - \ln N + 1),$$

and F^* is a function of N, T, g, and L.

If N is constant,

$$\Pi = -\left(\frac{\partial F^*}{\partial L}\right)_{T,g} = NkT\left(\frac{\partial \ln Z}{\partial L}\right)_{T,g},$$

and

$$\Gamma = \left(\frac{\partial F^*}{\partial g}\right)_{T,L} = -NkT\left(\frac{\partial \ln Z}{\partial g}\right)_{T,L}.$$

On carrying out the differentiations, we find

$$\Pi = \frac{Nmg}{\exp(mgL/kT) - 1}, \tag{12–34}$$

$$\Gamma = \frac{NkT}{g} - \frac{NmL}{\exp(mgL/kT) - 1}. \tag{12–35}$$

Thus the system has two equations of state, one expressing Π as a function of T, L, and g, and the other expressing Γ as a function of these variables.

The physical significance of Γ can be seen as follows. The gravitational potential energy E_p is

$$E_p = Y_2X_2 = g\Gamma,$$

and hence

$$\Gamma = \frac{E_p}{g}.$$

Thus Γ is the potential energy, per unit field intensity. The potential energy is therefore

$$E_p = g\Gamma = NkT - \frac{NmgL}{\exp(mgL/kT) - 1}. \tag{12–36}$$

The total energy E is

$$E = NkT^2\left(\frac{\partial \ln Z}{\partial T}\right)_{L,g} = \frac{5}{2}NkT - \frac{NmgL}{\exp(mgL/kT) - 1}, \tag{12–37}$$

and since $U = E - E_p$, it follows that

$$U = \frac{3}{2}NkT.$$

Hence the internal energy is the same as in the absence of a gravitational field and depends only on the temperature.

The entropy can be calculated from

$$S = \frac{E}{T} + Nk(\ln Z - \ln N + 1).$$

We next calculate the pressure P as a function of elevation. The number of molecules $\Delta\mathcal{N}_y$ in a macrolevel between y and $y + \Delta y$ is, from Eq. (12–16),

$$\Delta\mathcal{N}_y = \frac{N}{Z_y}\Delta\mathcal{G}_y \exp\left(\frac{-mgy}{kT}\right). \tag{12–38}$$

The volume of a thin cross section is $A\,\Delta y$, so the number of molecules per unit volume at a height y is

$$\mathbf{n}_y = \frac{\Delta\mathcal{N}_y}{A\,\Delta y}.$$

From the ideal gas law, the pressure P_y at this height is

$$P_y = \mathbf{n}_y kT.$$

It follows from the preceding three equations, after inserting the expressions for $\Delta\mathcal{G}_y$ and Z_y, that

$$P_y = \frac{Nmg}{A}\,\frac{\exp(-mgy/kT)}{1 - \exp(-mgL/kT)}.$$

At the bottom of the container, $y = 0$, and the pressure P_0 is

$$P_0 = \frac{Nmg}{A}\,\frac{1}{1 - \exp(-mgL/kT)}.$$

The pressure P_y can therefore be written more compactly as

$$P_y = P_0 \exp\left(\frac{-mgy}{kT}\right), \tag{12–39}$$

and the pressure decreases exponentially with elevation. Equation (12–39) is known as the *barometric equation* or the *law of atmospheres*. It can also be derived from the principles of hydrostatics and the equation of state of an ideal gas.

At the top of the container, $y = L$ and

$$P_L = \frac{Nmg}{A} \frac{1}{\exp(mgL/kT) - 1} = \frac{\Pi}{A}.$$

Therefore

$$\Pi = P_L A, \tag{12–40}$$

and the quantity Π is the force exerted against the piston at the top of the container. The work when the piston is displaced upward by an amount dL is

$$dW = \Pi \, dL = P_L A \, dL = P_L \, dV,$$

and the product $\Pi \, dL$ is the work when the gas expands.

In 1909 Perrin* used Eq. (12–39) in one of the earliest precision determinations of Avogadro's number N_A. Instead of gas molecules, he utilized particles of microscopic size suspended in a liquid of slightly smaller density, thus reducing the effective value of "g". The number of particles at different levels was counted with a microscope.

If $\Delta\mathcal{N}_1$ and $\Delta\mathcal{N}_2$ are the average numbers at heights y_1 and y_2, then

$$\frac{\Delta\mathcal{N}_1}{\Delta\mathcal{N}_2} = \exp\left[-\frac{mg(y_1 - y_2)}{kT}\right]. \tag{12–41}$$

All of the quantities in this equation can be measured experimentally with the exception of the Boltzmann constant k, so that the equation can be solved for k. Then N_A can be found, since k equals the universal gas constant R divided by N_A, and R is known from other experiments. Perrin concluded that the value of N_A lay between 6.5×10^{26} and 7.2×10^{26}, compared with the present best experimental value of 6.022×10^{26} molecules kilomole^{-1}.

12–5 THE PRINCIPLE OF EQUIPARTITION OF ENERGY

It will be recalled that the principle of equipartition of energy was introduced in Section 9–6 merely as an inference that might be drawn from some of the results of the kinetic theory of an ideal gas. We now show how this principle follows from the M-B or classical distribution function and what its limitations are.

The energy of a particle is in general a function of a number of different parameters. These might be the velocity components, the vertical elevation of the particles in a gravitational field, the angle that a molecular dipole makes with an electric field, and so on. Each of these parameters is called a *degree of freedom*. Let z represent any such parameter and $\epsilon(z)$ the energy associated with that parameter. If the energy can be expressed as a continuous function of the parameter, as in the preceding sections, the M-B and classical distribution function lead

* Jean Perrin, French physicist (1870–1942).

to the result that the average number of particles within a range Δz of the parameter is given by an expression of the form

$$\Delta \mathcal{N}_z = A \exp\left[\frac{-\epsilon(z)}{kT}\right] \Delta z,$$

where A is a constant independent of z. As examples, see Eq. (12–24) for the case in which z represents one of the rectangular components of the velocity, or Eq. (12–38) in which z represents the vertical coordinate y.

When the sum is replaced with an integral, the total number of particles, N, is given by

$$N = A\int \exp\left[\frac{-\epsilon(z)}{kT}\right] dz,$$

the limits of integration being over all values of z.

The total energy $E(z)$ associated with the parameter z is

$$E(z) = \int \epsilon(z)\, d\mathcal{N}_z = A\int \epsilon(z) \exp\left[-\frac{\epsilon(z)}{kT}\right] dz.$$

The mean energy $\bar{\epsilon}(z)$ of a single particle is

$$\bar{\epsilon}(z) = \frac{E(z)}{N}.$$

Now if the energy $\epsilon(z)$ is a *quadratic function* of z, that is, if it has the form $\epsilon(z) = az^2$, where a is a constant, and if the limits of z are from 0 to ∞, or from $-\infty$ to $+\infty$, then from Table 12–1,

$$\bar{\epsilon}(z) = \frac{\int az^2 \exp(-az^2/kT)\, dz}{\int \exp(-az^2/kT)\, dz} = \frac{1}{2} kT. \qquad (12\text{–}42)$$

That is, for every degree of freedom for which the conditions above are fulfilled, the mean energy per particle, in an assembly in equilibrium at the temperature T, is $kT/2$. This is the general statement of the equipartition principle.

The conditions above *are* fulfilled for the translational velocity components v_x, v_y, and v_z, since the energy associated with each is $mv_x^2/2$, $mv_y^2/2$, or $mv_z^2/2$ and the range of each is from $-\infty$ to $+\infty$. They are also fulfilled for the displacement x of a harmonic oscillator, since the potential energy associated with x is $Kx^2/2$, K being the force constant.

The conditions are *not* fulfilled for the vertical coordinate y of a gas in a gravitational field, where the potential energy is mgy; the mean *gravitational* potential energy is *not* $kT/2$. Neither are they fulfilled for the energy associated with molecular rotation, vibration, and electronic excitation, because of the quantized character of these energies, which can take on only certain discrete values and cannot be expressed as a *continuous* function of some coordinate. The energy associated with them is not a simple linear function of the temperature.

12–6 THE QUANTIZED LINEAR OSCILLATOR

We consider next an assembly of N identical linear oscillators, assumed distinguishable so that we can use Maxwell-Boltzmann statistics. The properties of such an assembly form the basis of the theory of the specific heat capacity of polyatomic gases and of solids.

A linear oscillator is a particle constrained to move along a straight line and acted on by a restoring force $F = -Kx$, proportional to its displacement x from some fixed point and oppositely directed. The equation of motion of the particle is

$$F = m\frac{d^2x}{dt^2} = -Kx,$$

where m is the mass of the particle. If displaced from its equilibrium position and released, the particle oscillates with simple harmonic motion of frequency ν, given by

$$\nu = \frac{1}{2\pi}\sqrt{K/m}.$$

The frequency depends only on K and m, and is independent of the amplitude x_m.

The energy ϵ of the oscillator is the sum of its kinetic energy $mv^2/2$ and its potential energy $Kx^2/2$. Since the total energy is constant, and the kinetic energy is zero when the displacement has its maximum value x_m, the potential energy at this displacement is equal to the total energy ϵ and hence

$$\epsilon = \frac{1}{2}Kx_m^2.$$

Thus the total energy is proportional to the *square* of the amplitude, x_m.

If the oscillators were completely independent, there could be no interchange of energy between them, and any given microstate of the assembly would continue indefinitely. We therefore assume that the interactions between the particles are large enough so that there can be sufficient exchanges of energy for the assembly to assume all possible microstates consistent with a given total energy, but small enough so that each particle can oscillate nearly independently of the others.

In classical mechanics, a particle can oscillate with any amplitude and energy. The principles of quantum mechanics, however, restrict the energy to some one of the set of values

$$\epsilon_j = \left(n_j + \frac{1}{2}\right)h\nu, \tag{12–43}$$

where $n_j = 0, 1, 2, \ldots$, and h is Planck's constant. An unexpected result is that the oscillator can never be in a state of zero energy, but that in the lowest level the energy is $h\nu/2$, in the next level it is $3h\nu/2$, and so on. The levels are nondegenerate; there is only one energy state in each level; and $g_j = 1$ in each level.

The quantum condition that the *energy* can have only some one of the set of values $[(n_j + 1/2)h\nu]$ is equivalent to the condition that the *amplitude* can have only some one of the set of values such that

$$x_m^2 = \left(n_j + \frac{1}{2}\right)\frac{h}{\pi}\sqrt{1/Km}.$$

Using Eq. (12–43), the partition function of the assembly is

$$Z = \sum_j \exp\left(\frac{-\epsilon_j}{kT}\right) = \sum_j \exp\left[-\left(n_j + \frac{1}{2}\right)\frac{h\nu}{kT}\right].$$

To evaluate the sum, let $z = h\nu/kT$ for brevity. Writing out the first few terms, we have

$$Z = \exp\left(-\frac{z}{2}\right) + \exp\left(-\frac{3z}{2}\right) + \exp\left(-\frac{5z}{2}\right) + \cdots$$

$$= \exp\left(-\frac{z}{2}\right)\{1 + \exp(-z) + [\exp(-z)]^2 + \cdots\}.$$

The sum in the preceding equation has the form of the infinite geometric series

$$1 + p + p^2 + \cdots,$$

which equals $1/(1 - p)$ as is readily verified by expanding the product $(1 - p) \times (1 + p + p^2 + \cdots)$. Therefore

$$Z = \exp\left(-\frac{z}{2}\right)\frac{1}{1 - \exp(-z)},$$

or

$$Z = \frac{\exp(-h\nu/2kT)}{1 - \exp(-h\nu/kT)}. \tag{12–44}$$

The temperature at which $kT = h\nu$ is called the characteristic temperature of the assembly and is represented by θ. Thus

$$k\theta = h\nu, \qquad \text{or} \qquad \theta = \frac{h\nu}{k}. \tag{12–45}$$

It follows that

$$\frac{h\nu}{kT} = \frac{\theta}{T},$$

and in terms of θ the partition function is

$$Z = \frac{\exp(-\theta/2T)}{1 - \exp(-\theta/T)}. \tag{12–46}$$

The value of the partition function at any temperature therefore depends, for a given assembly, on the ratio of the actual temperature T to the characteristic temperature θ, which thus provides a reference temperature for the assembly. The greater the natural frequency ν of the oscillators, the higher the characteristic temperature. Thus if the natural frequency is of the order of frequencies in the infrared region of the electromagnetic spectrum, say 10^{13} Hz,* then

$$\theta = \frac{h\nu}{k} = \frac{6.62 \times 10^{-34}\ \text{J s} \times 10^{13}\ \text{s}^{-1}}{1.38 \times 10^{-23}\ \text{J K}^{-1}} \simeq 500\ \text{K}.$$

An actual temperature T of 50 K is then approximately equal to $\theta/10$, and a temperature of 5000 K is approximately equal to 10 θ.

The average fractional number of oscillators in the jth energy level, from Eqs. (12–16) and (12–43) is

$$\frac{\bar{N}_j}{N} = \frac{1}{Z} \exp\left(-\frac{\epsilon_j}{kT}\right) = \frac{1}{Z} \exp\left(-\frac{\left(n_j + \frac{1}{2}\right)h\nu}{kT}\right).$$

Substituting Eq. (12–46) for Z and Eq. (12–45) for θ,

$$\frac{\bar{N}_j}{N} = \left[1 - \exp\left(\frac{-\theta}{T}\right)\right] \exp\left(-n_j \frac{\theta}{T}\right). \tag{12–47}$$

At any temperature T, the occupation number decreases exponentially with the quantum number n_j, and decreases more rapidly, the lower the temperature.

At the temperature at which $T = \theta$,

$$(\theta/T) = 1, \qquad \left[1 - \exp\left(\frac{-\theta}{T}\right)\right] = 0.632,$$

and

$$\frac{\bar{N}_j}{N} = 0.632 \exp(-n_j).$$

Thus for the four lowest energy levels, in which $n_j = 0, 1, 2$, and 3, we have

$$\frac{\bar{N}_0}{N} = 0.632, \qquad \frac{\bar{N}_1}{N} = 0.232, \quad \frac{\bar{N}_2}{N} = 0.085, \quad \frac{\bar{N}_3}{N} = 0.032.$$

About 63% of the oscillators are in the lowest energy level, about 23% in the next level, etc. Together, the four lowest levels account for about 98% of the oscillators.

It is left to the reader to show that when $T = \theta/2$,

$$\frac{\bar{N}_0}{N} = 0.865, \quad \frac{\bar{N}_1}{N} = 0.117, \quad \frac{\bar{N}_2}{N} = 0.016, \quad \frac{\bar{N}_3}{N} = 0.002.$$

---

* Heinrich R. Hertz, German physicist (1857–1894).

At this temperature, about 87% of the oscillators are in the lowest level, about 12% in the next level, etc. and almost all the particles are in the first four levels.

At a temperature $T = 2\theta$,

$$\frac{\overline{N}_0}{N} = 0.394, \quad \frac{\overline{N}_1}{N} = 0.239, \quad \frac{\overline{N}_2}{N} = 0.145, \quad \frac{\overline{N}_3}{N} = 0.088.$$

The first four levels then account for only about 86% of the oscillators, the remainder being distributed among the higher energy levels.

The lengths of the vertical lines in Fig. 12–14 represent the average fractional occupation numbers at the temperatures $T = \theta/2$, $T = \theta$, and $T = 2\theta$.

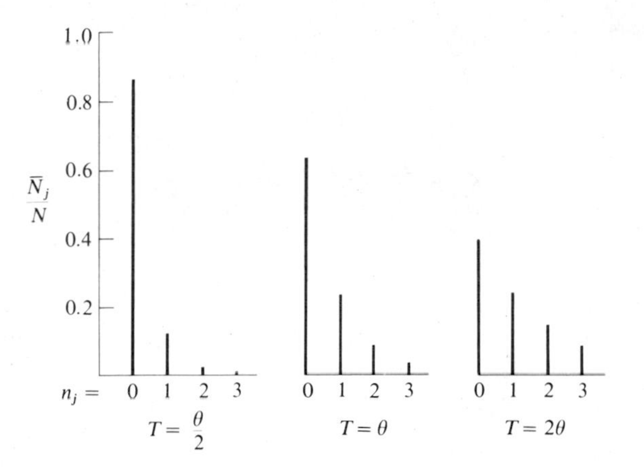

Fig. 12–14 The dependence on θ/T of the average fractional occupation number of the first four levels of a linear oscillator.

The total energy of the assembly, which in this case is its internal energy U, is

$$U = NkT^2 \frac{d \ln Z}{dT}$$

$$= Nk\theta\left[\frac{1}{\exp(\theta/T) - 1} + \frac{1}{2}\right]. \tag{12–48}$$

Thus for a given assembly of linear oscillators the internal energy is a function of temperature only. The heat capacity C_V of the assembly is

$$C_V = \frac{dU}{dT}$$

$$= Nk\left(\frac{\theta}{T}\right)^2 \frac{\exp(\theta/T)}{[\exp(\theta/T) - 1]^2}. \tag{12–49}$$

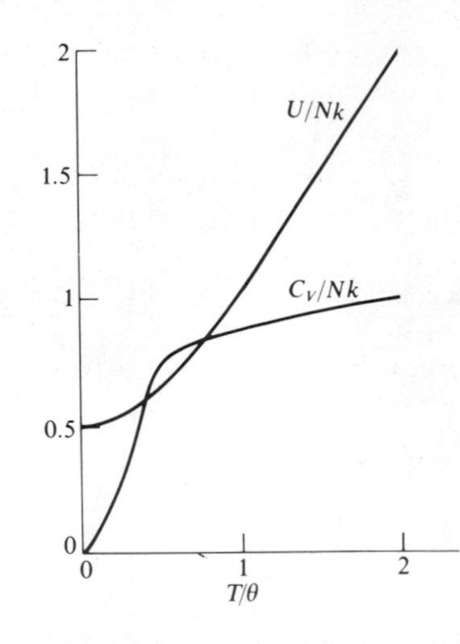

Fig. 12–15 The internal energy and heat capacity of an assembly of linear oscillators.

The curves in Fig. 12–15 are graphs of the internal energy U and of the heat capacity C_V (both divided by Nk) as functions of T/θ. The *ordinate* of the latter is proportional to the *slope* of the former.

As T approaches 0 K, very nearly all of the oscillators are in their lowest energy level with energy $h\nu/2$ and the total energy U approaches the zeropoint energy $Nh\nu/2$, or, $U/Nk \to 0.5$. The internal energy changes only slightly with changing temperature and the heat capacity approaches zero. The entropy of an assembly of linear oscillators also approaches zero as T approaches zero.

When $T \gg \theta$, $\theta/T \ll 1$, $\exp(\theta/T) - 1 \simeq \theta/T$, the term 1/2 is negligible compared with T/θ, and U approaches NkT. The mean energy per particle, U/N, approaches kT which is the value predicted by equipartition for an oscillator with two degrees of freedom (its position and its velocity). The internal energy increases nearly linearly with temperature and C_V approaches the constant value Nk.

12–7 SPECIFIC HEAT CAPACITY OF A DIATOMIC GAS

It was shown in Section 12–1 how the equation of state of a *monatomic* ideal gas, and its energy equation, could be derived by the methods of statistical thermodynamics. Consider next a gas whose molecules are *polyatomic*. If the energy of a molecule does not depend on the space coordinates x, y, and z of its center of mass, and if there is no mutual potential energy between molecules, the partition function will be directly proportional to the volume V, as in Eq. (12–6) for a monatomic gas. The Helmholtz function $F = -NkT(\ln Z - \ln N + 1)$ then has the same dependence on V as for a monatomic gas and the gas has the same equation of state, $PV = nRT$.

The specific heat capacity, however, will differ from that of a monatomic gas because a polyatomic molecule can have an "internal energy" of its own, made up of energy of rotation, vibration, and electronic excitation.

According to the classical equipartition principle, each degree of freedom associated with rotation and vibration shares equally with the three translational degrees of freedom, the mean energy of each being $kT/2$. The molal specific heat capacity at constant volume should equal $R/2$ for each degree of freedom and for a molecule with f degrees of freedom we should have $c_v = fR/2$, which should be constant independent of temperature.

This prediction is in good agreement with experiment for monatomic gases, for which there are three translational degrees of freedom only and for which c_v is very nearly equal to $3R/2$. At room temperature, however, the heat capacities of diatomic gases are nearly equal to $5R/2$, as if the molecules had an additional two degrees of freedom. Furthermore, the heat capacities are not constant, but vary with temperature and do not correspond to integral values of f.

A diatomic molecule can be considered to have the dumbbell-like structure of Fig. 9–5. In addition to the kinetic energy of translation of its center of mass, it may have energy of rotation about its center of mass and, since it is not a completely rigid structure, its atoms may oscillate along the line joining them. The rotational and vibrational energy are both quantized; and with each form of energy, as for an harmonic oscillator, there can be associated a characteristic temperature, θ_{rot} for rotation and θ_{vib} for vibration. The extent to which the rotational and vibrational energy levels are populated is determined by the ratio of the actual temperature T to the corresponding characteristic temperature. That is, the internal energies of rotation and vibration, and the corresponding specific heat capacities c_{rot} and c_{vib} are functions of the ratios T/θ_{rot} and T/θ_{vib}. We shall not give the precise form of this dependence, but simply state that the graphs of the specific heat capacities c_{rot} and c_{vib} have the same general form as the graph of c_v for an harmonic oscillator shown in Fig. 12–15. At very low temperatures, both heat capacities approach zero; at temperatures large compared to the characteristic temperatures, both approach the classical value Nk. Thus at sufficiently high temperatures the corresponding molal heat capacities approach the classical value R, as for a particle with two degrees of freedom.

What constitutes a "sufficiently" high temperature? This depends on the characteristic temperatures θ_{rot} and θ_{vib}. Table 12–2 lists some values of θ_{rot}. This temperature is inversely proportional to the moment of inertia of the molecule: the greater the moment of inertia, the lower the value of θ_{rot}. The highest value, about 86 K, is that for hydrogen, H_2, since its moment of inertia is smaller than for any other diatomic molecule. Molecules with one hydrogen atom form another group with values of θ_{rot} of approximately 20 K. For all others, the characteristic temperature is of the order of a few degrees or less. Thus "room temperature," say 300 K, is much greater than the characteristic temperature for rotation, and the molal specific heat capacity for rotation approaches the value R.

Table 12–2 Characteristic temperatures for rotation and vibration of diatomic molecules

Substance	θ_{rot}, (K)	θ_{vib}, (K)
H_2	85.5	6140
OH	27.5	5360
HCl	15.3	4300
CH	20.7	4100
CO	2.77	3120
NO	2.47	2740
O_2	2.09	2260
Cl_2	0.347	810
Br_2	0.117	470
Na_2	0.224	230
K_2	0.081	140

Table 12–2 also lists the characteristic temperatures θ_{vib} for the same molecules. These are all very much higher than the characteristic temperatures for rotation, which means that at room temperature, where $T \ll \theta_{\text{vib}}$, practically all molecules are in their lowest vibrational energy level and the specific heat capacity for vibration is practically zero. Only at much higher temperatures do the higher vibrational energy levels begin to be populated.

Thus at room temperature the specific heat capacities of most diatomic molecules have a contribution $3R/2$ for translation, plus R for rotation, making a total of $5R/2$ as is actually observed.

Figure 12–16 is a graph of experimental values of c_v/R for hydrogen, plotted as a function of temperature. (Hydrogen is the only diatomic gas that remains a gas down to low temperatures, of the order of 25 K.) At very low temperatures, c_v/R is equal to 3/2, the value for a *monatomic* gas. As the temperature is increased, c_v increases, and over a considerable range near room temperature c_v/R is about 5/2, which is the value (according to equipartition) if two degrees of freedom of rotation or vibration, but not both, are added to the translational degrees of freedom. Only at very high temperature does c_v/R approach 7/2, the value predicted by equipartition.

We can now understand in a general way the features of this graph. The characteristic temperatures for rotation and vibration, for hydrogen, are $\theta_{\text{rot}} =$ 85.5 K and $\theta_{\text{vib}} = 6140$ K. Below about 50 K, the temperature T is very much less than either characteristic temperature, and practically all molecules remain in their lowest energy states of rotation and vibration. The specific heat capacity is therefore the same as that of a monatomic gas, $3R/2$.

In the range from about 50 K to about 250 K, the temperature T is of the order of magnitude of θ_{rot} and the rotational states of higher energy begin to be populated. Above about 250 K, the molecules behave like classical rotators and make a contribution R to the specific heat capacity, which in this range equals $5R/2$. Starting at about 500 K, some molecules move to states of higher vibrational energy and c_v approaches the limiting classical value of $7R/2$.

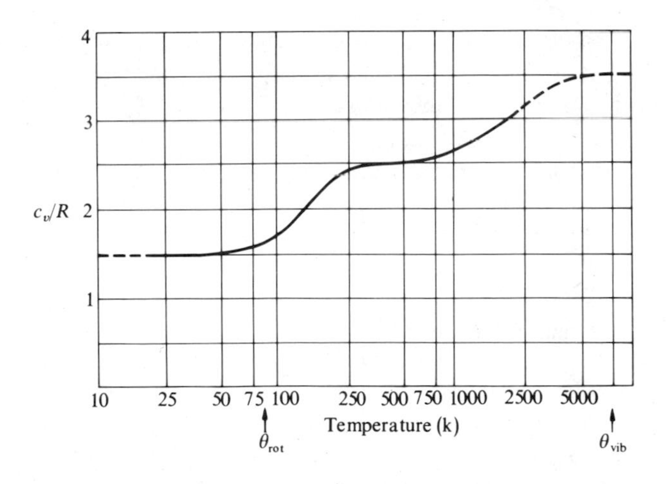

Fig. 12–16 Experimental values of c_v/R for hydrogen as a function of temperature plotted on a logarithmic scale.

Many important features of the general theory have been ignored in the (relatively) simple treatment of the problem given here. Some of these are: (a) the difference between the behavior of molecules such as H_2, whose atoms are alike, and those, such as NO, composed of unlike atoms; (b) the degeneracy of the rotational energy levels as a result of space quantization; (c) the energy associated with electronic excitation at high temperatures; (d) the coupling between rotational and vibrational states; and (e) the fact that the vibrations are not precisely simple harmonic. However, the exact theory is apparently so firmly established that specific heat capacities of gases can be computed theoretically, from optical measurements, more accurately than they can be measured experimentally by the technique of calorimetry.

PROBLEMS

12–1 In Section 12–1 the properties of a monatomic ideal gas were calculated using the classical distribution function. (a) Derive the equation of state and specific heat capacity of an ideal gas using, instead, the M-B distribution function. (b) Show that the M-B distribution function leads to an expression for the entropy of an ideal gas which is not extensive.

12–2 In a two-dimensional gas the molecules can move freely on a plane, but are confined within an area A. (a) Show that the partition function for a two-dimensional monatomic gas of N particles is given by

$$Z = \frac{A 2\pi m k T}{h^2}.$$

(b) Find the equation of state of the gas from its Helmholtz function.

12–3 Use the partition function of the previous problem to derive the heat capacity and entropy of a two-dimensional monatomic gas.

12–4 In Fig. 12–3, let $v_x = v_y = v_z = v_m$, and let $\Delta v_x = \Delta v_y = \Delta v_z = 0.01 v_m$. If $N =$ Avogadro's number, 6.02×10^{26} molecules, compute the average number of particles in each of the following elements of velocity space: (a) the slice of thickness Δv_y, (b) the rectangular parallelepiped common to two slices, (c) the volume element $\Delta v_x \, \Delta v_y \, \Delta v_z$, (d) the spherical shell of radius $\sqrt{3}\, v_m$ and thickness $0.01 v_m$.

12–5 (a) What is the "distance" v_y in Fig. 12–3, of a slice at right angles to the v_y-axis, if the slice contains one-half of the number of particles as a parallel slice of the same thickness at the origin. Express your answer in terms of v_m. (b) At what radial "distance" v from the origin in velocity space is the density ρ_v one-half as great as the origin.

12–6 Find the fractional number of molecules of a gas having (a) velocities with x-components between v_m and $1.01 v_m$, (b) speeds between v_m and $1.01 v_m$, (c) velocities with x-, y-, and z-components between v_m and $1.01 v_m$.

12–7 Show that $v_m = \sqrt{2kT/m}$.

12–8 (a) Compute to three significant figures the rms, average, and most probable speeds of an oxygen molecule at 300 K. (b) Compute the most probable speed of an oxygen molecule at the following temperatures: 100 K, 1000 K, 10,000 K.

12–9 Show that $(\overline{v^2}) - (\bar{v})^2 > 0$. This difference plays an important part in theory of fluctuations, and is the mean square deviation of the velocity from the average velocity.

12–10 Show that the average reciprocal speed $\overline{(1/v)}$ is given by $2/\sqrt{\pi}\, v_m = \sqrt{2m/\pi k T}$.

12–11 (a) Express Eq. (12–18) in terms of the kinetic energy $\epsilon (= mv^2/2)$ of the molecules. (b) Find the most probable and average energy of molecules having a distribution of speeds given by Eq. (12–18) and compare the results to $mv_m^2/2$ and $m\bar{v}^2/2$, respectively.

12–12 Show that the number of molecules with positive x-components of velocity less than some arbitrary value v is $\mathcal{N}_{0 \to x} = \frac{N}{2} \operatorname{erf}(x)$, where $x = v/v_m$ and erf (x) is the error function defined as

$$\text{erf}\,(x) = \frac{2}{\sqrt{\pi}}\int_0^x e^{-x^2}\,dx.$$

(b) Show that the number of molecules with positive x-components of velocity larger than the value v is $\mathcal{N}_{x\to\infty} = \frac{N}{2}\,[1 - \text{erf}\,(x)]$. Compute the fraction of molecules with x-components of velocity between (c) 0 and v_m, (d) v_m and ∞, (e) 0 and ∞, (f) $-v_m$ and $+v_m$. The value of erf (1) = 0.8427. (g) Illustrate your answers graphically in terms of the velocity distribution function.

12–13 Show that the number of molecules with *speeds* less than some arbitrary value v is given by

$$\mathcal{N}_{0\to x} = N\left[\text{erf}\,(x) - \frac{2}{\sqrt{\pi}}\,x\,e^{-x^2}\right],$$

where x and erf (x) are defined in the previous problem. (b) Show that the number of molecules with speeds greater than the arbitrary value is given by

$$\mathcal{N}_{x\to\infty} = N\left[1 - \text{erf}\,(x) + \frac{2}{\sqrt{\pi}}\,x\,e^{-x^2}\right].$$

Compute the fraction of molecules with speeds between (c) 0 and v_m, (d) v_m and ∞, and (e) 0 and ∞. (f) Illustrate your answers graphically in terms of the speed distribution function.

12–14 Show that v_{rms} for particles leaving a small hole in a furnace is given by $\sqrt{4kT/m}$.

12–15 Show that the number of molecules colliding with a surface of unit area per unit time, with components of velocity at right angles to the surface greater than some arbitrary value $v = xv_m$, is $[\mathbf{n}v_m \exp(-x^2)]/2\sqrt{\pi}$.

12–16 The oven in Fig. 12–10 contains bismuth at a temperature of 830 K, the drum is 10 cm in diameter and rotates at 6000 rpm. Find the distance between the points of impact of the molecules Bi and Bi_2 on the glass plate, G. Assume that all the molecules of each species escape the oven with the rms speed appropriate to that species.

12–17 A spherical bulb 10 cm in radius is maintained at a temperature of 27°C, except for one square centimeter, which is kept at a very low temperature. The bulb contains water vapor originally at a pressure of 10 Torr. Assuming that every water molecule striking the cold area condenses and sticks to the surface, how long a time is required for the pressure to decrease to 10^{-4} Torr?

12–18 A spherical bulb 10 cm in radius is pumped continuously to a high vacuum. In the bulb is a small vessel, closed except for a circular hole 0.2 mm in diameter located at the center of the bulb. The vessel contains mercury at 100°C, at which temperature its vapor pressure is 0.28 Torr. (a) Compute the average speed $\bar{v}$ of the molecules of mercury vapor in the small vessel. (b) Compute the rate of efflux of mercury through the hole, in milligrams hr^{-1}. (c) How long a time is required for 1 microgram of mercury to be

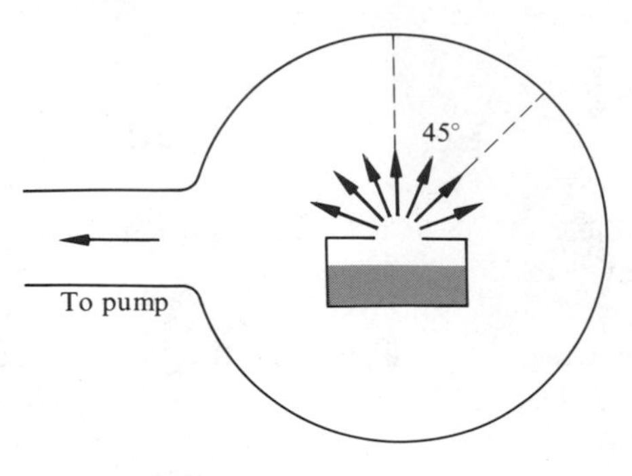

Figure 12–17

deposited on a square centimeter of the inner surface of the bulb, in a direction making an angle of 45° with the normal to the hole? (See Fig. 12–17.)

12–19 In a molecular beam experiment, the source is a tube containing hydrogen at a pressure $P_s = 0.15$ Torr and at a temperature $T = 400$ K. In the tube wall is a slit 30 mm × 0.025 mm, opening into a highly evacuated region. Opposite the source slit and 1 meter away from it is a second detector slit parallel to the first and of the same size. This slit is in the wall of a small enclosure in which the pressure P_d can be measured. When the steady state has been reached: (a) What is the discharge rate of the source slit in micrograms s^{-1}? (b) What is the rate of arrival of hydrogen at the detector slit, in micrograms s^{-1}, and in molecules s^{-1}? (c) How many molecules that will eventually reach the detector slit are in the space between source and detector at any instant? (d) What is the equilibrium pressure P_d in the detector chamber?

12–20 The distances OS and SD in the apparatus of Estermann, Simpson, and Stern in Fig. 12–11, are each 1 meter. Calculate the distance of the detector below the central position D, for cesium atoms having a speed equal to the rms speed in a beam emerging from an oven at a temperature of 460 K. Calculate also the "angle of elevation" of the trajectory. The atomic weight of cesium is 133.

12–21 The neutron flux across an area at the center of the Brookhaven reactor is about 4×10^{16} neutrons $m^{-2} s^{-1}$. Assume that the neutrons have a Maxwell-Boltzmann velocity distribution corresponding to a temperature of 300 K ("thermal" neutrons). (a) Find the number of neutrons per cubic meter. (b) Find the "partial pressure" of the neutron gas.

12–22 Derive Eq. (12–27) from Eq. (7–31) assuming $E_p = \Gamma g$, $Y_1 = \Pi$ and $X_1 = L$.

12–23 (a) Obtain the expressions for Z_v and Z_y given in Eqs. (12–31) and (12–32). (b) Obtain the expressions for Π and Γ given in Eqs. (12–34) and (12–35).

12–24 For the gas in a cylinder in a gravitational field, discussed in Section 12–4, show that as $g \to 0$, the number of molecules per unit volume approaches the constant value N/V, and hence is the same at all elevations. In other words, in the absence of a gravitational field the molecules of a gas are distributed *uniformly* throughout the volume of a container.

12–25 Show that the net downward force exerted on the container by the gas, in Section 12–4, equals the weight of the gas in the container.

12–26 If the height of the atmosphere is very large, show that (a) $\Pi = 0$, (b) $\Gamma = NkT/g$, (c) $E = \frac{5}{2}NkT$, (d) $dS = Nk[(5/2)(dT/T) - (dg/g)]$, and (e) that states at constant entropy are related by $T^{5/2}/g$ = constant.

12–27 (a) Calculate the fraction of hydrogen atoms which can be thermally ionized at room temperature. (b) At what temperature will e^{-1} of the atoms be ionized?

12–28 When a gas is whirled in a centrifuge, its molecules can be considered to be acted on by a radially outward centrifugal force of magnitude $m\omega^2 r$. Show that the density of the gas as a function of r varies as $\exp(m\omega^2 r^2/2kT)$.

12–29 Find the mean gravitational potential energy per molecule in an infinitely high isothermal atmosphere.

12–30 (a) Use the principle of equipartition of energy to find the total energy, the energy per particle, and the heat capacity of a system of N distinguishable harmonic oscillators in equilibrium with a bath at a temperature T. The kinetic energy of each oscillator is $m(v_x^2 + v_y^2 + v_z^2)/2$ and the potential energy is $K(x^2 + y^2 + z^2)/2$ where x, y, and z are the displacements from an equilibrium position. (b) Show that the expansivity of this system is zero because $\bar{x} = \bar{y} = \bar{z} = 0$.

12–31 A molecule consists of four atoms at the corners of a tetrahedron. (a) What is the number of translational, rotational, and vibrational degrees of freedom for this molecule? (b) On the basis of the equipartition principle, what are the values of c_v and γ for a gas composed of these molecules?

12–32 Using Eq. (11–62) derive (a) Eq. (12–48) and (b) Eq. (12–49). (c) Show that when $T \gg \theta$, C_V approaches Nk; and when $T \ll \theta$, C_V approaches zero as $e^{-\theta/T}$.

12–33 Calculate the average fractional number of oscillators in the jth energy level $\overline{N}_j/N$ for the four lowest energy levels when (a) $T = \theta/2$ and (b) $T = 2\theta$.

12–34 Make sketches of the average fractional number of oscillators in (a) the ground state, and (b) the first excited state, and (c) in the second excited state as a function of T/θ.

12–35 Making use of Eq. (11–66), show that the entropy of an assembly of quantized linear oscillators is

$$S = Nk\left\{\frac{\theta/T}{\exp(\theta/T) - 1} - \ln[1 - \exp(-\theta/T)]\right\},$$

where $\theta = h\nu/k$. (b) Show that S approaches zero as T approaches zero. (c) Why should Eq. (11–66) be used rather than Eq. (11–63)?

12–36 Consider 1000 diatomic molecules at a temperature $\theta_{\text{vib}}/2$. (a) Find the number in each of the three lowest vibrational states. (b) Find the vibrational energy of the system.

13

Applications of quantum statistics to other systems

13–1 THE EINSTEIN THEORY OF THE SPECIFIC HEAT CAPACITY OF A SOLID

In Section 9–8 and Fig. 3–10 it was shown that the specific heat capacity of many solids, at constant volume, approaches the Dulong-Petit value of $3R$ at high temperatures, but decreases to zero at very low temperatures. The first satisfactory explanation of this behavior was given by Einstein, who proposed that the atoms of a solid be considered in the first approximation as an assembly of quantized oscillators all vibrating with the same frequency ν. The principles of quantum mechanics had not been completely developed at the time this suggestion was made, and Einstein's original article assumed that the energy of an oscillator was given by

$$\epsilon_j = n_j h\nu.$$

The additional factor 1/2, which we introduced in Eq. (12–43), does not affect the method and we shall make use of the expressions already derived in Section 12–6. We must make one change, however. The atoms of a solid are free to move in *three* dimensions, not just one, so that an assembly of N atoms is equivalent to $3N$ linear oscillators. Then from Eq. (12–48), the internal energy U of a solid consisting of N atoms is

$$U = 3Nk\theta_{\text{E}}\left[\frac{1}{\exp(\theta_{\text{E}}/T) - 1} + \frac{1}{2}\right], \tag{13–1}$$

where the *Einstein temperature* θ_{E} is defined as

$$\theta_{\text{E}} \equiv \frac{h\nu}{k}. \tag{13–2}$$

The mean energy of an atom is

$$\bar{\epsilon} = \frac{U}{N} = 3k\theta_{\text{E}}\left[\frac{1}{\exp(\theta_{\text{E}}/T) - 1} + \frac{1}{2}\right],$$

and the specific heat capacity at constant volume is

$$c_v = 3R\left(\frac{\theta_{\text{E}}}{T}\right)^2 \frac{\exp(\theta_{\text{E}}/T)}{[\exp(\theta_{\text{E}}/T) - 1]^2}. \tag{13–3}$$

Figure 13–1 shows graphs of the dimensionless ratios $\bar{\epsilon}/k\theta_{\text{E}}$ and c_v/R, plotted as functions of T/θ_{E}. The ordinate of the latter curve, at any temperature, is proportional to the slope of the former. The general shape of the graph of c_v is in agreement with the experimental curve shown in Fig. 3–10. The value of θ_{E} (and hence of ν) for a particular substance is chosen so as to get the best fit between theoretical and experimental curves. However, it is not possible to find a value of θ_{E} that gives good agreement at both low and high temperatures.

When $T \gg \theta_{\text{E}}$, θ_{E}/T is small and c_v approaches the Dulong-Petit value,

$$c_v = 3R.$$

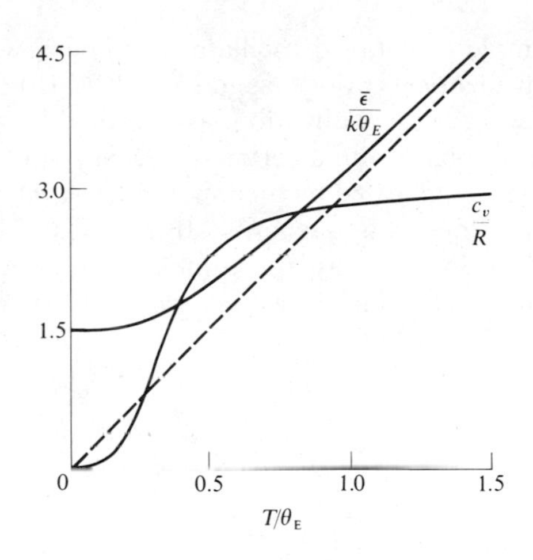

Fig. 13–1 Internal energy and specific heat capacity of a harmonic oscillator.

When $T \ll \theta_E$, the exponential term is large, we can neglect the 1 in the denominator, and

$$c_v = 3R\left(\frac{\theta_E}{T}\right)^2 \exp\left(-\theta_E/T\right).$$

(See Prob. 12–32.)

When T approaches zero the exponential term goes to zero more rapidly than $1/T^2$ goes to infinity, and c_v approaches zero in agreement with experiment and the third law. However, because of the rapid decrease of the exponential term, the theoretical values of c_v, at very low temperatures, decrease much more rapidly than the experimental values. Thus the Einstein theory, while it seems to indicate the correct approach to the problem, is evidently not the whole story.

13–2 THE DEBYE THEORY OF THE SPECIFIC HEAT CAPACITY OF A SOLID

The simple Einstein theory assumes that all atoms of a solid oscillate at the same frequency. Nernst and Lindemann* found empirically that the agreement between theory and experiment could be improved by assuming two groups of atoms, one oscillating at a frequency ν and the other at a frequency 2ν. This idea was extended by Born,† von Karman,‡ and Debye, who considered the atoms, not as isolated oscillators all vibrating at the same frequency, but as a system of coupled oscillators having a continuous spectrum of natural frequencies.

* Frederick A. Lindemann, First Viscount Cherwell, British physicist (1886–1957).
† Max Born, German physicist (1882–1970).
‡ Theodor von Kármán, Hungarian engineer (1881–1963).

As a simple example of coupled oscillators, suppose we have two identical particles connected by identical springs, as in Fig. 13–2. If both particles are given equal initial velocities in the same direction, as indicated by the upper arrows, the particles will oscillate in phase with a certain frequency ν_1. If the initial velocities are equal and opposite, as indicated by the lower arrows, the particles will oscillate out of phase but with a different frequency ν_2. If the initial velocities have arbitrary values, the resultant motion is a superposition of two oscillations of frequencies ν_1 and ν_2. The system is said to have two *natural frequencies.*

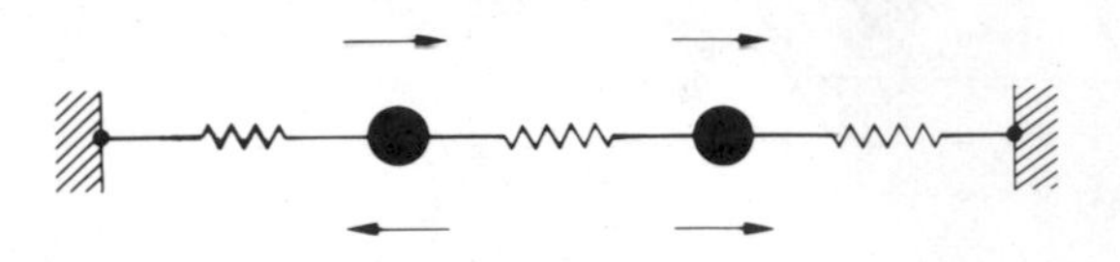

Fig. 13–2 Coupled oscillators.

Now suppose that the number of particles (and springs) is increased. It is no great task to calculate the natural frequencies when the number is small, but as the number is increased there are too many simultaneous equations to be solved. It turns out, however, that if there are N particles in the chain, the system will have N natural frequencies, whatever the value of N.

Now extend these ideas to three dimensions. A simple model of a crystal consists of a three-dimensional array of particles connected by springs, and such an array has $3N$ natural frequencies. Because of the impossibility of calculating these frequencies when N is as large as the number of molecules in a macroscopic crystal, Debye assumed that the natural frequencies of the atoms of a crystal would be the same as the frequencies of the possible stationary waves in a crystal if the crystal were a *continuous elastic solid.* This is a standard problem in the theory of elasticity, and we shall outline its solution without giving details. The procedure is closely analogous to that described in Section 11–2, except that we are now dealing with *real* elastic waves and not with the mathematical waves of wave mechanics.

As explained in Section 11–2, an elastic string of length L fixed at both ends, can oscillate in a steady state in any mode for which the wavelength λ is given by

$$\lambda = \frac{2L}{n},$$

where $n = 1, 2, 3, \ldots$, etc.

The fundamental equation of any sort of wave motion states that the speed of propagation c equals the product of the frequency ν and the wavelength λ:

$$c = \nu\lambda.$$

It follows that for any frequency ν, the number n is

$$n = \frac{2L}{c}\nu$$

and

$$n^2 = \frac{4L^2}{c^2}\nu^2.$$

The theory of elasticity leads to the result that the natural frequencies of stationary waves in an elastic solid in the form of a cube of side length L are given by the same equation except that the possible values of n^2 are

$$n^2 = n_x^2 + n_y^2 + n_z^2,$$

where n_x, n_y, and n_z are positive integers that can have the values 1, 2, 3, . . . , etc.

To find the number of waves in any frequency interval, or the *frequency spectrum*, we proceed in the same way as in Section 12–1 and Fig. 12–1. Let the numbers n_x, n_y, and n_z be laid off on three mutually perpendicular axes. Each triad of values determines a point in n-space, with corresponding values of n and of ν. Let $\mathscr{G}$ represent the total number of possible frequencies, up to and including that corresponding to some given n. This is equal to the number of points within an octant of a sphere of radius n, the volume of which is $(\pi/6)\,n^3$, and since $n = (2L/c)\nu$,

$$\mathscr{G} = \frac{4\pi}{3}\frac{L^3}{c^3}\nu^3.$$

But L^3 is the volume V of the cube, and it can be shown that regardless of the shape of the solid we can replace L^3 with V. Then

$$\mathscr{G} = \frac{4\pi}{3}\frac{V}{c^3}\nu^3. \tag{13–4}$$

However, three types of elastic waves can propagate in an elastic solid: a longitudinal or compressional wave (a sound wave) traveling with speed c_l, and two transverse or shear waves polarized in mutually perpendicular directions and traveling with a different speed c_t. The total number of possible stationary waves having frequencies up to and including some frequency ν is therefore

$$\mathscr{G} = \frac{4\pi}{3}V\left(\frac{1}{c_l^3} + \frac{2}{c_t^3}\right)\nu^3. \tag{13–5}$$

According to the Debye theory Eq. (13–5) can also be interpreted as describing the number of linear *oscillators* having frequencies up to and including the frequency ν. Thus, to be consistent with the notation of Section 12–2, $\mathscr{G}$ in Eq. (13–5) should be replaced by $\mathscr{N}$ and

$$\mathscr{N} = \frac{4\pi}{3}V\left(\frac{1}{c_l^3} + \frac{2}{c_t^3}\right)\nu^3. \tag{13–6}$$

If there were no upper limit to the frequency, the total number of oscillators would be infinite. But a crystal containing N atoms constitutes an assembly of $3N$ linear oscillators. Hence we assume that the frequency spectrum cuts off at a maximum frequency ν_m such that the total number of linear oscillators equals $3N$. Then setting $\mathcal{N} = 3N$ and $\nu = \nu_m$,

$$3N = \frac{4\pi}{3} V\left(\frac{1}{c_l^3} + \frac{2}{c_t^3}\right)\nu_m^3. \tag{13–7}$$

The wave speeds c_l and c_t can be calculated from a knowledge of the elastic properties of a given material and hence ν_m can be calculated from this equation. In a material like lead, which is easily deformed, the wave speeds are relatively small, while in a rigid material like diamond, the speeds are relatively large. Hence the value of ν_m for lead is much smaller than it is for diamond.

That there should be a maximum frequency of the stationary waves that can exist in a real solid can be seen as follows. For a single set of waves of speed c, the maximum frequency ν_m corresponds to a minimum wavelength $\lambda_{\text{min}} = c/\nu_m$, and Eq. (13–7) can be written

$$\lambda_{\text{min}} = \left(\frac{4\pi}{9}\right)^{1/3}\left(\frac{V}{N}\right)^{1/3}. \tag{13–8}$$

But (V/N) is the average volume per atom and the cube root of this, $(V/N)^{1/3}$, is of the order of the average interatomic spacing. Hence the structure of a real crystal (which is not a continuous medium) sets a limit to the minimum wavelength which is of the order of the interatomic spacing, as would be expected since shorter wavelengths do not lead to new modes of atomic motion. It follows from Eqs. (13–6) and (13–7) that

$$\mathcal{N} = \frac{3N}{\nu_m^3}\nu^3.$$

The number of linear oscillators having frequencies between ν and $\nu + \Delta\nu$ is then

$$\Delta\mathcal{N}_\nu = \frac{9N}{\nu_m^3}\nu^2\,\Delta\nu, \tag{13–9}$$

and the number per unit range of frequency is

$$\frac{\Delta\mathcal{N}_\nu}{\Delta\nu} = \frac{9N}{\nu_m^3}\nu^2. \tag{13–10}$$

Figure 13–3 is a graph of $\Delta\mathcal{N}_\nu/\Delta\nu$, plotted as a function of ν. The actual number $\Delta\mathcal{N}_\nu$ of oscillators of frequency between ν and $\nu + \Delta\nu$ is represented by the *area* of the shaded vertical strip, since the height of the strip is $\Delta\mathcal{N}_\nu/\Delta\nu$ and its width is $\Delta\nu$. This is in contrast to the Einstein model, in which *all* oscillators have the same frequency. The total area under the curve corresponds to the total number of linear oscillators, $3N$.

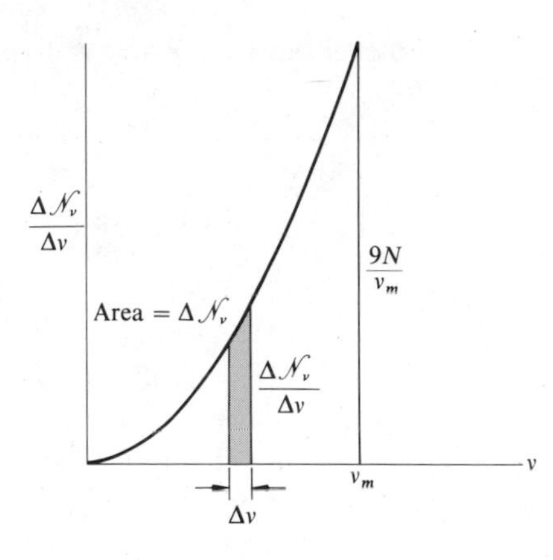

Fig. 13–3 The Debye frequency spectrum.

The oscillators of any frequency ν constitute a subassembly of linear oscillators all having the *same* frequency, as in the Einstein model. Then from Eq. (12–48) the internal energy ΔU_ν of the subassembly, replacing $3N$ with $\Delta \mathcal{N}_\nu$, is

$$\Delta U_\nu = \frac{9N}{\nu_m^3} \frac{h\nu^3}{\exp(h\nu/kT) - 1} \Delta\nu. \tag{13–11}$$

We omit the constant zero-point energy since this has no effect on the heat capacity.

The point of view thus far, in this section and in the preceding one, has been to consider the atoms of a crystal as distinguishable particles obeying the M-B statistics. An alternative approach is to consider the elastic waves themselves as the "particles" of an assembly. Each wave can also be considered as a particle called a *phonon*, and the assembly described as a *phonon gas*. Since the waves or phonons are indistinguishable, and there is no restriction on the number permitted per energy state, the assembly obeys the Bose-Einstein statistics.

We must, however, make one modification in the expression previously derived for the distribution function in this statistics. This is because the number N of waves, or phonons, in contrast to the number of atoms of a gas in a container of specified volume, cannot be considered one of the *independent* variables specifying the state of the assembly. If the assembly is a gas, we can arbitrarily fix both the volume V and the temperature T of a container, and still can introduce any arbitrary number N of molecules of gas into the container. But when the volume and temperature of a crystal are specified, the crystal itself, so to speak, determines the number of different waves, or phonons, that are equivalent to the oscillations of its molecules. Thus the crystal cannot be considered an *open* system for which N is an independent variable and the term $\mu\, dN$ does not appear in Eq. (11–22). This is equivalent to setting $\mu = 0$ and hence $\exp(\mu/kT) = 1$. The number of "particles" in a macrolevel between ϵ and $\epsilon + \Delta\epsilon$ is therefore

$$\Delta\mathcal{N} = \frac{\Delta\mathcal{G}}{\exp(\epsilon/kT) - 1}. \tag{13–12}$$

According to the principles of quantum mechanics, the energy of a wave (or phonon) of frequency ν is

$$\epsilon = h\nu,$$

where h is Planck's constant. Unlike a linear oscillator of frequency ν, which can have any one of the energies $(n_j + \frac{1}{2})h\nu$, where $n_j = 0, 1, 2, \ldots$, etc., a wave of frequency ν can have *only* the energy $h\nu$. Thus if a large amount of energy is associated with a given frequency, this simply means that a large number of waves, or phonons, all of the same energy, are present in an assembly.

An energy interval between ϵ and $\epsilon + \Delta\epsilon$ therefore corresponds to a frequency interval between ν and $\nu + \Delta\nu$. Thus the number of phonons with frequencies between ν and $\nu + d\nu$ is

$$\Delta\mathcal{N}_\nu = \frac{\Delta\mathcal{G}_\nu}{\exp(h\nu/kT) - 1}, \tag{13–13}$$

where $\Delta\mathcal{G}_\nu$ is the number of states having frequencies between $\nu + \nu + d\nu$.

The energy ΔU_ν of the waves in this frequency interval is

$$\Delta U_\nu = h\nu\,\Delta\mathcal{N}_\nu = \frac{h\nu\,\Delta\mathcal{G}_\nu}{\exp(h\nu/kT) - 1},$$

and comparison with Eq. (13–11) shows that

$$\Delta\mathcal{G}_\nu = \frac{9N}{\nu_m^3}\nu^2\,\Delta\nu, \tag{13–14}$$

which is the same as the expression for the number of *distinguishable* oscillators in this frequency interval. That is, the degeneracy $\Delta\mathcal{G}_\nu$ of a macrolevel is equal to the number of distinguishable oscillators in the same interval. Equation (13–13) can therefore be written

$$\Delta\mathcal{N}_\nu = \frac{9N}{\nu_m^3}\frac{\nu^2\,\Delta\nu}{\exp(h\nu/kT) - 1}. \tag{13–15}$$

There appears at first sight to be a discrepancy between the expression for $\Delta\mathcal{N}_\nu$ in the preceding equation and that in Eq. (13–9). However, the symbol $\Delta\mathcal{N}_\nu$ does not represent the same thing in the two equations. In Eq. (13–15), $\Delta\mathcal{N}_\nu$ is the number of indistinguishable waves (or phonons) having frequencies between ν and $\nu + \Delta\nu$, in a system obeying the B-E statistics. In Eq. (13–9), $\Delta\mathcal{N}_\nu$ is the number of distinguishable oscillators having frequencies in the same range, in a system obeying M-B statistics.

The total energy U of the assembly is now obtained by summing the expression for ΔU_ν over all values of ν from zero to ν_m, and after replacing the sum with an integral, we have

$$U = \frac{9N}{\nu_m^3}\int_0^{\nu_m} \frac{h\nu^3}{\exp(h\nu/kT) - 1}\, d\nu. \tag{13–16}$$

The *Debye temperature* θ_D is defined as

$$\theta_\text{D} \equiv \frac{h\nu_m}{k}, \tag{13–17}$$

and θ_D is proportional to the cut-off frequency ν_m. Some values are given in Table 13–1.

Table 13–1 Debye temperatures of some materials

Substance	θ_D(K)
Lead	88
Thallium	96
Mercury	97
Iodine	106
Cadmium	168
Sodium	172
Potassium bromide	177
Silver	215
Calcium	226
Sylvine (KCl)	230
Zinc	235
Rocksalt (NaCl)	281
Copper	315
Aluminum	398
Iron	453
Fluorspar (CaF_2)	474
Iron pyrites (FeS_2)	645
Diamond	1860

For convenience, we introduce the dimensionless quantities

$$x = \frac{h\nu}{kT}, \qquad x_m = \frac{h\nu_m}{kT} = \frac{\theta_\text{D}}{T}.$$

Then

$$U = 9NkT\left(\frac{T}{\theta_\text{D}}\right)^3 \int_0^{x_m} \frac{x^3\, dx}{\exp(x) - 1}. \tag{13–18}$$

This corresponds to Eq. (13–1) for the energy U according to the Einstein theory.

Consider first the high temperature limit, at which $x = h\nu/kT$ is small. Then $[\exp(x) - 1] \simeq x$ and the integral becomes

$$\int_0^{x_m} x^2\,dx = \frac{x_m^3}{3} = \frac{\theta_D^3}{3T^3}.$$

Then at high temperatures,

$$U = 3NkT, \qquad c_v = 3R,$$

in agreement with the Einstein theory and the Dulong-Petit law.

At intermediate and low temperatures, the value of the integral can be expressed only as an infinite series. To a good approximation, the upper limit of the integral when T is very small can be taken as infinity instead of x_m, since the integrand is small for values of x greater than x_m. The definite integral then equals $\pi^4/15$, and hence at low temperatures,

$$U = \frac{3}{5}\pi^4 NkT\left(\frac{T}{\theta_D}\right)^3;$$

and by differentiation,

$$c_v = \frac{12\pi^4}{5} R\left(\frac{T}{\theta_D}\right)^3. \qquad (13\text{–}19)$$

Equation (13–19) is known as the *Debye T^3 law*. According to this law, the heat capacity near absolute zero decreases with the *cube* of the temperature, instead of *exponentially* as in the Einstein theory. The decrease is therefore less rapid and the agreement with experiment is much better. Although the Debye theory is based on an analysis of elastic waves in a homogeneous, isotropic, continuous medium, experimental values of the specific heat capacity of many crystalline solids are in good agreement with the Debye theory at temperatures below $\theta_D/50$, or when $T/\theta_D < 0.02$. As the temperature is increased, the specific heat capacity increases somewhat faster than the theory would predict. There is recent experimental evidence that amorphous materials do not appear to follow the Debye T^3 law even at temperatures below $\theta_D/100$, or when $T/\theta_D < 0.01$.

The heat capacity at any temperature can be calculated by evaluating the integral in Eq. (13–18), which gives the internal energy as a function of T, and differentiating the result with respect to T. As in the Einstein theory, the result is a function of T/θ_D only, and hence a *single* graph represents the temperature variation of c_v for *all* substances. The curve in Fig. 13–4 (what can be seen of it) is a graph of c_v/R, plotted as a function of T/θ_D, and the points are experimental values for a variety of materials.

It will be seen from Fig. 13–4 that roughly speaking, when T/θ_D is greater than 1, or when the actual temperature exceeds the Debye temperature, the system behaves "classically" and c_v is nearly equal to the "classical" or "non-quantum" value $3R$. When the actual temperature is less than the Debye temperature,

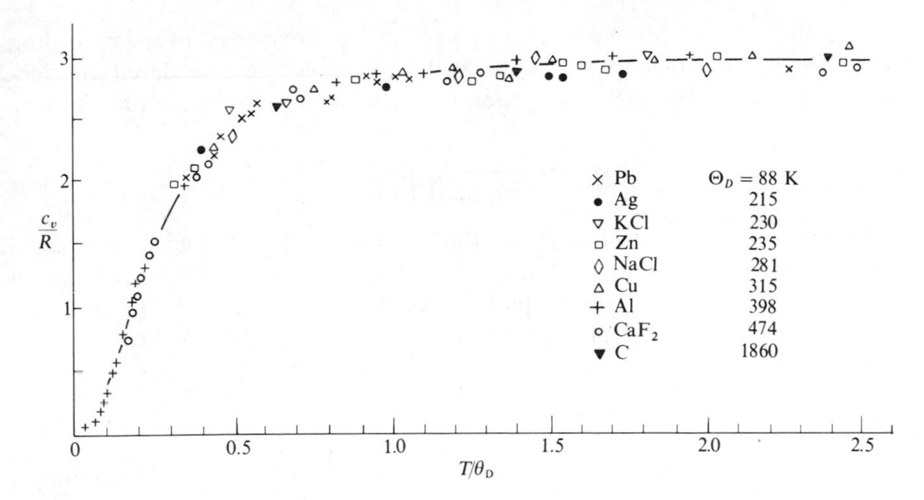

Fig. 13–4 Specific heat capacities of various solids as functions of T/θ_D.

quantum effects become significant and c_v decreases to zero. Thus for lead, with a Debye temperature of only 88 K, "room temperature" is well above the Debye temperature, while diamond, with a Debye temperature of 1860 K, is a "quantum solid" even at room temperature.

At intermediate temperatures there is good agreement between values of the specific heat capacity calculated by the Einstein and by the Debye theories. This agreement might be expected, since the Dulong-Petit theory is a first approximation that works at high temperatures. The Einstein theory is a second approximation which works for high and intermediate temperatures. The Debye theory is a third approximation that works at low temperatures when other effects do not dominate.

13–3 BLACKBODY RADIATION

The thermodynamics of blackbody radiation was discussed in Section 8–7 and we now consider the statistical aspects of the problem. The radiant energy in an evacuated enclosure whose walls are at a temperature T is a mixture of electromagnetic waves of all possible frequencies from zero to infinity, and it was the search for a theoretical explanation of the observed energy distribution among these waves that led Planck to the postulates of quantum theory.

To apply the methods of statistics to a batch of radiant energy, we consider the waves themselves as the "particles" of an assembly. Each wave can be considered a particle called a *photon* and the assembly can be described as a *photon gas*. Because the photons are indistinguishable and there is no restriction on the number per energy state, the assembly obeys the Bose-Einstein statistics.

The problem is very similar to that of a phonon gas discussed in the preceding section. The number of photons in the enclosure cannot be considered an independent variable and the B-E distribution function reduces to the simpler form,

$$\Delta \mathcal{N}_\nu = \frac{\Delta \mathcal{G}_\nu}{\exp(h\nu/kT) - 1}.$$

There is, however, a difference in the expression for the degeneracy $\Delta\mathcal{G}_\nu$. As we showed in the preceding section, the degeneracy of a macrolevel, in an assembly of waves (or photons) is equal to the possible number $\Delta\mathcal{G}_\nu$ of stationary waves in the frequency interval from ν to $\nu + \Delta\nu$. Let us return to Eq. (13–5),

$$\mathcal{G} = \frac{4\pi}{3}\frac{V}{c^3}\nu^3,$$

where $\mathcal{G}$ is the number of stationary waves with frequencies up to and including ν. Electromagnetic waves are purely transverse and there can be *two* sets of waves, polarized in mutually perpendicular planes and both traveling with the speed of light c. Also, since empty space has no structure, there is no upper limit to the maximum possible frequency. Then interpreting $\mathcal{G}$ as the total number of possible energy states of all frequencies up to and including ν, we have

$$\mathcal{G} = \frac{8\pi}{3}\frac{V}{c^3}\nu^3.$$

The degeneracy $\Delta\mathcal{G}_\nu$ is therefore

$$\Delta\mathcal{G}_\nu = \frac{8\pi V}{c^3}\nu^2\,\Delta\nu,$$

and the number of waves (or photons) having frequencies between ν and $\nu + \Delta\nu$ is

$$\Delta\mathcal{N}_\nu = \frac{8\pi V}{c^3}\frac{\nu^2}{\exp(h\nu/kT) - 1}\Delta\nu. \tag{13–20}$$

The energy of each wave is $h\nu$, and after dividing by the volume V, we have for the energy per unit volume, in the frequency range from ν to $\nu + \Delta\nu$, or the *spectral energy density* $\Delta\mathbf{u}_\nu$,

$$\Delta\mathbf{u}_\nu = \frac{8\pi h}{c^3}\frac{\nu^3}{\exp(h\nu/kT) - 1}\Delta\nu. \tag{13–21}$$

This equation has the same *form* as the experimental law (Planck's law) given in Section 8–7, and we now see that the experimental constants c_1 and c_2 in Eq. (8–50) are related to the fundamental constants h, c, and k, by the equations

$$c_1 = \frac{8\pi h}{c^3}, \qquad c_2 = \frac{h}{k}. \tag{13–22}$$

When numerical values of h, c, and k are inserted in these equations, the calculated values of c_1 and c_2 agree exactly with their experimental values, within the limits of experimental error.

At a given temperature T, and at high frequencies for which $h\nu \gg kT$, the exponential term is large; we can neglect the 1; and

$$\Delta \mathbf{u}_\nu \simeq \frac{8\pi h}{c^3}\nu^3 \exp(-h\nu/kT)\,\Delta\nu. \tag{13–23}$$

An equation of this form had been derived by Wien* before the advent of quantum theory and it is known as *Wien's law*. It is in good agreement with experiment at high frequencies but in very poor agreement at low frequencies.

However, at low frequencies for which $h\nu \ll kT$, $[\exp(h\nu/kT) - 1]$ is very nearly equal to $h\nu/kT$ and

$$\Delta \mathbf{u}_\nu \simeq \frac{8\pi kT}{c^3}\nu^2\,\Delta\nu. \tag{13–24}$$

This equation had been derived by Rayleigh† and Jeans,‡ also before the quantum theory, and had been found to agree with experiment at low, but not at high, frequencies. That it cannot be correct in general can be seen by noting that as the frequency becomes very high, the predicted energy density approaches infinity. (This result is sometimes referred to as the "ultraviolet catastrophe.")

It is interesting to note that Planck's first approach to the problem was purely empirical. He looked for an equation having a mathematical form such that it would reduce to the Wien equation when $h\nu/kT$ was large, and to the Rayleigh-Jeans equation when $h\nu/kT$ was small. He found that Eq. (13–21) had this property, and his search for a theoretical explanation of the equation led to the development of quantum theory.

Figure 13–5 shows graphs of the dimensionless quantity $\dfrac{\Delta \mathbf{u}_\nu}{\Delta\nu}\left(\dfrac{c^3h^2}{8\pi k^3T^3}\right)$, plotted as a function of the dimensionless quantity $h\nu/kT$. The solid curve is a graph of Planck's law, and the dotted curves are, respectively, graphs of the Rayleigh-Jeans law, applicable when $h\nu \ll kT$, and of Wien's law, applicable when $h\nu \gg kT$.

The total energy density $\mathbf{u}_\nu$, including all frequencies, can now be found by summing $\Delta\mathbf{u}_\nu$ over all values of ν from zero to infinity, since there is no limit to the maximum value of ν. Replacing the sum with an integral, we have

$$\mathbf{u}_\nu = \frac{8\pi h}{c^3}\int_0^\infty \frac{\nu^3}{\exp(h\nu/kT) - 1}\,d\nu;$$

* Wilhelm Wien, German physicist (1864–1928).

† John W. Strutt, Lord Rayleigh, English physicist (1842–1919).

‡ Sir James H. Jeans, English mathematician (1877–1946).

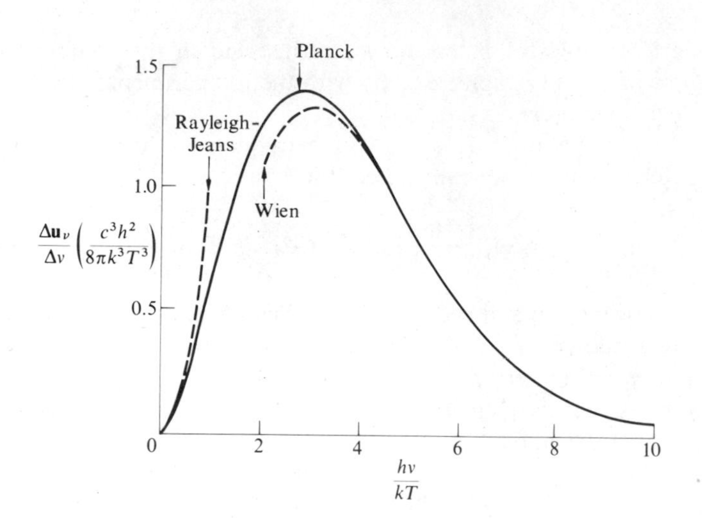

Fig. 13–5 Graphs of Planck's law, Wien's law, and the Rayleigh-Jeans law.

or, if we define a dimensionless variable $x = h\nu/kT$,

$$\mathbf{u}_\nu = \frac{8\pi k^4}{c^3h^3} T^4 \int_0^\infty \frac{x^3\, dx}{\exp(x) - 1}.$$

The value of the definite integral is $\pi^4/15$, so

$$\mathbf{u}_\nu = \frac{8\pi^5 k^4}{15c^3h^3} T^4 = \sigma T^4, \tag{13–25}$$

where

$$\sigma = \frac{8\pi^5 k^4}{15c^3h^3}. \tag{13–26}$$

Equation (13–25) is the same as *Stefan's law* (Eq. (8–54)); and when the values of k, c, and h are inserted in Eq. (13–26), the calculated and experimental values of σ agree exactly, within the limits of experimental error.

Thus quantum theory and the methods of statistics provide a theoretical basis for the form of Planck's law, and relate the experimental constants c_1, c_2, and σ to the fundamental constants h, c, and k. Expressions for the internal energy, the entropy, and the Helmholtz and Gibbs functions of blackbody radiant energy were derived by the principles of thermodynamics in Section 8–7 and need not be repeated here. It will be recalled that the Gibbs function $G = 0$, which might also have been taken as a justification for setting $\mu = 0$ in the B-E distribution function.

13–4 PARAMAGNETISM

We now consider the statistics of a paramagnetic crystal. The properties of such crystals are chiefly of interest in the region of extremely low temperatures, of the order of a few kelvins or less. A number of simplifying assumptions will be made, but the procedure is the same as in more complicated cases.

A typical paramagnetic crystal is chromium potassium sulfate, $Cr_2(SO_4)_3 \cdot K_2SO_4 \cdot 24H_2O$. Its paramagnetic properties are due solely to the chromium atoms, which exist in the crystal as ions, Cr^{+++}. Every electron in an atom has not only an electric charge but also a magnetic moment μ_B of 1 Bohr* magneton, equal (in MKS units) to 9.27×10^{-24} A m^2, as if the electron were a tiny sphere of electric charge spinning about an axis. In most atoms, the *resultant* magnetic moment of the electrons is zero, but the chromium ion Cr^{+++} has three uncompensated electron spins and a magnetic moment of $3\mu_B$.

For every chromium ion there are 2 sulfur atoms, 1 potassium atom, 20 oxygen atoms, and 24 hydrogen atoms, making a total of 47 other particles which are nonmagnetic. The magnetic ions are therefore so widely separated that there is only a small magnetic interaction between them.

It was shown in Section 8–8 that the thermodynamic properties of a paramagnetic crystal could be calculated from a knowledge of the quantity $F^* = E - TS$. Using the methods of statistics, the expression for F^* can be derived in terms of the temperature T and the parameters that determine the energy levels of the atoms in the crystal. Because the atoms can be labeled according to the positions they occupy in the crystal lattice, the system obeys M-B statistics, and as usual the first step is to calculate the partition function Z, defined as

$$Z = \sum_j \Delta \mathscr{G}_j \exp - \frac{\epsilon_j}{kT}.$$

Because of their oscillatory motion, the molecules have the same set of vibrational energy levels as those of any solid, and the total vibrational energy constitutes the internal energy U_{vib}. In addition, the small interaction between the magnetic ions, and their interactions with the electric field set up by the remainder of the lattice, gives rise to an additional internal energy (of the ions only) which we write as U_{int}. Finally, if there is a magnetic field in the crystal, set up by some external source, the ions have a magnetic potential energy which, like the gravitational potential energy of particles in a gravitational field, is a joint property of the ions and the source of the field and cannot be considered an *internal* energy. The total magnetic potential energy is E_p.

The vibrational energy levels, the levels associated with internal magnetic and electrical interactions, and the potential energy levels are all independent. The partition function Z, as in the case of a gas in a gravitational field, can be expressed

* Niels H. D. Bohr, Danish physicist (1885–1962).

as the product of independent partition functions which we write as Z_{vib}, Z_{int}, and $Z_{\mathscr{H}}$. Thus

$$Z = Z_{\text{vib}} Z_{\text{int}} Z_{\mathscr{H}}.$$

The magnetic ions constitute a *subassembly*, characterized by the partition functions Z_{int} and $Z_{\mathscr{H}}$ only, and they can be considered independently of the remainder of the lattice, which can be thought of simply as a container of the subassembly. Although the energy U_{int} and the partition function Z_{int} play important roles in the complete theory, we shall neglect them and consider that the total energy of the subassembly is its potential energy E_{p} only. Thus we consider only the partition function $Z_{\mathscr{H}}$.

As shown in Appendix E, the potential energy of an ion in a magnetic field of intensity $\mathscr{H}$ is $-\mu\mathscr{H} \cos\theta$, where μ is the magnetic moment of the ion and θ the angle between its (vector) magnetic moment and the direction of the field. For simplicity, we consider only a subassembly of ions having a magnetic moment of 1 Bohr magneton μ_{B}. The principles of quantum mechanics restrict the possible values of θ, for such an ion, to either zero or 180°, so that the magnetic moment is either parallel or antiparallel to the field. (Other angles are permitted if the magnetic moment is greater than μ_{B}). The corresponding values of $\cos\theta$ are then $+1$ and -1, and the possible energy levels are $-\mu_{\text{B}}\mathscr{H}$ and $+\mu_{\text{B}}\mathscr{H}$. The energy levels are nondegenerate; there is only one state in each level, but there is no restriction on the number of ions per state. The partition function $Z_{\mathscr{H}}$ therefore reduces also to the sum of two terms:

$$Z_{\mathscr{H}} = \exp\left(\frac{\mu_{\text{B}}\mathscr{H}}{kT}\right) + \exp\left(\frac{-\mu_{\text{B}}\mathscr{H}}{kT}\right) = 2\cosh\frac{\mu_{\text{B}}\mathscr{H}}{kT}, \tag{13–27}$$

since by definition the hyperbolic cosine is given by

$$\cosh x = \tfrac{1}{2}[\exp(x) + \exp(-x)].$$

Let $N_{\uparrow}$ and $N_{\downarrow}$ represent respectively the number of ions whose moments are aligned parallel and antiparallel to the field $\mathscr{H}$. The corresponding energies are $\epsilon_{\uparrow} = -\mu_{\text{B}}\mathscr{H}$ and $\epsilon_{\downarrow} = \mu_{\text{B}}\mathscr{H}$. The average occupation numbers in the two directions are then

$$\bar{N}_{\uparrow} = \frac{N}{Z}\exp\frac{-\epsilon_{\uparrow}}{kT}, \qquad \bar{N}_{\downarrow} = \frac{N}{Z}\exp\frac{\epsilon_{\downarrow}}{kT}.$$

The excess of those ions in the parallel, over those in the antiparallel alignment, is

$$\bar{N}_{\uparrow} - \bar{N}_{\downarrow} = \frac{N}{Z}\left[\exp\left(-\frac{\epsilon_{\uparrow}}{kT}\right) - \exp\left(\frac{\epsilon_{\downarrow}}{kT}\right)\right] = \frac{N}{Z}2\sinh\frac{\mu_B\mathscr{H}}{kT},$$

which reduces to

$$\bar{N}_{\uparrow} - \bar{N}_{\downarrow} = N\tanh\frac{\mu_{\text{B}}\mathscr{H}}{kT}. \tag{13–28}$$

The net magnetic moment M of the crystal is the product of the magnetic moment μ_B of each ion and the excess number of ions aligned parallel to the field. Then

$$M = (\bar{N}_\uparrow - \bar{N}_\downarrow)\mu_B = N\mu_B \tanh \frac{\mu_B \mathscr{H}}{kT}. \tag{13–29}$$

This is the *magnetic equation of state* of the crystal, expressing the magnetic moment M as a function of $\mathscr{H}$ and T. Note that M depends only on the ratio $\mathscr{H}/T$.

The equation of state can also be derived as follows. The function F^* is

$$F^* = -NkT \ln Z = -NkT \ln \left[2 \cosh \frac{\mu_B \mathscr{H}}{kT}\right]. \tag{13–30}$$

The magnetic moment M, which in this case corresponds to the extensive variable X, is

$$M = -\left(\frac{\partial F^*}{\partial \mathscr{H}}\right)_T = N\mu_B \tanh \frac{\mu_B \mathscr{H}}{kT}. \tag{13–31}$$

In strong fields and at low temperatures, where $\mu_B \mathscr{H} \gg kT$, tanh $(\mu_B \mathscr{H}/kT)$ approaches 1 and the magnetic moment approaches the value

$$M = N\mu_B. \tag{13–32}$$

But this is simply the *saturation magnetic moment* M_{sat}, which would result if all ionic magnets were parallel to the field.

At the other extreme of weak fields and high temperatures, $\mu_B \mathscr{H} \ll kT$, tanh $(\mu_B \mathscr{H}/kT)$ approaches $\mu_B \mathscr{H}/kT$, and Eq. (13–31) becomes

$$M = \left(\frac{N\mu_B^2}{k}\right)\frac{\mathscr{H}}{T}. \tag{13–33}$$

But this is just the experimentally observed *Curie law*, stating that in weak fields and at high temperatures the magnetic moment is directly proportional to $(\mathscr{H}/T)$, or

$$M = C_C \frac{\mathscr{H}}{T}, \tag{13–34}$$

where C_C is the Curie constant. The methods of statistics therefore not only lead to the Curie law, but they also provide a theoretical value of the Curie constant, namely,

$$C_C = \frac{N\mu_B^2}{k}. \tag{13–35}$$

Workers in the field of paramagnetism customarily use cgs units. The unit of magnetic intensity is 1 oersted* [(1 Oe) equal to 10^{-4} A m^2.] The Bohr magneton is

$$\mu_B = 0.927 \times 10^{-20} \text{ erg Oe}^{-1},$$

---

* Hans C. Oersted, Danish physicist (1777–1851).

and the Boltzmann constant is

$$k = 1.38 \times 10^{-16} \text{ erg K}^{-1}.$$

If the number of particles is Avogadro's number N_A, equal to 6.02×10^{23} cgs units, the Curie constant as given by Eq. (13–33) is

$$C_C = \frac{N_A \mu_B^2}{k} = 0.376 \text{ cm}^3 \text{ K mole}^{-1}.$$

The complete theory leads to the result that for chromium ions Cr^{+++}, of magnetic moment $3\mu_B$, the value of C_C is 5 times as great, or

$$C_C = 5 \times 0.376 = 1.88 \text{ cm}^3 \text{ K mole}^{-1}.$$

The experimentally measured value is

$$C_C = 1.84 \text{ cm}^3 \text{ K mole}^{-1}$$

in good agreement with the predictions of quantum theory.

The ratio M/M_{sat} is

$$\frac{M}{M_{sat}} = \tanh \frac{\mu_B \mathscr{H}}{kT}. \tag{13–36}$$

Figure 13–6 is a graph of the *magnetization curve* of the system, in which the ratio M/M_{sat} is plotted as a function of $\mu_B \mathscr{H}/kT$. The magnetization curve represents the balance struck by the system between the *ordering* effect of the external field $\mathscr{H}$, which is to align all ionic magnets in the direction of the field, and the *disordering* effect of thermal agitation, which increases with temperature. In weak

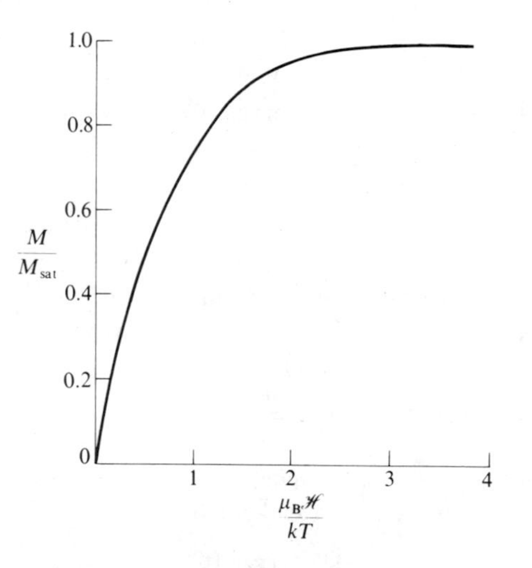

Fig. 13–6 Magnetization curve of a paramagnetic crystal.

fields the values of the two energy levels are nearly equal, both are nearly equally populated, and the *resultant* magnetic moment is very small. In strong fields, the difference between the energy levels is large, the ordering effect predominates, and nearly all magnets line up in the lower energy level where they have the same direction as $\mathscr{H}$.

It will be seen from Fig. 13–6 that saturation, as predicted by quantum theory, is very nearly attained when $\mu_B\mathscr{H}/kT = 3$, or when

$$\frac{\mathscr{H}}{T} = \frac{3k}{\mu_B} = 45 \text{ kOe K}^{-1}.$$

Hence, if $T = 300$ K, a field of 13.5×10^6 Oe would be required for saturation. On the other hand, if the temperature is as low as 1 K, a field of 4.5×10^4 Oe would produce saturation, and at a temperature of 0.1 K, a field of only 4.5×10^3 Oe would be required. (Modern superconducting electromagnets can produce magnetic intensities up to 1.5×10^5 Oe.)

We now calculate the other thermodynamic properties of the system. The total energy E, which in this case is the potential energy E_p, is

$$E = E_p = NkT^2\left(\frac{\partial \ln Z_{\mathscr{H}}}{\partial T}\right)_{\mathscr{H}}$$

$$= -Nk\left(\frac{\mu_B\mathscr{H}}{k}\right)\tanh\frac{\mu_B\mathscr{H}}{kT}. \tag{13–37}$$

Comparison with Eq. (13–29) shows that the potential energy is

$$E_p = -\mathscr{H}M. \tag{13–38}$$

The potential energy is negative because of our choice of reference level; that is, the potential energy of a magnetic dipole is set equal to zero when the dipole is at right angles to the field.

The heat capacity at constant $\mathscr{H}$ is

$$C_{\mathscr{H}} = \left(\frac{\partial E}{\partial T}\right)_{\mathscr{H}}$$

$$= Nk\left(\frac{\mu_B\mathscr{H}}{kT}\right)^2 \operatorname{sech}^2\frac{\mu_B\mathscr{H}}{kT}. \tag{13–39}$$

Figure 13–7 shows graphs of E_p and $C_{\mathscr{H}}$, (both divided by Nk) as functions of $kT/\mu_B\mathscr{H}$. The curves differ from the corresponding curves for the internal energy and heat capacity of an assembly of harmonic oscillators because there are only two permitted energy levels and the energy of the subassembly cannot increase indefinitely with increasing temperature.

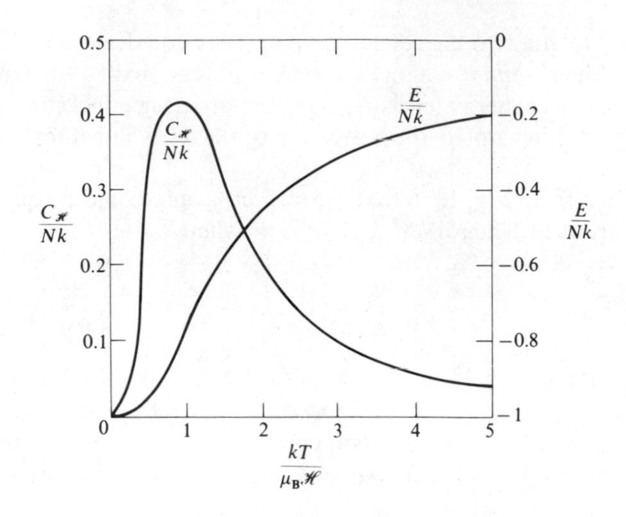

Fig. 13–7 The specific potential energy and specific heat capacity at constant magnetic intensity, both divided by Nk, for a paramagnetic crystal as a function of $kT/\mu_B H$.

Let us compare the heat capacity $C_{\mathscr{H}}$ of the magnetic ion subassembly with the heat capacity C_V of the entire crystal. Let $T = 1$ K and $\mathscr{H} = 10^4$ Oe. Then

$$\frac{kT}{\mu_B \mathscr{H}} \simeq 1.5, \qquad \operatorname{sech}^2 \frac{\mu_B \mathscr{H}}{kT} \simeq 0.81,$$

and by Eq. (13–39),

$$C_{\mathscr{H}} \simeq Nk(1.5)^{-2} \times 0.81 \simeq 0.36\, Nk.$$

Assuming there are 50 nonmagnetic particles for every magnetic ion, and taking a Debye temperature of 300 K as a typical value, we have from the Debye T^3 law,

$$C_V \simeq Nk(50) \times \frac{12\pi^4}{5}\left(\frac{1}{300}\right)^3,$$

$$\simeq 0.5 \times 10^{-5}\, Nk.$$

At this temperature, then, the heat capacity of the magnetic ions is about 100,000 times as great as the vibrational heat capacity of the crystal lattice. Much more energy is required to orient the ionic magnets than to increase the vibrational energy of the molecules of the lattice. It is this energy of orientation which allows the cooling of the lattice during the process of adiabatic demagnetization described in Section 8–8.

The entropy of the subassembly can now be calculated from the equation $F^* = E - TS$. From Eqs. (13–30) and (13–37) we have

$$S = \frac{E - F^*}{T} = Nk\left[\ln\left(2\cosh\frac{\mu_B \mathscr{H}}{kT}\right) - \frac{\mu_B \mathscr{H}}{kT}\tanh\frac{\mu_B \mathscr{H}}{kT}\right]. \qquad (13\text{–}40)$$

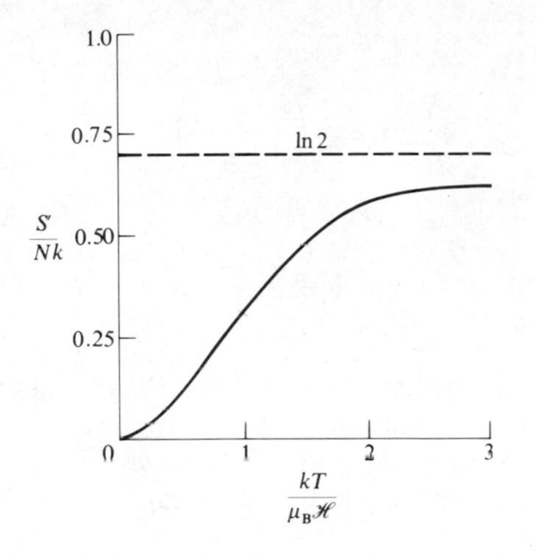

Fig. 13–8 The entropy of a paramagnetic crystal.

Figure 13–8 is a graph of S/Nk, plotted as a function of $kT/\mu_B\mathscr{H}$. At a given value of $\mathscr{H}$, S approaches zero as T approaches zero. At this temperature, all dipoles are in their lower energy state; there is only one possible microstate; and $S = k \ln \Omega = k \ln 1 = 0$. At the other limit, when $kT \gg \mu_B\mathscr{H}$,

$$\cosh (\mu_B\mathscr{H}/kT) \to 1, \ (\mu_B\mathscr{H}/kT) \to 0, \ \tanh (\mu_B\mathscr{H}/kT) \to 1,$$

and $S \to Nk \ln 2$. The entropy is also a function of $(\mathscr{H}/T)$ only. In a reversible adiabatic demagnetization, S and hence $(\mathscr{H}/T)$ remains constant. Thus as $\mathscr{H}$ is decreased, T must decrease also in agreement with the thermodynamic result.

13–5 NEGATIVE TEMPERATURES

Consider again a system with just two possible magnetic energy levels, in which the magnetic moment μ_B of a particle can be either parallel or antiparallel to a magnetic intensity $\mathscr{H}$. The energy of the lower level, in which μ_B is parallel to $\mathscr{H}$, is $\epsilon_1 = -\mu_B\mathscr{H}$; and that of the upper level, in which μ_B is opposite to $\mathscr{H}$, is $\epsilon_2 = +\mu_B\mathscr{H}$. In the equilibrium state at a temperature T, the average occupation numbers of the levels are

$$\bar{N}_1 = \frac{N}{Z} \exp\left(\frac{-\epsilon_1}{kT}\right),$$

$$\bar{N}_2 = \frac{N}{Z} \exp\left(\frac{-\epsilon_2}{kT}\right).$$

The ratio $\bar{N}_1/\bar{N}_2$ is

$$\frac{\bar{N}_1}{\bar{N}_2} = \exp\left(\frac{\epsilon_2 - \epsilon_1}{kT}\right)$$

or

$$T = \frac{1}{k}\left[\frac{\epsilon_2 - \epsilon_1}{\ln \bar{N}_1 - \ln \bar{N}_2}\right], \tag{13–41}$$

and we can consider this as the equation *defining* T, in terms of ϵ_1, ϵ_2, $\bar{N}_1$, and $\bar{N}_2$. If $\epsilon_2 > \epsilon_1$ and $\bar{N}_1 > \bar{N}_2$, the right side of the equation is positive and T is positive. The situation can be represented graphically as in Fig. 13–9(a), in which the lengths of the heavy lines correspond to the average occupation numbers $\bar{N}_1$ and $\bar{N}_2$.

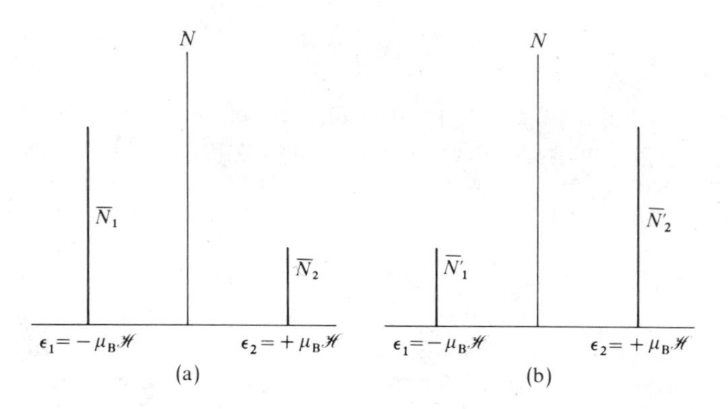

Fig. 13–9 (a) In the state of stable equilibrium the occupation number $\bar{N}_1$ of the level of lower energy is larger than the occupation number $\bar{N}_2$ of the level of higher energy. (b) Population inversion immediately after the magnetic intensity $\mathscr{H}$ has been reversed.

Now suppose the direction of the magnetic intensity is suddenly reversed. Those magnetic moments that were parallel to the original field, and in the state of lower energy ϵ_1, are opposite to the new field and are now in the higher-energy state, while those that were opposite to the original field, and in the higher-energy state ϵ_2, are parallel to the new field and are now in the lower-energy state. Eventually, the moments in the higher-energy state will flop over to the new lower-energy state, but immediately after the field has been reversed, and before any changes in occupation numbers have taken place, the situation will be that depicted in Fig. 13–9(b). The average occupation number $\bar{N}'_2$ of the new upper state is the same as the number $\bar{N}_1$ in the original lower state, and the occupation number $\bar{N}_1'$ of the new lower state is the same as the number $\bar{N}_2$ in the original upper state. We say there has been a *population inversion*. Then if we consider that the temperature of the system is *defined* by Eq. (13–41), and if T' is the temperature corresponding to Fig. 13–9(b),

$$T' = \frac{1}{k}\left[\frac{\epsilon_2 - \epsilon_1}{\ln N_1' - \ln N_2'}\right]. \tag{13–42}$$

Since $\bar{N}_2'$ is greater than $\bar{N}_1'$, the denominator on the right side of the equation is negative and T' is *negative*.

Negative temperatures can be looked at from another viewpoint. At a temperature $T = 0$, all magnets are in their lower-energy states. As the temperature is increased, more and more magnets move to the state of higher energy and when $T = +\infty$, both states are equally populated. Then one might say that if the number in the higher state is even *greater* than that in the lower state, as it is when there is a population inversion, the temperature must be *hotter* than infinity. We thus have the paradoxical result that a system at a negative temperature is even hotter than at an infinite temperature.

In paramagnetic substances, the interactions between the ionic magnets and the lattice are so great that the substance cannot exist in a state of population inversion for an appreciable time. However, it was found by Pound, Purcell, and Ramsey, in 1951, that the *nuclear* magnetic moments of the lithium atoms in LiF interact so slowly with the lattice that a time interval of several minutes is required for equilibrium with the lattice to be attained, a time long enough for experiments to be made showing that a population inversion actually existed.

13–6 THE ELECTRON GAS

The most important example of an assembly obeying the Fermi-Dirac statistics is that of the free electrons in a metallic conductor. We assume that each atom in the crystal lattice parts with some (integral) number of its outer valence electrons and that these electrons can move freely throughout the metal. There is, of course, an electric field within the metal due to the positive ions and which varies widely from point to point. On the average, however, the effect of this field cancels out except at the surface of the metal where there is a strong localized field (or *potential barrier*) that draws an electron back into the metal if it chances to make a small excursion outside the surface. The free electrons are therefore confined to the interior of the metal in much the same way that gas molecules are confined to the interior of a container. We speak of the electrons as an *electron gas*.

The degeneracies of the energy levels are the same as those of free particles in a box, with one exception. There are two sets of electrons in a metal, identical except that they have oppositely directed spins. The Pauli exclusion principle, instead of asserting that there can be no more than one particle per state, now permits *two* electrons per state provided they have oppositely directed spins. This is equivalent to doubling the number of states in a macrolevel, or the degeneracy $\Delta\mathscr{G}$ of the macrolevel, and permitting only *one* electron per state. Hence, instead of Eq. (12-17) we have

$$\Delta\mathscr{G}_v = \frac{8\pi m^3 V}{h^3}\, v^2\, \Delta v.$$

It will be more useful to express the degeneracy in terms of the kinetic energy $\epsilon = \frac{1}{2}mv^2$. Then since

$$v^2 = \frac{2\epsilon}{m}, \qquad v = \left(\frac{2}{m}\right)^{1/2}\epsilon^{1/2}, \qquad \Delta v = \frac{1}{2}\left(\frac{2}{m}\right)^{1/2}\epsilon^{-1/2}\,\Delta\epsilon,$$

it follows that

$$\Delta\mathscr{G}_\epsilon = 4\pi V\left(\frac{2m}{h^2}\right)^{3/2}\epsilon^{1/2}\,\Delta\epsilon. \tag{13-43}$$

If for brevity we set

$$A \equiv 4\pi V\left(\frac{2m}{h^2}\right)^{3/2}, \tag{13-44}$$

then

$$\Delta\mathscr{G}_\epsilon = A\epsilon^{1/2}\,\Delta\epsilon. \tag{13-45}$$

The degeneracy therefore increases with the square root of the energy. Then from the F-D distribution function, Eq. (11-40), the average number $\Delta\mathscr{N}$ of electrons in a macrolevel is

$$\Delta\mathscr{N} = \frac{\Delta\mathscr{G}_\epsilon}{\exp\left[(\epsilon - \mu)/kT\right] + 1} = A\,\frac{\epsilon^{1/2}}{\exp\left[(\epsilon - \mu)/kT\right] + 1}\,\Delta\epsilon. \tag{13-46}$$

The chemical potential μ can be evaluated from the requirement that $\sum \Delta\mathscr{N} = N$, where N is the total number of electrons. Replacing the sum with an integral, we have

$$N = A\int_0^\infty \frac{\epsilon^{1/2}}{\exp\left[(\epsilon - \mu)/kT\right] + 1}\,d\epsilon.$$

The integral cannot be evaluated in closed form and the result can be expressed only as an infinite series. The result, first obtained by Sommerfeld,* is

$$\mu = \epsilon_F\left[1 - \frac{\pi^2}{12}\left(\frac{kT}{\epsilon_F}\right)^2 + \frac{\pi^4}{80}\left(\frac{kT}{\epsilon_F}\right)^4 + \cdots\right]. \tag{13-47}$$

The quantity ϵ_F is a constant for a given metal and is called the *Fermi energy*. As we shall show, ϵ_F is a function of the number of electrons per unit volume, N/V,

* Arnold J. W. Sommerfeld, German physicist (1868–1951).

so the preceding equation expresses μ as a function of T and N/V. When $T = 0$, $\mu^0 = \epsilon_F$. The distribution function at $T = 0$ is then

$$\Delta\mathcal{N}^0 = \frac{\Delta\mathcal{G}_\epsilon}{\exp[(\epsilon - \epsilon_F)/kT] + 1}. \tag{13–48}$$

The significance of the Fermi energy ϵ_F can be seen as follows. In all levels for which $\epsilon < \epsilon_F$, the difference $(\epsilon - \epsilon_F)$ is a negative quantity, and at $T = 0$,

$$\frac{\epsilon - \epsilon_F}{kT} = -\infty.$$

The exponential term in Eq. (13–48) is then zero and in all levels for which $\epsilon < \epsilon_F$,

$$\Delta\mathcal{N}^0 = \Delta\mathcal{G}_\epsilon = A\epsilon^{1/2}\,\Delta\epsilon. \tag{13–49}$$

That is, the average number of electrons in a macrolevel equals the number of states in the level, and all levels with energies less than ϵ_F are fully occupied with their quota of one electron in each state.

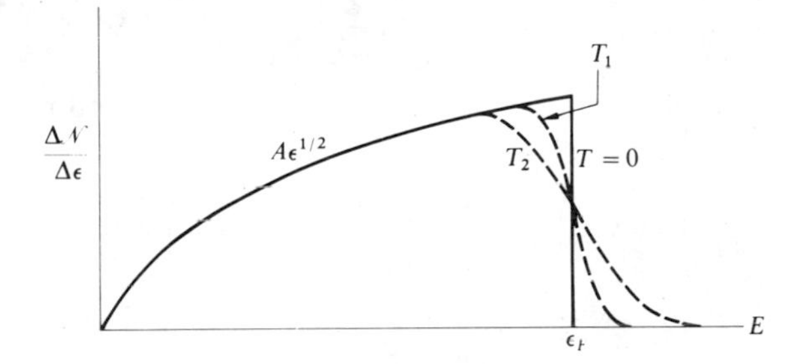

Fig. 13–10 Graphs of the distribution function of the free electrons in a metal, at $T = 0$ and at two higher temperatures T_1 and T_2.

In all levels for which $\epsilon > \epsilon_F$, the term $(\epsilon - \epsilon_F)$ is positive. Hence at $T = 0$ the exponential term equals $+\infty$ and $\Delta\mathcal{N}^0 = 0$. Thus there are no electrons in these levels and the Fermi energy ϵ_F is *the maximum energy of an electron at absolute zero*. The corresponding level is called the *Fermi level*.

The solid curve in Fig. 13–10 is a graph of the number of electrons per unit energy interval, $\Delta\mathcal{N}^0/\Delta\epsilon = A\epsilon^{1/2}$, at $T = 0$. The curve extends from $\epsilon = 0$ to $\epsilon = \epsilon_F$, and is zero at all energies greater than ϵ_F.

An expression for the Fermi energy can now be obtained from the requirement that $\sum \Delta\mathcal{N}^0 = N$. Replacing the sum with an integral, introducing the distribution function at $T = 0$, and integrating over all levels from zero to ϵ_F, we have

$$N = A\int_0^{\epsilon_F} \epsilon^{1/2}\,d\epsilon = \frac{2}{3}A\epsilon_F^{3/2};$$

or, after inserting the expression for A,

$$\epsilon_F = \frac{h^2}{8m}\left(\frac{3N}{\pi V}\right)^{2/3}. \qquad (13\text{–}50)$$

Thus as stated earlier, ϵ_F is a function of the number of electrons per unit volume, N/V, but is independent of T.

As a numerical example, let the metal be silver, and since silver is monovalent we assume one free electron per atom. The density of silver is 10.5×10^3 kg m^{-3}, its atomic weight is 107, and the number of free electrons per cubic meter, N/V, equals the number of atoms per cubic meter which is 5.86×10^{28}. The mass of an electron is 9.11×10^{-31} kg and $h = 6.62 \times 10^{-34}$ J s. Then

$$\epsilon_F = 9.1 \times 10^{-19}\text{ J} = 5.6\text{ eV}.$$

The total energy U of the electrons is

$$U = \sum \epsilon\, \Delta\mathcal{N}, \qquad (13\text{–}51)$$

or, replacing the sum with an integral,

$$U = A\int_0^\infty \frac{\epsilon^{3/2}}{\exp[(\epsilon - \mu)/kT] + 1}\, d\epsilon.$$

Again, the integral cannot be evaluated in closed form and must be expressed as an infinite series. The result is

$$U = \frac{3}{5}N\epsilon_F\left[1 + \frac{5\pi^2}{12}\left(\frac{kT}{\epsilon_F}\right)^2 - \frac{\pi^4}{16}\left(\frac{kT}{\epsilon_F}\right)^4 + \cdots\right]. \qquad (13\text{–}52)$$

When $T = 0$,

$$U^0 = \frac{3}{5}N\epsilon_F. \qquad (13\text{–}53)$$

It is left as a problem to show that the same result is obtained if one inserts in Eq. (13–51) the expression for the distribution function at $T = 0$, and integrates from $\epsilon = 0$ to $\epsilon = \epsilon_F$.

The mean energy per electron at absolute zero is

$$\overline{\epsilon^0} = \frac{U^0}{N} = \frac{3}{5}\epsilon_F.$$

Hence for silver,

$$\overline{\epsilon^0} = \frac{3}{5} \times 5.6\text{ eV} \approx 3.5\text{ eV}.$$

The mean kinetic energy of a gas molecule at room temperature is only about 0.03 eV, and the temperature at which the mean kinetic energy of a gas molecule is 3.5 eV is nearly 28,000 K. Hence the mean kinetic energy of the electrons in a metal, even at absolute zero, is much greater than that of the molecules of an ordinary gas at temperatures of many thousand kelvins.

At a temperature of 300 K, and for silver for which $\epsilon_F = 9.1 \times 10^{-19}$ J,

$$\frac{kT}{\epsilon_F} = \frac{1.38 \times 10^{-23} \times 300}{9.1 \times 10^{-19}} = 4.58 \times 10^{-3}.$$

Thus at this temperature the terms in powers of (kT/ϵ_F), in the series expansion in Eq. (13–47), are all very small and to a good approximation one can consider that $\mu = \epsilon_F$ at any temperature.

The dotted curves in Fig. 13–10 are graphs of the distribution function $\Delta\mathcal{N}/\Delta\epsilon$, at higher temperatures T_1 and T_2, where $T_2 > T_1$. It will be seen that the occupation numbers change appreciably with increasing temperature, only in those levels near the Fermi level. The reason for this is the following. Suppose the energy U of the metal is gradually increased from its value U^0 at $T = 0$, thus gradually raising its temperature. In order to accept a small amount of energy, an electron must move from its energy level at $T = 0$ to a level of slightly higher energy. But except for those electrons near the Fermi level, all states of higher energy are fully occupied so that only those electrons near the Fermi level can move to a higher level when the temperature is increased. With increasing temperature, those levels just below the Fermi level become gradually depleted, electrons at still lower levels can move to those that have been vacated, and so on.

For the particular level at which $\epsilon = \mu$, the quantity $(\epsilon - \mu) = 0$, and at any temperature above $T = 0$, the exponential term in the distribution function equals 1, and the occupation number is

$$\Delta\mathcal{N} = \tfrac{1}{2}\Delta\mathcal{G}_\epsilon.$$

If the temperature is not too great, then to a good approximation $\mu = \epsilon_F$ and to this approximation we can say that at any temperature above $T = 0$, the Fermi level is 50% occupied.

The heat capacity at constant volume, C_V, is given by

$$C_V = \left(\frac{\partial U}{\partial T}\right)_V,$$

and from Eq. (13–52),

$$C_V = \frac{\pi^2}{2}\left(\frac{kT}{\epsilon_F}\right)Nk\left[1 - \frac{3\pi^2}{10}\left(\frac{kT}{\epsilon_F}\right)^2 + \cdots\right]. \tag{13–54}$$

If the temperature is not too great, we can neglect terms in powers of (kT/ϵ_F) higher than the first, and to this approximation

$$C_V = \frac{\pi^2}{2}\left(\frac{kT}{\epsilon_F}\right)Nk. \tag{13–55}$$

Replacing Nk with nR, where n is the number of moles, and dividing both sides by n, we have for the molal specific heat capacity of the free electrons in a metal,

$$c_v = \frac{\pi^2}{2}\left(\frac{kT}{\epsilon_F}\right)R, \tag{13–56}$$

which is zero at $T = 0$ and which increases linearly with the temperature T. For silver at 300 K, using the value of (kT/ϵ_F) previously calculated,

$$c_v = 2.25 \times 10^{-2}R.$$

The molal specific heat capacity of a monatomic ideal gas, on the other hand, is

$$c_v = \frac{3}{2} R.$$

Thus although the mean kinetic energy of the electrons in a metal is very much larger than that of the molecules of an ideal gas at the same temperature, the energy *changes* only very slightly with changing temperature and the corresponding heat capacity is extremely small. This result served to explain what had long been a puzzle in the electron theory of metallic conduction. The observed molal specific heat capacity of metallic conductors is not very different from that of nonconductors, namely, according to the Dulong-Petit law, about $3R$. But the free electrons, if they behaved like the molecules of an ideal gas, should make an additional contribution of $3R/2$ to the specific heat capacity, resulting in a value much larger than that actually observed. The fact that only those electrons having energies near the Fermi level can *increase* their energies as the temperature is increased leads to the result above, namely, that the electrons make only a negligible contribution to the heat capacity.

To calculate the entropy of the electron gas, we make use of the fact that in a reversible process at constant volume, the heat flow into the gas when its temperature increases by dT is

$$dQ_r = C_V\, dT = T\, dS;$$

and hence at a temperature T the entropy is

$$S = \int_0^T \frac{dQ_r}{T} = \int_0^T \frac{C_V}{T}\, dT.$$

Inserting the expression for C_V from Eq. (13–54) and carrying out the integration, we obtain

$$S = Nk\frac{\pi^2}{2}\left(\frac{kT}{\epsilon_F}\right)\left[1 - \frac{\pi^2}{10}\left(\frac{kT}{\epsilon_F}\right)^2 + \cdots\right] \tag{13–57}$$

Hence the entropy is zero at $T = 0$, as it must be since there is only one possible microstate at $T = 0$ and at this temperature $\Omega = 1$, $S = k \ln \Omega = 0$.

The Helmholtz function F is

$$F = U - TS,$$

and from the expressions derived above for U and S,

$$F = \frac{3}{5} N\epsilon_F\left[1 - \frac{5\pi^2}{12}\left(\frac{kT}{\epsilon_F}\right)^2 + \cdots\right]. \tag{13–58}$$

The pressure P of the electron gas is given by

$$P = -\left(\frac{\partial F}{\partial V}\right)_T,$$

and since

$$\epsilon_F = \frac{h^2}{8m}\left(\frac{3N}{V}\right)^{2/3},$$

it follows that

$$P = \frac{2}{5}\frac{N\epsilon_F}{V}\left[1 + \frac{5\pi^2}{12}\left(\frac{kT}{\epsilon_F}\right)^2 + \cdots\right]. \tag{13–59}$$

This is the equation of state of the electron gas, expressing P as a function of V and T.

Comparison with Eq. (13–52) shows that the pressure is two-thirds of the energy density

$$P = \frac{2}{3}\frac{U}{V}.$$

For silver, $N/V \simeq 6 \times 10^{28}$ electrons per cubic meter and $\epsilon_F \simeq 10 \times 10^{-19}$ J. Hence at absolute zero,

$$P \simeq \tfrac{2}{5} \times 6 \times 10^{28} \times 10 \times 10^{-19} \simeq 24 \times 10^9 \text{ N m}^{-2}$$

$$\simeq 240{,}000 \text{ atm!}$$

In spite of this tremendous pressure, the electrons do not all evaporate spontaneously from the metal because of the potential barrier at its surface.

PROBLEMS

13–1 (a) Show that the entropy of an assembly of N Einstein oscillators is given by

$$S = 3Nk\left\{\frac{\theta_E/T}{\exp(\theta_E/T) - 1} - \ln\left[1 - \exp(-\theta_E/T)\right]\right\}.$$

(b) Show that the entropy approaches zero as T approaches zero and (c) that the entropy approaches $3Nk[1 + \ln(T/\theta_E)]$ when T is large. (d) Make a plot of S/R versus T/θ_E.

13–2 (a) From Fig. 3–10 find the characteristic Einstein temperature θ_E for copper such that the Einstein equation for c_v agrees with experiment at a temperature of 200 K. (b) Using this value of θ_E, calculate c_v at 20 K and 1000 K and compare with the experimental values. (c) Make a sketch of θ_E versus temperature so that the Einstein equation for c_v will yield the experimental values.

13–3 The characteristic Debye temperature for diamond is 1860 K and the characteristic Einstein temperature is 1450 K. The experimental value of c_v for diamond, at a temperature of 207 K, is 2.68×10^3 J kilomole^{-1} K^{-1}. Calculate c_v at 207 K from the Einstein and Debye equations and compare with experiment.

13–4 (a) Show that the heat capacity of a one-dimensional array of N coupled linear oscillators is given by

$$C_V = 3Nkx_m^{-1}\int_0^{x_m} \frac{x^2 e^x\, dx}{(e^x - 1)^2},$$

where $x = h\nu/kT$, and it is assumed that both longitudinal and transverse waves can propagate along the array. (b) Evaluate this expression for C_V in the low and high temperature limits.

13–5 To show that the Debye specific heat capacity at low temperature can be determined from measurements of the velocity of sound, (a) show that

$$\theta_D = \frac{hc}{k}\left(\frac{3N}{4\pi V}\right)^{1/3},$$

where

$$\frac{1}{c^3} = \frac{1}{3}\left(\frac{1}{c_l^3} + \frac{2}{c_t^3}\right);$$

and (b) show that the specific heat per kilogram c_v is

$$c_v = \frac{16\pi^5}{5}\frac{k^4}{h^3}\frac{T^3}{\rho c^3} = 1.22 \times 10^{11}\frac{T^3}{\rho c^3},$$

where ρ is the density of the material. (c) Calculate the average value of the sound velocity in copper. For copper, ρ is approximately 9000 kg m^{-3} and $c_v = 0.15$ J kg^{-1} K^{-1} at 5 K. (d) Calculate a value for θ_D and for ν_m for copper. (e) Calculate the value of λ_{min} and compare to the interatomic spacing, assuming that copper has a cubic structure.

13–6 Calculate values (a) for c_1 and c_2 of Eq. (13–22) and (b) for the Stefan-Boltzmann constant σ.

13–7 (a) Show that for electromagnetic radiation the energy per unit volume in the wavelength range between λ and $\lambda + d\lambda$ is given by

$$d\mathbf{u}_\lambda = \frac{8\pi hc}{\lambda^5}\frac{d\lambda}{\exp(hc/\lambda kT) - 1}.$$

(b) Show that the value of λ for which $\Delta\mathbf{u}_\lambda$ is a maximum is given by $\lambda_m T = 2.9 \times 10^{-3}$ m K. This is known as Wien's displacement law. (c) Calculate λ_m for the earth, assuming the earth to be a blackbody.

13–8 (a) Show that Wien's law can be derived by assuming that photons obey M-B statistics. (b) Show that Wien's law results in a total energy density which is nearly the same as that derived in Section 13–3.

13–9 If the magnetic moment $\mu = g\mu_B$ of an atom is large enough, there will be $2J + 1$ possible angles θ between the magnetic moment and the applied magnetic intensity $\mathcal{H}$ corresponding to magnetic levels having energies $\epsilon_J = m_J\mu\mathcal{H}$ where m_J has values between $-J$ and $+J$. (a) Show that $Z_{\mathcal{H}}$ will be given by

$$Z = \frac{\sinh\dfrac{(2J+1)\mu\mathcal{H}}{2kT}}{\sinh\dfrac{\mu\mathcal{H}}{2kT}}. \qquad (13\text{–}60)$$

[*Hint:* See the derivation of Eq. (12–44).] (b) Show that the net magnetic moment of the system is given by

$$M = N\mu\left[\frac{(2J+1)}{2}\coth(2J+1)\frac{\mu\mathscr{H}}{2kT} - \coth\frac{\mu\mathscr{H}}{2kT}\right].$$

This is called the Brillouin* function. (c) Show that the net magnetic moment follows Curie's law in the limit of high temperatures and low fields. (d) In the limit of low temperature and high fields, show that all the dipoles are aligned. (e) Show that the expression for the net magnetic moment derived in part (b) reduces to Eq. (13–29) when $2J + 1 = 2$ and $g = 2$.

13–10 Use Eq. (13–60) of the previous problem to calculate the entropy of N distinguishable magnetic dipoles. Evaluate the expression in the limit of high and low temperatures and make a graph of the entropy as a function of T and $\mathscr{H}$.

13–11 A paramagnetic salt contains 10^{25} magnetic ions per cubic meter, each with a magnetic moment of 1 Bohr magneton. Calculate the difference between the number of ions aligned parallel to the applied intensity of 10 kOe and that aligned antiparallel at (a) 300 K, (b) 4 K, if the volume of the sample is 100 cm³. Calculate the magnetic moment of the sample at these two temperatures.

13–12 Use the statistical definitions of work, total energy, and net magnetic moment to show that the work of magnetization is given by $dW = -\mathscr{H}\,dM$. [*Hint:* See Section 3–13.]

13–13 Derive expressions for the magnetic contribution to the entropy and the heat capacity at constant magnetic intensity $\mathscr{H}$ for the system discussed in Section 13–4. Sketch curves of these properties as a function of $\mathscr{H}/T$.

13–14 Calculate the mean speed, the root-mean-square speed and the mean-reciprocal speed in terms of $v_F \equiv (2\epsilon_F/m)^{1/2}$ for an electron gas at 0 K.

13–15 (a) Show that the average number of electrons having speeds between v and $v + dv$ is given by

$$\Delta\mathscr{N}_v = \frac{8\pi m^3 V}{h^3}\,\frac{v^2\,\Delta v}{\exp\left[(\frac{1}{2}mv^2 - \mu)/kT\right] + 1}.$$

(b) Sketch $\Delta\mathscr{N}_v/\Delta v$ as a function of v^2 at $T = 0$ K.

13–16 (a) Calculate ϵ_F for aluminum assuming 3 electrons per aluminum atom. (b) Show that for aluminum at 1000 K, μ differs from ϵ_F by less than 0.01%. (c) Calculate the electronic contribution to the molal specific heat capacity of aluminum at room temperature and compare it to $3R$. (The density of aluminum is 2.7×10^3 kg m^{-3} and its atomic weight is 27.)

13–17 The Fermi velocity is defined as $v_F \equiv (2\epsilon_F/m)^{1/2}$ and the Fermi temperature as $T_F \equiv \epsilon_F/k$. (a) Calculate values of the Fermi velocity, momentum, and temperature for electrons in silver. (b) Determine the magnitude of the second term in Eqs. (13–47), (13–52), (13–54), (13–57), (13–58), and (13–59) at room temperature. (c) At what temperature does the second term contribute approximately a 1% correction in the above equations?

---

* Leon N. Brillouin, French physicist (1889–1969).

13–18 Find the mean energy per electron by substituting the expression for $\Delta\mathcal{N}^0$ into Eq. (13–51).

13–19 Derive Eqs. (13–57), (13–58), and (13–59).

13–20 In a one-dimensional electron gas $\Delta\mathcal{G}_\epsilon = \frac{2L}{h}\sqrt{2m/\epsilon}\,\Delta\epsilon$ where L is the length of the sample of N electrons, (a) Sketch $\mathcal{N}^0(\epsilon)$ as a function of ϵ. (b) Show that $\epsilon_F = \frac{h^2N^2}{32mL^2}$. (c) Find the average energy per electron at 0 K.

13–21 (a) Use the data shown in Fig. 7–7 to determine the Fermi energy of liquid He^3 which can also be considered as a gas of particles obeying Fermi-Dirac statistics. (b) Determine the Fermi velocity and temperature for He^3. (See Problem 13–17).

13–22 The free electrons in silver can be considered an electron gas. Calculate the compressibility and expansivity of this gas and compare them to the experimental values for silver of 0.99×10^{-11} m^2 N^{-1} and 56.7×10^{-6} K^{-1}, respectively.

Appendix

A

Selected differentials from a condensed collection of thermodynamic formulas by P. W. Bridgman

Any partial derivative of a state variable of a thermodynamic system, with respect to any other state variable, a third variable being held constant [for example, $(\partial u/\partial v)_T$] can be written, from Eq. (4–20), in the form

$$(\partial u/\partial v)_T = \frac{(\partial u/\partial z)_T}{(\partial v/\partial z)_T}$$

where z is any arbitrary state function. Then if one tabulates the partial derivatives of all state variables with respect to an arbitrary function z, any partial derivative can be obtained by dividing one tabulated quantity by another. For brevity, derivatives of the form $(\partial u/\partial z)_T$ are written in the table below in the symbolic form $(\partial u)_T$. Then, for example,

$$\left(\frac{\partial u}{\partial v}\right)_T = \frac{(\partial u)_T}{(\partial v)_T} = \frac{T(\partial v/\partial T)_P + P(\partial v/\partial P)_T}{-(\partial v/\partial P)_T} = \frac{T\beta}{\kappa} - P,$$

which agrees with Eq. (6–9). Ratios (not derivatives) such as $d'q_P/dv_P$ can be treated in the same way. For a further discussion, see *A Condensed Collection of Thermodynamics Formulas* by P. W. Bridgman (Harvard University Press, 1925), from which the table below is taken.

P constant	*T* constant
$(\partial T)_P = 1$	$(\partial P)_T = -1$
$(\partial v)_P = (\partial v/\partial T)_P$	$(\partial v)_T = -(\partial v/\partial P)_T$
$(\partial s)_P = c_P/T$	$(\partial s)_T = (\partial v/\partial T)_P$
$(\partial q)_P = c_P$	$(\partial q)_T = T(\partial v/\partial T)_P$
$(\partial w)_P = P(\partial v/\partial T)_P$	$(\partial w)_T = -p(\partial v/\partial P)_T$
$(\partial u)_P = c_P - P(\partial v/\partial T)_P$	$(\partial u)_T = T(\partial v/\partial T)_P + P(\partial v/\partial P)_T$
$(\partial h)_P = c_P$	$(\partial h)_T = -v + T(\partial v/\partial T)_P$
$(\partial g)_P = -s$	$(\partial g)_T = -v$
$(\partial f)_P = -s - P(\partial v/\partial T)_P$	$(\partial f)_T = P(\partial v/\partial P)_T$

h constant

$(\partial P)_h = -c_P$

$(\partial T)_h = v - T(\partial v/\partial T)_P$

$(\partial v)_h = -c_P(\partial v/\partial P)_T - T(\partial v/\partial T)_P^2 + v(\partial v/\partial T)_P$

$(\partial s)_h = vc_P/T$

$(\partial q)_h = vc_P$

$(\partial w)_h = -P[c_P(\partial v/\partial P)_T + T(\partial v/\partial T)_P^2 - v(\partial v/\partial T)_P]$

g constant

$(\partial P)_g = s$

$(\partial T)_g = v$

$(\partial v)_g = v(\partial v/\partial T)_P + s(\partial v/\partial P)_T$

$(\partial s)_g = \frac{1}{T}[vc_P - sT(\partial v/\partial T)_P]$

$(\partial q)_g = -sT(\partial v/\partial T)_P + vc_P$

$(\partial w)_g = P[v(\partial v/\partial T)_P + s(\partial v/\partial P)_T]$

s constant

$(\partial P)_s = -c_P/T$

$(\partial T)_s = -(\partial v/\partial T)_P$

$(\partial v)_s = -\frac{1}{T}[c_P(\partial v/\partial P)_T + T(\partial v/\partial T)_P^2]$

$(\partial q)_s = 0$

$(\partial w)_s = -\frac{P}{T}[c_P(\partial v/\partial P)_T + T(\partial v/\partial T)_P^2]$

$(\partial u)_s = \frac{P}{T}[c_P(\partial v/\partial P)_T + (T\,\partial v/\partial T)_P^2]$

$(\partial h)_s = -vc_P/T$

$(\partial g)_s = -\frac{1}{T}[vc_P - sT(\partial v/\partial T)_P]$

$(\partial f)_s = \frac{1}{T}[Pc_P(\partial v/\partial P)_T + PT(\partial v/\partial T)_P^2 + sT(\partial v/\partial T)_P]$

v constant

$(\partial P)_v = -(\partial v/\partial T)_P$

$(\partial T)_v = (\partial v/\partial P)_T$

$(\partial s)_v = \frac{1}{T}[c_P(\partial v/\partial P)_T + T(\partial v/\partial T)_P^2]$

$(\partial q)_v = c_P(\partial v/\partial P)_T + T(\partial v/\partial T)_P^2$

$(\partial w)_v = 0$

$(\partial u)_v = c_P(\partial v/\partial P)_T + T(\partial v/\partial T)_P^2$

$(\partial h)_v = c_P(\partial v/\partial P)_T + T(\partial v/\partial T)_P^2 - v(\partial v/\partial T)_P$

$(\partial g)_v = -v(\partial v/\partial T)_P - s(\partial v/\partial P)_T$

$(\partial f)_v = -s(\partial v/\partial P)_T$

B

The Lagrange method of undetermined multipliers

In an algebraic equation such as

$$ax + by = 0, \tag{B-1}$$

one is accustomed to consider one of the variables, say x, as the *independent* variable and the other variable, y, as the *dependent* variable. The equation is then considered as imposing a relation between the dependent and independent variables in terms of the coefficients a and b, namely, in this case, $y = -(a/b)x$.

Suppose, however, that *both* x and y are *independent* variables. Then y may have *any* value regardless of the value of x, and we can no longer *require* that $y = -(a/b)x$. The equation $ax + by = 0$ can be satisfied for *any* pair of variables x and y only if $a = 0$, $b = 0$.

Suppose next that x and y are not completely independent but must also satisfy a *condition equation*, which we take, for example, as

$$x + 2y = 0. \tag{B-2}$$

What can we now say about the coefficients a and b in Eq. (B–1)? One procedure is to consider Eq. (B–1) and the condition equation (B–2) as a pair of simultaneous linear equations. We solve Eq. (B–2) for x and substitute in Eq. (B–1):

$$x = -2y$$

$$a(-2y) + by = 0,$$

$$b = 2a. \tag{B-3}$$

Then Eq. (B–1) is satisfied for any pair of values of a and b that satisfy Eq. (B–3), *provided* the values of x and y satisfy the condition equation (B–2).

If the number of independent variables and condition equations is small, the procedure above is adequate. But when these numbers become very large, there are too many simultaneous equations to solve. In this case, we use the Lagrange* method of undetermined multipliers. Each condition equation is multiplied by an undetermined constant λ. If there are k condition equations, there are k such

* Joseph L. Lagrange, French mathematician (1736–1813).

multipliers: $\lambda_1, \lambda_2, \ldots, \lambda_k$. In our problem there is only one such equation and one multiplier λ. Then from Eq. (B–2),

$$\lambda x + 2\lambda y = 0. \tag{B–4}$$

Now add this to Eq. (B–1), giving

$$(a + \lambda)x + (2\lambda + b)y = 0. \tag{B–5}$$

Now assign a value to λ such that the coefficient of either x or y is zero. If we choose x, then

$$(a + \lambda) = 0; \qquad \lambda = -a. \tag{B–6}$$

Equation (B–5) then reduces to

$$(2\lambda + b)y = 0, \tag{B–7}$$

which contains only *one* of the variables. But since either one of the variables can be considered independent, Eq. (B–7) is satisfied only if

$$(2\lambda + b) = 0; \qquad b = -2\lambda. \tag{B–8}$$

Then from Eqs. (B–6) and (B–8) we have

$$b = 2a, \tag{B–9}$$

which is the same as Eq. (B–3).

In effect, the use of Lagrange multipliers leads to an equation, Eq. (B–5), which has the same property as if both x and y were independent, since the coefficient of each is zero.

We now use the Lagrange method of undetermined multipliers to explain how Eqs. (8–29), the equations of phase equilibrium, are a necessary consequence of Eq. (8–27), which expresses the condition that the Gibbs function shall be a minimum, subject to the condition equations (8–28). If the values of the $dn_i^{(j)}$'s in Eq. (8–27) were completely independent, the equation could be satisfied for an arbitrary set of the $dn_i^{(j)}$'s only if the coefficient of each were zero. The method of undetermined multipliers takes the condition equations into account so as to eliminate some of the terms in Eq. (8–27) to obtain an equation in which the remaining $dn_i^{(j)}$'s are independent, so that the coefficient of each can be set equal to zero. The procedure is as follows.

We multiply the first of the condition equations (8–28) by a constant λ_1 whose value for the present is undetermined. The second equation is multiplied by a second constant λ_2, the next by a constant λ_3, and so on. These equations are then added to Eq. (8–27). The result is the equation

$$\begin{aligned}
&(\mu_1^{(1)} + \lambda_1)\, dn_1^{(1)} + (\mu_1^{(2)} + \lambda_1)\, dn_1^{(2)} + \cdots + (\mu_1^{(\pi)} + \lambda_1)\, dn_1^{(\pi)} \\
&\quad + (\mu_2^{(1)} + \lambda_2)\, dn_2^{(1)} + (\mu_2^{(2)} + \lambda_2)\, dn_2^{(2)} + \cdots + (\mu_2^{(\pi)} + \lambda_2)\, dn_2^{(\pi)} \\
&\quad \vdots \\
&\quad + (\mu_k^{(1)} + \lambda_k)\, dn_k^{(1)} + (\mu_k^{(2)} + \lambda_k)\, dn_k^{(2)} + \cdots + (\mu_k^{(\pi)} + \lambda_k)\, dn_k^{(\pi)} = 0.
\end{aligned} \tag{B–10}$$

The total number of $dn_i^{(j)}$'s in this equation is $k\pi$, one for each of the k constituents in each of the π phases. For any constituent i, arbitrary values may be assigned to the dn_i's in all phases but one, making a total of $(\pi - 1)$ arbitrary values. The remaining dn_i then takes up the slack, since

$$\sum_{j=1}^{j=\pi} dn_i^{(j)} = 0.$$

Then since there are k constituents, the total number of $dn_i^{(j)}$'s which can be given arbitrary values, or the number that are independent, is $k(\pi - 1) = k\pi - k$. Let us therefore assign values to the (as yet) undetermined multipliers, such that for each constituent i, in some one of the phases j, the sum $(\mu_i^{(j)} + \lambda_i) = 0$. For example, let us select phase 1 and assign a value to λ_1 such that in phase 1

$$(\mu_1^{(1)} + \lambda_1) = 0, \qquad \text{or} \qquad \mu_1^{(1)} = -\lambda_1.$$

Then the product $(\mu_1^{(1)} + \lambda_1)\, dn_1^{(1)}$ is zero regardless of the value of $dn_1^{(1)}$ and this term drops out of the sum in Eq. (B–10). In the same way, we let

$$(\mu_2^{(1)} + \lambda_2) = 0, \qquad \text{or} \qquad \mu_2^{(1)} = -\lambda_2,$$

and so on for each of the k constituents. This reduces the number of $dn_i^{(j)}$'s in Eq. (B–1) by k, leaving a total of $k\pi - k$. But since this is the number of $dn_i^{(j)}$'s that can be considered independent, it follows that the coefficient of each of the remaining $dn_i^{(j)}$'s must be zero. Therefore for *any* constituent i in *any* phase j,

$$\mu_i^{(j)} = -\lambda_i.$$

Therefore the chemical potential of any constituent i has the same value $-\lambda_i$ in all phases, which leads to the equations of phase equilibrium, Eqs. (8–29). Note that the values of the λ_i's themselves need not be known; the only significant aspect is that the values of the chemical potentials of every phase are *equal*, whatever these values may be.

One can consider that, *in effect*, the method of Lagrange multipliers makes *all* the $dn_i^{(j)}$'s in Eq. (B–10) independent, since the coefficient of each is zero, but the coefficients are zero for different reasons. In phase 1, the coefficients are zero because we assigned values to the λ's to make them zero. In the other phases, the coefficients are zero because the remaining $dn_i^{(j)}$'s are independent.

The choice of phase 1 in the preceding argument was not essential; we could equally well have started with any other phase and, in fact, could have selected different phases for each constituent. In any case, we would eliminate the same number k of $dn_i^{(j)}$'s from Eq. (B–10), and the remainder would be independent.

C

Properties of factorials

In the derivations of the distribution functions of particles obeying the various statistics, many properties of the factorial are used. In this appendix we derive these properties by investigating the gamma function $\Gamma(s)_x$. Stirling's approximation for calculating factorials of large numbers is also developed.

The factorial of a positive integer n is written $n!$ and defined as

$$n! = n(n-1)(n-2)\cdots 1. \tag{C–1}$$

From this definition it follows that

$$(n+1)! = (n+1)n!. \tag{C–2}$$

Equation (C–2) can be used to determine $0!$ and $(-n)!$

If $n = 0$, Eq. (C–2) gives $1! = (0!)$ and

$$0! = 1. \tag{C–3}$$

If $n = -1$, Eq. (C–2) results in the expression $0! = 0(-1)!$. Since $0! = 1$, we can take $(-1)!$ to be ∞, that is

$$(-1)! = \infty. \tag{C–4}$$

However, this involves division by zero which is undefined mathematically. The gamma function is an expression for values of n which may not be integer, which yields Eqs. (C–1) to (C–3) for integer n. In the limit that n approaches -1, the gamma function approaches ∞.

Integrals of the form

$$f(s) = \int_0^\infty a(t)e^{-st}\,dt$$

are called Laplace* transforms. They are very useful in many branches of science and engineering. The gamma function is a Laplace transform in which $s = 1$ and $a(t) = t^n$ where n need not be an integer. Thus

$$f(1) = \Gamma(n+1) \equiv \int_0^\infty t^n e^{-t}\,dt. \tag{C–5}$$

* Marquis de Pierre S. Laplace, French mathematician (1749–1827).

For $n \geq -1$, integration by parts yields

$$\int_0^\infty t^n e^{-t}\,dt = -t^n e^{-t}\Big|_0^\infty + n\int_0^\infty t^{n-1}e^{-t}\,dt.$$

The first term on the right is zero at both limits since e^{-t} approaches zero faster than t^n approaches infinity at the upper limit. Then

$$\int_0^\infty t^n e^{-t}\,dt = n\int_0^\infty t^{n-1}e^{-t}\,dt$$

or

$$\Gamma(n+1) = n\Gamma(n). \tag{C–6}$$

The gamma function can be successively integrated by parts so that

$$\Gamma(n+1) = n(n-1)(n-2)\cdots 1,$$

and if n is an integer

$$\Gamma(n+1) = n!. \tag{C–7}$$

If $n = 0$, the gamma function can be integrated directly and

$$\Gamma(1) = \int_0^\infty e^{-t}\,dt = 1.$$

Since by Eq. (C–7), $\Gamma(1) = 0!$,

$$0! = 1, \tag{C–8}$$

in agreement with Eq. (C–3).

The integral of Eq. (C–5) diverges if $n \leq -1$, but by rewriting Eq. (C–6) as

$$n^{-1}\Gamma(n+1) = \Gamma(n), \tag{C–9}$$

the definition of $\Gamma(n)$ can be extended to negative integers. If $0 < n < 1$, $\Gamma(n)$ can be determined from Eq. (C–9). Using this recursion formula again, the values of $\Gamma(n)$ for $-1 < n < 0$ can be found from the values for $\Gamma(n)$ when $0 < n < 1$, and so on. Thus $\Gamma(n)$ is determined for all noninteger values of n.

However, since $\Gamma(1) = 1$ the method fails for $n = 0$, since division by zero is undefined. Thus

$$\lim_{n\to 0}\Gamma(n) = \lim_{n\to 0} n^{-1}\Gamma(n+1) = \pm\infty. \tag{C–10}$$

Similar behavior is found for all negative integers.

For small values of n the factorial can be evaluated by direct computation. However, it is often necessary to evaluate $n!$ for large values of n. The factorial of a large number can be found with sufficient precision by Stirling's approximation which we now derive.

The natural logarithm of factorial n is

$$\ln(n!) = \ln 2 + \ln 3 + \cdots + \ln n.$$

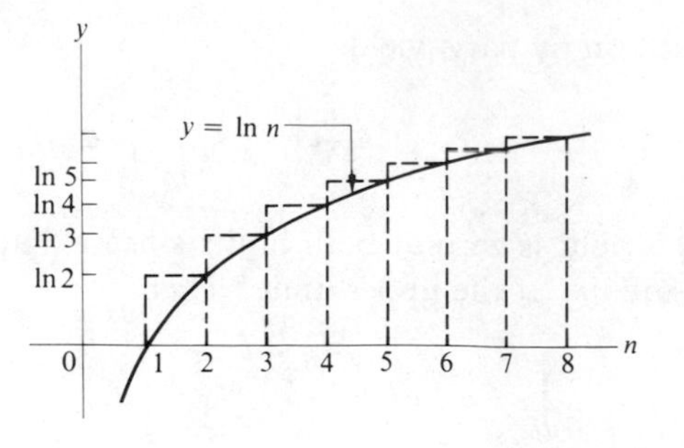

Fig. C–1 A graph of ln n as a function of n.

This is exactly equal to the area under the step curve shown by dotted lines in Fig. C–1, between $n = 1$ and $n = n$, since each rectangle is of unit width and the height of the first is ln 2, that of the second is ln 3, etc. This area is approximately equal to the area under the smooth curve $y = \ln n$ between the same limits, provided n is large. For small values of n the step curve differs appreciably from the smooth curve, but the latter becomes more and more nearly horizontal as n increases. Hence approximately, for large n,

$$\ln (n!) = \int^{n} \ln n \, dn.$$

Integration by parts gives

$$\ln (n!) = n \ln n - n + 1,$$

and if n is large we may neglect the 1, so finally

$$\ln (n!) = n \ln n - n. \tag{C–11}$$

This is Stirling's approximation.

An exact analysis leads to the following infinite series.

$$n! = \sqrt{2\pi n}\left(\frac{n}{e}\right)^n\left[1 + \frac{1}{12n} + \frac{1}{288n^2} - \frac{139}{51840n^3} + \cdots\right]. \tag{C–12}$$

If all terms in the series except the first are neglected, we obtain

$$\ln (n!) = \tfrac{1}{2} \ln 2\pi + \tfrac{1}{2} \ln n + n \ln n - n. \tag{C–13}$$

If n is very large compared with unity, the first two terms of this expression are negligible also, and we get Eq. (C–11).

D

An alternative derivation of distribution functions

At the end of Section 11–5, it was noted that when the number of particles in an assembly becomes large, the occupation numbers of the levels in the most probable macrostate are very nearly the same as the average occupation numbers for the assembly. This is not only true for particles obeying B–E statistics, but it holds equally well for the other statistics. Thus when the system is in equilibrium, the distribution of particles among levels can also be determined from the occupation numbers of the macrostate with the maximum thermodynamic probability, subject to the constraints that the total energy and the total number of particles of the assembly is constant.

When one looks at a large number of identical assemblies, one macrostate occurs the most often. The assumption is that this macrostate is the distribution of particles among levels for the system in equilibrium. Therefore the properties of the system are determined by the distribution of particles among levels that has the maximum thermodynamic probability. In the text we assume that the properties of the system are determined by the average occupation numbers of the levels. In the limits of large numbers of particles both methods lead to the same distribution functions, as we shall show.

We now describe the conventional procedure for calculating occupation numbers in the most probable macrostate, or, the most probable occupation numbers. If we let $\mathscr{W}^*$ represent the thermodynamic probability of the most probable macrostate, the entropy S is set proportional to the logarithm of $\mathscr{W}^*$, that is,

$$S = k_B \ln \mathscr{W}^*.$$

To find the most probable macrostate, we use the usual criterion for the maximum value of a function, namely, that its first variation is equal to zero. (Strictly speaking, it should also be shown that this leads to a *maximum* value and not to a *minimum*.) We shall illustrate by considering the Maxwell-Boltzmann statistics, although the same procedure can be followed in the other statistics as well.

In the M-B statistics, the thermodynamic probability of a macrostate is given by Eq. (11–21),

$$\mathscr{W}_{\text{M-B}} = N! \prod_j \frac{g_j^{N_j}}{N_j!} \tag{D–1}$$

Instead of maximizing $\mathscr{W}$, it is simpler to maximize $\ln \mathscr{W}$, since if $\mathscr{W}$ is a maximum, its logarithm is a maximum also. Then considering the thermodynamic probability of the most probable macrostate,

$$\ln \mathscr{W}^* = \ln N! + \sum_j N_j \ln g_j - \sum_j \ln N_j!. \tag{D–2}$$

We assume that $N \gg 1$, and that in any level j, $N_j \gg 1$, so that we can use the Stirling approximation (see Appendix C), and

$$\ln N! = N \ln N - N,$$

$$\ln N_j! = N_j \ln N_j - N_j.$$

Then

$$\ln \mathscr{W}^* = N \ln N - N + \sum_j N_j \ln g_j - \sum_j N_j \ln N_j + \sum_j N_j.$$

But $\sum N_j = N$, so

$$\ln \mathscr{W}^* = N \ln N + \sum_j N_j \ln g_j - \sum_j N_j \ln N_j = N \ln N - \sum_j N_j \ln \frac{g_j}{N_j}. \tag{D–3}$$

Now compare this macrostate with a neighboring macrostate in which the occupation numbers are slightly different. Let the occupation number of any level j differ from its most probable value by δN_j. Since $\delta N_j \ll N_j$, we can use the methods of differential calculus, considering δN_j as a mathematical differential. The differential of $\ln \mathscr{W}^*$ is then, since N and g_j are constants,

$$\delta \ln \mathscr{W}^* = \sum_j \ln g_j \, \delta N_j - \sum_j \delta N_j - \sum_j \ln N_j \, \delta N_j. \tag{D–4}$$

Since the total number of particles is the same in the two macrostates, any increases in the occupation numbers of some levels must be balanced by decreases in the occupation numbers of other levels, and hence $\sum_j \delta N_j = 0$. Since $\ln \mathscr{W}^*$ is to be a maximum, we set $\delta \ln \mathscr{W}^* = 0$. Then

$$\sum_j \ln \frac{g_j}{N_j} \, \delta N_j = 0,$$

or

$$\left(\ln \frac{g_1}{N_1}\right) \delta N_1 + \left(\ln \frac{g_2}{N_2}\right) \delta N_2 + \cdots = 0. \tag{D–5}$$

If the N_j's were *independent*, then as explained in Appendix B, this equation could be satisfied only if the coefficient of each δN_j were zero. But the δN_j's are not independent. We have shown above that

$$\delta N = \sum_j \delta N_j = 0; \tag{D-6}$$

and since the total energy $U = \sum_j \epsilon_j N_j$ is the same in both macrostates, any increase in energy resulting from an increase in the occupation number of a level must be balanced by a decrease in the energy of other levels and a second condition equation is

$$\delta U = \sum_j \epsilon_j \, \delta N_j = 0. \tag{D-7}$$

We therefore use the Lagrange method of undetermined multipliers described in Appendix B. Multiply the first condition equation, Eq. (D–6), by a constant which for later convenience we write as $\ln \alpha$, multiply the second by a constant $-\beta$, and add these products to Eq. (D–5), obtaining

$$\sum_j \left(\ln \frac{g_j}{N_j} + \ln \alpha - \beta \epsilon_j \right) \delta N_j = 0.$$

In effect, the δN_j's are now independent and the coefficient of each must be zero. Hence for any level j,

$$\ln \frac{g_j}{N_j} + \ln \alpha - \beta \epsilon_j = 0, \tag{D-8}$$

or

$$N_j = \alpha g_j \exp(-\beta \epsilon_j), \tag{D-9}$$

which is the distribution function for the most probable occupation numbers, expressed in terms of the constants α and β.

Now sum the preceding equation over all j's, and let

$$Z = \sum_j g_j \exp(-\beta \epsilon_j)$$

where Z is the single particle partition function described in Section 11–14. Then since $\sum_j N_j = N$, it follows that

$$\alpha = \frac{N}{Z}, \tag{D-10}$$

and from Eq. (D–9),

$$\frac{N_j}{g_j} = \frac{N}{Z} \exp(-\beta \epsilon_j). \tag{D-11}$$

To evaluate the constant β, we insert in Eq. (D–3) the expression for $\ln(g_j/N_j)$ from Eq. (D–11), and set $S = k_B \ln \mathscr{W}^*$, giving

$$S = k_B \left[N \ln N - \sum_j N_j \ln N + \sum_j N_j \ln Z + \beta \sum_j \epsilon_j N_j \right],$$

or

$$S = N k_B \ln Z + \beta k_B U. \tag{D-12}$$

If the energy levels are functions of the volume V (or some other extensive parameter), then Z is a function of β and V and has the same value in two equilibrium states in which the values of β and V are the same. The entropy difference S between the states, since $\ln Z$ is a constant, is

$$\Delta S = \beta k_{\mathrm{B}} \Delta U. \tag{D-13}$$

From the principles of thermodynamics, the entropy difference between two equilibrium states at the same temperature and volume is

$$\Delta S = \frac{\Delta U}{T}.$$

It follows that $\beta k_{\mathrm{B}} = 1/T$, or

$$\beta = \frac{1}{k_{\mathrm{B}} T}. \tag{D-14}$$

Hence Eq. (D–12) can be written

$$S = N k_{\mathrm{B}} \ln Z + \frac{U}{T}, \tag{D-15}$$

and

$$F = U - TS = -N k_{\mathrm{B}} T \ln Z. \tag{D-16}$$

The chemical potential μ is

$$\mu = \left(\frac{\partial F}{\partial N}\right)_{T,V} = -k_{\mathrm{B}} T \ln Z, \tag{D-17}$$

and hence

$$\ln Z = \frac{-\mu}{k_{\mathrm{B}} T}, \qquad \frac{1}{Z} = \exp \frac{\mu}{k_{\mathrm{B}} T}. \tag{D-18}$$

The distribution function, from Eq. (D–11), can now be written as

$$\frac{N_j}{g_j} = N \exp \frac{\mu - \epsilon_j}{k_{\mathrm{B}} T}. \tag{D-19}$$

Comparison with Eq. (11–44) shows that the distribution function for the *most probable* occupation numbers is given by the same expression as that for the *average* occupation numbers.

One objection to the conventional procedure is that if an N_j is calculated from the preceding equation, the value obtained is not necessarily an *integer*, while the actual occupation number of a level is necessarily integral. If we consider the right side of Eq. (D–19) does give the correct values of the average occupation numbers, this equation can be interpreted to mean that the occupation numbers in the most probable macrostate are the nearest integer to their values averaged over all macrostates. Since the occupation numbers are all very large, the "nearest integer" will differ by only a *relatively* small amount from the average.

A more serious objection is the following. One of the terms in the expression for the thermodynamic probability of a macrostate in the Fermi-Dirac statistics is $(g_j - N_j)!$. If $\ln (g_j - N_j)!$ is evaluated by the Stirling approximation, and the procedure above is followed, one does obtain the same expression for the most probable occupation numbers as that for their average values. But in the F-D statistics, the difference $(g_j - N_j)$ is not necessarily a large number and may in fact be zero if a level is fully occupied. The use of the Stirling approximation to evaluate $\ln (g_j - N_j)!$ is therefore questionable, even if it leads to the right answer. The procedure followed in Section 11–10, however, does not require the use of Stirling's approximation and is valid provided only that the N_j's themselves are large numbers.

E

Magnetic potential energy

Each magnetic ion in a paramagnetic crystal is a small permanent magnet and is equivalent to a tiny current loop as in Fig. E–1. The ion has a magnetic moment μ, which if the ion actually did consist of a current I in a loop of area A, would equal (in the system of units we are using) the product IA. The moment can be represented by a vector perpendicular to the plane of the loop.

If the moment vector makes an angle θ with the direction of an external magnetic field of intensity $\mathscr{H}$, a torque τ of magnitude $\mu\mathscr{H} \sin\theta$ is exerted on the loop, in such a direction as to align the magnetic moment in the same direction as $\mathscr{H}$. In Fig. E–1, this torque is clockwise. In the usual sign convention, the angle θ is considered positive when measured *counterclockwise* from the direction of θ, so we should write

$$\tau = -\mu\mathscr{H} \sin\theta. \tag{E–1}$$

If the loop is given a small counterclockwise displacement, so that the angle θ increases by $d\theta$, the work of this torque is

$$dW = \tau\, d\theta = -\mu\mathscr{H} \sin\theta\, d\theta.$$

The increase in magnetic potential energy of the loop, $d\epsilon_{\mathrm{p}}$, is defined as the *negative* of this work, just as the increase in gravitational potential energy of a body of mass m, when it is lifted vertically in a gravitational field of intensity g, is the negative of the work of the downward gravitational force $-mg$ exerted on it. Hence

$$d\epsilon_{\mathrm{p}} = \mu\mathscr{H} \sin\theta\, d\theta. \tag{E–2}$$

The total change in potential energy when the angle θ is increased from θ_1 to θ_2 is

$$\epsilon_{\mathrm{p}_2} - \epsilon_{\mathrm{p}_1} = \mu\mathscr{H} \int_{\theta_1}^{\theta_2} \sin\theta\, d\theta = \mu\mathscr{H}(\cos\theta_1 - \cos\theta_2).$$

Let us take the reference level of potential energy as that at which the moment is at right angles to the field, where $\theta = 90°$ and $\cos\theta = 0$. Hence if we set

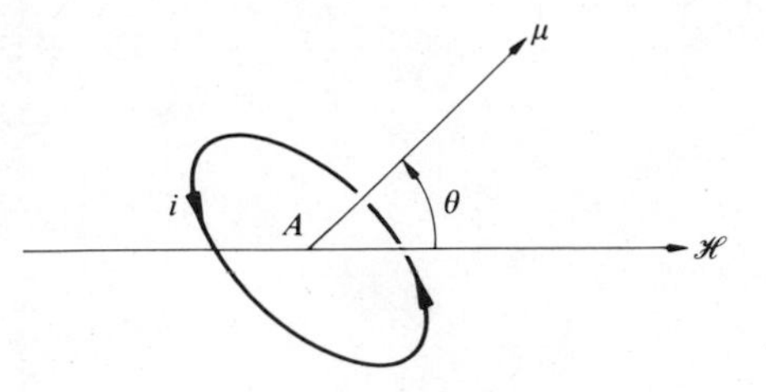

Fig. E–1 A magnetic ion of magnetic moment μ is equivalent to a small current loop.

$\theta_1 = 90°$ and $\epsilon_{p_1} = 0$, and let ϵ_{p_2} and θ_2 refer to any arbitrary angle θ,

$$\epsilon_p - 0 = \mu\mathscr{H}(0 - \cos\theta),$$

and

$$\epsilon_p = -\mu\mathscr{H}\cos\theta. \tag{E–3}$$

When the angle θ is less than 90°, as in Fig. E–1, $\cos\theta$ is positive and the potential energy ϵ_p is negative. That is, the potential energy is less than that in the reference level. When θ is greater than 90°, $\cos\theta$ is negative and ϵ_p is positive.

Let $\Delta\mathscr{N}_\theta$ be the number of atomic magnets whose moments make angles with the field between θ and $\theta + \Delta\theta$. Each of these has a component moment in the direction of the field of $\mu\cos\theta$, and the moment due to these is

$$\Delta M = \Delta\mathscr{N}_\theta\,\mu\cos\theta.$$

The total moment M of the entire crystal is

$$M = \sum\Delta\mathscr{N}_\theta\,\mu\cos\theta. \tag{E–4}$$

In the same way, the total potential energy E_p of the crystal is

$$E_p = -\sum\Delta\mathscr{N}_\theta\,\mu\mathscr{H}\cos\theta.$$

It follows from the two preceding equations that

$$E_p = -\mathscr{H}M. \tag{E–5}$$

Answers to Problems

Chapter 1

1-1 (a) no; (d) yes.

1-2 (a) extensive; (d) intensive.

1-3 (a) 10^3 kg m^{-3}; (b) 10^{-3} m^3 kg^{-1}; (c) 18×10^{-3} m^3 kilomole^{-1}; (d) 1.29 kg m^{-3}, 0.775 m^3 kg^{-1}, 22.4 m^3 kilomole^{-1}.

1-4 About 100 Torr.

1-5 (b) 1.01×10^5 N m^{-2}.

1-6 (a) 4.

1-7 (c) decrease.

1-8 153 K, 185 K, 193 K, 197 K.

1-9 (a) 328 K; (b) 6.84 cm; (c) no.

1-10 (a) $a = 1.55 \times 10^{-3}$, $b = -115$; (b) 112 degrees; (c) 5.97 cm.

1-11 (a) 73.3; (b) 26.7 degrees.

1-12 (a) 672; (b) 180 degrees.

1-13 (a) $A = 3.66 \times 10^{-4}$ atm K^{-1}, $B = 321$ degrees, $C = 3.66 \times 10^{-3}$ K^{-1}; (b) 130 degrees; (c) 0.12 atm; (d) $-\infty$.

1-14 (a)

t(°C)	−100	0	200	400	500
$\mathscr{E}$(mV)	−60	0	60	40	0

;

(b) $a = 2.5$ degrees m V^{-1}, $b = 0$;

(c)

t(°C)	−100	0	200	400	500
t^*(deg)	−150	0	150	100	0

.

1-15 (a) −195.80°C; (b) 139.23 R; (c) −320.44°F.

1-16 (a) 14.20 kelvins; (b) 14.20 deg C; (c) 25.56 rankines; (d) 25.56 deg F.

1-17 (a) no; (b) yes.

1-21 (a) reversible isobaric process; (b) quasistatic isothermal process; (c) irreversible (adiabatic) compression; (d) irreversible isochoric process; (e) reversible isothermal process; (f) irreversible adiabatic process.

Chapter 2

2-2 (a) 5.7×10^{-2} m^3 kilomole^{-1}; (b) 8.8 kilomoles; (c) 5.3 kilomoles.

2-3 (a) $A = P_1^2/RT_1$; (c) 800 K.

2-4 (a) 0.25 m; (b) 500 Torr.

2–5 (a) 456 K.

2–6 0.18 m.

2–7 8.66*d*.

2–9 (a) 300 K; (b) 6.24 m^3 $kilomole^{-1}$; (c) 750 K, 120 K; (d) 10 m^3; (e) 8 kg.

2–10 (a) 0.308 kilomoles; (b) 9.86 kg; (c) 3.96×10^6 N m^{-2}; (d) 0.277 kilomoles.

2–11 (a) 1 m^3; (b) 150 K; (c) 200 K, 0.67 m^3; (d) 225 K, 0.75 m^3.

2–13 (b) 0.06, 0.22, 0.51.

2–14 (a) 4.87×10^7 N m^{-2}; (b) 5.10×10^7 N m^{-2}; (c) 8.31×10^3 and 8.70×10^3 J kilomole^{-1} K^{-1}.

2–19 6.5×10^7 N m^{-2}.

2–23 (a) $\beta = (v - b)/vT$, $\kappa = (v - b)^2/RTv$.

2–25 $v = v_0 \exp(aT^3/P)$, $a/b = \frac{1}{3}$.

2–26 (a) $L_0\alpha$; (b) $L_0(YA)^{-1}$; (c) $-\Delta\mathscr{F}/\alpha YA$.

2–27 (a) 2.88×10^5 N; (b) 6 m.

2–29 (a) 0.031 m^3 $kilomole^{-1}$; (b) 0.042 m^3 $kilomole^{-1}$.

2–30 (b) 0.270.

2–32 $[(v - b)(vRT + a)]/T[a(v - b) - v^2RT]$.

2–33 (a) $R/(v - b)$; (b) $R/(v - b)$; (c) $[\exp(-a/vRT)](v - b)^{-1}(R + a/vT)$.

2–35 (b) $10^{-12}(6.4 + 3.3 \times 10^{-3}T)$ m^2 N^{-1}; (c) -3.3×10^{-15} m^2 N^{-1} K^{-1}; (d) 5.2×10^{-3}.

Chapter 3

3–1 1.69×10^6 J.

3–2 1.91×10^5 J.

3–3 $-3nRT_1/8$.

3–4 2.03 J.

3–5 1.13 J.

3–6 (b) Work on the gas; (c) 8.15×10^4 J, 0.434 J; (d) 0.4 m^3, 1.44×10^{-6} m^3.

3–7 (a) $W = RT \ln [(v_2 - b)/(v_1 - b)] + a[(1/v_2) - (1/v_1)]$; (b) 4.26×10^6 J; (c) 4.3×10^6 J.

3–8 (b) $d'W = nR\,dT + nRT\,dP/P$.

3–9 (a) $d'W = -\mathscr{F}L_0(d\mathscr{F}/YA + \alpha\,dT)$; (b) $W_{\mathscr{F}} = -\mathscr{F}L_0\alpha(T_2 - T_1)$; (c) $W_T = -L_0(\mathscr{F}_2^2 - \mathscr{F}_1^2)/2YA$.

3–10 (a) $d'W = -C_C\mathscr{H}\,d\mathscr{H}/T + C_C\mathscr{H}^2\,dT/T^2$; (b) $W_{\mathscr{H}} = -C_C\mathscr{H}^2(1/T_f - 1/T_i)$; (c) $W_T = -(C_C/2T)(\mathscr{H}_f^2 - \mathscr{H}_i^2)$.

3–11 $-3(\mu_0 V + C_C/T)\mathscr{H}_i^2/2$.

3–13 -2.03×10^3 J.

3–14 (a) -3.11×10^5 J; (b) -4.32×10^5 J; (c) 150 K; (d) 1.25×10^5 N m^{-2}.

3–16 $W_a = 0$, $W_b = 11.2 \times 10^5$ J, $W_c = -8.08 \times 10^5$ J, $W_{abca} = 3.12 \times 10^5$ J.

3–17 (a) 6×10^6 J; (b) clockwise.

3–18 (a) 2.51×10^{-8} J; (b) counterclockwise.

3–19 $C_C\mathscr{H}^2/T$.

3–22 2.8×10^4 J.

3–26 (a) 60 J; (b) 70 J are liberated; (c) $Q_{a\text{-}d} = 50$ J, $Q_{b\text{-}d} = 10$ J.

3–27 (a) $\Delta U_{a\text{-}b} = Q_{a\text{-}b} = 100$ J, $\Delta U_{b\text{-}c} = 900$ J, $-\Delta U_{c\text{-}a} = 1000$ J, $W_{a\text{-}b} = Q_{c\text{-}a} = \Delta U_{\text{cycle}} = 0$, $Q_{\text{cycle}} = W_{\text{cycle}} = -500$ J.

3–28 (a) $Q = n[a(T_2 - T_1) + b(T_2^2 - T_1^2) + c(1/T_2 - 1/T_1)]$;
(b) $\bar{c}_P = a + b(T_2 + T_1) - c/T_1T_2$; (c) 24.0×10^3 and 26.0×10^3 J kilomole^{-1} K^{-1}.

3–29 (a) 0.589 J kilomole^{-1} K^{-1}; (b) 73.6 J kilomole^{-1} K^{-1}; (c) 1850 J; (d) 37.3 J kilomole^{-1} K^{-1}.

3–30 (a) 118 J; (b) 124 J; (c) 118 J.

3–31 (a) $C = \mathscr{P}\dfrac{dt}{dT}$.

3–32 (b) 1.39×10^4 J.

3–33 (a) 1.24×10^5 J; (b) 4000 J; (c) 1.16×10^5 J.

3–35 (a) -5.35×10^4 J; (b) $W_c = -5.25 \times 10^4$ J; (c) $W_d = -0.98 \times 10^3$ J.

3–36 (a) -3.6×10^5 J kg^{-1}; (b) -4.22×10^5 J kg^{-1}.

Chapter 4

4–2 (a) a.

4–3 (b) $5/[3(T_f + T_i)]$.

4–4 (a) $a = 24.0$ J kilomole^{-1} K^{-1}, $b = 6.9 \times 10^{-3}$ J kilomole^{-1} K^{-2}; (b) 2.03×10^4 J kilomole^{-1}.

4–7 (a) 27×10^3: 4.02×10^{-2}; (b) $\frac{5}{2}R:R$; (c) 0.60; (d) almost all.

4–8 (b) $a + R$.

4–11 (a) $q_{a\text{-}c\text{-}b} = 19\ RT_1/2$, $q_{a\text{-}d\text{-}b} = 17\ RT_1/2$, $q_{a\text{-}b} = 9\ RT_1$; (b) 3 R.

4–16 $\Delta T = (2n_An_B - n_A^2 - n_B^2)a/c_vV(n_A + n_B)2$.

4–18 (a) $a/(c_vv^2)$; (b) $c_vT - 2a/v - RTv/(v - b)$;
(c) $[2av(v - b)^2 - RTv^3b]/c_P[RTv^3 - 2a(v - b)^2]$.

4–21 (a) $nc_vT_0/2$; (b) $3T_0/2$; (c) $5.25\ T_0$; (d) $4.75\ nc_vT_0$.

4–22 885 K.

4–23 (a) $W_T = -3.46 \times 10^5$ J, $W_S = -2.5 \times 10^5$ J; (b) $W_T = -3.46 \times 10^5$ J; $W_S = -4.43 \times 10^5$ J.

4–24

Process	V_c(m^3)	T_f(K)	W(J)	Q(J)	ΔU(J)
a	32	400	6.73×10^6	6.73×10^6	0
b	13.9	174	2.74×10^6	0	-2.74×10^6
c	32	400	0	0	0

4–25 (b)

Process	ΔT(K)	ΔV(m^3)	ΔP(atm)	W(J)	Q(J)
T = const	0	22.4	−0.5	1.57×10^6	1.57×10^6
P = const	273	44.8	0	2.27×10^6	5.68×10^6
V = const	−438	0	−0.401	0	-5.45×10^6
$Q = 0$	165	−67.2	0.901	-2.04×10^6	0
Cycle	0	0	0	1.8×10^6	1.8×10^6

Process	ΔU(J)	ΔH(J)
T = const	0	0
P = const	3.41×10^6	5.68×10^6
V = const	-5.45×10^6	-9.09×10^6
$Q = 0$	2.04×10^6	3.41×10^6
Cycle	0	0

4–26 (a) $T(v - b)^{c_v/R}$ = constant, $(P + a/v^2)(v - b)^{(c_v+R)/R}$ = constant, (b) $W = c_v(T_i - T_f) + (a/v_f - a/v_i)$.

4–30 (a) 900 Calories; (b) 1600 Calories; (c) 300 and 400 Calories.

4–31 (b) lower T_1.

4–32 $\eta c = T_2/T_1$.

4–33 73 K, 230 K.

4–34 (a) 0.25, 3; (b) 0.167, 5.

4–36 (a) 2.34×10^5 watts; (b) 5.5; (c) 1.52×10^8 J; (d) 6.06×10^7 J.

4–37 13.6

4–38 3.1 watts, about 0.3%.

Chapter 5

5–1 83.3 K and 166.6 K.

5–3 (a) 12.2 J K^{-1}; (b) 6.06×10^3 J K^{-1}.

5–4 (a) $Q_{a\text{-}b}$ = 2192 J, $Q_{b\text{-}c}$ = 10,966 J, $Q_{c\text{-}d}$ = −6576 J, $Q_{d\text{-}a}$ = −5480 J; (b) 0.996×10^5 N m^{-2}; (c) $S_{a\text{-}b}$ = 5.54 J K^{-1}, $S_{b\text{-}c}$ = 11.0 J K^{-1}, $S_{c\text{-}d}$ = −5.54 J K^{-1}, $S_{d\text{-}a}$ = −11.0 J K^{-1}.

5–5 (a) 0; (b) 0.167 J K^{-1}.

5-6 293 J K^{-1}.

5-7 (a) 1200 J absorbed at 300 K, 200 J given up at 200 K; (b) -3 J K^{-1}, -1 J K^{-1}, 4 J K^{-1}; (c) 0.

5-8 (a) 777 J K^{-1}; (b) -777 J K^{-1}.

5-9 (a) 0.171 J K^{-1}; (b) -0.171 J K^{-1}.

5-10 (a) $am \ln (T_2/T_1) + bm(T_2 - T_1)$; (b) 2.47×10^4 J kilomole^{-1} K^{-1}.

5-13 (a) engine; (b) 250R J, -100R J; (c) 0.6; (d) 0.667.

5-15

	ΔS_{body}(J K^{-1})	ΔS_{res}(J K^{-1})	ΔS_{u}(J K^{-1})
(a)	6.93	-5.0	1.93
(b)	11.0	-6.67	4.33
(c)	-6.93	20.0	13.1

5-16 (a) $\Delta S_{H_2O} = 1300$ J K^{-1}, $\Delta S_{\text{res}} = -1120$ J K^{-1}, $\Delta S_{\text{u}} = 180$ J K^{-1}; (b) $\Delta S_{H_2O} =$ 1300 J K^{-1}, $\Delta S_{\text{res}} = -1210$ J K^{-1}, $S_{\text{u}} = 90$ J K^{-1}.

5-17 290 K, 190 J K^{-1}.

5-20 (c) T_f of part (b).

5-22 $-0.555\, RT_1 \leq w_g \leq 0$, $0 \leq \Delta u \leq 0.555\, RT_1$, $0 \leq \Delta S \leq 0.693\, R$.

5-27 No.

Chapter 6

6-1 (a) $P\kappa v - T\beta v$; (c) 0.

6-2 (a) 3360 J kilomole^{-1} K^{-1}; (b) 0.135.

6-3 (a) R; (b) $R \ln v/v_0$.

6-4 (a) $\Delta S = 3\, aVT + \dfrac{1}{T}\int P\, dV +$ constant; (c) $\kappa(T)$.

6-7 (a) $-(T\beta - 1)/\kappa$; (c) 0.

6-17 5.68 J K^{-1}.

6-18 (b) $(c_v + R)(T - T_0) + h_0$, $c_P(T - T_0) + h_0$.

6-19 (a) $a + bT - R$; (b) $s = a \ln (T/T_0) + b(T - T_0) - R \ln P/P_0 + s_0$, $h = a(T - T_0) + b(T^2 - T_0^2)/2 + h_0$; (c) $(a - R)(T - T_0) + b(T^2 - T_0^2)/2 + u_0$.

6-20 (a) 3.73×10^6 J; (b) 1.15×10^4 J K^{-1}.

6-22 (a) -4.6 J kg^{-1}; (b) -155 J kg^{-1}; (d) 0.394 K.

6-25 (a) $\sim$ 0.22 K; (b) 0; (c) $\sim$ 3.5 K.

6-26 (a) -253 J; (b) -253 J; (c) -91 J.

6-28 (a) $\eta = 0$, $\mu = -b/c_P$; (b) $\eta = -b/c_v$, $\mu = 0$.

6-29 -2.1 K.

6–30 (a) $\Delta T = 0$, $\Delta s = 1.91 \times 10^4$ J kilomole^{-1} K^{-1}; (b) $\Delta T = -146$ K, $\Delta s = 6.1 \times 10^3$ J kilomole^{-1} K^{-1}.

6–32 19.3 atm.

6–33 (a) 0.02 K atm^{-1}; (b) 0.098 K atm^{-1}; (c) -0.27 K, 12.3 K.

6–34 35.3 K.

6–38 $(\partial V/\partial M)_{S,T} = MV/C_C nR$

Chapter 7

7–6 (a) $P(v + A) = RT$, $s = -R \ln (P/P_0) + A'P$; (b) $h = P(A'T - A)$, $u = T(A'P - R)$, $f = RT[\ln (P/P_0) - 1]$; (c) $c_P = PA''T$, $c_v = 2A'P + A''TP - \dfrac{P^2A'^2}{R} - R$; (d) $\kappa = \dfrac{RT}{P(RT - AP)}$, $\beta = (R - A'P)/(RT - AP)$; (e) $\mu = (A - A'T)/PA''T$.

7–22 (b) -10^3 J, -50 J K^{-1}, -1.5×10^4 J, -1.48×10^4 J, -800 J, 3.6 J K^{-1}.

7–23 (a) -1.35×10^7 N m^{-2} K^{-1}; (b) 268 atm; (c) 1.31×10^6 N m^{-2} K^{-1}; (d) 24.6 atm.

7–25 (a) 200 K, 1.01 atm; (b) $l_{13} = 0.492$ J kilomole^{-1}, $l_{23} = 0.328$ J kilomole^{-1}, $l_{12} = 0.164$ J kilomole^{-1}.

7–27 (a) -0.15 K.

Chapter 8

8–1 (a) $(n_A + n_B)R \ln 2$.

8–2 (a) $\frac{1}{6}, \frac{1}{3}, \frac{1}{2}$; (b) $\frac{1}{3}, \frac{2}{3}$, 1 atm; (c) $-1.5 + 10^7$ J; (d) $+5 \times 10^4$ J K^{-1}.

8–5 (a) 2; P and T.

8–6 (c) K is not a function of P and $K = e^{-\Delta G^\circ/RT}$.

8–7 2.

8–8 (a) $A - T$, % Cd; $B - T$; $C - T$, % Cd; $D - 0$; $E - T$, % Cd; (c) $k = 22{,}000$ K kg kilomole^{-1}.

8–10 (a) 1.28×10^{-2} Torr; (b) 76.3 atm.

8–12 (a) 0.146 J m^{-2}; (b) $\lambda = 0.085$ J m^{-2}, $c_A = 6.82 \times 10^{-5}$ J m^{-2} K^{-1}, s $= 2.28 \times 10^{-4}$ J m^{-2} K^{-1}; (c) 2.5 K.

8–13 (a) $d\sigma(A_2 - A_1)$; (b) $\lambda(A_2 - A_1)$.

8–15 (a) $c_{\mathscr{F}} - c_l = l\alpha^2T/\kappa$, $\alpha = \dfrac{1}{l}\left(\dfrac{\partial l}{\partial T}\right)_{\mathscr{F}}$, $\kappa = \dfrac{1}{l}\left(\dfrac{\partial l}{\partial \mathscr{F}}\right)_T$; (b) $c_{\mathscr{F}}/c_l = \kappa/\kappa_s$.

8–17 (a) 4.2×10^{-2} J K^{-1}; (b) 12.6 J; (c) 20.3 J; (d) -7.7 J.

8–18 (c) $\Delta G = -20.3$ J, $\Delta H = -7.74$ J.

8–19 -228×10^6 J.

8–20 (b) $\dfrac{d\mathbf{u}}{3}(V_2 - V_1)$; (c) $\dfrac{4}{3}\mathbf{u}(V_2 - V_1)$.

8–22 (a) 378 K; (b) 2.04×10^{-6} and 5.14×10^{-6} N m^{-2}.

8–24 (a) $b\mathscr{H}/T$.

8–25 (a) -0.815 J; (b) -1.63 J, -1.63 J, 0, -0.815 J; (c) 100 Oe; (d) 7.93×10^4.

8–28 (g) 8.00×10^5 J; (h) -2.02×10^5 J, -3.96; (i) 5.98×10^5 J; (j) 0.300; (k) 5.48×10^5 J, 0.275.

8–29 (a) 220 Btu lbm^{-1}; (b) 70 Btu lbm^{-1}.

8–30 (c) $c = 8.7$.

Chapter 9

9–1 (a) 3.2×10^{19} molecules; (b) 3.2×10^{10} molecules.

9–2 3300 Å.

9–3 (a) 6.9×10^{-6}; (b) same as (a).

9–4 (a) 1.7×10^{-5}; (b) 2.8×10^{-3}.

9–5 (a) 0.01; (b) 1.7×10^{-7}, 2.8×10^{-5}; (c) $1.64 \times 10^{18}\, v_0$ molecules $\text{m}^{-2}\,\text{s}^{-1}$, 9.4×10^{20} molecules $\text{m}^{-2}\,\text{s}^{-1}$.

9–6 (a) 20 m s^{-1}, 20 m s^{-1}; (b) 12.5 m s^{-1}, 14.6 m s^{-1}; (c) 10 m s^{-1}, 12.2 m s^{-1}; (d) 10 m s^{-1}, 14.1 m s^{-1}; (e) 11.5 m s^{-1}, 12.7 m s^{-1}.

9–7 (c) $2\, v_0/3$; (d) $0.707\, v_0$.

9–8 (b) $3.4 \times 10^{20}\, v_0$ molecules $\text{m}^{-2}\,\text{s}^{-1}$; (c) $4.5 \times 10^{24}\, v_0$ molecules $\text{m}^{-2}\,\text{s}^{-1}$.

9–10 Force per unit length $= \mathbf{n}' m\bar{v}^2/2$.

9–11 (a) 1360 m s^{-1}; (b) 2400 K; (c) 0.31 eV.

9–12 (a) 2.9×10^{23} impacts s^{-1}; (b) 120 m.

9–13 (a) 7.2; (b) 1.22×10^{-3} atm.

9–14 (a) 2×10^{19} molecules cm^{-2}; (b) 3.3×10^{23} impacts s^{-1}; (c) same as (b); (d) mean energy is about 0.1 of heat of vaporization per molecule.

9–15 (a) 9.4×10^{-6} g $\text{cm}^{-2}\,\text{s}^{-1}$; (b) about the same.

9–16 $2.77\, V/\bar{v}A$.

9–17 (a) 10^{17} molecules; (b) 1.6×10^{-3} Torr.

9–18 $P_l = \dfrac{P_0}{2}\,[1 + \exp(-\bar{v}At/2V)]$.

9–20 (b) $v_{\text{rms}} \propto P^{1/5}$.

9–21 (a) 3 translational, 3 rotational, and $2(3N - 6)$ vibrational; (b) $9R$, 1.11.

9–23 (a) 1.5×10^3 J; (b) 1.36×10^3 m s^{-1}.

Chapter 10

10–3 (a) 3.2×10^{-19} m^2; (b) 5.2×10^{-8} m; (c) 9.6×10^9 s^{-1}.

10–4 $l \propto P^{-1}$, $z \propto P$.

10–5 (a) 5×10^{-10} m; (b) 7.9×10^{-19} m^2; (c) 7.9×10^5 m^{-1}; (d) 7.9×10^5 m^{-1}; (e) 0.45; (f) 0.88×10^{-6} m; (g) 1.3×10^{-6} m.

10–6 4.4×10^{-5} s.

10–7 (a) 3.2; (b) 0.05.

10–8 (a) 3.7×10^3; (b) 1.35×10^3; (c) 1.8×10^2; (d) 1.8×10^3; (e) 7.4×10^2; (f) 1.5×10^2; (g) $= 0$.

10–9 (a) 6; (b) 6; (c) 6.

10–10 1.2×10^{-10} m.

10–11 (a) 10 cm; (b) 61 μA.

10–12 (a) 4.9×10^{-10} m; (b) 160; (c) 2.7 m s^{-1}; (d) 160; (e) 48.

10–13 (a) 7.2×10^{-14} s; (b) 7.78×10^{-9} m, 34 atomic distances; (c) 0.2; (d) 632 s.

10–15 4×10^{-4}.

10–16 (a) $\eta/\sqrt{T_{avg}} = 9.6 \times 10^{-7}$ N s m^{-2} $K^{-1/2}$; (b) 4.2×10^{-10} m; (c) 2.8×10^{-10} m, 2.1×10^{-10} m.

10–17 (a) $\lambda \propto T^{1/2}$; (b) 0.058 J K^{-1} m^{-1} s^{-1}.

10–18 (a) 2.52×10^{-4} m^2 s^{-1}, 1.03×10^{-4} m^2 s^{-1}; (b) $D \propto T^{3/2}m^{1/2}P^{-1}$.

10–19 (a) -1.22×10^{25} (molecules m^{-3}) m^{-1}; (b) $(2.32 \times 10^{23} + 4.75 \times 10^{15})$ molecules s^{-1}; (c) $(2.32 \times 10^{23} - 4.75 \times 10^{15})$ molecules s^{-1}; (d) 9.50×10^{15} molecules s^{-1}, 0.70 μg s^{-1}.

10–20 (a) 1.26×10^{-5} N s m^{-2}; (b) 0.98×10^{-5} m^2 s^{-1}; (c) 9.1×10^{-3} J m^{-1} s^{-1} K^{-1}.

Chapter 11

11–3 10^8.

11–4 (b) 45, 50, 120, 75, 60, 100.

11–5 (a) 5; (b) 5, 4, 3, 2, 1; (c) 16, 32, 24, 8, 4; (d) 15, 84.

11–6 (a) 6.55×10^8; (b) 1.52×10^{32}; (c) 21.

11–8 (b) 2427; (c) 3.68, 1.79, 0.838, 0.394, 0.189, 0.078, 0.035; (d) 7.00.

11–9 (a) 14 macrostates; (d) 2.584, 1.585, 0.877, 0.485, 0.250, 0.135, 0.058, 0.027.

11–10 (b) 6; (c) 36.

11–12 (a) 8 macrostates; (d) 2.278, 1.722, 1.056, 0.667, 0.222, 0.056.

11–13 (d) 2.500, 1.591, 0.955, 0.530, 0.265, 0.114, 0.0378, 0.0075.

11–14 (a) $\delta N_1 = \delta N_2 = 1$; (b) 5.55×10^{12}, 4.13×10^{12}.

11–15 (a) 729; (b) 60; (c) 6ϵ; (d) 126.

11–16

	Macrostates	1	2	3	4	5	6
$\mathscr{W}_k$	B-E	45	50	120	75	60	100
	F-D	0	0	60	0	12	6
	M-B	4,500	2,400	10,800	400	3,840	4,320

	j	4	3	2	1
$\bar{N}_j$	B-E	0.744	1.333	2.100	0.822
	F-D	0.769	1.385	1.923	0.923
	M-B	0.861	1.362	1.694	1.083

11–17 0.423 k_B, 0.797 k_B, 0.539 k_B.

11–18 (b) 0.395 k_B.

11–19 (b) 3; (c) 195; (d) 2.923, 1.385, 0.462, 0.231; (f) $-2.06\ k_B$.

11–21 (b) 12; (c) 2.75, 1.50, 0.75; (e) $-1.81\ k_B$.

11–27 (b) 8505; (c) 2.86, 1.43, 0.571, 0.143; (d) $-3.4\ k_B$.

11–29 (a) 4×10^{-3} eV, 6.51 k_B; (b) 127 K; (c) $2 + \exp(-23.2/T)$; (d) 4×10^{-3} eV, 4.43 k_B, 14.4 K, $1 + 2\exp(-23.2/T)$.

11–34 (a) $1 + \exp(-\epsilon/k_BT)$; (b) $[1 + \exp(-\epsilon/k_BT)]^{-1}$, $[1 + \exp(\epsilon/k_BT)]^{-1}$; (c) $N\epsilon[1 + \exp(\epsilon/k_BT)]^{-1}$; (d) $Nk_B \ln[1 + \exp(-\epsilon/k_BT)] + N\epsilon\{T[1 + \exp(\epsilon/k_BT)]\}^{-1}$; (e) $Nk_B(\epsilon/k_BT)^2 \exp(\epsilon/k_BT)[1 + \exp(\epsilon/k_BT)]^{-2}$.

11–35 (a) $E = -\mu\mathscr{H}_0N/6$, $M = N\mu/3$; (b) $\Delta E = \mu\mathscr{H}_0N/12$, $\Delta M = 0$; (c) $\Delta M = \mu N/3$.

11–36 (b) $U = 0$, $S = N(\mu\mathscr{H}/2T)\tanh(\mu\mathscr{H}/k_BT) + Nk_B \ln 2\cosh(\mu\mathscr{H}/k_BT)$, $F^* = -Nk_BT \ln 2\cosh(\mu\mathscr{H}/k_BT)$, $M = -N(\mu/2)\tanh(\mu\mathscr{H}/2k_BT)$; (c) $N\tanh(\mu\mathscr{H}/2k_BT)$, $N[1 - \tanh(\mu\mathscr{H}/2k_BT)]$.

Chapter 12

12–1 $S = Nk\left[\ln V + \frac{3}{2}\ln T + \frac{3}{2}\ln\frac{2\pi mk}{h^2} + \frac{3}{2}\right]$.

12–2 (b) $\mathscr{F} = NkT/A$.

12–3 (a) $C_V = Nk$; (b) $S = Nk\left[2 + \ln\frac{A2\pi mkT}{Nh^2}\right]$.

12–4 (a) 1.25×10^{24} molecules; (b) 2.6×10^{21} molecules; (c) 5.4×10^{18} molecules; (d) 2.0×10^{24} molecules.

12–5 (a) 0.83 v_m; (b) 0.83 v_m.

12–6 (a) 2.08×10^{-3}; (b) 8.3×10^{-3}; (c) 9×10^{-9}.

12–8 (a) $v_m = 394$ m s^{-1}; $\bar{v} = 445$ m s^{-1}, $v_{rms} = 482$ m s^{-1}; (b) 227 m s^{-1}, 719 m s^{-1}, 2270 m s^{-1}.

12–11 (b) $\epsilon_m = kT/2$, $\bar{\epsilon} = 3kT/2$.

12–12 (c) 0.421; (d) 0.079; (e) 0.500; (f) 0.843.

12–13 (c) 0.573; (d) 0.427; (e) 1.00.

12–16 3.6×10^{-3} m.

12–17 3.26 s.

12–18 (a) 198 m s^{-1}; (b) 13.5 mg hr^{-1}; (c) 118 s.

12–19 (a) 5.81 μg s^{-1}; (b) 3.49×10^{11} molecules s^{-1}; 1.17 $\mu\mu$g s^{-1}; (c) 1.36×10^8 molecules; (d) 3.26×10^{-8} Torr.

12–20 0.086 mm, 2.5×10^{-3} deg.

12–21 (a) 6.34×10^{13} neutrons m^{-3}; (b) 2.63×10^{-7} N m^{-2}.

12–27 (a) 10^{-228}; (b) 1.57×10^4 K.

12–29 kT.

12–30 (a) $3kT$.

12–31 (a) 12; (b) $9kT$, 1.11.

12–36 (a) 865, 117, 16; (b) 149 $k\theta_{vib}$.

Chapter 13

13–2 (a) 246 K; (b) 172 J kilomole^{-1} K^{-1}, 24.9 × 10^{3} J kilomole^{-1} K^{-1}.

13–3 1.12 × 10^{3} J kilomole^{-1} K^{-1}, 2.66 × 10^{-3} J kilomole^{-1} K^{-1}.

13–4 $C_V = (6Nk^2/h\nu_m)T$, $C_V = 3Nk$.

13–5 (c) 2.24 × 10^{3} m s^{-1}; (d) 292 K, 6.1 × 10^{12} Hz; (e) 3.69 × 10^{-10} m, 2.27 × 10^{-10} m.

13–6 (a) 6.17 × 10^{-64} J s^{4} m^{-3}, 4.8 × 10^{-11} K s; (b) 7.62 × 10^{-16} J m^{-3} K^{-4}.

13–7 (c) 10^{-5} m.

13–11 (a) 2.24 × 10^{18} atoms; (b) 1.66 × 10^{20} atoms; (c) 2.08 × 10^{-2} Oe cm^{2}, 1.54 Oe cm^{2}.

13–13 $S = N\mu_B\mathscr{H} \tanh(\mu_B\mathscr{H}/kT)/T - Nk \ln 2 \cosh(\mu_B\mathscr{H}/kT)$, $C_V = Nk(\mu_B\mathscr{H}/kT)^2 \tanh(\mu_B\mathscr{H}/kT)$.

13–14 0.75 v_F, 0.77 v_F, 1.5 v_F^{-1}.

13–16 18.7 × 10^{-19} J; (c) 1.09 × 10^{-2} R.

13–17 (a) 1.4 × 10^{6} m s^{-1}, 1.3 × 10^{-24} kg m s^{-1}, 6.5 × 10^{4} K; (b) 8.9 × 10^{-5}, 6.4 × 10^{-5}, 2.1 × 10^{-5}, 8.9 × 10^{-5}, 8.9 × 10^{-5}; (c) 3200 K.

13–20 (c) $\epsilon_F/3$.

13–21 (a) 2.13 × 10^{-5} eV; (b) 2.46 K, 116 m s^{-1}.

13–22 2.81 × 10^{-11} m^{2} N^{-1}; 3.4 × 10^{-7} K^{-1}.

Index

Constants and Conversion Factors

The constants given below are based on values reported by B. N. Taylor, W. H. Parker, and D. N. Langenberg in *Reviews of Modern Physics 41* (1969); 375. This paper contains additional digits and also standard deviation uncertainties.

Avogadro's Number N_A	6.0222×10^{26} molecules kilomole^{-1}
Boltzmann's constant k	1.3806×10^{-23} J K^{-1} 8.6171×10^{-5} eV K^{-1}
Gas constant R	8.3143×10^{3} J kilomole^{-1} K^{-1} 8.2056×10^{-2} m^3 atm kilomole^{-1} K^{-1}
Temperature of the triple point of H_2O T_3	273.16 K = 0.01°C
Normal atmospheric pressure P	1.01325×10^{5} N m^{-2} 760 Torr
Normal specific volume of ideal gas v	22.4136 m^3 kilomole^{-1}
Electronic charge e	1.6022×10^{-19} C
Electron mass m	9.1096×10^{-31} kg
Bohr magneton μ_B	9.2741×10^{-24} A m^2
Faraday constant	9.6487×10^{7} C kilomole^{-1}
Permeability constant μ_o	$4\pi \times 10^{-7}$ H m^{-1}
Permittivity constant ϵ_o	8.8542×10^{-12} C^2 N^{-1} m^{-2}
Velocity of light c	2.9979×10^{8} m s^{-1}
Planck's constant h	6.6262×10^{-34} J s^{-1}
Stefan-Boltzmann constant $\frac{\sigma c}{4}$	5.6696×10^{-8} W m^{-2} K^{-4}
Atomic mass unit amu	1.6605×10^{-27} kg
Standard acceleration due to gravity g	9.80665 m s^{-2}